NONLINEARITY AND FUNCTIONAL ANALYSIS

Lectures on Nonlinear Problems
in Mathematical Analysis

This is a volume in
PURE AND APPLIED MATHEMATICS

A Series of Monographs and Textbooks

Editors: SAMUEL EILENBERG AND HYMAN BASS

A list of recent titles in this series appears at the end of this volume.

NONLINEARITY AND FUNCTIONAL ANALYSIS

Lectures on Nonlinear Problems in Mathematical Analysis

Melvyn S. Berger

Department of Mathematics
University of Massachusetts
Amherst, Massachusetts

ACADEMIC PRESS New York San Francisco London 1977

A Subsidiary of Harcourt Brace Jovanovich, Publishers

ACADEMIC PRESS, INC.
111 Fifth Avenue, New York, New York 10003

United Kingdom Edition published by
ACADEMIC PRESS, INC. (LONDON) LTD.
24/28 Oval Road, London NW1

Library of Congress Cataloging in Publication Data

Berger, Melvyn, Date
Nonlinearity and functional analysis.

(Pure and applied mathematics)
Bibliography: p.
Includes index.
1. Mathematical analysis. 2. Nonlinear theories. I. Title.
QA300.B458 515′.7 76-26039
ISBN 0-12-090350-4

8

Transferred to digital printing 2005

To the memory of my father,
Abraham Berger

How manifold are Thy works, O Lord! In wisdom hast Thou made them all.

PSALMS 104

The approach to a more profound knowledge of the basic principles of physics is tied up with the most intricate mathematical methods.

A. EINSTEIN

All depends, then, on finding out these easier problems, and on solving them by means of devices as perfect as possible and of concepts capable of generalization.

D. HILBERT

CONTENTS

PART III ANALYSIS IN THE LARGE

Chapter 5 Global Theories for General Nonlinear Operators

Chapter 6 Critical Point Theory for Gradient Mappings

PREFACE

For many decades great mathematical interest has focused on problems associated with linear operators and the extension of the well-known results of linear algebra to an infinite-dimensional context. This interest has been crowned with deep insights, and the substantial theory that has been developed has had a profound influence throughout the mathematical sciences. However when one drops the assumption of linearity, the associated operator theory and the many concrete problems associated with such a theory represent a frontier of mathematical research. Nonetheless, the fundamental results so far obtained in this direction already form a deep and beautiful extension of this linear theory. Just as in the linear case, these results were inspired by and are highly relevant to concrete problems in mathematical analysis. The object of the lectures represented here is a systematic description of these fundamental nonlinear results and their applicability to a variety of concrete problems taken from various fields of mathematical analysis.

Here I use the term ''mathematical analysis'' in the broadest possible sense. This usage is in accord with the ideas of Henri Poincaré (one of the great pioneers of our subject). Indeed, by carefully scrutinizing the specific nonlinear problems that arise naturally in the study of the differential geometry of real and complex manifolds, classical and modern mathematical physics, and the calculus of variations, one is able to discern recurring patterns that inevitably lead to deep mathematical results.

From an abstract point of view there are basically two approaches to the subject at hand. The first, as mentioned above, consists of extending specific results of linear functional analysis associated with the names of Fredholm, Hilbert, Riesz, Banach, and von Neumann to a more general nonlinear context. The second approach consists of viewing the subject matter as an infinite-dimensional version of the differential geometry of manifolds and mappings between them. Obviously, these approaches are closely related, and when used in conjunction with modern topology, they form a mode of mathematical thought of great power.

Finally, over and above these two approaches, there are phenomena that are genuinely both nonlinear and infinite dimensional in character. A framework for understanding such facts is still evolving.

The material to be described is divided into three parts, with each part containing two chapters. Part I is concerned first with the motivation and preliminary mathematical material necessary to understand the context of later developments in the book and second with providing a rudimentary calculus and classification of nonlinear operators. Part II deals with local analysis. In Chapter 3, I treat various infinite-dimensional extensions of the classical inverse and implicit function theorems as well as Newton's, the steepest descent and majorant methods for the study of operator equations. In Chapter 4, I turn attention to those parameter-dependent perturbation phenomena related to bifurcation and singular perturbation problems. In this chapter the use of topological ("transcendental") methods makes its first decisive appearance. The third and final part of the book describes analysis in the large and shows the necessity of combining concrete analysis with transcendental methods. Chapter 5 develops global methods that are applicable to general classes of operators. In particular, it treats the various theories and applications of the mapping degree and its recent extensions involving higher homotopy groups of spheres as well as linearization and projection methods. Chapter 6 describes the calculus of variations in the large and its current developments in modern critical point theory. This material evolves naturally from minimization and isoperimetric problems involving critical points of higher type.

A main object of the text is the application of the abstract results obtained to resolve interesting problems of geometry and physics. The applications represented have been chosen with regard both to their intrinsic interest and to their relation to the abstract material presented in the text. In many cases the specific examples require an extension of theory and so serve as a motive force for further developments. It is hoped that the deeper and more complicated applications included will enhance the value and interest of this rapidly developing subject.

Moreover, I have chosen a few nonlinear problems as models for our abstract developments. These include

(i) the determination of periodic solutions for systems of nonlinear ordinary differential equations,

(ii) Dirichlet's problem for various semilinear elliptic partial differential equations,

(iii) the differential geometric problem of determining the "simplest" metric on a given compact manifold (simplest meaning constant curvature here),

(iv) the structure of the solutions of von Kármán's equations of nonlinear elasticity.

All these models illustrate the need for new theoretical developments and more subtle and incisive methods of study. In addition, the classical nature of these problems indicates the tremendous scope for research on the abstract essentials of less classical nonlinear problems.

Many of the abstract results and applications described in the text are of recent origin, but I hope nonetheless that they form a unified pattern of development that differs from existing monographs on the subject. The choice of subject matter presented here has been highly subjective, and, in order to keep the text to a reasonable number of pages, many important topics have been treated only superficially, if at all. Thus material dealing with ordered Banach spaces, variational inequalities, convex analysis, monotone mappings, and parabolic and hyperbolic partial differential equations has been largely avoided. Moreover, these topics have been well covered by a number of recent monographs and survey articles. In a somewhat different vein, I have avoided applications too special to illustrate the general principles addressed. An example is two-point boundary value problems for a single second-order nonlinear differential equation. Such problems can (for example) be successfully treated by phase-plane methods. Finally, the recent "Euclidean" field theory methods of modern physics have shown that nonlinear hyperbolic systems can often be treated in terms of the nonlinear elliptic boundary value problems treated here.

This book has been written over a period of years, so that various kinds of misprints inevitably arise. I ask the reader to inform me of any such misprint, so that an errata list may be prepared. Yet, I hope the material described here has sufficient coherence, intrinsic interest, and attractiveness to provide the reader with a framework for further excursions into nonlinear analysis.

Many interesting nonlinear problems and illustrative examples have had to be deleted from the text to keep the book within manageable size. I hope in the near future to complete another volume containing these items as well as instructive but more routine problems. This volume will also contain a more complete bibliography.

Finally, I would like to thank all of those who helped in producing this book. They include D. Westreich, R. Plastock, E. Podolak, J. vande Koppel, T. Goldring, S. Kleiman, A. Steif, S. Nachtigall, M. Schechter, L. E. Fraenkel, S. Karlin, W. B. Gordon, A. Wightman, and last but by no means least, my editors at Academic Press. This book could not have been written without the generous financial support of the Air Force Office of Scientific Research and the National Science Foundation. To both organizations I extend my hearty thanks.

NOTATION AND TERMINOLOGY

Ω	an open subset of real N-dimensional Euclidean space $\mathbb{R}^N$
$\mathfrak{M}^N$	an N-dimensional smooth manifold
$x = (x_1, \ldots x_N)$	the Cartesian coordinates of a point in $\mathbb{R}^N$
$D_j = \partial / \partial x_j$	the elementary partial derivative operators acting on functions defined on Ω
$\alpha = (\alpha_1, \alpha_2, \ldots, \alpha_N)$	multi-index
$\|\alpha\|$	$\Sigma_{i=1}^N \alpha_i$
D^α	$\Pi_{i=1}^N D_i^{\alpha_i}$
$F(x, D^\beta u, D^m u)$	a differential operator of order m depending explicitly on the elementary higher order differential operators D^α of order m with $\|\beta\| < m$
X	some linear vector space of functions
linear operator F	$F(\alpha f + \beta g) = \alpha F(f) + \beta F(g)$ for each f, $g \in X$, and scalars α, β
nonlinear operator F	an operator F that is not necessarily linear
quasilinear differential operator $F(x, D^\beta f, D^\alpha f)$	F is a linear differential operator when regarded as a function of the elementary differential operators $D^\alpha f$ of order m alone
differential equation defined on Ω	an equation between two differential operators which must hold at each point of Ω
classical solution	a (sufficiently smooth) function (defined on Ω), which satisfies the equation at each point of Ω
$\|x\|$	length of a vector $x \in \mathbb{R}^N$
$\|\|u\|\|$	the norm of an element u of a Banach space X
absolute constant	used in connection with the inequality $F(x) \leq cG(x)$ to mean that the c does not depend on x as x varies
seminorm	a nonnegative function g defined on X such that $g(\alpha x) = \|\alpha\| g(x)$ and $g(x+y) \leq g(x) + g(y)$

SUGGESTIONS FOR THE READER

The present book is intended as a synthesis between certain aspects of mathematical analysis and other areas of science. Such a synthesis requires much motivation and a creative approach usually not found in textbooks.

Thus those parts of Chapters 1 and 2 that provide background material and preliminary information need not be read straight through. Rather the reader is encouraged to skip around to find bits of knowledge that excite his interest and to pursue these directly into the later Chapters. When necessary, the reader should return to Part I to pick up necessary information. Reading this book is not intended to be a linear experience!

Chapter 3 is intended to be abstract in nature and to help develop a facility in utilizing the "functional analysis" language. The first three sections of Chapter 3 form a necessary prerequisite for all that follows. In contrast, Chapter 4 is applications-oriented throughout. Indeed, a proper understanding of parameter-dependent local analysis requires careful thinking about specific classic model problems. Again the reader can choose only those applications that fit his or her interest.

Part III can be read in separate pieces. Chapter 5, for example, contains three separate lines of development: Section 5.1, Section 5.2, and Sections 5.3–5.5 (a deep study requiring, of course, a blending of each strand). Similarly Chapter 6 divides naturally in three parts: Sections 6.1–6.2, Sections 6.3–6.4, and Sections 6.5–6.7. The first two parts do not make use of topological methods, but such methods are essential for the third.

The reader is expected to have some prerequisite knowledge of conventional linear functional analysis, ordinary and partial differential equations. Some acquaintance with undergraduate physics and differential geometry will be helpful in understanding the applications. These applications are treated rather tersely and with varying degrees of thoroughness. A comparison with more detailed and traditional treatments of each application will prove helpful. My idea has been to provide a sense of the scope, utility, and diversity of the subject matter without obscuring the key ideas.

PART I

PRELIMINARIES

Many problems arising naturally in differential geometry and mathematical physics as well as in many other areas of science involve the study of solutions of systems of nonlinear differential equations. However, since most of these systems are "nonintegrable," in the sense that their solutions cannot be written in closed form, classical methods of studying such systems generally fail. Thus new methods of study are required. In recent years a new approach to these problems has proved both relatively successful and straightforward. The approach consists essentially in this: A given problem is reformulated in the language of function spaces; this abstracted problem is then analyzed as completely as possible by the methods of functional analysis, and the results obtained are then retranslated into statements concerning the original problem. The generality thus attained is important in several respects. First, a given problem is stripped of extraneous data, so that the analytic core of the problem is revealed. Secondly, seemingly diverse problems are shown to be specializations of the same theoretical ideas. Finally, the abstract structures that lie at the foundation of the study of novel nonlinear phenomena can be clearly aşcertained. In the sequel, we shall describe this circle of ideas as well as the resulting interplay between (nonlinear) functional analysis and concrete problems.

The aim of Part I The subject matter to be discussed is distinguished from many other mathematical areas by the mixing of various "structures" inherent in even the simplest examples. Consequently, although most of the problems presented here are easily stated, the number of prerequisites necessary to the adequate understanding of the solution to a given problem may be quite large. Thus the aim of Part I is fourfold:

(i) to set out these prerequisites in a systematic way;

(ii) to motivate the various specific problems to be studied in the sequel;

(iii) to indicate the steps necessary to reformulate a specific problem in terms of appropriate abstract nonlinear operators;

(iv) to develop an elementary calculus for these abstract operators.

The first two points are treated in Chapter 1, while the last two are covered in Chapter 2.

CHAPTER 1

BACKGROUND MATERIAL

This chapter is divided into six sections. The first two sections list a number of classical geometric and physical nonlinear problems, as well as the typical difficulties encountered in studying such problems. Next we summarize the results from linear functional analysis that will be useful in the sequel. Then we review the regularity results for linear elliptic partial differential equations that have proven invaluable for the successful application of functional analysis to the nonlinear problems discussed in the first section. Finally we survey the basic facts concerning mappings between finite-dimensional spaces (and, in particular, results from topology) that will be needed in the text.

1.1 How Nonlinear Problems Arise

Before commencing a systematic study of nonlinearity, it is of interest to mention some important sources of the problems, some of which will be discussed in the sequel. Three classic sources of nonlinear problems are mentioned below: first, differential-geometric problems, in which nonlinearity enters naturally via curvature considerations; next, mathematical problems of classical and modern physics; and finally problems of the calculus of variations involving nonquadratic functionals. Of course these sources are not exhaustive, and the mathematical aspects of fields such as economics, genetics, and biology offer entirely new nonlinear phenomena (see the Notes at the end of this chapter).

1.1A Differential-geometric sources

Differential-geometric problems associated with the effects of curvature are a rich and historic source of nonlinear differential systems. The following examples indicate their scope:

(I) Geodesics on manifolds Consider the simple hypersurface S defined by setting $S = \{x | x \in \mathbb{R}^N, f(x) = 0\}$, where $f(x)$ is a C^2 real-valued function defined on $\mathbb{R}^N$ such that $|\nabla f| \neq 0$ on S. The geodesics on S are characterized as curves $g = x(t)$ on S that are critical points of the arc length functional. Geometrically, they are characterized by the property that the principal normal of g coincides with the normal of S. Analytically, geodesics are found as solutions of the following system of N ordinary differential equations:

$$(1.1.1) \qquad \text{(a)} \quad x_{tt} + \mu(t)\, \nabla f(x) = 0, \qquad \text{(b)} \quad f(x(t)) = 0,$$

where $\mu(t)$ is some real-valued function of t.

Apart from a few exceptional cases this system depends *nonlinearly* on $x(t)$. For example, let us determine the function $\mu(t)$ of (1.1.1) in terms of f. In fact, if we differentiate the relation $f(x(t)) = 0$ twice with respect to t, we find

$$H(f)x_t \cdot x_t + \nabla f \cdot x_{tt} = 0,$$

where $H(f)$ denotes the Hessian matrix $(\partial^2 f/\partial x_i\, \partial x_j)$ of f. Thus (1.1.1) implies that

$$\mu(t) = \{H(f)x_t \cdot x_t\}|\nabla f|^{-2}.$$

Consequently we find $|\nabla f|^2 \mu(t) = H(f)x_t \cdot x_t$. Thus the system (1.1.1) is nonlinear in x unless S is either a sphere so that $\mu(t)$ is a constant, or a hyperplane in which case $\mu(t) \equiv 0$. If S is an ellipsoid, Jacobi showed that the resulting system (1.1.1) could be explicitly solved in terms of elliptic functions. However, such integrable systems are rare; and the study of geodesics for hypersurfaces differing only slightly from an ellipsoid requires new and quite refined methods of study. More generally, if $(\mathfrak{M}^N, g)$ denotes an N-dimensional differentiable manifold with the Riemannian metric

$$ds^2 = \sum_{i,j=1}^{N} g_{ij}(x)\, dx_i\, dx_j,$$

then the geodesics on $(\mathfrak{M}^N, g)$ can be found as the solutions of the nonlinear system

$$(1.1.2) \qquad \ddot{x}_i + \sum_{j,k} \Gamma^i_{jk} \dot{x}_j \dot{x}_k = 0 \qquad (i, j, k = 1, 2, \ldots, N),$$

where Γ^i_{jk} denotes the so-called Christoffel symbol of the second kind. These symbols can be computed in terms of the functions g_{ij} and their derivatives. Thus the geodesics on $(\mathfrak{M}^N, g)$, defined by the equation (1.1.2), are *directly* related to the intrinsic metric and consequently to the curvature properties of $(\mathfrak{M}^N, g)$.

The study of the solutions of (1.1.2) in terms of the geometry and topology of $(\mathfrak{M}^N, g)$ has been a motive force in the discovery of many new global methods for studying nonlinear systems. These methods, applicable to both nonlinear ordinary and partial differential equation systems, will be discussed in Part III.

(II) Minimal surfaces Two-dimensional analogues of geodesics are minimal surfaces, that is, critical points of the area integral and so surfaces with zero mean curvature. A classic problem with regard to minimal surfaces, due to Plateau, can be phrased as follows: Given a closed Jordan curve γ in $\mathbb{R}^3$, find a (smooth) area-minimizing minimal surface $\mathbb{S}$ spanning γ. When the surface $\mathbb{S}$ can be represented as $z = z(x, y)$, the function z satisfies the nonlinear partial differential equation

$$(1.1.3) \qquad z_{xx}(1 + z_y^2) - 2z_{xy}z_x z_y + z_{yy}(1 + z_x^2) = 0.$$

If the surface $\mathbb{S}$ is represented parametrically by $\{w | w_i = w_i(u, v)$ $(i = 1, 2, 3)$ with u, v isothermal coordinates$\}$, the equation (1.1.3) becomes *almost linear*. Indeed, the vector $\mathrm{w} = (w_1, w_2, w_3)$ must satisfy the much simpler relations

$$(1.1.4a) \qquad \Delta \mathbf{w} = 0,$$

$$(1.1.4b) \qquad \mathbf{w}_u^2 = \mathbf{w}_v^2, \qquad \mathbf{w}_u \cdot \mathbf{w}_v = 0,$$

where $\mathbf{w}_u$ and $\mathbf{w}_v$ denote the partial derivatives of the vector $\mathbf{w}$ with respect to u and v, respectively. We shall derive these last relations in Chapter 6 and in addition solve Plateau's problem for a rectifiable Jordan curve there.

An important distinction between geodesics and minimal surfaces is the following observation of H. A. Schwarz. The length of a rectifiable curve γ can be found by approximating γ by sufficiently small straight line segments. However, the area of a surface $\mathbb{S}$ (of finite area $A(\mathbb{S})$) cannot necessarily be found by approximating $\mathbb{S}$ by polyhedra. Generally, the areas of the approximating polyhedra converge to a number larger than $A(\mathbb{S})$. This fact serves as motivation for the notion of lower semicontinuity introduced in Chapter 6.

Interesting facts about minimal surfaces can be observed by finding a minimal surface (of least area) between two parallel circles C_1, C_2 (a distance h apart) in $\mathbb{R}^3$, whose centers lie on a line perpendicular to the planes of C_1 and C_2. For h sufficiently small, the surface of smallest area spanning C_1 and C_2 is a catenoid formed by revolving a catenary about the line $r = 0$. Its equation is $r = k_1 \cosh[(z - k_2)/k_1]$, where the constants k_1 and k_2 are so chosen that the catenoid bounds C_1 and C_2. Actually, there will be two distinct catenoids spanning C_1 and C_2, one of which will be the desired minimal surface of smallest area. Now if h is sufficiently large, no catenoid spanning C_1 and C_2 will exist, and the minimal surface of smallest area spanning C_1 and C_2 will consist of two disconnected surfaces, one spanning C_1, the other spanning C_2. Such facts demonstrate the interesting "discontinuity" and "symmetry-breaking" phenomena inherent in the study of the solutions of nonlinear problems.

(III) Uniformization of Riemann surfaces Let $F(w, z)$ be an irreducible polynomial in the complex variables w and z with constant complex coefficients, then uniformization theory is concerned with finding a representation of the points of $F(w, z) = 0$ in the form $z = z(t)$, $w = w(t)$, with the (global) parameter t varying over a simply connected domain of the complex plane. Poincaré and Klein succeeded in reducing the proof of the existence of such a parametrization to the following differential geometric problem:

(Π) Let $(\mathfrak{M}^2, g)$ denote a compact smooth two-dimensional manifold with Riemannian metric

$$g = ds^2 = \sum_{i,j} g_{ij}\, dx_i\, dx_j \qquad (i, j = 1, 2).$$

Does $\mathfrak{M}^2$ possess another Riemannian metric $\bar{g}$, conformally equivalent to g, such that the Gaussian curvature of $(\mathfrak{M}^2, \bar{g})$ is constant?

This reduction is accomplished as follows. On an appropriate compact Riemann surface S the relation $F(w, z) = 0$ can be written $w = f(z)$. Now if we can represent S as the quotient of a domain D in the complex plane by a discontinuous group Γ (acting without fixed points), then the canonical surjection mapping σ: $D \to D/\Gamma = S$ is easily shown to be analytic and single-valued. Thus, $z = \sigma(t)$ and $w = f(\sigma(t))$ for $t \in D$ determines the desired uniformization. Now this representation of S and D/Γ (up to conformal equivalence) is precisely the content of the Clifford–Klein space problem for two-dimensional manifolds of constant Gaussian curvature.

In order to solve (Π), we first recall that two metrics g and $\bar{g}$ are conformally equivalent (apart from a diffeomorphism) if there is a smooth function σ defined on $\mathfrak{M}^2$ such that $\bar{g} = e^{2\sigma}g$. (Also, conformally equivalent metrics represent the same complex-analytic structure on $\mathfrak{M}^2$.) Now the problem (Π) can be reformulated in terms of nonlinear differential systems as follows: Let (u, v) denote isothermal parameters on $\mathfrak{M}^2$. Then from elementary differential geometry we observe that the Gaussian curvature K of $(\mathfrak{M}^2, g)$ with respect to $ds^2 = \lambda(u, v)\{du^2 + dv^2\}$ can be written as

$$K = e^{-2\sigma}\{(\log \lambda)_{uu} + (\log \lambda)_{vv}\},$$

and (after a short calculation) the Gaussian curvature $\bar{K}$ with respect to $\bar{g}$ can be written as

$$\bar{K} = e^{-2\sigma}\{K - \Delta\sigma\},$$

where Δ denotes the Laplace–Beltrami operator on $(\mathfrak{M}^2, g)$. Thus, if $\bar{K}$ is constant, we find that the required conformal mapping σ is a solution of the nonlinear elliptic partial differential equation

$$(1.1.5) \qquad \Delta\sigma - K(x) + \bar{K}e^{2\sigma} = 0.$$

The existence of a smooth solution of this equation is discussed in Part III. This provides an approach to uniformization theory independent of covering space notions.

In 1900, Hilbert posed the problem of extending this uniformization theory to algebraic relations between three or more complex variables. However, despite the efforts of many distinguished scientists and the achievement of a number of partial results, this problem of Hilbert is still unresolved.

(iv) Metrics with prescribed curvature properties The problem (Π) described above has many interesting generalizations. The extensions we have in mind consist in either supposing that the manifold in question $\mathfrak{M}^N$ has dimension $N > 2$, or finding a metric g on $\mathfrak{M}^2$ with a prescribed curvature function $K(x)$ (or both). The immediate difficulties with this program are easily identified. First, all two-dimensional Riemannian manifolds can be viewed as one-dimensional complex manifolds. Thus in extending (Π), one must be careful to distinguish between the real and complex differentiable structures on manifolds. Secondly, the notion of a scalar Gaussian curvature function has several distinct generalizations for higher dimensional manifolds $\mathfrak{M}^N$ with $N > 2$. The simplest scalar function is the so-called scalar curvature function $R(x)$ associated with a Riemannian metric g. More generally, the Ricci tensor $R_{ij}(x)$ and the set of "sectional curvatures" are equally justifiable generalizations. Finally, if we seek a Riemannian metric $\bar{g}$ conformally equivalent to a given metric g on $\mathfrak{M}^2$ with prescribed curvature $K(x)$, we must bear in mind that $\bar{g} = e^{2\sigma}g$ for some smooth function $\sigma(x)$ *only up to a diffeomorphism*. Thus, in problem (Π), if we allow the curvature function $K(x)$ to be variable, we need only solve the equation

$$(1.1.6) \qquad \Delta\sigma - K(x) + \bar{K}(\tau(x))e^{2\sigma} = 0,$$

where τ is *any* diffeomorphism of $\mathfrak{M}^2$ onto itself.

In the sequel, we shall study the following generalization of (Π):

(Π_N) Find a metric $\tilde{g}$ conformally equivalent to g on a compact manifold $(\mathfrak{M}^N, g)$ with a C^∞ prescribed scalar curvature function $\tilde{R}(x)$.

We shall moreover specialize (Π_N) by (i) letting $N = 2$, or (ii) letting $\tilde{R}(x) = \text{constant}$. This latter problem for $N > 2$ was discussed by H. Yamabe in 1960, but is still not completely resolved.

In order to answer (Π_N), we recall the following formula for the change of scalar curvature under a conformal deformation $\tilde{g} = e^{2\sigma}g$:

$$\tilde{R} = e^{-2\sigma}\left\{ R - 2(N-1)\left[\Delta\sigma + \left(\frac{N}{2} - 1 \right)|\nabla\sigma|^2 \right] \right\}.$$

For $N = 2$, this formula reduces to (1.1.5) upon setting $R = 2K$. However, for $N > 2$, this equation behaves quite differently from (1.1.5). Indeed, in this case, upon setting $u = \exp(\frac{1}{2}N - 1)\sigma$ we find that u satisfies the nonlinear equation

(1.1.6′) $\qquad c(N)\,\Delta u - R(x)u + \tilde{R}u^{b(N)} = 0,$

where $b(N) = (N+2)/(N-2)$ and $c(N) = 4(N-1)/(N-2)$. Thus, for $N > 2$, we must find a *strictly positive* smooth function u defined on $(\mathfrak{M}^N, g)$ that satisfies (1.1.6).

The problem (Π_N) can be substantially sharpened by restricting attention to complex manifolds $(\mathfrak{M}^N, g)$ and using the complex structure so defined on $\mathfrak{M}^N$ to compute the "Hermitian scalar curvature" of the conformally deformed metric $\tilde{g} = e^{2\sigma}g$ (see Chapter 6, Section 2). Indeed, in the Hermitian case, the radical change in the above formula for the deformed scalar curvature, as the dimension changes, *does not occur.* Moreover, we shall find complex analytic obstructions to solving (Π_N) that have no real analogue for $N > 2$.

(v) Mapping properties of holomorphic functions Nonlinear partial differential equations analogous to (1.1.5) arise quite naturally in the geometric study of holomorphic functions of one complex variable. Let f be a holomorphic mapping of the unit disk D equipped with the Poincaré metric $ds^2 = (1 - z\bar{z})^{-2}\,dz\,d\bar{z}$ into the extended complex plane. Let the metric $dS^2 = df\,d\bar{f}$ be defined on $f(D)$, and set $e^u = (ds/dS)^2$. Then the Laplace–Beltrami operator Δ relative to dS^2 can be written $\Delta u = -\frac{1}{4}(\partial^2 u/\partial f\,\partial\bar{f})$ and a short computation shows that u satisfies the equation

$$\Delta u = 2e^u.$$

This equation is independent of f and was used by Poincaré in his studies on automorphic functions and later by F. Nevanlinna in his differential geometric proof of the value distribution of meromorphic functions.

(vi) Deformations of complex structures Deformations of the complex structure on a compact manifold $\mathfrak{M}'$ of complex dimension 1 were first studied by Riemann in 1857. He found that the number $m(\mathfrak{M}')$ of independent complex parameters on which the deformation depends can be completely described in terms of the Euler characteristic of $\mathfrak{M}'$ (or equivalently, in terms of the genus of the Riemann surface associated with $\mathfrak{M}'$). Riemann called $m(\mathfrak{M}')$ the number of moduli, and the study of these complex parameters has occupied the attention of a great many researchers down to the present day. Thus two Riemann surfaces may be topologically equivalent, but the "moduli" of these surfaces determine their analytic equivalence.

For higher dimensional complex manifolds $\mathfrak{M}^n$, the analogous deformation problem is less well understood, and (in contrast to the one-dimensional case) has a *highly nonlinear* character. To illustrate this point, let $\mathfrak{M}^n$ be given a complex structure V_0 with suitable distinguished complex coordinates $z_1, \ldots, z_n$. Let $\tilde{V}$ be another complex structure underlying $\mathfrak{M}^n$ with local coordinates $y_1, \ldots, y_n$ and such that the (1, 0) forms can be written

$$dy_j = dz_j + \sum_k \varphi_{kj}\, d\bar{z}_k \qquad \text{with } \varphi_{kj} \text{ small.}$$

Then $\tilde{V}$ is said to be an almost complex structure near the complex structure V_0. Then $\tilde{V}$ will define a true complex structure on $\mathfrak{M}^n$ if and only if the vector-valued (0, 1) form

$$\omega = \omega_{\bar{k}}\, d\bar{z}^k$$

satisfies an appropriate integrability condition. This condition takes the form of the *nonlinear* partial differential equation

$$\text{(1.1.7)} \qquad \bar{\partial}\omega = [\omega, \omega].$$

Here the linear differential operator $\bar{\partial}$ is the canonical operator that maps vector-valued (0, 1) forms into vector-valued (0, 2) forms by the rule

$$\text{(1.1.8)} \qquad \bar{\partial}(a\, d\bar{z}^{s_1} \wedge \cdots \wedge d\bar{z}^{s_q}) = \sum_k \frac{\partial a}{\partial \bar{z}_k}\, d\bar{z}_k \wedge d\bar{z}^{s_1} \wedge \cdots \wedge d\bar{z}^{s_q},$$

and the bracket $[\omega, \omega]$ is a certain bilinear vector-valued (0, 2) form. Consequently, to study the complex structures on $\mathfrak{M}^n$ near V_0, we need only study the solutions ω of (1.1.7) that lie sufficiently close to zero. This notion distinguishes a nonlinear aspect of the deformation problem if $n \geqslant 2$ since for $n = 1$ almost all complex structures are automatically integrable, the bracket $[\omega, \omega]$ vanishes identically and consequently the system (1.1.7) is linear. For a further discussion, the reader is referred to Chapter 4 and the references quoted there. As we shall see, the resolution of this deformation problem is greatly simplified by the introduction of Hilbert space techniques together with notions of "bifurcation" theory.

1.1B Sources in mathematical physics

Another equally rich source of nonlinear differential systems can be found in the basic problems of mathematical physics. The following general examples are among those discussed in the sequel.

I Classical mathematical physics

(I) Newtonian mechanics of particles Consider a system of N particles p_i of mass m_i $(i = 1, \ldots, N)$ moving in $\mathbb{R}^3$ subject to forces derived from a potential function $U(x_1, \ldots, x_{3N})$. The motions of these particles are found as solutions of the differential system

$$m_i \ddot{x}_i + \frac{\partial U}{\partial x_i} = 0 \qquad (i = 1, \ldots, 3N). \tag{1.1.9}$$

This system is clearly nonlinear when U is not a quadratic function of its arguments, but depends on its arguments to some higher order.

Now a fundamental problem for classical mechanics is the determination of the periodic motions of (1.1.9) for various reasonable potential functions U. The importance of periodic motions resides in their observation for many diverse natural phenomena governed by equations of the form (1.1.9). Moreover, Poincaré has conjectured that the periodic solutions of (1.1.9) (for appropriately restricted U) are "dense" in the set of all solutions. Here, *density* means given any solution $x(t)$, then there is a periodic solution differing only slightly from $x(t)$ for a given length of time.

When the forces acting on the particles are of a purely gravitational nature, Newton's law of gravitation implies

$$U(x_1, \ldots, x_N) = \sum_{k<j} \frac{m_k m_j}{|x_k - x_j|} \qquad (j, k = 1, \ldots, N),$$

and the resulting system (1.1.9) represents the governing equations for the classically formidable N-body problem. In celestial mechanics, the well-known two-body or Kepler problem can be solved quite explicitly and is consequently important since many problems of astronomy can be regarded as perturbations of it. Indeed, one such perturbation, the well-known restricted three-body problem, was considered by Poincaré to be typical of general dynamical systems. As an even simpler example, the equations of motion governing an autonomous perturbation $\epsilon f(\mathbf{x})$ of the Kepler problem can be written

$$\ddot{\mathbf{x}} + \frac{\mathbf{x}}{|\mathbf{x}|^3} + \epsilon f(\mathbf{x}) = 0, \qquad \mathbf{x} \in \mathbb{R}^N. \tag{1.1.10}$$

At $\mathbf{x} = 0$, the term $\mathbf{x}/|\mathbf{x}|^3$ has a singularity. The removal of the difficulties inherent in this fact has given rise to a rather elaborate "regularization" theory, in which one avoids analysis of (1.1.10) near $\mathbf{x} = 0$ by appropriate changes of coordinates in (1.1.10) on a fixed energy surface (cf. Chapter 6). The system (1.1.10) is then reduced to the form

$$\ddot{\mathbf{y}} + \operatorname{grad} W(\mathbf{y}) = 0, \qquad \tfrac{1}{2}\dot{\mathbf{y}}^2 + W(\mathbf{y}) = \text{const}.$$

where $W(\mathbf{y})$ is a smooth function vanishing at $\mathbf{y} = 0$ to the second order.

Classical methods for studying the periodic solutions of such nonlinear systems as (1.1.9) often break down due to "resonance" effects (among others). This fact has given rise to many new attempts to utilize topological methods in studying such problems, and we shall discuss this topic in Chapters 4 and 6.

(II) Elasticity A deformable body B is called elastic if it may be deformed by the application of a given class of forces to part of B, but returns to its original state after the forces are withdrawn. The simplest classical formulation of elasticity is based on the two assumptions of a linear stress–strain law (Hooke's law) and of infinitesimally small displacements. These assumptions imply linear governing equations. However, if one takes into account possible large deformations produced but retains Hooke's law, the resulting equations governing the equilibrium states of the elastic body B are nonlinear. Thus, the equations governing the equilibrium states of a one-dimensional elastic body B (a rod) under the action of compressive forces of magnitude λ applied to its ends can be written as the boundary value problem

(1.1.11) $$w_{ss} + \lambda w(1 - w_s^2)^{1/2} = 0, \qquad w(0) = w(1) = 0.$$

Here w is a measure of the horizontal deflection produced in B by the compressive force. This classic system is known as the *elastica* problem of Euler since it was solved completely by him in 1744. Its two-dimensional analogue (discussed by von Kármán in 1910) concerns the deformations produced in a two-dimensional elastic body B of arbitrary shape $\Omega \subset \mathbb{R}^2$ (a thin elastic plate) acted on by compressive forces of magnitude λ on its boundary. This problem is considerably more difficult than the one-dimensional case. By utilizing the modern techniques described in the sequel, research has only recently begun to give an adequate mathematical treatment of the problem. The resulting deformations are governed by a system of two coupled partial differential equations, the so-called von Kármán equations, defined on Ω, and can be written (after suppressing certain physical parameters)

(1.1.12) $$\Delta^2 F = -\tfrac{1}{2}[w, w], \qquad \epsilon^2 \Delta^2 w = [f, w],$$

$$\left.\begin{aligned} D^\alpha F|_{\partial\Omega} &= \lambda\psi_0, \\ D^\alpha w|_{\partial\Omega} &= 0, \end{aligned}\right\} \quad |\alpha| \leqslant 1,$$

where Δ^2 denotes the biharmonic operator and

$$[f, g] = f_{xx}g_{yy} + f_{yy}g_{xx} - 2f_{xy}g_{xy}.$$

Here ϵ^2 is a measure of the thickness of the thin plate, w represents the vertical deflection of B from its undeformed state, and F represents the Airy stress function from which all the stress components of the deformation can be found. Although the deformations predicted from (1.1.11) can be found explicitly in terms of elliptic functions, the integration of (1.1.12), in general, can be understood only by a careful qualitative analysis using the methods to be described in the sequel. The equations (1.1.12) possess many subtle properties, and we shall use these in the sequel as a specific, nontrivial example of our theoretical developments.

(iii) Ideal incompressible fluids The velocity distribution u of an ideal incompressible fluid is governed by the (nonlinear) Euler equations of motion and the equation of continuity. Denoting the velocity components by u_i, the density of the fluid by ρ, and the pressure by p, and assuming the fluid is acted on by forces F_i, these equations are

$$(1.1.13) \qquad \frac{\partial u_i}{\partial t} + \sum_j u_j \frac{\partial u_i}{\partial x_j} = -\frac{1}{\rho}\frac{\partial p}{\partial x_i} + F_i \qquad (i = 1, 2, 3)$$

$$(1.1.14) \qquad \operatorname{div} \mathbf{u} = 0.$$

Assuming that (i) the flow is irrotational, so that the velocity vector is the gradient of a velocity potential ζ; (ii) gravity is the only force acting on the fluid; (iii) the flow is stationary; and (iv) the flow is two dimensional; the system (1.1.13) has the first integral

$$(1.1.15) \qquad \tfrac{1}{2}\left(\zeta_{x_1}^2 + \zeta_{x_2}^2\right) + gx_2 = \text{const.} \qquad \text{on} \quad \partial\Gamma,$$

and (1.1.14) becomes

$$(1.1.16) \qquad \Delta\zeta = 0 \qquad \text{on} \quad \Gamma.$$

The nonlinear aspect of this problem is *twofold*. The boundary $\partial\Gamma$ of Γ is unknown, and the boundary condition imposed on $\partial\Gamma$ is nonlinear. Solutions of the system (1.1.15)–(1.1.16) have great importance in the theory of water waves, a subject renowned for its many interesting local and global nonlinear phenomena (see Section 5.5).

The vortex motions of an ideal incompressible fluid (whose study was initiated by Helmholtz in 1858) exhibit particular striking nonlinear phenomena. Consider, for example, vortex rings of a permanent form that can be observed in such a fluid. By a vortex ring we mean a continuous axisymmetric solenoidal vector field q defined on $\mathbb{R}^3$ and a subset Σ of $\mathbb{R}^3$ (homeomorphic to a solid torus) such that (taking axes fixed in Σ), both q and Σ do not vary with time, the vorticity $\omega = \operatorname{curl} q$ vanishes outside Σ (but not in Σ) and moreover satisfies the Euler equations of motion (1.1.13) and the appropriate boundary condition at infinity. In the sequel (Section 6.4) we shall derive and study the following semilinear elliptic partial differential equation for the "Stokes stream function" ψ associated with q:

$$\psi_{rr} - \frac{1}{r}\psi_r + \psi_{zz} = \begin{cases} 0 & \text{in} \quad \mathbb{R}^3 - \Sigma, \\ -\lambda r^2 f(\psi) & \text{in} \quad \Sigma. \end{cases} \tag{1.1.17}$$

Here the given function f governs the distribution of vorticity in Σ, while both ψ and its gradient must be continuous across the boundary of Σ, $\partial\Sigma$. Again, as in (1.1.15)–(1.1.16), the equation (1.1.17) is nonlinear in two respects: both ψ and Σ must be determined from it and the appropriate boundary conditions. Classically, two extreme explicit solutions of (1.1.17) were known. The problem of finding a one-parameter family of vortex rings joining these extremes requires global methods and will be discussed in Chapter 6. (See Fig. 1.1.)

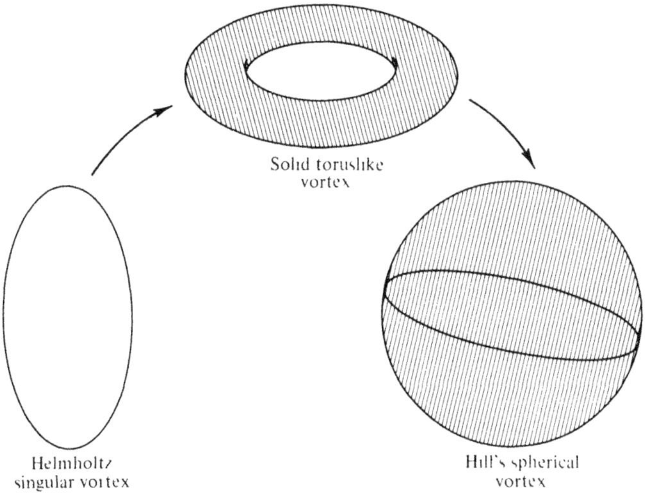

FIG. 1.1 Vortex ring distributions in $\mathbb{R}^3$ illustrating the intermediate solid toruslike vortex rings of varying cross sections interpolating between the classical Helmholtz singular vortex ring and Hill's spherical vortex.

(iv) Viscous incompressible fluids The equations governing the velocity distribution of a viscous incompressible fluid, the so-called Navier–Stokes equations, are

$$\frac{\partial u_i}{\partial t} + \sum_j u_j \frac{\partial u_i}{\partial x_j} = -\frac{1}{\rho}\frac{\partial p}{\partial x_i} + \nu\,\Delta u_i + F_i \qquad (i = 1, 2, 3), \tag{1.1.18}$$

$$\operatorname{div} \mathbf{u} = 0, \tag{1.1.19}$$

and so differ from the Euler equations only by the addition of the terms

$\nu\, \Delta u_i$. Here ν is a measure of the viscosity of the fluid. If the fluid under consideration occupies a region Ω with boundary $\partial\Omega$, one generally adds a homogeneous or inhomogeneous boundary condition $u_i|_{\partial\Omega} = g_i$, which implies in the homogeneous case that the velocity of u at $\partial\Omega$ vanishes. These equations describe a vast range of observed hydrodynamic behavior for both large and small values of the viscosity ν. It is still an open problem to show that the complicated phenomena of turbulence can be described on the basis of the nonlinearity of these equations, although as we shall see in Chapter 4, the onset of turbulence, via the appearance of secondary solutions, can be rigorously established in many instances, as a bifurcation phenomenon.

II Contemporary mathematical physics

(I) Theory of quantum fields The principle of superposition definitely rules out the appearance of nonlinear equations governing elementary quantum-mechanical phenomena. However, as soon as one considers the interaction of various quantum fields, the possibility of nonlinear equations of motion reappears. In fact, in recent years, some success has been achieved by the study of the following Lorentz-invariant nonlinear Klein–Gordan equation

$$\zeta_{tt} - \Delta\zeta = -m^2\zeta + F(|\zeta|^2)\zeta \tag{1.1.20}$$

and its generalizations. Here ζ is a complex-valued "wave" function.

An interesting nonlinear problem in quantum field theory, posed by A. Wightman, concerns the notions of dynamical instability and broken symmetry for model theories. In a simple (mean field) approximation (and rewritten in terms of contemporary "Euclidean field theory") such questions can be studied relative to the equation

$$\Delta u - m^2 u + P'(u) = f_\infty \qquad \text{(defined on } \mathbb{R}^N\text{)}, \tag{1.1.21}$$

where f_∞ is regarded as a constant given a priori, and $P'(u)$ is a polynomial-like function of u such that $P(u) \to -\infty$ as $|u| \to \infty$. The mathematical problem to be studied concerns the minima of the associated functional

$$\mathcal{G}_f(u) = \int_{\mathbb{R}^3} \{ \tfrac{1}{2}(|\nabla u|^2 + m^2u^2) - P(u) + f_\infty u \}\, dV \tag{1.1.22}$$

and the question of its uniqueness and existence relative to various choices of the constant f_∞ and the zeros of the function $-m^2u + P'(u)$. The idea is that by perturbing the right-hand side of (1.1.21) by various perturbations of f_∞, different unique absolute minima of (1.1.22) result, and these may help account for the appearance of the various so-called strange particles occurring in contemporary field theory. (For more details see Section 6.2.) See Fig. 1.2 for a pictorial view of this approach to dynamic instability.

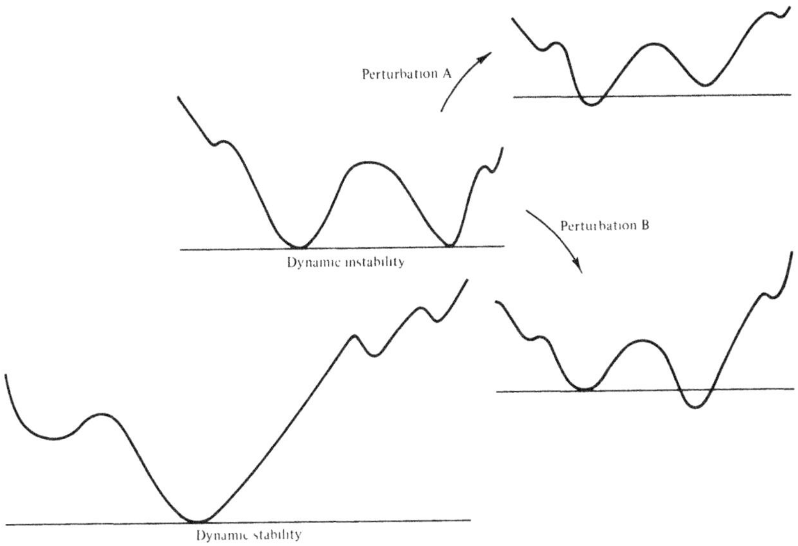

FIG. 1.2 Qualitative behavior of dynamic stability versus dynamic instability in quantum field theory models.

(ii) Relativistic theory of gravitation According to the general theory of relativity, space–time is represented by a four-dimensional (normal hyperbolic) manifold (V^4, g) that has an indefinite metric $ds^2 = g_{\alpha\beta}\, dx^\alpha\, dx^\beta$ $(\alpha, \beta = 1, 2, 3, 4)$. Gravitational effects on test particles and light rays are described by saying that the motions of such entities are geodesics with respect to the metric g. The ten components $g_{\alpha\beta}$ of the metric g are not arbitrary in Einstein's theory, but must satisfy certain nonlinear partial differential equations. For free space, these equations can be written as

$$(1.1.23) \qquad R_{\alpha\beta} = 0,$$

where $R_{\alpha\beta}$ is the Ricci tensor with respect to the metric $g_{\alpha\beta}$. A time-independent, radially symmetric solution of this quasilinear system can be found, which, in addition, has the property that it is asymptotic at large distances to the Lorentz metric $ds^2 = dt^2 - dx_1^2 - dx_2^2 - dx_3^2$. This solution, the so-called Schwarzschild metric, can be written in spherical polar coordinates as

$$(1.1.24) \qquad ds^2 = \left(1 - \frac{2m}{r}\right) dt^2 - \frac{dr^2}{1 - (2m/r)} - r^2\{d\theta^2 + \sin^2\theta\, d\varphi^2\}.$$

However, in general, the nonlinearity of the system (1.1.23) and its generalizations has caused great difficulties in exploring the further implications of Einstein's theory. An important new development centers around "Euclidean gravity," obtained by analytically continuing the usual nonlinear hyperbolic equations to nonlinear elliptic ones.

(III) Phase transitions in solids One of the most intriguing nonlinear problems in contemporary mathematical physics is the theory of phase transitions. From the point of view of thermodynamics, the sharp transition of a substance from one state to another can be understood in terms of the Gibbs internal energy function $U(x_1, \ldots, x_{n+1})$, where the variables x_i stand for appropriate generalized coordinates. For a broad class of systems, the function U is homogeneous of degree 1, so it is customary to normalize U by setting $x_{n+1} = 1$ and scaling U accordingly. Moreover, a fundamental physical postulate of thermodynamics states that the stable equilibrium state of the system described by U is the minimum of U. Now the stability of the system at a given equilibrium state $\bar{x}$ is expressed by the definiteness of the quadratic form $\sum_{i,j} U_{x_i x_j}(\bar{x})\xi_i\xi_j$; if this form is indefinite, then the state $\bar{x}$ cannot exist in a homogeneous form but breaks up into two or more phases, each of which satisfies the stability condition. In applications, states with semidefinite forms are of special interest and correspond to generalized critical points of the system. A deeper study of the phase transition problem going beyond a thermodynamic interpretation has been established by Onsager for a particular model, the two-dimensional Ising model. This was accomplished by explicit computation, and it is an important unresolved problem to study such phase transitions successfully by qualitative methods, especially in three dimensions.

1.1C Sources from the calculus of variations

A third source of nonlinear differential systems is intimately connected with the formal development of the calculus of variations. Indeed, the characterization of physical and geometric entities by extremal principles is a basic goal of a large portion of scientific thought and, for example, has survived the transition from classical to modern physics. In mathematical terms, if $u(x)$ is a stationary value of some functional defined over a domain Ω,

$$\mathcal{I}_0(w) = \int_\Omega F(x, w, Dw, \ldots, D^m w)$$

(relative to a sufficiently large class of admissible functions $\mathcal{C}$), then $u(x)$ satisfies the Euler–Lagrange differential equation

$$(1.1.25) \qquad \mathcal{I}_0'(w) \equiv \sum_{|\alpha| \leq m} (-1)^{|\alpha|} D^\alpha F_\alpha(x, w, Dw, \ldots, D^m w) = 0.$$

Here we have used the notation $\partial F/\partial w_\alpha = F_\alpha$. If the dependence of F on w

is higher than quadratic, this differential equation will be *nonlinear* in w (in general).

More generally, if $u(x)$ is a stationary value of the functional $\mathcal{I}_0(w)$ subject to the integral constraints

$$\mathcal{I}_i(w) = \int_\Omega G_i(x, w, \ldots, D^k w) \qquad (i = 1, \ldots, N),$$

then $u(x)$ will satisfy the equation

$$(1.1.26) \qquad \sum_{j=0}^{N} \lambda_j \mathcal{I}_j'(w) = 0 \qquad (j = 0, 1, \ldots, N),$$

where the λ_j are real numbers not all zero, and each $\mathcal{I}_j'(w)$ has the form (1.1.25).

Boundary value problems associated with equations of the form (1.1.25) recur throughout mathematical physics. Thus it is natural to make a careful mathematical analysis of the partial differential equations defined by (1.1.25)–(1.1.26), and, in fact, we shall do so in the sequel.

As well as satisfying the Euler–Lagrange equations (1.1.25), an extremal $u(x)$ of $\mathcal{I}_0(w)$ also may satisfy *natural boundary conditions*. In particular, if the admissible class $\mathcal{C}$ does not restrict the boundary behavior of the admissible competing functions, we may derive extra restrictions on an extremal by considering general variations of boundary values. Thus for the integral

$$\mathcal{I}(y) = \int_0^1 F(x, y, y') \, dx$$

the "natural boundary condition" at $x = 0$ and $x = 1$ is easily determined to be $\partial F(x, w(x), w'(x))/\partial y' = 0$. Moreover the following questions arise in this connection:

Does the fact that a nonlinear system is derivable from a variational principle affect the nature of its solutions? Can this variational principle be utilized in a qualitative study of these solutions? As we shall see in the sequel, the answer to these questions is a resounding yes for many of the classical problems of geometry and physics just described. Moreover, this affirmative answer also holds for general classes of nonlinear systems. The elaboration of these facts is one of the key motifs of this monograph. It is remarkable that this idea has not been utilized in most classical approaches to nonlinear problems, despite the widespread formal use of variational principles in deriving the governing equations of nonlinear systems.

Furthermore, the solutions of many nonlinear problems arising as the stationary points of some functional $\mathcal{I}(w)$ are *saddle points* of this functional. Thus classical methods of the calculus of variations become inapplicable since these deal primarily with absolute minima. The study of vortex rings of permanent form and periodic motions of Hamiltonian systems illustrates this situation. One method of treating this situation, used in the sequel, is the conversion of the problem into an isoperimetric variational problem by a clever choice of a constraint C. (See Fig. 1.3.)

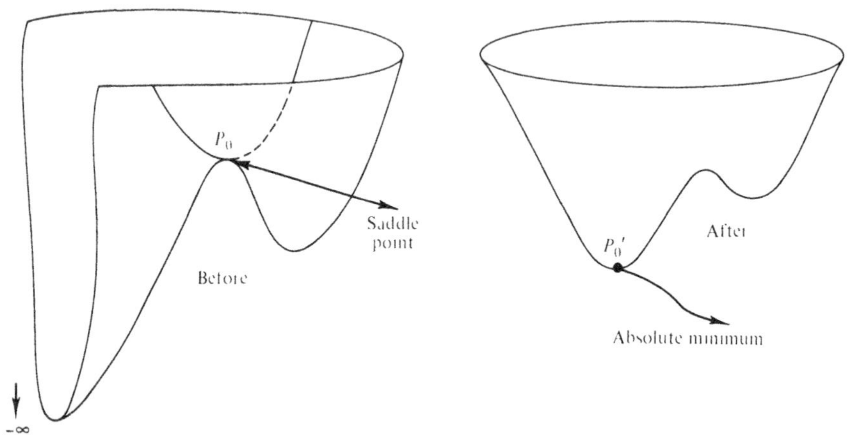

FIG. 1.3 Graphs of a typical energy functional before and after the imposition of a constraint C, illustrating the conversion of a saddle point to an absolute minimum.

1.2 Typical Difficulties Encountered

Nonlinear systems of the type considered in Section 1.1 possess certain general properties and peculiarities that make their study both difficult and interesting. Actually, one can distinguish two classes of difficulties, one inherent in the nonlinear system itself, the other related to the manner in which such systems are generally studied. We illustrate these below by citing simple examples.

1.2A Inherent difficulties

(I) Nonuniqueness This property is perhaps the most characteristic feature of nonlinear systems. For example, the boundary value problem for the system

(1.2.1) $$y_{tt} + 2y^3 = 0,$$

(1.2.2) $$y(0) = y(A) = 0$$

has a *countably infinite number of distinct solutions* $y_N(t)$ such that $\sup_{[0,A]} |y_N(t)| \to \infty$ as $N \to \infty$. Indeed, solutions $y(t)$ of this system passing through (0, 0) must satisfy $t = \int_0^y (c^4 - y^4)^{-1/2}\, dy$, so that $y(t)$ is a periodic function (varying between the limits $\pm c$) and of period $T = 2c^{-1}\int_{-1}^{1}(1 - t^4)^{-1/2}\, dt = (2/c)\beta$ (say). In order that $y(A) = 0$, it must be that for some integer N, $T = A/N$; so that for $c = (2N/A)\beta$ ($N = 1, 2, \ldots$), the system (1.2.1)–(1.2.2) possesses the desired solutions $y_N(t)$. Because of this nonuniqueness, the study of the *stability* of solutions of nonlinear systems assumes great importance.

(ii) Singularities The solutions of many nonlinear systems may develop singularities although the systems themselves have smooth coefficients. Consider, for example, the (Navier–Stokes-like) nonlinear boundary value problem $\ddot{y} + y\dot{y} = 0$, $y(0) = y(A) = 0$. One readily verifies by direct integration that this system possesses no solutions without singularities however small the number A is chosen.

(iii) Critical dependence on parameters The structure of the solutions of nonlinear systems containing explicit parameters often change radically with the parameter. Thus the system

(1.2.3) $$\ddot{y} + \tfrac{1}{2}\lambda e^{y} = 0,$$

(1.2.4) $$y(0) = y(1) = 0$$

has (a) no (smooth) solution for $\lambda > \beta$ (a certain positive real number), (b) exactly one (smooth) solution for $\lambda = \beta$, and (c) precisely two solutions for $0 < \lambda < \beta$. These facts can be proved by direct integration. Indeed, for $\lambda \geqslant 0$, the solutions $y(t)$ of (1.2.3) passing through the origin satisfy

$$t = \lambda^{-1/2}\int_0^y (c - e^s)^{-1/2}\, ds, \qquad \text{where} \quad c = 1 + \lambda^{-1}y^2(0);$$

which shows that the curve $y = y(t)$ increases to a maximum at $y(t_1) = \log c$. Since the curve $y(t)$ is symmetric with respect to $t = t_1$, we require that

$$1 = 2t_1 = 2\lambda^{-1/2}\int_0^{\log c} \frac{ds}{\sqrt{c - e^s}} = 2\lambda^{-1/2}\int_1^c \frac{du}{u\sqrt{c - u}}\,.$$

Hence, after a simple computation, one finds $\cosh^2(\sqrt{\lambda c}/4) = c$; and the solutions of this transcendental equation imply (a)–(c) above.

Furthermore, as first noted by Poincaré, nonlinear differential systems $\mathbb{S}_\lambda$ (depending explicitly on a parameter λ) exhibit a large variety of bifurcation phenomena. That is, there exist certain values of λ, λ_c say, such that $\mathbb{S}_\lambda$ possesses at least two distinct curves of solutions $y_0(\lambda)$ and $y_1(\lambda)$ with $y_0(\lambda) \to y_1(\lambda)$ as $\lambda \to \lambda_c$. As an example, consider the system

(1.2.5) $$\ddot{y} + \lambda^2\{y - y^3\} = 0, \qquad y(0) = y(1) = 0.$$

Using the Jacobi elliptic function $\operatorname{sn}(\xi, k)$ with quarter-period $K(k)$, one finds that (1.2.5) can be integrated directly. Indeed, setting

$$y_n(x) = \left(\frac{2k^2}{1+k^2} \right)^{1/2} \operatorname{sn}(2nK(k)x, k), \qquad 0 \leqslant k < 1, \quad n = 1, 2, \ldots,$$

with $\lambda = 2n(1+k^2)^{1/2}K(k) \geqslant n\pi$, one finds that for $m\pi < \lambda < (m+1)\pi$, the set of solutions of (1.2.5) is $0, \pm y_1, \ldots, \pm y_m$. Also as $\lambda \to m\pi$, $y_m(x) \to 0$ uniformly on $[0, 1]$. Consequently, the system (1.2.5) exhibits "bifurcation" phenomena at the values $\lambda^2 = m^2\pi^2$.

In the investigation of systems with features akin to (1.2.5), Poincaré introduced the notion of *exchange of stability*. We say that a solution $y(\lambda)$ of (1.2.5) is stable if it minimizes

$$\mathfrak{L}(y) = \int_0^1 [\tfrac{1}{2}\dot{y}^2 - \lambda^2(\tfrac{1}{2}y^2 - \tfrac{1}{4}y^4)].$$

With this definition, it follows that $y_0(\lambda) \equiv 0$ is stable for $0 \leqslant \lambda^2 \leqslant \pi^2$, but only $y_1(\lambda)$ is stable for all $\lambda^2 > \pi^2$. Thus as λ^2 increases through π^2, the property of stability is exchanged between the families $y_0(\lambda)$ and $y_1(\lambda)$.

(iv) Dimension versus nonlinear growth For certain differential systems $\mathbb{S}$ defined on a domain Ω, there is a striking relationship between the growth of the nonlinear terms in $\mathbb{S}$, the dimension of the set Ω, and the existence of singularity-free solutions of $\mathbb{S}$ defined on Ω. As an example, we shall show that *the equation*

$$(1.2.6) \qquad \Delta u - u + |u|^\sigma u = 0$$

has no nontrivial smooth solutions defined on $\mathbb{R}^N (N > 2)$ *with* $|u| \to 0$ *as* $|x| \to \infty$, *for* $\sigma \geqslant 4/(N-2)$. (In Chapter 6, we shall show that (1.2.6) has smooth solutions for $0 < \sigma < 4/(N-2)$.) The nonexistence of smooth solutions of (1.2.6) with $\sigma \geqslant 4/(N-2)$ is obtained by proving:

(∗) If $u(x)$ is a solution of the equation

$$(1.2.7) \qquad \Delta u + f(u) = 0 \quad \text{on} \quad \mathbb{R}^N \quad (N > 2)$$

with $|u| \to 0$ as $|x| \to \infty$, then

$$\left(\frac{2N}{N-2} \right) \int_{\mathbb{R}^N} F(u)\, dx = \int_{\mathbb{R}^N} uf(u)\, dx,$$

where $F_u(u) = f(u)$.

Proof: Assuming the truth of (∗) for the time being, we find that if u is a smooth solution of (1.2.6) with

$$f(u) = -u + |u|^\sigma u \qquad \text{and} \qquad F(u) = -\frac{u^2}{2} + \frac{1}{\sigma+2}|u|^\sigma u^2,$$

then (∗) implies that

$$\left(\frac{2N}{N-2} - 1 \right) \int_{\mathbb{R}^N} u^2\, dx = \left(\frac{2N}{(N-2)(\sigma+2)} - 1 \right) \int_{\mathbb{R}^N} |u|^\sigma u^2\, dx.$$

Thus for $u \neq 0$, $\sigma < 4/(N-2)$. To prove (*), we note that a solution $u(x)$ of (1.2.7) is a critical point of the functional

$$\mathcal{G}(u(x)) = \int_{\mathbb{R}^N} [\tfrac{1}{2} |\nabla u|^2 - F(u)]\, dx,$$

so that we must have $(d/dk)\mathcal{G}(u(kx))|_{k=1} = 0$. Making the change of variables $y = kx$ in $\mathcal{G}(u(x))$, we find

$$(\dagger) \qquad 0 = -\left(\frac{N-2}{2}\right)\int_{\mathbb{R}^N} |\nabla u|^2 + N\int_{\mathbb{R}^N} F(u).$$

On the other hand, multiplying the equation $\Delta u + f(u) = 0$ by $u(x)$ and integrating by parts, we find

$$(\dagger\dagger) \qquad \int_{\mathbb{R}^N} |\nabla u|^2 = \int_{\mathbb{R}^N} f(u)u.$$

Combining (†) and (††) gives (*).

(v) Decay at infinity Another special property of nonlinear systems, exemplified in (1.2.6), can be described roughly as the *amplification of decay* of solutions of nonlinear elliptic differential systems defined on unbounded domains. Indeed, if $u(x)$ is a solution of (1.2.6), it can also be regarded as a solution of the linear system

$$(1.2.8) \qquad \Delta u - u + p(x)u = 0, \qquad |u| \to 0 \quad \text{as} \quad |x| \to \infty,$$

where $p(x) = |u|^\sigma \to 0$ as $|x| \to \infty$. Then, using well-known results from the linear theory of elliptic partial differential equations, one finds that $|u(x)| = O(|x|^{-\beta})$ ($\beta > 0$ say) as $|x| \to \infty$. Iterating this procedure, one eventually finds that $|u(x)| = O(e^{-\gamma|x|})$ as $|x| \to \infty$ for some constant $\gamma > 0$.

(vi) Symmetric causes may not produce symmetric effects A thin circular elastic plate is clamped along its edge and subjected to an axisymmetric uniform compression (of large magnitude) there. One observes that the elastic plate deforms into a new stable *nonaxisymmetric* equilibrium state under this force. The deformations observed are governed by a system of nonlinear partial differential equations due to von Kármán (1.1.12), and it is the nonlinearity of these equations that gives rise to this unusual effect since the associated linear theory would predict an axisymmetric equilibrium state.

1.2B Nonintrinsic difficulties

All the above properties were connected with the intrinsic behavior of solutions of nonlinear differential systems. We now consider those difficulties associated with particular methods of studying nonlinear systems.

(i) Inadequacy of linearization procedures The most well-known of these methods is a process called "linearization" in which higher order terms in the nonlinear system are totally disregarded locally (in the immediate vicinity of the origin (say)). Such a procedure may well yield incorrect results. Indeed, consider the structure of the nontrivial periodic solutions (near the origin) of the system

$$(1.2.9) \qquad \text{(a)} \quad x_{tt} + x - y^3 = 0, \qquad \text{(b)} \quad y_{tt} + y + x^3 = 0.$$

The linearized system $x_{tt} + x = 0$, $y_{tt} + y = 0$ possesses a four-parameter family of periodic solutions, whereas the system (1.2.9) has only the trivial periodic solution $x \equiv y \equiv 0$. To prove this last statement, suppose the system (1.2.9) had a β-periodic solution $(x(t), y(t))$. Then multiplying (1.2.9a) by $y(t)$, (1.2.9b) by $x(t)$, subtracting, and integrating by parts (over a period), we find $\int_0^\beta [x^4(t) + y^4(t)]\, dt = 0$, so that $x(t) \equiv y(t) \equiv 0$. This procedure of linearization can be somewhat generalized by the so-called conjugacy theory of mappings. For example, if one is given the first-order system of nonlinear ordinary differential equations

$$(1.2.10) \qquad z_t = Az + f(z), \qquad \text{where} \quad |f(z)| = o(|z|),$$

one could attempt to find a C^1 (or even continuous) change of coordinates $z = \varphi(\xi)$ such that locally near $z = 0$, the new system could be written in its linearized form $\xi_t = A\xi$. Clearly, if one converts (1.2.9) to the form (1.2.10) by setting $x_t = u, y_t = w$, and $z = (x, y, u, w)$, such a transformation ϕ cannot exist. For then nontrivial periodic solutions in ξ (near the origin) would correspond to nontrivial periodic solutions in z (near the origin) and thus to nontrivial periodic solutions $(x(t), y(t))$ for (1.2.9).

(ii) Small-divisor problems Another property (historically known as *small-divisor problems*) is a consequence of proving convergence via the Cauchy majorant method. In this method, the solutions of a large class of nonlinear differential systems are constructed as *formal* power (or Fourier) series in a parameter μ (say), and one attempts to prove the convergence of the resulting series for a large class of values of μ. Thus, for example, suppose one wishes to find a 2π-periodic solution $g(z)$ of the difference equation

$$(1.2.11) \qquad g(z + 2\pi\mu) - g(z) = f(z),$$

where $f(z)$ is 2π-periodic. If $f(z) = \sum_{n>0} f_n e^{inz}$ is the Fourier series of f, then the Fourier coefficients for $g = \sum_{n>0} g_n e^{inz}$ are $g_n = f_n[e^{2\pi i\mu n} - 1]^{-1}$. For rational μ some of these denominators vanish, but for irrational μ, $|e^{2\pi i\mu n} - 1|$ can be arbitrarily small. Nonetheless, as can be shown, for almost all μ (in the sense of Lebesgue measure) (1.2.11) has a unique solution $g(z)$ which is somewhat less smooth than the function $f(z)$.

(iii) Asymptotic solutions On the other hand, in certain problems involving power series, the convergence of these formally constructed series $\xi_N(\mu)=\sum_{n=0}^{N}\xi_n\mu^n$ $(N=1,2,\ldots)$ may be difficult (if not impossible) to determine as $N\to\infty$. Yet these series may be "asymptotic" to a given solution $\xi(\mu)$ of the nonlinear system, for μ small, in the sense that $\mu^{-N}\|\xi(\mu)-\xi_N(\mu)\|\to 0$ as $\mu\to 0$, with N fixed. Thus, for example, consider the system

$$(1.2.12)\qquad \mu^2 x_{tt}-x=e^{-t^2},\qquad x\to 0\quad\text{as}\quad |t|\to\infty.$$

With $q(t)=e^{-t^2}$, the sequence

$$x_{2N}(t,\mu)=q(t)+\mu^2 q^{(2)}(t)+\cdots+\mu^{2N}(-1)^N q^{(2N)}(t)$$

is asymptotic to the unique solution of (1.2.12) for μ sufficiently small, although the associated power series has radius of convergence zero. See Section 4.4.

(iv) Lack of a priori bounds Many methods of studying a nonlinear differential system $\mathcal{S}$ are based on finding bounds for all its possible solutions, depending on the special form of $\mathcal{S}$. These "a priori" bounds assert the existence of certain universal constants, bounding some measure of the size of any solution $u(x)$ of $\mathcal{S}$. For linear systems of the form $Lu=g$, the nonexistence of such a priori bounds indicates that the equation $Lu=g$ may not be solvable *for all smooth functions* g. However, for nonlinear systems, this fact may be quite false. Consider, for example,

$$(1.2.13)\qquad y_{tt}+y^3=g(t),\qquad y_t(0)=y_t(1)=0$$

which is solvable for arbitrary $g(t)\in C[0,1]$. Yet, as in the discussion of (1.2.6) there can be no a priori estimate for a solution y of (1.2.13) in terms of g since for $g\equiv 0$, (1.2.13) possesses solutions $u(t)$ of arbitrary amplitude, i.e., for which $\sup_{[0,1]}|u(t)|$ can be arbitrarily large.

(v) Resonance phenomena An interesting nonlinear effect related to small divisor problems is associated with the term *resonance phenomena*. These phenomena are well illustrated by the problem of "normal modes" for a nonlinear Hamiltonian system near a point of equilibrium. A system of N coupled linear oscillators is governed by the system of linear ordinary differential equations

$$(1.2.14)\qquad \ddot{\mathbf{x}}+A\mathbf{x}=\mathbf{0}\qquad\text{where}\quad \mathbf{x}(t)\in\mathbb{R}^N.$$

Here the matrix A is assumed self-adjoint and nonsingular with eigenvalues $0<\lambda_1^2\leqslant\lambda_2^2\leqslant\cdots\leqslant\lambda_N^2$. Such a system has N linearly independent periodic solutions $\mathbf{x}_j(t)$ with minimal periods $2\pi/\lambda_j$ $(j=1,\ldots,N)$, called "normal modes." These solutions can be found explicitly by diagonalizing A, so that (1.2.14) becomes uncoupled. Moreover, every solution of the system (1.2.14) is a superposition of these fundamental normal modes. It is of great importance to study the behavior of these normal modes (near

$\mathbf{x} = 0$) under a nonlinear Hamiltonian perturbation $\nabla V(\mathbf{x})$, where $|\nabla V(\mathbf{x})| = o(|\mathbf{x}|)$ for $|\mathbf{x}|$ sufficiently small. Thus, the new nonlinear Hamiltonian system can now be written

$$(1.2.15) \qquad \ddot{\mathbf{x}} + A\mathbf{x} + \nabla V(\mathbf{x}) = \mathbf{0}.$$

Following the Cauchy majorant method it is customary to seek solutions $\mathbf{x}(t)$ with period λ near the jth normal mode $\mathbf{x}_j(t)$ in the form

$$\mathbf{x}(s) = \epsilon \mathbf{x}_j(s) + \sum_{n=2}^{\infty} \boldsymbol{\alpha}_n(s)\epsilon^n,$$

where $\lambda s = t$ and ϵ is sufficiently small,

$$\lambda = 2\pi/\lambda_j + \sum_{n=1}^{\infty} \beta_n \epsilon^n.$$

However, in order to prove that the resulting power series converges, it is necessary to make a severe assumption on the eigenvalues λ_j, namely $\lambda_k/\lambda_j \neq$ integer (for $k = 1, 2, \ldots, N$ with $k \neq j$). Indeed, for any integer n, terms of the form $(\lambda_k - n\lambda_j)^{-1}$ appear in the formal series expressions for the coefficients α_n and β_n. These irrationality conditions or resonance conditions thus seem essential for the permanence of the jth normal mode under nonlinear perturbation. This is indeed the case for non-Hamiltonian arbitrary perturbations. However, *by restricting our class of perturbations to Hamiltonian ones* (as defined above) a quite different fact is true; viz. these irrationality conditions are *unnecessary* for the preservation of normal modes. This interesting fact will be described in Chapter 4 as a bifurcation phenomenon. See Fig. 1.4 for an illustration of this situation.

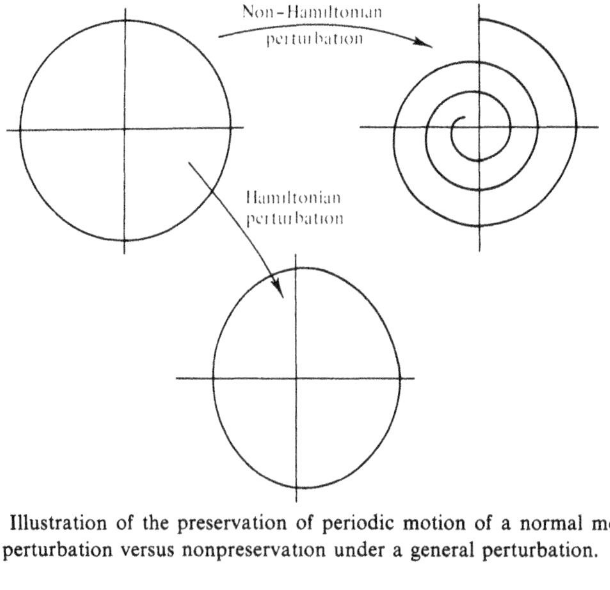

FIG. 1.4 Illustration of the preservation of periodic motion of a normal mode under a Hamiltonian perturbation versus nonpreservation under a general perturbation.

1.3 Facts from Functional Analysis

Many of the problems concerning nonlinear systems mentioned so far can be reduced to the solution of an infinite number of equations (albeit nonlinear) in as many unknowns. Thus it is natural to attempt to extend the basic concepts of the functional analysis of linear operators to this broader context. Here we summarize those basic concepts and results from classical functional analysis needed in the sequel.

Essentially, the elementary facts we need in the sequel concern (i) geometrical properties of Banach and Hilbert spaces, (ii) properties of bounded linear functionals and operators on a Banach space, (iii) facts concerning compactness in Banach spaces, and (iv) explicit examples of (i)–(iii) for certain standard Banach spaces. For complete proofs and references of the results mentioned here, see the bibliographic notes at the end of this chapter.

1.3A Banach and Hilbert spaces

A Banach space $(X, \|\ \|)$ is a normed vector space (over the real or complex numbers) that is complete with respect to the metric $d(x, y) = \|x - y\|$. In the sequel, we shall be concerned primarily with such spaces and the (geometrically simpler) special case of Hilbert spaces. Recall that a Hilbert space H is a vector space with a positive-definite inner product $(\ ,\ )$ that defines a Banach space upon setting $\|x\|^2 = (x, x)$ for $x \in H$. Therefore, in a Hilbert space H, orthogonal vectors and the orthogonal complement $M^{\perp}$ of a subset M of H ($M^{\perp} = \{x \mid x \in H, (x, y) = 0$ for all $y \in M\}$) are defined as in the finite-dimensional case. (Strong) convergence (or convergence in norm) of a sequence $\{x_n\}$ to x in a Banach space X means that $\|x_n - x\| \to 0$ as $n \to \infty$. Thus, for example, $f(x) = \|x\|$ is a continuous function with respect to convergence in norm. A seminorm defined on a Banach space X is a nonnegative real-valued functional $|x|$ defined on H that satisfies the properties $|\alpha x| = |\alpha|\,|x|$ and $|x + y| \leqslant |x| + |y|$. A seminorm f is compact relative to $\|\ \|$ if every bounded sequence in X has a convergent subsequence in $|\ |$.

Closed linear subspaces of Banach (Hilbert) spaces are again Banach (Hilbert) spaces. A similar result holds for finite direct sums; and if X is the direct sum of X_1 and X_2, we write $X = X_1 \oplus X_2$. However the rather intricate geometry of general Banach spaces can be seen by noting that there are closed subspaces M of a Banach space X for which no closed subspace N of X exists satisfying $X = M \oplus N$. Fortunately, that situation does not occur if (i) X is a Hilbert space, (ii) $\dim M < \infty$, or (iii) $\operatorname{codim} M < \infty$. In fact, if X_1 is any closed subspace of a Hilbert space X, then $X = X_1 \oplus X_1^{\perp}$.

If a normed vector space X has two norms $\| \|_1$ and $\| \|_2$, we call these norms equivalent if there are positive constants α and β such that $\alpha\|x\|_1 \leq \|x\|_2 \leq \beta\|x\|_1$ for all $x \in X$. Much simplification generally results from a judicious choice of equivalent norms in the problems we shall discuss.

The pleasant geometric properties of Hilbert space may be attributed to the so-called parallelogram law. In fact, a Banach space $(X, \| \|)$ is a Hilbert space if and only if the parallelogram law holds, i.e., for each $u, v \in X$,

(1.3.1) $$\|u+v\|^2 + \|u-v\|^2 = 2\{\|u\|^2 + \|v\|^2\}.$$

This law can be generalized to a useful class of Banach spaces as in the next definition. A Banach space $(X, \| \|)$ is called *uniformly convex* if for all $\epsilon > 0$ and u, v of norm 1 in X with $\|u - v\| > \epsilon$, there is a $\delta = \delta(\epsilon)$ independent of u, v such that $\|\frac{1}{2}(u+v)\| \leq 1 - \delta$. Such spaces show many of the useful geometric properties of Hilbert spaces. Thus, for example,

(1.3.2) Let M be a closed convex subset of a uniformly convex Banach space X, and let u be a point of $X - M$. Then the distance $d(u, M)$ is attained by one and only one point $m \in M$.

We say that a Banach space Y is *imbedded* in X and write $X \supset Y$ if (i) the elements of Y are also elements of X and (ii) (strong) convergence of a sequence $\{u_n\}$ in Y also implies (strong) convergence of $\{u_n\}$ in X. This implies the existence of an absolute constant $c > 0$ such that $\|u\|_X \leq c\|u\|_Y$ for each $u \in Y$. Y is called compactly imbedded in X if in addition to (i) and (ii) bounded subsets in Y are compact in X.

In many problems of analysis it is useful to consider those one-parameter families of Banach spaces X_α where the parameter α varies over the positive integers or real numbers with the property that $X_{\alpha_1} \subset X_{\alpha_2}$ for $\alpha_2 < \alpha_1$. Such families are called scales of Banach spaces, or a Banach scale.

A metric space is called separable if it possesses a countable dense subset. In most of the specific problems we consider, the Banach spaces involved are in fact separable. Linear subspaces of a separable Banach space X are separable, as are quotients of X by closed linear subspaces. Any separable Hilbert space possesses a countable orthonormal basis and consequently all such spaces are isometric.

1.3B Some useful Banach spaces

Let Ω be a domain in $\mathbb{R}^N$. Then the following special Banach spaces will prove important in the sequel.

(I) Spaces of continuously differentiable functions Let m be a nonnegative integer and α a multi-index. Then

$$C^m(\bar{\Omega}) = \{f \mid D^\alpha f \text{ is continuous on } \bar{\Omega} \text{ for } |\alpha| \leqslant m\}$$

is a Banach space with respect to the norm

$$\|f\|_m = \sum_{|\alpha| \leqslant m} \sup_{\Omega} |D^\alpha f|.$$

Clearly the spaces $C^m(\bar{\Omega})$ form a Banach scale as m varies over the positive integers.

Unfortunately, these spaces are often inconvenient for many problems in analysis. For example, in potential theory if Δ denotes the Laplacian on $\mathbb{R}^N$, the simple equation $\Delta u = f(x)$ is not necessarily solvable in a domain $\Omega \subset \mathbb{R}^N$ $(N > 1)$ for arbitrary $f \in C^0(\bar{\Omega})$.

This difficulty can be overcome by defining

(II) Spaces of Hölder continuous functions Let α be a positive number, $0 < \alpha < 1$. Then a function $u(x)$ is said to satisfy a Hölder condition with exponent α in Ω if for $x \neq y$

$$H_\alpha(u) = \sup_{x, y \in \Omega} \frac{|u(x) - u(y)|}{|x - y|^\alpha} < \infty \qquad (0 \leqslant \alpha \leqslant 1).$$

The set $C^{m,\alpha}(\bar{\Omega}) = \{u \mid u \in C^m(\bar{\Omega}),\ H_\alpha(D^\beta u) < \infty,\ |\beta| = m\}$ is a Banach space with respect to the norm

$$\|f\|_{m,\alpha} = \|f\|_m + \sup_{|\beta| = m} H_\alpha(D^\beta f).$$

Hence for $\alpha = 0$, the norms $\|f\|_{m,0}$ and $\|f\|_m$ are equivalent. This fact is useful if one notes that for fixed m, the space $C^{m,\alpha}(\bar{\Omega})$ form a Banach scale as α varies over $[0, 1]$. These spaces solve the potential theory problem mentioned above since if $f \in C^{0,\alpha}(\bar{\Omega})$, the Poisson equation $\Delta u = f$ always has a solution $u \in C^{2,\alpha}(\bar{\Omega})$ with $0 < \alpha < 1$.

(III) Spaces of μ-integrable functions Let $(\Omega, \mathfrak{B}, \mu)$ be a measure space defined on Ω, and let p be a positive number with $1 \leqslant p < \infty$. We denote by

$$L_p(\Omega, \mu) \equiv \left\{f \mid \int_\Omega |f|^p \, d\mu < \infty, f\ \mu\text{-measurable}\right\}.$$

If one identifies functions differing only on a set of μ-measure zero, then $L_p(\Omega, \mu)$ is a Banach space with respect to the norm

$$\|f\|_{L_p} = \left\{\int_\Omega |f|^p \, d\mu\right\}^{1/p}.$$

Clearly $L_2(\Omega, \mu)$ is a Hilbert space with respect to the inner product $(f, g)_{L_2} = \int_\Omega fg \, d\mu$. With $\|f\|_{L_\infty} = \text{ess sup}_\Omega |f|$,

$$L_\infty(\Omega, \mu) \equiv \{f \mid \|f\|_\infty < \infty, f \ \mu\text{-measurable}\}$$

is a Banach space. Generally, we shall suppose $(\Omega, \mathcal{B}, \mu)$ is the Lebesgue measure space defined on Ω, and we write $d\mu = dx$.

The relations among the L_p spaces, as the number p varies, will play an important role in the sequel. In particular, we note the following three important inequalities:

(a) *Hölder's inequality* If $f_i \in L_{p_i}(\Omega, \mu)$ and $1 = \sum_{i=1}^k 1/p_i$, then $\prod_{i=1}^k f_i$ is μ-integrable and

$$\int_\Omega \left(\prod_{i=1}^k f_i \right) d\mu \leqslant \prod_{i=1}^k \|f_i\|_{L_{p_i}}. \tag{1.3.3}$$

Hence for $\mu(\Omega) < \infty$ and $f \in L_p(\Omega, \mu)$,

$$\|f\|_{L_r} \leqslant [\mu(\Omega)]^{r^{-1}-p^{-1}} \|f\|_{L_p} \qquad \text{for} \quad r \leqslant p. \tag{1.3.4}$$

Thus for $\mu(\Omega) < \infty$, the $L_p(\Omega, \mu)$ spaces form a Banach scale for $p \in [1, \infty)$.

(b) Moreover, if $f \in L_p \cap L_{p+\beta}$ for $\beta > 0$, then $f \in L_{p+t\beta}(\Omega, \mu)$ for $0 \leqslant t \leqslant 1$ and $\psi(s) = \log \|f\|_{L_s}$ is convex in s for $s \in [p, p+\beta]$. **if $\mu(\Omega) = 1$.**

(c) *Clarkson's inequalities* Let $f, g \in L_p(\Omega, \mu)$ with $p^{-1} + q^{-1} = 1$, then for $p \geqslant 2$:

$$\|\tfrac{1}{2}(f+g)\|_{L_p}^p + \|\tfrac{1}{2}(f-g)\|_{L_p}^p \leqslant \tfrac{1}{2}\{\|f\|_{L_p}^p + \|g\|_{L_p}^p\}. \tag{1.3.5}$$

An easy immediate consequence of (1.3.5) is that for $2 \leqslant p < \infty$, $L_p(\Omega, \mu)$ is uniformly convex. An inequality, analogous to (1.3.5) also holds for L_p, $1 < p < 2$, so that $L_p(\Omega, \mu)$ is also uniformly convex for $1 < p < 2$.

(iv) (Sobolev) spaces of functions with generalized L_p derivatives In many problems involving differential operators, it is convenient to incorporate the L_p norms of the derivative of a function into a Banach norm. To accomplish this, consider the functions in the class $C^\infty(\Omega)$. For any number $p \geqslant 1$ and integer $m \geqslant 0$, we take the closure of $C^\infty(\Omega)$ with respect to the norm

$$\|u\|_{m,p} = \left\{ \sum_{|\alpha| \leqslant m} \|D^\alpha u\|_{L_p}^p \right\}^{1/p}. \tag{1.3.6}$$

The resulting Banach space is called the Sobolev space $W_{m,p}(\Omega)$. Now $W_{0,p}(\Omega) = L_p(\Omega)$, for fixed p, the Banach spaces $W_{m,p}(\Omega)$ are uniformly convex for $1 < p < \infty$ and clearly form a Banach scale as m varies over the nonnegative integers. Furthermore, for $p = 2$, $W_{m,2}(\Omega)$ is a Hilbert space with respect to the inner product

$$(u, v)_{m,2} = \sum_{|\alpha| \leqslant m} \int_\Omega D^\alpha u \cdot D^\alpha v \, dx.$$

The spaces $W_{m,p}(\Omega)$ can be modified to incorporate boundary conditions. Thus the closure of $C_0^\infty(\Omega)$ in $W_{m,p}(\Omega)$, denoted $\mathring{W}_{m,p}(\Omega)$, contains functions whose derivatives up to order m vanish in the "generalized sense," on the boundary $\partial\Omega$ of Ω. If Ω is a bounded domain, a basic inequality of Poincaré implies the existence of an absolute constant $k(\Omega)$ such that for $\mathring{W}_{1,p}(\Omega)$ $(1 < p < \infty)$,

(1.3.7) $$\|u\|_{L_p} \leqslant k(\Omega)\|\nabla u\|_{L_p}.$$

Consequently, for bounded domains in $\mathring{W}_{m,p}(\Omega)$, the "short" norm given by

$$\|u\|_{m,p} = \left\{ \sum_{|\alpha| = m} \int_\Omega |D^\alpha u|^p \right\}^{1/p}$$

is equivalent to the norm given by (1.3.6). This is easily seen by iterating (1.3.7).

The Sobolev spaces $W_{m,p}$ can be defined on Riemannian manifolds $(\mathfrak{M}^N, g)$ by using local coordinate neighborhoods and partitions of unity. For example, the space $W_{1,2}(\mathfrak{M}^N, g)$ can be defined as the closure of C^∞ $(\mathfrak{M}^N, g)$ in the norm

$$\|u\|_{1,2} = \int_{\mathfrak{M}^N} \{u^2 + |\nabla u|^2\} \, dV_g,$$

where in terms of local coordinates we have

$$\int_{\mathfrak{M}^N} |\nabla u|^2 \, dV_g = \sum_{i,j=1}^{N} \int_{\mathfrak{M}^N} g^{ij}(x) \frac{\partial u}{\partial x_i} \frac{\partial u}{\partial x_j} \, dV_g.$$

All these integrals are independent of the local coordinate neighborhoods and partitions of unity used to define them.

The Sobolev spaces $W_{m,p}(\Omega)$ can be extended to cover "negative" orders of differentiation. If $m \geqslant 0$, $1 < p < \infty$, and $p^{-1} + q^{-1} = 1$, then taking derivatives in the distribution sense,

$$W_{-m,p}(\Omega) = \left\{ u \mid u = \sum_{|\alpha| \leqslant m} D^\alpha g_\alpha, \, g_\alpha \in L_p(\Omega) \right\}$$

and is a Banach space with respect to the norm

$$\|u\|_{-m,p} = \sup \sum_{|\alpha| \leqslant m} \int_\Omega (-1)^{|\alpha|} g_\alpha \, D^\alpha f \, dx$$

over the class of functions f in $W_{m,q}$ with norm 1.

The Sobolev spaces are important in potential theory due to the property that if any $f \in L_p(\Omega)$, $1 < p < \infty$, the Dirichlet problem for the equation $\Delta u = f$ has a "generalized" solution $u \in \mathring{W}_{2,p}(\Omega)$.

1.3C Bounded linear functionals and weak convergence

A bounded linear functional $h(x)$ defined on a Banach space X is a linear mapping $X \to R^1$ such that $|h(x)| \leqslant K\|x\|_X$ for some constant K independent of $x \in X$. Clearly $h(x)$ is continuous with respect to convergence in norm. Furthermore, the set of all bounded linear functionals on X, denoted X^* is called the conjugate space of X. It is a Banach space with respect to the norm $\|h\| = \sup |h(x)|$ over the sphere $\|x\|_X = 1$. Bounded linear functionals on X are also written $x^*(x)$, where $x^* \in X^*$ and x varies over X. If $(X^*)^* = X$, then the space X is called reflexive, and such spaces share many special geometric properties with Hilbert space. In particular, all uniformly convex spaces are reflexive. The following well-known results concerning the extension of bounded linear functionals will be important later on in the text.

(1.3.8) **Hahn–Banach Theorem** Let $p(x)$ be a seminorm defined on a Banach space X, M a linear subspace of X, and $f(x)$ a linear functional defined on M with $|f(x)| \leqslant p(x)$ for $x \in M$. Then f can be extended to a bounded linear function $F(x)$ on X with $|F(x)| \leqslant p(x)$.

This result has the following consequences.

(1.3.9) (i) Let $h(x)$ be a bounded linear functional on a closed linear subspace M of X. Then $h(x)$ can be extended to a bounded linear functional $F(x)$ on X with $\|h\|_M = \|F\|_X$.

(ii) If $x_0 \in X$, then there is a linear functional $h \in X$ of norm one with $h(x_0) = \|x_0\|$. Thus $\|x\|_X = \sup h(x)$ over the unit sphere $\|h\|_{X^*} = 1$.

(iii) Let σ be a nonempty convex open subset of X, and M a linear subspace not intersecting σ. Then there is a closed proper subspace of X containing M and intersecting σ.

Furthermore, it is useful to have a concrete representation for an arbitrary linear functional on a given Banach space X. In this connection, we note

(1.3.10) (i) **Riesz Representation Theorem** Let X be a Hilbert space. Then any bounded linear functional $h(x)$ defined on X can be uniquely written as $h(x) = (x, y)$ for some $y \in X$.

(ii) $L_p^*(\Omega, \mu) = L_q(\Omega, \mu)$ for $1 < p < \infty$ and $p^{-1} + q^{-1} = 1$. Thus, for such values of p, $L_p(\Omega, \mu)$ is reflexive.

(iii) $\mathring{W}^*_{m,p}(\Omega) = W_{-m,q}(\Omega)$, m an integer, $1 < p < \infty$, and $p^{-1} + q^{-1} = 1$. Thus for such values of m and p, $\mathring{W}_{m,p}(\Omega)$ is reflexive.

In order to discuss various types of convergence in a Banach space X, the bounded linear functionals on X prove exceptionally useful. A sequence $x_n \in X$ converges weakly to an element $x \in X$ if $h(x_n) \to h(x)$ for each $h \in X^*$. Weak convergence is novel only for infinite-dimensional Banach spaces X since it coincides with the norm topology when $\dim X < \infty$. Weak convergence has the following fundamental properties.

(1.3.11) (i) Weak limits are unique if they exist

(ii) If $x_n \to x$ strongly in X, then $x_n \to x$ weakly in X.

(iii) If $x_n \to x$ weakly in X, then $\{\|x_n\|\}$ is uniformly bounded and $\|x\| \leqslant \underline{\lim} \|x_n\|$.

(iv) If $x_n \to x$ weakly, a convex combination of the x_n converges strongly to x.

(v) Let X be a reflexive Banach space, then X is (sequentially) weakly complete.

(vi) In a uniformly convex Banach space X, $x_n \to x$ weakly and $\|x_n\| \to \|x\|$ imply that $x_n \to x$ strongly.

In many nonlinear problems of analysis to be studied in the sequel, it is important to determine precise conditions for weakly convergent sequences to be strongly convergent. The simplest nontrivial result is (1.3.11(vi)) above.

1.3D Compactness

In discussing problems involving Banach spaces X of infinite dimension, one attempts to find subsets of X that share the main properties of closed bounded sets in finite-dimensional vector spaces. To this end, one says that a set M of a Banach space X is compact if M is closed (in the norm topology) and such that every sequence in M contains a strongly convergent subsequence. One then defines the notion of weak sequential compactness analogously. Results useful in this connection are:

(1.3.12) (i) In a compact set M of a Banach space X, weak and strong convergence coincide.

(ii) The closed convex hull of a compact subset of a Banach space is again compact.

(iii) A bounded set in a reflexive Banach space X is weakly sequentially compact.

(iv) If the sphere $\partial\sigma_n = \{x \mid \|x\| = n, x \in X\}$ in a Banach space X is compact, $\dim X < \infty$.

A mapping A with range a Banach space Y and domain a Banach space X is called compact if $A(\sigma)$ is relatively compact in Y for all bounded sets σ in X. Such compact mappings have separable ranges since $A(X) = A(\bigcup_{n=1}^{\infty}\sigma_n) \subset \bigcup_{n=1}^{\infty} A(\sigma_n)$. Since each set $A(\sigma_n)$ is separable, $A(X)$ is itself separable. A closer study of such operators is fundamental for our study and will be made in Sections 1.4E and 2.4.

It is extremely important to determine specific criteria for compactness in the special Banach spaces introduced in Section 1.2. Two well-known criteria are:

(1.3.13) **Arzela–Ascoli Theorem** If Ω is bounded, then a set S of $C(\Omega)$ is conditionally compact if and only if the set S is bounded in the sup norm and the elements of S are equicontinuous.

(1.3.14) **M. Riesz–Tamarkin Theorem** If Ω is bounded and $1 \leqslant p < \infty$, then a set S of $L_p(\Omega)$ is conditionally compact if and only if (i) the set S is bounded in the L_p norm, and (ii) equicontinuous in the L_p norm (i.e., $\|f(x+y) - f(x)\|_{L_p} \to 0$ as $|y| \to 0$ uniformly for $f \in S$).

Immediate consequences of (1.3.13) and (1.3.14) that will be useful later are:

(1.3.15) If Ω is a bounded domain, then bounded sets in the $C^{m,\alpha}(\Omega)$ norm are conditionally compact in the $C^{m',\alpha'}(\Omega)$ norm, provided $m \geqslant m'$ and $\alpha \geqslant \alpha'$, with at least one of the inequalities proper.

(1.3.16) **Rellich's Lemma** If Ω is a bounded domain, $1 \leqslant p < \infty$, and $m \geqslant 1$, then sets that are bounded in the $\mathring{W}_{m,p}(\Omega)$ norm are conditionally compact in the $\mathring{W}_{m-1,p}(\Omega)$ norm.

The extensions of the above results to *general* unbounded domains $\Omega \subset \mathbb{R}^N$ is important for many of the problems to be discussed. Some typical results in this direction are:

(1.3.14′) The Riesz–Tamarkin theorem (1.3.14) is valid for unbounded domains $\Omega \subset \mathbb{R}^N$ provided we add to the conditions (i) and (ii):

(iii) S is equismall at infinity (i.e., $\lim_{R\to\infty} \|f\|_{L_p(\Omega - \{x \mid |x| \leqslant R\})} = 0$ uniformly for $f \in S$).

(1.3.16′) The imbedding of $W_{m,p}(\Omega) \to W_{m-1,p}(\Omega)$ is compact if and only if $\operatorname{vol}[\Omega \cap \{y \mid |y - x| < 1\}] \to 0$ as $|x| \to \infty$.

1.3E Bounded linear operators

A linear operator L with domain X and range contained in Y (X, Y Banach spaces) is *bounded* if there is a constant K independent of $x \in X$

such that $\|Lx\|_Y \leqslant K\|x\|_X$ for all $x \in X$. Such operators are continuous with respect to *both strong and weak topologies defined on* X. The set of such maps for fixed X, Y is again a Banach space, denoted $L(X, Y)$, with respect to the norm $\|L\| = \sup \|Lx\|_Y$ for $\|x\|_X = 1$. Any bounded linear operator $L \in L(X, Y)$ has an adjoint $L^* \in L(Y^*, X^*)$ that is uniquely defined by setting $L^*g = f$, where $f(x) = g(Lx)$ for every bounded linear functional $f \in X^*$. Thus $\|L^*\| = \|L\|$; and for two operators $L_1, L_2 \in L(X, Y)$,

$$(\alpha L_1 + \beta L_2)^* = \alpha L_1^* + \beta L_2^* \qquad \text{and} \qquad (L_1 L_2)^* = L_2^* L_1^*.$$

The resolvent set $\rho(L)$ of an operator $L \in L(X, X)$ is the set of all scalars λ such that $L - \lambda I$ has a bounded inverse. All other scalars λ comprise the spectrum of L, denoted $\sigma(L)$. A number $\lambda \in \sigma(L)$ is an *eigenvalue* of L if $\text{Ker}(L - \lambda I) \neq \{0\}$, and a nonzero element $x \in X$ is an eigenvector of L corresponding to the eigenvalue λ if $x \in \text{Ker}(L - \lambda I)$. The set of such eigenvalues is called the point spectrum of L. The essential spectrum of a bounded linear operator L, $\sigma_e(L)$, consists of those numbers $\lambda \in \sigma(L)$ that cannot be removed from the spectrum by the addition to L of a compact linear operator C. It turns out that $\lambda \notin \sigma_e(L)$ is equivalent to the fact that $\lambda I - L$ has a closed range, finite-dimensional kernel, and cokernel with $\dim \text{Ker}(\lambda I - L) = \dim \text{coker}(\lambda I - L)$.

Some specific linear operators that play important roles in the sequel will be the following:

(1.3.17) **Sobolev's Integral Operator** Let Ω be a bounded domain in $\mathbb{R}^N$, and λ a positive number. Then the linear operator defined by

$$Sf(x) = \int_\Omega \frac{f(y)\, dy}{|x - y|^\lambda}$$

has the following properties:

(i) For $f \in L_p(\Omega)$ and $\lambda < N(1 - 1/p)$, S is a bounded linear operator of $L_p(\Omega) \to C^{0,\mu}(\bar{\Omega})$, where $\mu = \min(1, N(1 - 1/p) - \lambda)$.

(ii) For $\lambda > N(1 - 1/p)$, S is a bounded linear operator of $L_p(\Omega) \to L_r(\Omega)$ for $r < Np/\{N - (N - \lambda)p\}$.

(1.3.18) **Calderon–Zygmund Singular Integral Operator** For $x \in \mathbb{R}^N$: let $K(x) = \omega(x)/|x|^N$, where $\omega(x)$ is a positive C^∞ function on $\mathbb{R}^N - \{0\}$ such that $\int_{|x|=1} \omega(x)\, dS = 0$. The linear convolution operator $Lu = K * u$ is then a bounded linear mapping from $L_p(\mathbb{R}^N) \to L_p(\mathbb{R}^N)$ for $1 < p < \infty$.

(1.3.19) **Korn–Lichtenstein Theorem** If $K(x)$ is a function with the properties described in (1.3.18) above, and $u \in C^{0,\alpha}(\mathbb{R}^N)$ $(0 < \alpha < 1)$ has compact support, then the convolution $Lu = K * u$ is a bounded linear mapping of $C^{0,\alpha}(\mathbb{R}^N) \to C^{0,\alpha}(\mathbb{R}^N)$ for $0 < \alpha < 1$.

In the sequel we also use the following basic results on the properties of general bounded linear operators.

(i) On inverse operators

(1.3.20) **Banach's Theorem** Suppose $L \in L(X, Y)$ is injective and surjective, then L has a *bounded* linear inverse $L^{-1} \in L(Y, X)$.

(1.3.21) **Lax–Milgram Lemma** If X is a Hilbert space and $L \in L(X, X)$ is such that for all $x \in X$ there is an absolute constant β with $|(Lx, x)| \geqslant \beta \|x\|^2$, then L has a bounded inverse and $\|L^{-1}\| \leqslant 1/\beta$.

(1.3.22) If $L \in L(X, Y)$, L has a bounded inverse L^{-1} if and only if L^* has a bounded inverse $(L^*)^{-1}$; and then $(L^{-1})^* = (L^*)^{-1}$.

(ii) Mapping properties of linear operators

(1.3.23) **Open Mapping Theorem** Operators $L \in L(X, Y)$ map open sets of X onto open sets of Y, provided they are surjective.

(1.3.24) **Closed Range Theorem** Suppose $L \in L(X, Y)$ and L^* is injective and has a closed range. Then the range of L is Y. Moreover, any operator $L \in L(X, Y)$ has a closed range if and only if there is an absolute constant C such that

$$\|Lx\| \geqslant C\, d(x, \operatorname{Ker} L)$$

where Ker L denotes the null space of L.

(1.3.25) **Uniform Boundedness Theorem** If $L_n \in L(X, Y)$ and $\lim_{n\to\infty} L_n x$ exists for each $x \in X$, then $\{\|L_n\|\}$ is uniformly bounded, and there is a bounded operator $L = \lim_{n\to\infty} L_n$ such that $L_n x \to Lx$ for all $x \in X$.

(iii) Projection and imbedding operators

(1.3.26) **Projection Operators** An operator $P \in L(X, X)$ is called a projection if $P^2 = P$; and if $R(P)$ and $R(I - P)$ denote the ranges of P and $I - P$, respectively, then $X = R(P) \oplus R(I - P)$. Conversely if $X = M \oplus N = \{x \mid x = m + n,\ m \in M,\ n \in N,\ x \in X\}$, with M or N finite dimensional and $M \cap N = \varnothing$, the map $Qx = m$ is a projection.

(1.3.27) **Imbedding Operators** If the Banach space X is imbedded in the Banach space Y, the linear mapping i: $X \to Y$ defined by setting $i(x) = x$ is called an imbedding operator. Since $X \subset Y$, the operator i is continuous, and so $i \in L(X, Y)$. Moreover, if X is compactly imbedded in Y, the operator i is also compact (See Section 1.3F(i)).

The following results show how new inequalities can be derived from the properties of imbedding operators.

(1.3.28) **Lions' Lemma** Let X_1, X_2, X_3 be three Banach spaces satisfying the imbedding relations $X_1 \subset X_2 \subset X_3$. Suppose the imbedding $X_1 \to X_2$ is compact. Then given any $\epsilon > 0$, there is a $K(\epsilon) > 0$ such that for all $y \in X_1$

$$\|y\|_{X_2} \leqslant \epsilon \|y\|_{X_1} + K(\epsilon)\|y\|_{X_3}.$$

Proof: Assume the inequality is false. Thus there is a sequence $\{y_n\}$ in X_1 such that $\|y_n\|_{X_2} \geqslant \epsilon \|y_n\|_{X_1} + n\|y_n\|_{X_3}$. Setting $v_n = y_n / \|y_n\|_{X_1}$, we obtain

(1.3.29) $$\|v_n\|_{X_1} = 1 \quad \text{and} \quad \|v_n\|_{X_2} \geqslant \epsilon + n\|v_n\|_{X_3}.$$

By the properties of the imbeddings $X_1 \subset X_2 \subset X_3$, there is a subsequence of v_n, which we relabel v_n, such that $v_n \to v$ strongly in X_2 and X_3. On the other hand, using (1.3.29) we must have both $\|v\|_{X_3} = 0$ and $\|v\|_{X_2} \geqslant \epsilon > 0$, a contradiction.

1.3F Special classes of bounded linear operators

(I) Compact linear operators A linear operator $C \in L(X, Y)$ is called compact if for any bounded set $B \subset X$, $C(B)$ is conditionally compact in Y. Bounded linear mappings with *finite-dimensional ranges* are automatically compact; and conversely, if X and Y are Hilbert spaces, then a compact linear mapping C is the uniform limit of such mappings. The theory of compact linear operators is highly developed and its principal results can be summarized as follows:

(1.3.30) Let $C \in L(X, X)$ be compact, and set $L = I + C$. Then (a) L has closed range, (b) dim Ker L = dim coker $L < \infty$, (c) there is a finite integer β such that $X = \text{Ker}(L^\beta) \oplus \text{Range}\ (L^\beta)$ and L is a linear homeomorphism of Range(L^β) onto itself, (d) the spectrum of C, $\sigma(C)$, consists of a countable set of eigenvalues λ_N. These eigenvalues, with the possible exception of zero, are isolated and of finite multiplicity.

(1.3.31) A compact linear operator $C \in L(X, Y)$ maps weakly convergent sequences in X into strongly convergent sequences in Y (i.e., C is necessarily a *completely continuous operator*). Conversely, if X and Y are Hilbert spaces, any completely continuous linear operator is compact.

The set of compact operators, $K(X, Y)$ defined in $L(X, Y)$ has the following properties:

(1.3.32) The set $K(X, Y)$ is closed in the uniform operator topology of $L(X, Y)$. Moreover, $C \in K(X, Y)$ if and only if $C^* \in K(Y^*, X^*)$. The set $K(X, Y)$ is a closed two-sided ideal in the normed ring $L(X, Y)$.

Generally speaking, a bounded linear operator C between function spaces X and Y is compact if the functions in the range of C are "smoother" than the functions of X. Thus, for example,

(1.3.33) The Sobolev integral operator defined by

$$Sf(x) = \int_\Omega \frac{f(y)\,dy}{|x-y|^\lambda} \qquad (\Omega \text{ a bounded domain in } \mathbb{R}^N)$$

is compact as a linear operator from $L_p(\Omega)$ to $C^{0,\alpha}(\overline{\Omega})$ or $L_r(\Omega)$ as in (1.3.17 (i), (ii)).

(1.3.34) The imbedding operator $i : C^{\mu,\alpha}(\overline{\Omega}) \to C^{\mu',\alpha'}(\overline{\Omega})$ is compact provided Ω is a bounded domain and $\mu \geqslant \mu'$, $\alpha \geqslant \alpha'$ (with at least one of the inequalities being strict).

(1.3.35) The imbedding operator $i : \mathring{W}_{m,p}(\Omega) \to \mathring{W}_{m-1,p}(\Omega)$ is compact provided Ω is bounded.

This result is also true for the imbedding $i : W_{m,p}(\Omega) \to W_{m-1,p}(\Omega)$, provided the boundary of the bounded domain Ω is sufficiently regular.

(ii) Fredholm operators and their generalizations An operator $L \in L(X, Y)$ is called Fredholm if (a) the range of L is closed in Y, and (b) the subspaces Ker L and coker L are finite dimensional. The set of Fredholm maps contained in $L(X, Y)$ is denoted by $\Phi(X, Y)$. $\Phi(X, Y)$ can be shown to be an open subset of $L(X, Y)$. The index of a Fredholm map L, ind L, can be defined by either of the formulas

(1.3.36) $$\operatorname{ind} L = \dim \ker L - \dim \operatorname{coker} L = \dim \ker L - \dim \ker L^*$$

and can be shown to be invariant under compact perturbations and perturbations by elements of $L(X, Y)$ of sufficiently small norm. Thus, the index is constant on the connected components of $\Phi(X, Y)$. Moreover, if $A \in \Phi(X, Y)$ and $B \in \Phi(Y, Z)$, $BA \in \Phi(X, Z)$ and ind $BA =$ ind $B +$ ind A. The subset of Fredholm maps of index k is denoted by $\Phi_k(X, Y)$.

By virtue of (1.3.30), compact perturbations of the identity are Fredholm operators of index zero, and conversely, (see (1.3.38) below) any Fredholm map $L \in \Phi_0(X, Y)$ differs from a compact perturbation of the identity only by a linear homeomorphism in $L(X, Y)$. Examples of Fredholm maps of arbitrary index on any separable Hilbert space can be constructed by considering forward and backward shift operators.

It is useful to extend the concept of a Fredholm operator L by requiring that L have closed range but allowing dim coker $L = \infty$. Such operators are called semi-Fredholm and are denoted by $\Phi_+(X, Y)$. Moreover, such operators can be characterized by:

(1.3.37) An operator $L \in L(X, Y)$ is semi-Fredholm if and only if there is a compact seminorm $|\ |$ with respect to $\|\ \|_X$ such that for all $x \in X$,

$$\|Lx\| + |x| \geq c\|x\|,$$

where c is a positive absolute constant.

Thus L is Fredholm if and only if the result (1.3.37) holds for L and its adjoint L^*.

Basic properties of Fredholm and semi-Fredholm operators are:

(1.3.38) (i) If $L \in \Phi_k(X, Y)$ for $k \geq 0$, there is a compact linear mapping C of arbitrarily small (but nonzero) norm and finite rank such that $L = L_0(I + C)$, where $L_0 \in L(X, Y)$ is surjective and dim Ker $L_0 = k$.

(ii) If $L \in \Phi(X, Y)$, then there are closed linear subspaces X_0 of X, Y_0 of Y and an operator $L_0 \in L(X, Y)$ such that:

(1) If $\hat{L}$ denotes the restriction of L to X_0 and P is a projection of Y onto coker L, $\hat{L}$ is invertible and $L_0 = \hat{L}^{-1}(I - P)$ is a two-sided inverse of L modulo compact operators;

(2) $X = X_0 + \text{Ker } L$ and $Y = Y_0 + \text{coker } L$; and in fact

(3) $L_0 L$ is a projection of X onto X_0, while LL_0 is a projection of Y onto Y_0.

(iii) For $L \in \Phi_+(X, Y)$, dim Ker L is upper semicontinuous in the sense that if $\|B\|$ is sufficiently small dim Ker$(L + B) \leq$ dim Ker L.

(iv) If $X = Y$ is a Hilbert space and $L \in \Phi(X, Y)$, the linear equation $Lx = y$ is solvable if and only if y is orthogonal to Ker L^*.

(III) Self-adjoint operators defined on a Hilbert space H An operator $L \in L(H, H)$ is called self-adjoint if $(Lx, y) = (x, Ly)$ for each x, $y \in H$. The structure of such operators has been studied intensively since the fundamental research of Hilbert. The following results are used in the sequel:

(1.3.39) **Bilinear Forms and Self-Adjoint Operators** If $\zeta(x,y)$ is a bounded symmetric bilinear functional, there is a unique self-adjoint operator $L \in L(H, H)$ such that $\zeta(x,y) = (Lx,y)$. L is compact if and only if $\zeta(x,y)$ is continuous (in x and y) with respect to weak convergence.

(1.3.40) **The Spectrum of Self-Adjoint Operators**

(i) The spectrum of a self-adjoint operator L is contained in the interval $[m, M]$ of the real axis where $m = \inf(Lx, x)$ over $\partial\Sigma_1 = \{x \mid \|x\| = 1\}$ and $M = \sup(Lx, x)$ over $\partial\Sigma_1$. Moreover, the numbers $m, M \in \sigma(L)$, and

$$\|L\| = \max(|m|, |M|) = \sup_{\|x\|=1} |(Lx, x)|.$$

(ii) For a self-adjoint operator L, a number $\lambda \notin \sigma(L)$ if and only if there is an absolute positive constant C such that $\|Lx - \lambda x\| \geqslant C\|x\|$ for all $x \in H$. Furthermore, the essential spectrum of a self-adjoint operator L, $\sigma_e(L)$, consists of those numbers of $\sigma(L)$ that are not eigenvalues of finite multiplicity.

(iii) If L is a self-adjoint compact operator, then $\sigma(L)$ consists of at most a countably infinite number of real eigenvalues $\{\lambda_k\}$. This set is discrete except possibly at $\lambda = 0$. Furthermore, the multiplicity of $\lambda_k = \dim \operatorname{Ker}(L - \lambda_k I) < \infty$ and $Lx = \sum_{k=1}^{\infty}\lambda_k(x, x_k)x_k$, where $\{x_k\}$ is an orthonormal sequence of eigenvectors and λ_k is repeated $\dim \operatorname{Ker}(L - \lambda_k I)$ times.

(iv) The eigenvalues of a self-adjoint compact operator L can be characterized by minimax principles. In particular, if the positive eigenvalues λ_k^+ are ordered in decreasing order (with multiplicities repeated)

(1.3.41) $$\lambda_k^+ = \min_{\pi_{k-1}} \max_{x \in \pi_{k-1}} (Lx, x)/\|x\|^2,$$

where π_{k-1} denotes an arbitrary linear subspace of H of *codimension* $k - 1$. Alternatively,

(1.3.42) $$\lambda_k^+ = \max_{P_k} \min_{x \in P_k} (Lx, x)/\|x\|^2,$$

where P_k denotes an arbitrary linear subspace of H of dimension k.

(v) An isolated eigenvalue λ_0 of finite multiplicity of a self-adjoint operator L has stability properties under "analytic perturbations" $L(\epsilon)$ (i.e., $L(\epsilon) = L + \sum_{n=1}^{\infty}\epsilon^n L^{(n)}$ for ϵ so small and $L^{(n)}$ self-adjoint with $\|L(\epsilon)\| < \infty$). Indeed, for $|\epsilon|$ sufficiently small and $\lambda \in \Delta = (\lambda_0 + \beta, \lambda_0 - \alpha)$, where α and β are chosen so that Δ contains no other spectral value of L, there exist μ convergent real power series

$$\lambda_i(\epsilon) = \lambda_0 + \sum_{j=1}^{\infty} \epsilon^j \lambda_i^{(j)}$$

$$x_i(\epsilon) = x_i + \sum_{j=1}^{\infty} \epsilon^j x_i^{(j)} \qquad (i = 1, \ldots, \mu)$$

such that the spectrum of L in Δ consists of the eigenvalues $\lambda_i(\epsilon)$ with associated orthonormal eigenvectors $x_i(\epsilon)$.

(vi) A positive self-adjoint operator L (i.e., an operator with $\sigma(L) \subset [0,\infty)$) has a unique positive self-adjoint square root $L^{1/2}$ such that $(Lx, x) = \|L^{1/2}x\|^2$.

(1.3.43) **Projection Operators** Let M be a closed subspace of H. If the elements $\{x\}$ of $H = M \oplus M^{\perp}$ are denoted $x = m + m^{\perp}$, then the linear map $Px = m$ is called the orthogonal projection of H on M.

(i) Any orthogonal projection map P on a closed subspace M of H is self-adjoint with $P^2 = P$, $\|P\| = 1$, and $\|(I - P)x\| = d(x, M)$. Conversely, any self-adjoint operator Q with $Q^2 = Q$ is the orthogonal projection of H onto $Q(H)$.

(ii) An orthogonal projection operator P of H onto M is compact if and only if $\dim M < \infty$.

Remark on the Laplace–Beltrami operator: The properties of bounded self-adjoint linear operators will be of great importance since they recur throughout geometry and mathematical physics relative to appropriate Hilbert space structures. Thus the Laplace–Beltrami operator Δ acting on smooth functions defined on a compact Riemannian manifold $(\mathfrak{M}, g)$ and given in local coordinates by the formula

$$\Delta u = \sum_{i,j=1}^{N} \frac{1}{\sqrt{|g|}} \frac{\partial}{\partial x_i} \left\{ \sqrt{g}\, g^{ij} \frac{\partial u}{\partial x_j} \right\},$$

where $|g| = \det(g_{ij})$ and $g_{ij} g^{jk} = \delta_{ik}$, can be extended to a bounded self-adjoint linear operator L mapping $W_{1,2}(\mathfrak{M}, g)$ into itself by setting (and integrating by parts)

$$\zeta(u, v) = \int v\, \Delta u\, dV_g = -\sum_{i,j=1}^{N} \int g^{ij} \frac{\partial u}{\partial x_i} \frac{\partial v}{\partial x_j}\, dV_g .$$

Indeed, by the Cauchy-Schwarz inequality, $|\zeta(u, v)| \leqslant K\|u\|_{1,2}\|v\|_{1,2}$ for each $u, v \in C^{\infty}(\mathfrak{M}, g)$ and some absolute constant $K > 0$. Thus $\zeta(u, v)$ is a bounded bilinear functional, and so by (1.3.39) there is a unique bounded self-adjoint operator L mapping $W_{1,2}(\mathfrak{M}, g)$ into itself such that $(Lu, v) = \zeta(u, v)$.

Similar results hold for any "formally self-adjoint" differential operator of the form

$$\mathfrak{L}u = \sum_{|\alpha|, |\beta| \leqslant m} (-1)^{|\alpha|} D^{\alpha}\{a_{\alpha\beta}(x)\, D^{\beta} u\}$$

(where $a_{\alpha\beta} = a_{\beta\alpha}$ are bounded measurable functions) defined on a bounded domain $\Omega \subset \mathbb{R}^N$ taken together with appropriate boundary conditions on the boundary $\partial\Omega$ of Ω. For the simplest case, the so-called Dirichlet boundary conditions $D^{\alpha}u|_{\partial\Omega} = 0$, $|\alpha| \leqslant m - 1$, for $u, v \in C_0^{\infty}(\Omega)$, we set

$$\zeta(u, v) = \int_{\Omega} v \mathfrak{L} u = \sum_{|\alpha|, |\beta| \leqslant m} \int_{\Omega} a_{\alpha\beta}(x)\, D^{\alpha}u\, D^{\beta}v$$

after numerous integrations by parts. As before, by (1.3.39), there is a unique self-adjoint operator $L \in L(\mathring{W}_{m,2}(\Omega), \mathring{W}_{m,2}(\Omega))$ such that $\zeta(u, v) = (Lu, v)$.

1.4 Inequalities and Estimates

In order to carry out the program described at the beginning of Part I, one associates with a given nonlinear differential system $\mathfrak{S}$ a mapping $f(\mathfrak{S})$ acting between suitably chosen Banach spaces. The key point here is that in this association $f(\mathfrak{S})$ should retain the major *qualitative* properties of the system $\mathfrak{S}$. These qualitative feature of $f(\mathfrak{S})$ include such properties as boundedness, continuity, and compactness, and can be established only by making essential use of inequalities and estimates of the type presented below.

We shall describe only two classes of results since they will be the main analytical facts necessary in the sequel. The first class (described in Sections 1.4A–1.4C) are the so-called calculus inequalities for the Sobolev spaces $W_{m,p}(\Omega)$. These inequalities describe in precise terms the connection between the sizes of:

(i) the L_p norms of the generalized derivatives of a function f,
(ii) the L_p norms defined on a domain $\Omega \subset \mathbb{R}^N$ of the function f itself,
(iii) the pointwise behavior of the function f and its derivatives,
(iv) the dimension of the set Ω, and
(v) bounded sets in $W_{m,p}(\Omega)$ when regarded as subsets of "larger" Banach spaces $X \supset W_{m,p}(\Omega)$.

The second class (described in Section 1.4D–1.4E) can be described as estimates for the solutions of linear elliptic partial differential equations of two types:

(i) L_p estimates for $(1 < p < \infty)$ (i.e., estimates of the solutions in an integral or "averaged" norm), and
(ii) pointwise estimates in the Hölder spaces $C^{m,\alpha}(\overline{\Omega})$ $(0 < \alpha < M)$ (i.e., estimates in the usual pointwise sense).

Remark on pointwise versus integral estimates: The close relationship between the pointwise and integral estimates in both these cases is crucial for the resolution of the concrete nonlinear problems we shall discuss. Thus, for example, many nonlinear problems first written down in the classical pointwise form can be naturally reformulated and solved in a Hilbert space context. Then it is necessary to ensure that this "Hilbert space" solution yields a solution to the actual nonlinear problem as stated in terms of smooth functions. It is exactly at this point that the estimates to be described prove invaluable since the Hilbert space norms we utilize are inevitably integral norms and the estimates thus provide pointwise information to be obtained from integral data.

1.4A The spaces $W_{1,p}(\Omega)$ $(1 \leqslant p < \infty)$

We first consider the simple case $\Omega \equiv \mathbb{R}^N$ and functions of compact support.

(1.4.1) **Theorem** Suppose $u \in W_{1,p}(\mathbb{R}^N)$ has compact support in $\mathbb{R}^N$, then

(i) $p > N$ and $N + \mu p < p$ imply $u \in C^{0,\mu}(\mathbb{R}^N)$ and

(1.4.2) $$\|u\|_{C^{0,\mu}(\mathbb{R}^N)} \leqslant \text{const.}\, \|\nabla u\|_{L_p(\mathbb{R}^N)},$$

where the constant depends on the support of u but not on u itself;

(ii) $p \leqslant N$ and $(N-p)r < Np$ imply $u \in L_r(\mathbb{R}^N)$ and

(1.4.3) $$\|u\|_{L_r(\mathbb{R}^N)} \leqslant \text{const.}\, \|\nabla u\|_{L_p(\mathbb{R}^N)}$$

where again the constant depends on the support of u but not on u itself.

The proof of these inequalities follows immediately from the boundedness properties of the Sobolev integral operator since for any $u \in C_0^\infty(\mathbb{R}^N)$

$$(1.4.4) \qquad |u(x)| \leq \frac{1}{\omega_N} \int \frac{|\nabla u|}{|x-y|^{N-1}}\, dy, \qquad \text{where} \quad \omega_N = \operatorname{vol}(|x| \leq 1).$$

This last inequality follows by noting that for any fixed direction,

$$u(x) = -\int_0^\infty \frac{\partial u}{\partial r}(x + r\omega)\, dr,$$

so that integrating over all directions ω gives

$$u(x) = -\frac{1}{\omega_N} \int r^{1-N} \frac{\partial u(y)}{\partial r}\, dy.$$

Remark on limiting cases: Two limiting cases will be of interest in the sequel. In each case the result given in (1.4.1) can be sharpened in an important way. In the first case, we suppose that $p = N$ so that (1.4.3) implies $u \in L_p(\mathbb{R}^N)$ for every finite p. Examples show that $W_{1,p}(\mathbb{R}^N) \not\subset L_\infty(\mathbb{R}^N)$. Thus it is natural to inquire about the exact integrability properties of e^{u^α} as α varies over $(0, \infty)$.

In the second case, we suppose $p < N$ but $r = Np/(N-p)$. In this case, we shall see that a sharpened analogue of (1.4.4) holds with the constant *independent* of the size of the support of the given function u. The plausibility of this latter fact by dimensional analysis is given in the Notes at the end of this chapter. However, we shall also see that this limiting case loses a certain compactness property. This "loss of compactness" will play a critical role in solving certain problems of mathematical physics and differential geometry to be discussed in Chapter 6.

(1.4.1′) **Theorem** Let $u \in W_{1,p}(\mathbb{R}^N)$ have compact support in $\mathbb{R}^N$.

(i) $p < N$ and $r = Np/(N-p)$ imply $u \in L_r(\mathbb{R}^N)$ and

$$(1.4.5) \qquad \|u\|_{L_r(\mathbb{R}^N)} \leq c_{r,N} \|\nabla u\|_{L_p(\mathbb{R}^N)} \qquad \text{with} \quad c_{r,N} = \frac{p}{2\sqrt{N}} \left(\frac{N-1}{N-p} \right).$$

(ii) $p = N$ implies there are positive numbers c_1, c_2 depending only on N such that

$$(1.4.6) \qquad \int \exp\left(c_1 \frac{|u|}{\|\nabla u\|_{L_p}} \right)^{N/(N-1)} \leq c_2 \mu(\operatorname{supp}(u)).$$

Proof of (1.4.5): We first prove the result for $u \in C_0^\infty(\mathbb{R}^N)$ and $p = 1$; the result for general $p < N$ then follows immediately from Hölder's inequality by applying the special case to $v = u^\sigma$ with $\sigma = ((N-1)/(N-p))p$. For $p = 1$, we first observe that

$$|u(x)| \leq \tfrac{1}{2} \int_{-\infty}^\infty \left| \frac{\partial u}{\partial x_i} \right| dx_i = \tfrac{1}{2} I_i \quad \text{(say)}.$$

Multiplying these inequalities together, we find $|2u(x)|^{N/(N-1)} \leqslant (I_1 I_2 \cdots I_N)^{1/(N-1)}$. Now we integrate this last inequality successively over the variables $x_1, x_2, \ldots, x_N$ and apply Hölders inequality (1.3.3) at each stage. Thus in the first step, we find upon setting $I_{ij} = \int_{-\infty}^{\infty}\int_{-\infty}^{\infty}|\partial u/\partial x_i|\, dx_i\, dx_j$,

$$\int_{-\infty}^{\infty}|2u(x)|^{N/(N-1)}\, dx_1 \leqslant I_1^{1/(N-1)}\int_{-\infty}^{\infty}(I_2 \cdots I_N)^{1/(N-1)}\, dx_1 \leqslant I_1^{1/(N-1)}(I_{21} \cdots I_{N1})^{1/(N-1)}.$$

So that at the last stage, since the geometric mean is at most equal to the root mean square,

$$\int_{\mathbb{R}^N}|2u(x)|^{N/(N-1)}\, dx_1 \cdots dx_N \leqslant \left\{\prod_{i=1}^{N}\int_{\mathbb{R}^N}\frac{\partial u}{\partial x_i}\, dx_1 \cdots dx_N\right\}^{1/(N-1)},$$

which implies

$$\|u\|_{N/(N-1)} \leqslant \frac{1}{2\sqrt{N}}\int_{\mathbb{R}^N}|\nabla u|\, dx_1 \cdots dx_N \leqslant \|\nabla u\|_{L_1}.$$

Proof of (1.4.6) The proof is based on keeping a careful bound on the absolute constant in the inequality (1.4.5) as a function of p. In fact, we can prove (from (1.4.4)) that for a function $u \in C_0^\infty(\Omega)$,

$$\int_\Omega |u|^{Np/(N-1)} \leqslant c_0(c_1\|\nabla u\|_{0,N})^{Np/(N-1)}p^p\,\mu(\Omega).$$

From this estimate, we easily establish that

$$\int \exp\left\{\frac{u}{c_1\|\nabla u\|_{L_N}}\right\}^{N/(N-1)} = \sum_{p=0}^{\infty}\int\frac{1}{p!}\left(\frac{u}{c_1\|\nabla u\|_{L_N}}\right)^{pN/(N-1)}$$

$$\leqslant \text{const.}\,\mu(\Omega)\sum_{p=0}^{\infty}\left(\frac{c_2}{c_1}\right)^{pN/(N-1)}\frac{p^p}{p!}.$$

This series converges for c_2/c_1 sufficiently small, and consequently, for some constants c_1, c_2, the inequality is established.

Remark on relation of inequalities to the calculus of variations: The inequalities (1.4.2)–(1.4.5) just described can be reformulated in terms of isoperimetric variational problems. Indeed, (1.4.3), for example, can be phrased as follows: Consider the functions Σ in $W_{1,p}(\Omega)$ for which $\|u\|_{1,p} = 1$, then (i) for which numbers r is $\sup_{u\in\Sigma}\|u\|_{L_r} < \infty$, and (ii) for which numbers r and which domains Ω is $\sup_{u\in\Sigma}\|u\|_{L_r}$ *attained* by an element of Σ. The results (1.4.1)–(1.4.1′) are definitive answers for (i). To answer (ii), the compact imbedding result (described below) is crucial. Indeed, since $W_{1,r}(\Omega)$ is compactly imbedded in $L_r(\Omega)$, (1.3.31) implies that the functional $G(u) = \|u\|_{L_r}$ is continuous with respect to weak convergence in $W_{1,p}(\Omega)$. Thus if $\alpha = \sup_{u\in\Sigma}\|u\|_{L_r} < \infty$ and $u_n \in \Sigma$, $\|u_n\|_{L_r} \to \alpha$, $\{u_n\}$ will have a weakly convergent subsequence with weak limit $\bar{u}$ such that $\|\bar{u}\|_{L_r} = \alpha$. Moreover, $\|\bar{u}\|_{1,p} = 1$ since otherwise $0 < \|\bar{u}\|_{1,p} < 1$ by (1.3.11), and so for some $t > 1$, $\|t\bar{u}\|_{L_r} > \alpha$, while $\|t\bar{u}\|_{1,p} = 1$, a contradiction. Thus the following result is of great importance:

(1.4.7) **Kondrachov Compactness Theorem and Extensions** Suppose S is a set of functions in $W_{1,p}(\mathbb{R}^N)$ with a common compact support and such that the set $\{\|u\|_{1,p} | u \in S\}$ is uniformly bounded. Then S is conditionally compact in (i) $C^{0,\mu}(\mathbb{R}^N)$ for $N + \mu p < p$, and (ii) $L_r(\mathbb{R}^N)$ for $r < Np/(N-p)$. In addition, in the exceptional case (iii) $N = p$, the functional $\int_{\mathbb{R}^N} e^{ku}$ is continuous with respect to weak convergence in S, while (iv) for $r = Np/(N-p)$ the set S is not necessarily conditionally compact.

Proof of (i): (i) follows immediately from the Arzela–Ascoli theorem, the boundedness of the Sobolev integral operator (1.3.33), and the fact that for bounded domains $C^{0,\alpha}(\Omega) \subset C^{0,\alpha'}(\Omega)$ is a *compact* imbedding for $\alpha' < \alpha$.

Proof of (ii): For $W_{1,p} \to L_p$ the result can easily be derived from the M. Riesz–Tamarkin theorem. For r in the open interval $(p, Np/(N-p))$, we use (1.3.35) and the logarithmic convexity property of the L_p norm stated in Section 1.3B. Indeed, if $\{f_n\}$ is any weakly convergent sequence in L_r, $\{f_n\}$ is strongly convergent in L_p, and setting $p_* = Np/(N-p)$ for some $0 < \alpha < 1$,

$$\|f_n - f_m\|_{L_r}^r \leqslant \|f_n - f_m\|_{L_p}^{\alpha p} \|f_n - f_m\|_{L_{p^*}}^{(1-\alpha)p_*}.$$

Consequently, $\{f_n\}$ is a convergent sequence in L_r, as required. Clearly this implies S is conditionally compact.

Proof of (iii): Clearly by (1.3.16) if $u_n \to u$ weakly in $W_{1,N}(\mathbb{R}^N)$ and $\{u_n\}$ have the same compact support, then $u_n \to u$ in measure. Consequently, to show that $\int_{\mathbb{R}^N} e^{(ku_n)} \to \int_{\mathbb{R}^N} e^{(ku)}$, it suffices to observe that by (1.4.6), for some constant c, $\{\int_{\mathbb{R}^N} exp(cu_n^2)\}$ is uniformly bounded; and so the result follows by Lebesgue integration theory.

In order to find Sobolev inequalities valid for any domain $\Omega \subset \mathbb{R}^N$, with constant *independent* of the support of the functions involved, we prove the following extension of (1.4.2).

(1.4.8) Suppose $1 < p < N$ and $|\nabla u| \in L_p(\mathbb{R}^N)$, while $u \in L_s(\mathbb{R}^N)$. Then (i) $u \in L_r(\mathbb{R}^N)$ for any real r in the closed interval $[s, Np/(N-p)]$, and moreover, for some $\alpha \in [0, 1]$ there is an absolute constant C_α such that

(1.4.9) $$\|u\|_{L_r} \leqslant C_\alpha \|u\|_{L_s}^\alpha \|\nabla u\|_{L_p}^{1-\alpha},$$

where

$$\frac{1}{r} = \alpha\left(\frac{1}{p} - \frac{1}{N}\right) + \frac{1-\alpha}{s}.$$

(ii) Consequently the inequality (1.4.3) holds with the absolute constants independent of the support of u, provided $\|\nabla u\|_{L_p}$ is replaced by $\|u\|_{1,p}$ on the right-hand side of each.

Proof of (i): The inequality (1.4.9); follows from convexity properties of L_p norms, (1.4.2), and (1.3.16). The relationship among α, r, p, N, and s is determined by dimensional analysis.

Proof of (ii): Set $s = p$ in (1.4.9) and use the fact that

$$\|u\|_{L_p}^\alpha \|\nabla u\|_{L_p}^{1-\alpha} \leqslant c_p \|u\|_{1,p}$$

where c_p depends only on p.

(1.4.1′) **Corollary** For arbitrary $u \in W_{1,p}(\mathbb{R}^N)$, the inequalities (1.4.2)–(1.4.3) hold with the constants independent of the support of u as well as of u itself, provided $\|\nabla u\|_{L_p}$ is replaced by $\|u\|_{W_{1,p}}$.

(1.4.10) **Compactness Theorem** Suppose S is a bounded set of functions in $W_{1,p}(\mathbb{R}^N)$ possessing a common compact support in $\mathbb{R}^N$. Then in accord with (1.4.7), S is conditionally compact in

(i) $C^{0,\mu}(\mathbb{R}^N)$ for $N + \mu p < p$,
(ii) $L_r(\mathbb{R}^N)$ for $r < Np/(N-p)$.

Moreover, the functional $\int_\Omega e^{ku} = \mathcal{G}_k(u)$ is continuous with respect to weak convergence in $W_{1,N}(\Omega)$ for Ω a bounded domain in $\mathbb{R}^N$ for each $k > 0$.

(1.4.11) **Corollary** All the results (1.4.1)–(1.4.7) hold if the space $W_{1,p}(\mathbb{R}^N)$ is replaced by $\mathring{W}_{1,p}(\Omega)$ for any bounded domain $\Omega \subset \mathbb{R}^N$. More generally, if $p < N$, $\nabla u \in L_p(\mathbb{R}^N)$ and $u \in L_s$, then $u \in L_r$ for $r \in [1/s, Np/(N-p)]$ and

$$\|u\|_{L_r} \leqslant \text{const.}\ \|u\|_{L_s}^{\alpha} \|\nabla u\|_{L_p}^{1-\alpha}$$

for some $\alpha \in [0, 1]$.

1.4B The spaces $W_{m,p}(\mathbb{R}^N)$ and $\mathring{W}_{m,p}(\Omega)$ ($m \geqslant 1$, an integer, and $1 \leqslant p < \infty$)

The following results can be proven by iterating the results of Section 1.4A.

(1.4.12) **Theorem** Suppose $u \in W_{m,p}(\mathbb{R}^N)$ has compact support in $\mathbb{R}^N$, then:

(i) $mp > N$ and $N + p(\alpha + \mu) < mp$ imply $u \in C^{\alpha,\mu}(\mathbb{R}^N)$ and

(1.4.13) $$\|u\|_{C^{\alpha,\mu}} \leqslant \text{const.}\ \|D^m u\|_{L_p},$$

where the constant depends on supp(u) but not on u itself;

(ii) $mp \leqslant N$ and $(N - \beta p)r < Np$ imply $D^{m-\beta}u \in L_r(\mathbb{R}^N)$ and

(1.4.14) $$\|D^{m-\beta}u\|_{L_r} \leqslant \text{const.}\ \|D^m u\|_{L_p},$$

where the constant again depends on supp(u) but not on u itself;

(iii) $mp < N$ and $r = Np/(N - \beta p)$ imply $D^{m-\beta}u \in L_r(\mathbb{R}^N)$ and

(1.4.15) $$\|D^{m-\beta}u\|_{L_r} \leqslant c(\beta, N, m)\|D^m u\|_{L_p},$$

where the constant depends only on the numbers r and N but not on u or supp(u).

(1.4.16) **Corollary** For arbitrary $u \in W_{m,p}(\mathbb{R}^N)$, the inequalities (1.4.13)–(1.4.15) hold with the constants independent of supp(u) as well as u itself, provided $\|D^m u\|_{L_p}$ is replaced by $\|u\|_{W_{m,p}}$. More generally, if $u \in L_s(\mathbb{R}^N)$, $D^m u \in L_p(\mathbb{R}^N)$, then $D^{m-\beta}u \in L_r(\mathbb{R}^N)$ for $r < Np/(N - m\beta)$ and

(1.4.17) $\|D^{m-\beta}u\|_{L_r} \leqslant \text{const.}\ \|D^m u\|^{\alpha}_{L_p} \|u\|^{1-\alpha}_{L_s}, \qquad \alpha \in [0, 1].$

(1.4.18) **Corollary** Suppose S is a bounded set of functions in $W_{m,p}(\mathbb{R}^N)$ possessing a common compact support in $\mathbb{R}^N$. Then, assuming the appropriate restrictions as stated in (1.4.12), S is conditionally compact in

(i) $C^{\alpha,\mu}(\mathbb{R}^N)$ for $Np + (\alpha + \mu)p < mp$

(ii) $W_{m-\beta,r}(\mathbb{R}^N)$ for $r < Np/(N - \beta p)$.

(1.4.19) **Corollary** All the results (1.4.12)–(1.4.18) hold if the space $W_{m,p}(\mathbb{R}^N)$ is replaced by $\mathring{W}_{m,p}(\Omega)$ for any domain $\Omega \subset \mathbb{R}^N$.

Remark on arbitrary domains: For arbitrary domains $\Omega \subset \mathbb{R}^N$, the above calculus inequalities hold, provided the boundary of Ω, $\partial\Omega$, possesses appropriate smoothness properties. This result can be obtained from the so-called Calderon extension theorem that shows that for $\partial\Omega$ sufficiently smooth, there is a bounded linear transformation E from $W_{m,p}(\Omega)$ into $W_{m,p}(\mathbb{R}^N)$ such that for each $u \in W_{m,p}(\Omega)$, the restriction of Eu to Ω coincides with u.

1.4C Estimates for linear elliptic differential operators

The estimates in question are of two types:

(1) L_p estimates ($1 < p < \infty$), i.e., estimates of an operator in the integral or averaged sense, and

(2) pointwise estimates of Schauder type in the Hölder spaces $C^{m,\mu}$, $0 < \mu < 1$.

Linear differential operators $P(x, D) = \sum_{|\alpha| \leqslant m} a_\alpha(x)\, D^\alpha$ are often classified by their principal part $P_m(x, D) = \sum_{|\alpha| = m} a_\alpha(x)\, D^\alpha$. In particular, we can associate a homogeneous polynomial $P_m(x, \xi)$ in the indeterminate ξ with $P_m(x, D)$ by setting

$$P_m(x, \xi) = \sum_{|\alpha| = m} a_\alpha(x)\xi^\alpha.$$

Then $P(x, D)$ is elliptic in a domain $\Omega \subset \mathbb{R}^N$ if the polynomial $P_m(x, \xi)$ is definite in ξ for $x \in \Omega$. This clearly implies that m is an even integer. The operator $P(x, D)$ is uniformly elliptic if there is a constant $c > 0$ such that $P_m(x, \xi) \geqslant c|\xi|^m$. A differential operator $P(x, D)$ is said to be written in divergence form if

(1.4.20) $$P(x, D) = \sum_{|\alpha|,\, |\beta| \leqslant m} D^\alpha \{a_{\alpha\beta}(x)\, D^\beta\}.$$

With these preliminaries we are now in a position to state the estimates in question

(1.4.21) **(I) Gårding's inequality** Let $P(x, D)$ be a uniformly elliptic differential operator in divergence form defined on a bounded domain Ω of $\mathbb{R}^N$ with bounded measurable coefficients, but with the top order coefficient $a_{\alpha\beta}(x)$ for $|\alpha| = |\beta| = m$ uniformly continuous. Then for $u \in \mathring{W}_{m,2}(\Omega)$,

(1.4.22) $$(P(x, D)u, u)_{L_2(\Omega)} \geqslant c_1\|u\|^2_{\mathring{W}_{m,2}(\Omega)} - c_2\|u\|_{L_2(\Omega)},$$

where $c_1 > 0$ and c_2 are constants independent of u.

Proof for $m = 2$: The second order case $m = 2$ is easy, while the details of the more difficult $m > 2$ case are given in many texts and will thus not be reproduced here.

$$(P(x, D)u, u)_{L_2} = \sum_{|\alpha|, |\beta| = 1} \int a_{\alpha\beta}(x)\, D^\alpha u\, D^\beta u - R(x, u, D^\alpha u)$$

where $R(x, u, D^\alpha u)$ is a bilinear form in u and $D^\alpha u$ with $|\alpha| = 1$, with measurable, bounded coefficients. By virtue of Hölder's inequality, we may suppose for any $\epsilon > 0$ and some absolute constant $c(\epsilon)$, that

$$R(x, u, D^\alpha u) \leqslant \epsilon\|\nabla u\|^2_{0,2} - c(\epsilon)\|u\|^2_{0,2}.$$

Consequently, the result follows from the uniform ellipticity of $P(x, D)$.

(II) L_p estimates for solutions of linear elliptic equations Let L be an elliptic differential operator of order $2m$ defined on a bounded domain $\Omega \subset \mathbb{R}^N$. For simplicity, we suppose the coefficients of L are C^∞ functions and the boundary of Ω, $\partial\Omega$, is also of class C^∞. By a solution u of the elliptic boundary value problem

(1.4.23) $$Lu = f, \qquad D^\alpha u|_{\partial\Omega} = 0, \qquad f \in L_p(\Omega)$$

we mean a function $u \in L_p(\Omega)$ such that

(1.4.24) $$\int uL^*\varphi = \int f\varphi \qquad \text{for all} \quad \varphi \in C_0^\infty(\Omega),$$

where L^* is the formal adjoint of L (see Section 1.5).

(1.4.25) **Theorem** Let $1 < p < \infty$, and suppose u is a solution of (1.4.23) in the sense (1.4.24), then $u \in W_{2m,p}$ and

(1.4.26) $$\|u\|_{2m,p} \leqslant c_1\|Lu\|_{0,p} + c_2\|u\|_{0,p},$$

where the positive constants c_1, c_2 are independent of u. Furthermore, if $\operatorname{Ker} L = 0$, then $c_2 = 0$.

(III) $C^{m,\alpha}$ estimates of Schauder type

(1.4.27) **Theorem** Under the same hypothesis for L and Ω, suppose

$u \in C^{0,\alpha}(\Omega)$ is a solution of (1.4.23) with $f \in C^{\alpha}$ in the sense (1.4.24). Then $u \in C^{2m,\alpha}(\Omega)$ and

$$\|u\|_{C^{2m,\alpha}(\Omega)} \leqslant c_1\{\|Lu\|_{C^{0,\alpha}} + c_2\|u\|_{C^{0,\alpha}}\}. \tag{1.4.28}$$

Idea of the proofs for $L = \Delta$ on $\mathbb{R}^N$: We consider Green's function $G(x, y)$ for the Laplace operator Δ on $\mathbb{R}^N$:

$$G(x, y) = \begin{cases} -\omega_N^{-1}(N-2)^{-1}|x-y|^{2-N} & \text{for} \quad N > 2, \\ -(2\pi)^{-1} \ln |x-y| & \text{for} \quad N = 2. \end{cases}$$

We observe first that for f of compact support and $\Delta u = f$,

$$u = \int_{\mathbb{R}^N} G(x, y) f(y)\, dy = h * f,$$

so that (formally)

$$D_i D_j u = (D_i D_j h) * f. \tag{1.4.29}$$

Next we remark that the Calderon–Zygmund inequality (1.3.18) is applicable to this last convolution equation since for all $N \geqslant 2$, $D_j h$ is smooth on $\mathbb{R}^N - \{0\}$ and homogeneous of degree $1 - N$, and hence is a Calderon–Zygmund kernel function. Thus by (1.3.18), we find that for any fixed $p \in (1, \infty)$,

$$\|u\|_{2,p} \leqslant \text{const.}\ \|f\|_{L_p} \leqslant \text{const.}\ \|\Delta u\|_{L_p}.$$

Next, to derive the Schauder-type estimate, we apply (1.3.19) to the equation (1.4.29) and again observe that the kernel $D_i D_j h$ is a Calderon–Zygmund kernel function. Thus

$$\|u\|_{C^{2,\alpha}} \leqslant \text{const.}\ \|f\|_{0,\alpha} \leqslant \text{const.}\ \|\Delta u\|_{0,\alpha}.$$

For the proof of the general case, the reader is referred to the papers by Agmon *et al.* (1959, 1964).

1.5 Classical and Generalized Solutions of Differential Systems

The notion of generalized differentiation allows one to distinguish between classical solutions of a differential system $\mathfrak{S}$ and the solutions satisfying $\mathfrak{S}$ in some averaged (i.e., integral) sense. Thus, for example, a locally integrable function u is a "distribution" solution of the linear differential equation of order m, $Lu \equiv \sum_{|\alpha|=m} a_\alpha(x)\, D^\alpha u = f$ on a domain $\Omega \subset \mathbb{R}^N$ if for all test functions $\varphi \in C_0^\infty(\Omega)$,

$$\int_\Omega uL^*\varphi = \int_\Omega f\varphi, \qquad \text{where} \quad L^* = \sum_{|\alpha|=m} (-1)^\alpha D^\alpha a_\alpha. \tag{1.5.1}$$

On the other hand, a "*classical*" *solution* of $Lu = f$ on Ω is generally a function u, m-times continuously differentiable on Ω, that satisfies $Lu = f$ at each point $x \in \Omega$.

This broadened concept of a solution has proved extremely useful in the theory of linear partial differential equations. Moreover, generally for linear elliptic operators L, the class of distribution solutions of $Lu = f$ coincides with the class of classical solutions provided f and the coefficients $a_\alpha(x)$ are sufficiently smooth.

1.5A Weak solutions in $W_{m,p}$

In general, distributions cannot be multiplied. Thus for nonlinear problems, another type of generalized solution, "weak solution" (intermediate between the notion of classical and distribution solution), is important. We say that u is a weak solution in $W_{m,p}$ over the open set $\Omega \subset \mathbb{R}^N$ of the Dirichlet problem

$$(1.5.2) \qquad \mathcal{C}(u) = \sum_{|\alpha| \leq m} (-1)^{|\alpha|} D^\alpha A_\alpha(x, u, \ldots, D^m u) = 0, \quad D^\alpha u|_{\partial\Omega} = 0,$$

if $u \in W_{m,p}(\Omega)$ and

$$(1.5.3) \qquad \sum_{|\alpha| \leq m} \int_\Omega A_\alpha(x, u, \ldots, D^\beta u)\, D^\alpha \varphi = 0 \quad \text{for all} \quad \varphi \in C_0^\infty(\Omega).$$

Partial differential equations of the form (1.5.2) arise naturally from problems in the calculus of variations as discussed in Section 1.1C. Integration by parts in (1.5.3) shows immediately that if a weak solution of $\mathcal{C}(u) = 0$ is sufficiently smooth, it is a solution of $\mathcal{C}(u) = 0$ in the classical pointwise sense (provided of course that the functions A_α are smooth). Thus a classical solution of the Dirichlet problem for $\mathcal{C}(u) \equiv 0$ is a solution in the weak sense. As is well known, the converse statement is false (even for linear equations).

Example Let $\Omega = \{x | x = (x_1, x_2, \ldots, x_N),\ |x| < 1,\ x \in \mathbb{R}^N,\ N > 2\}$. In Ω the Dirichlet problem for the equation (*) $\Delta u = (N + 2)x_1 x_2 |x|^{-2}$ has the unique weak solution $u(x) = x_1 x_2 \log |x| \in \mathring{W}_{1,2}(\Omega)$. Since this weak solution $u(x)$ is not continuous at $x = 0$, the Poisson equation (*) has no classical solution $w(x)$. For if $w(x)$ existed, it would necessarily coincide with $u(x)$, since the only generalized harmonic function vanishing on $\partial\Omega$ is identically zero (Weyl's lemma).

The notion of weak solution just introduced has proven highly successful in the systematic study of nonlinear differential equations since it conveniently splits the study of the solutions of such systems into two parts: one part concerned with the existence and nature of weak solutions, another concerned exclusively with the smoothness of such weak solutions. Furthermore, the structure of these weak solutions can generally be reformulated in terms of abstract operators (generally nonlinear) acting between suitable Banach spaces. Thus powerful results of functional analysis become applicable in the study of nonlinear differential equations.

1.5B Regularity of weak solutions for semilinear elliptic systems

By enlarging the class of possible solutions for a system $\mathbb{S}$, extraneous and "unreal" solutions may be introduced into consideration. This contingency must be excluded. Thus any discussion of weak solutions must take up the problem of proving that such generalized solutions are sufficiently smooth to be solutions in the classical pointwise sense. Such questions are referred to as regularity theory. The simplest (and also one of the most useful) regularity theories involves boundary value problems for semilinear elliptic differential equations.

Two key facts are used. First, the known generalized solution and the linearity of the top order part of the given differential operator are used to regard the regularity problem as one for a *linear* inhomogeneous equation with the inhomogeneous term in some L_p class. Secondly, the nonlinearity of the problem is used *to iterate* the gain in smoothness obtained from the linear regularity theory. By "recycling" the smoother generalized solution, $u(x)$ say, back into the inhomogeneous term, we find (by virtue of the Sobolev inequalities) that this term is an element of some new L_r class with $r > p$. This, in turn, yields an additional improved smoothness for $u(x)$ via the linear regularity theory, and so on.

The ideas used in studying this regularity theory can be clearly seen in the study of the following semilinear Dirichlet problem. Let L be a linear elliptic differential operator of order $2m$ with smooth (say C^∞) coefficients defined on a bounded domain $\Omega \subset \mathbb{R}^N$, so that

$$Lu = \sum_{|\alpha, |\beta| \leqslant m} (-1)^{|\alpha|} D^\alpha(a_{\alpha\beta}(x) D^\beta u).$$

Then we suppose $u \in \mathring{W}_{m,2}(\Omega)$ is a weak solution of the system

$$(1.5.4) \qquad Lu = f(x, u) \qquad \text{in } \Omega;$$

$$(1.5.5) \qquad D^\alpha u|_{\partial\Omega} = 0, \qquad |\alpha| \leqslant m - 1.$$

By (1.5.3), this means that for all $\varphi \in C_0^\infty(\Omega)$,

$$(1.5.6) \qquad \sum_{|\alpha|, |\beta| \leqslant m} \int_\Omega a_{\alpha\beta}(x) D^\beta u \, D^\alpha \varphi = \int_\Omega f(x, u)\varphi.$$

Thus, if $f(x, u)$ is a C^1 function of its arguments (say) and if one knows a priori that $|u| \leqslant$ const. on Ω, one can immediately conclude that $u \in C^{2m,\alpha}$. Indeed, u can be regarded as a weak solution of the *linear* inhomogeneous equation $Lu = g(x)$ in Ω, where $g(x) = f(x, u(x)) \in L_\infty(\Omega)$. Thus by (1.4.25), $u \in W_{2m,p}$ for any $1 < p < \infty$. For sufficiently large p, by the Sobolev imbedding theorem, $g(x) = f(x, u(x)) \in C^{0,\alpha}(\Omega)$. Consequently, by (1.4.27), $u \in C^{2m,\alpha}(\Omega)$. Hence one achieves a regularity result in this case by an iterative argument based on the L_p and Schauder estimates for linear elliptic equations, combined with the Sobolev imbedding theorems (1.4.12). This argument can be sharpened as follows:

(1.5.7) Suppose $f(x, u)$ is a Lipschitz continuous function of x and u satisfying the growth condition

$$(1.5.8) \qquad |f(x, u)| \leqslant k\{1 + |u|^{\sigma}\}, \qquad 0 < \sigma < \frac{N + 2m}{N - 2m}$$

for $|u|$ sufficiently large. Then any weak solution of (1.5.4) is a classical solution in Ω and at all sufficiently smooth portions of $\partial\Omega$. Conversely, if $\sigma > (N + 2m)/(N - 2m)$, (1.5.4) may possess weak solutions that are not continuous in Ω.

We simplify the proof of this result by supposing $f(u) = k\{u^{\sigma}\}$ and postpone justification until Section 2.2, where a study of the simple composition operator $f(u) = f(x, u)$ is given. To prove the second part of the theorem, we let $L = \Delta$, the Laplace operator, and observe that if $r = |x|$ and $\Omega = \{x \mid |x| < 1\}$, then $r^{\alpha} \in W_{1,2}(\Omega)$ for $\alpha > 1 - N/2$. Moreover, a simple computation shows that $u = r^{\alpha}$ satisfies the equation (except at $x = 0$)

$$\Delta u + K(\alpha, N)u^{(\alpha-2)/\alpha} = 0, \qquad \text{where} \quad K(\alpha, N) = \alpha(2 - \alpha - N).$$

Thus for $0 > \alpha > 1 - N/2$, $v = r^{\alpha} - 1$ will be a weak solution in $\mathring{W}_{1,2}(\Omega)$ of the Dirichlet problem for an equation of the type (1.5.4), and the resulting nonlinear term $f(x, v)$ will satisfy a growth condition (1.5.8) with $\sigma = (\alpha - 2)/\alpha = 1 - 2/\alpha > (N + 2)/(N - 2)$. Finally, we observe that at $x = 0$, $v = r^{\alpha} - 1$ has a singularity and so is not continuous in Ω.

We return to the first part and assume $f(u) = ku^{\sigma}$ with $\sigma < (N + 2m)/(N - 2m)$. Then to show the weak solution $u \in \mathring{W}_{1,2}(\Omega)$ is actually smooth enough to be a classical solution, we use an iterative argument called a bootstrapping procedure; i.e., we gradually increase the regularity properties of $u(x)$ by first showing that for any compact subdomain Ω' of Ω, $u \in_{2m,p}(\Omega')$ for *some* p and then show that $u \in W_{2m,p'}$ for any finite $\tilde{p}$. Then as soon as $N < 2m\tilde{p}$, Sobolev's imbedding theorem implies $u \in C_{0,\alpha}(\Omega')$. Thus $f(x, u(x)) \in C_{0,\alpha}(\Omega')$. Consequently, the Schauder regularity theorem (1.4.27) implies that $u \in C_{2m,\alpha}(\Omega')$.

In case $N \leqslant 2m$, the argument is easy since the Sobolev imbedding theorem implies that $|u|^{\sigma} \in L_p(\Omega')$ for any finite p. Therefore, since $u(x)$ is known, we can regard the equation $Lu = ku^{\sigma}(x)$ as an *inhomogeneous linear* elliptic partial differential equation in u; whence the L_p regularity theorem (1.4.25) implies $u(x) \in W_{2m,p}(\Omega')$ for any finite p'.

In case $N > 2m$, we first show that $u \in W_{2m,p}(\Omega')$ with $p = 2N(1 + \epsilon)/(N + 2m)$ for some $\epsilon > 0$. To this end, we first observe that by the Sobolev imbedding theorem $u \in \mathring{W}_{m,2}(\Omega)$ implies $u \in L_p(\Omega)$ for $p = 2N/(N - 2m)$. Hence $k|u|^{\sigma} \in L_s(\Omega)$ for $s = p/\sigma$. Since $\sigma < (N + 2m)/(N - 2m)$, for some $\epsilon > 0$, $s = 2N(1 + \epsilon)/(N + 2m)$. Then, as in the preceding paragraph, we may consider the equation $Lu = k\{u^{\sigma}\}$ as a linear inhomogeneous elliptic equation for u and again (1.4.25) implies $u \in W_{2m,s}(\Omega')$ with $s = 2N(1 + \epsilon)/(N + 2m)$. Now we show that $u \in W_{2m,s_1}(\Omega')$ with $s_1 > s$. Since $u \in W_{2m,s}(\Omega')$, the Sobolev imbedding theorem shows that $u \in L_{p1}(\Omega')$ for $\rho_1 = Ns/(N - 2ms)$. Thus $k|u|^{\sigma} \in L_{s_1}$ for $s_1 = \rho_1/\sigma$. Now to show that u has improved regularity, we note that

$$\frac{s_1}{s} = \frac{\rho_1}{p} = \frac{(Ns/2N)(N - 2m)}{N - 2ms}.$$

After a short calculation, we find that $s_1/s = (1+\epsilon)(N-2m)/(N-2m-4m\epsilon) > 1+\epsilon$. Consequently, the L_p regularity theorem for inhomogeneous elliptic equations implies not only that $u \in W_{2m,s_1}$ with $s_1 > s$, but also that after a finite number of repetitions of this last argument, $u \in W_{2m,\bar{s}}$ for $\bar{s}$ arbitrarily large. Thus the desired result is attained.

The equation (1.5.4) is semilinear since the operator L is linear. The regularity theory of quasilinear elliptic equations of the form (1.5.2) is much more difficult unless either $m = 1$ (i.e., for second-order equations) or $N = 1$ (i.e., for ordinary differential equations). For second order ordinary differential equations, the following regularity result suffices for many applications.

(1.5.9) Suppose $p > 1$ and $u(x) \in \mathring{W}_{1,p}(a,b)$ satisfies the following integral identity for all $\varphi \in \mathring{W}_{1,p}(a,b)$

(1.5.10) $$\int_a^b \{F_z(x,u,u_x)\varphi_x + F_y(x,u,u_x)\varphi\}\,dx = 0,$$

where $F(x,y,z)$ is a C^2 function of its arguments. Then if $F_{zz}(x,y,z) \neq 0$, $u(x) \in C^2(a,b)$.

Proof: By a simple application of Sobolev's inequality, we note that the function $\int_a^x F_y(x,\tilde{u},\tilde{u}_x) = G(x)$ is Lipschitz continuous. Integrating the second term in the braces in (1.5.10) by parts we find

$$\int_a^b \{F_z(x,\tilde{u},\tilde{u}_x) - G(x)\}\zeta_x\,dx = 0.$$

Now, ζ_x is an arbitrary bounded measureable function with $\int_a^b \zeta_x\,\mathrm{dx} = 0$, so that (after a possible redefinition of $\tilde{u}(x)$ on a set of measure zero)

(1.5.11) $$F_z(x,\tilde{u},\tilde{u}_x) = G(x) + \text{const}.$$

Since $F_{zz} > 0$, the finite-dimensional implicit function theorem can be used to solve $\tilde{u}_x(x)$ in terms of $\tilde{u}(x)$ and $G(x)$, so that $\tilde{u}_x$ is Lipschitz continuous. Hence $G(x)$ must be continuously differentiable, and again (5.1.11) implies that $\tilde{u}_{zz}$ is continuous, as required.

The regularity results for quasilinear second-order elliptic partial differential equations are quite difficult to establish and need not detain us at this stage. In the first place these results are the principal object of study of a number of excellent recent monographs (see the bibliographic Notes at the end of this chapter) and secondly they are not needed for the major part of our study. Indeed, by and large, most of the nonlinear problems in mathematical physics and differential geometry that we shall discuss involve only semilinear equations, for which simple results analogous to (1.5.7) suffice.

1.6 Mappings between Finite-Dimensional Spaces

A large portion of the study of linear systems is based on knowledge derived from the theory of finite-dimensional vector spaces and linear mappings between them. Consequently, it is natural to base a study of general (nonlinear) systems on those ideas that extend from linear algebra to a nonlinear context. In this section we mention some results in this connection needed in the sequel. References for a full discussion and proofs will be given in the bibliographic notes at the end of this chapter.

1.6A Mappings between Euclidean spaces

Let Ω denote an open set in $\mathbb{R}^N$, and f a smooth mapping (of class C^p say): $\Omega \to \mathbb{R}^M$. Then one attempts to determine the mapping properties of f by studying the derivative of f, $f'(x)$ (i.e., the $N \times M$ matrix $(D_j f_i(x)$ where $f = (f_1, \ldots, f_N))$. Thus, if at a point x_0, $\operatorname{rank}(f'(x_0)) = M$, f maps a small neighborhood of x_0 onto a small neighborhood of $f(x_0)$. Such a point x_0 is called regular with respect to f. The complement in Ω, i.e., the set $\mathcal{C} = \{x \mid x \in \Omega, \operatorname{rank} f'(x) < M\}$, is called the critical set and a point $x \in \Omega$ is called a critical point. The set $\mathcal{C}$ is closed in Ω since if $x_n \to x$ in Ω, $\operatorname{rank} f'(x_n) \geqslant \operatorname{rank} f(x)$. The following additional results concerning the set C are important.

(1.6.1) Let Ω be an open subset of $\mathbb{R}^N$. Then:

(i) Sard's Theorem If $f(x)$ is a p-times continuously differentiable mapping of Ω into $\mathbb{R}^m$, then the critical values $f(\mathcal{C})$ have measure zero in $\mathbb{R}^m$, provided $N - m + 1 \leqslant p$.

(ii) A. Morse's Theorem If $F(x)$ is a N-times continuously differentiable real-valued function defined on Ω, and $\mathcal{C}$ denotes the critical points of $F(x)$, $F(\mathcal{C})$ has measure zero in $\mathbb{R}^1$.

These two results have numerous applications in analysis, and in Chapter 3 we shall discuss infinite-dimensional analogues of these facts.

For complex analytic mappings f defined on a bounded domain Ω of $\mathbb{C}^N$ into $\mathbb{C}^M$, many additional mapping properties of f are known. Thus

(1.6.2) (i) For $N = M$, then z_0 is a singular point of f if and only if f is not one-to-one near z_0.

(ii) If z_0 is a point on the set $S = \{z \mid f(z) = p\}$ in a small neighborhood U of z_0, $S \cap U$ consists of a finite number of irreducible components $\{V_i\}$ each of which is either a point or contains an analytic (nontrivial) curve. Moreover, if $V_i \neq V_j, V_i$ contains an analytic curve not contained in V_j.

(iii) If $S = \{z \mid f(z) = p\}$ is compact, S consists of a finite number of points.

Sard's theorem can be used to define the degree of a continuous mapping f: $\Omega \to \mathbb{R}^N$. This integer provides an "algebraic" count of the number of solutions of the equation $f(x) = p$ in Ω for $p \in \mathbb{R}^N$ provided $f(x) \neq p$ on $\partial\Omega$ and Ω is bounded. Our definition is given in three parts:

(i) Suppose f is a C^1 mapping of $\Omega \to \mathbb{R}^N$ and the rank of $f'(x)$ is N, so that whenever $f(x_0) = p$, the Jacobian determinant of f at x_0 $|J_f(x_0)| \neq 0$. Then we define the degree of f at p relative to Ω as

$$d(f, p, \Omega) = \sum_{f(x)=p} \operatorname{sgn} |J_f(x)|.$$

This sum is finite by virtue of the compactness of $\bar{\Omega}$, and the inverse function theorem.

(ii) Suppose now that f is known to be a C^1 mapping of $\Omega \to \mathbb{R}^N$. Then by (1.6.1(i)) we can find a sequence of regular points $\{p_n\}$ (with respect to (f, Ω)) such that $p_n \to p$ in $\mathbb{R}^N$. We then define the degree of f at p as

$$d(f, p, \Omega) = \lim_{n\to\infty} d(f, p_n, \Omega).$$

(iii) Finally, if f is only known to be continuous in Ω, there is a sequence of C^1 mappings $f_n \to f$ uniformly on $\bar{\Omega}$, and we set

$$d(f, p, \Omega) = \lim_{n\to\infty} (f_n, p, \Omega).$$

The function $d(f, p, \Omega)$ can be shown to be well defined in (ii) and (iii), and the limits exist and are independent of the approximating sequences. The basic properties of the degree function follow readily from this definition. For a bounded domain $D \subset \mathbb{R}^N$, they are

(1.6.3) (i) (*boundary value dependence*) $d(f, p, D,)$ is uniquely determined by the action of $f(x)$ on ∂D.

(ii) (*homotopy invariance*) Suppose $H(x, t) = p$ has no solution $x \in \partial D$ for any $t \in [0, 1]$, then $d(H(x, t), p\ D)$ is constant independent of $t \in [0, 1]$ provided $H(x, t)$ is a continuous function of x and t.

(iii) (*continuity*) $d(f, p, D)$ is a continuous function of $f \in C(\bar{D})$ (with respect to uniform convergence) and of $p \in D$.

(iv) $d(f, p, D) = d(f, p', D)$ for p and p' in the same component of $\mathbb{R}^N - f(\partial D)$.

(v) (*domain decomposition*) If $\{D_i\}$ is a finite collection of disjoint open subsets of D and $f(x) \neq p$ for $x \in (\bar{D} - \cup_i D_i)$, then $d(f, p, D) = \sum_i d(f, p, D_i)$.

(vi) (*Cartesian product formula*) If $p \in D \subset \mathbb{R}^n$, $p' \in D' \subset \mathbb{R}^m$ and $f\colon D \to \mathbb{R}^n$, $g\colon D' \to \mathbb{R}^m$, then $d((f, g), (p, p'), D \times D') = d(f, p, D) \cdot d(g, p', D')$ provided the right-hand side is defined.

(vii) If $f(x) \neq p$ in $\bar{D}$, then $d(f, p, D) = 0$.

(viii) (*odd mappings*) Let D be a symmetric domain about the origin, and $f(-x) = -f(x)$ on ∂D, with $f\colon D \to \mathbb{R}^n$, and $f(x) \neq 0$ on ∂D, then $d(f, 0, D)$ is an odd integer.

(ix) If $d(f, p, D) \neq 0$, then the equation $f(x) = p$ has solutions in D.

(x) Let f be a complex analytic mapping of a neighborhood U of the origin of $\mathbb{C}^n$ into itself with $f(0) = 0$. If the origin is an isolated zero of f, and the Jacobian determinant det $|J_f(0)| = 0$, then $d(f, 0, U') \geqslant 2$ for any sufficiently small open neighborhood U' of the origin.

The degree of a mapping just described is most useful in establishing qualitative properties of continuous mappings. As a simple example, we prove

(1.6.4) **Brouwer Fixed Point Theorem** Let f be a continuous mapping of the unit ball $\sigma = (x|\ |x| \leqslant 1)$ in $\mathbb{R}^N$ into itself. Then f has at least one fixed point in σ.

Proof: We show f has a fixed point either on the boundary of σ or in the interior. Suppose f has no fixed point on the boundary $\partial\sigma$ of σ, then the degree $\delta = d(x - f(x), 0, |x| < 1)$ is defined. In this case we show $\delta = 1$, by use of the homotopy $h(x, t) = x - tf(x)$, $t \in [0, 1]$, joining the identity mapping to $x - f(x)$. Since f maps σ into itself, $|f(x)| \leqslant 1$; and so the equation $h(x, t) = 0$ has no solutions on the boundary of σ, $\partial\sigma$. Indeed, if the equation had solutions on $\partial\sigma$, $t = 1$ and f would, of necessity, have a fixed point on $\partial\sigma$. Thus by the homotopy invariance property of degree mentioned in (1.6.3), $\delta = d(x, 0, |x| < 1) = 1$ (by definition). Thus f has a fixed point in σ by (1.6.3(ix)) and so the result is established.

In Chapter 5, we shall investigate the extensions of the degree to classes of mappings between Banach spaces. Moreover, this extension is used there to solve many problems of analysis.

1.6B Homotopy invariants

Homotopy groups Let $M(X, Y)$ denote the set of continuous mappings between the topological spaces X and Y. Two mappings $f, g \in M(X, Y)$ are called homotopic if there is a one-parameter family of mappings $f_t \in M(X, Y)$ depending continuously on $t \in [0, 1]$ and joining f and g, i.e., such that $f_0 = f$ while $f_1 = g$. One easily verifies that homotopy is an equivalence relation, and so partitions $M(X, Y)$ into a set of homotopy classes, written $[X, Y]$. Then one attempts to obtain information about these homotopy classes and, if possible, to determine the classes $[X, Y]$ in terms of the topological properties of X and Y.

Accordingly, an algebraic structure is introduced into $[X, Y]$. Thus if $X = S^1$ denotes the interval $[0,1]$ with end points identified, one fixes a base point $y_0 \in Y$ and restricts attention to mappings $f \in [S^1, Y]$ with $f(0) = y_0$. For two such mappings $f, g \in [S^1, Y]$, one defines $[f]\cdot[g] = [f\cdot g]$, where

$$f\cdot g(s) = \begin{cases} g(2s) & \text{for } 0 \leqslant s \leqslant \frac{1}{2}, \\ f(2s-1) & \text{for } \frac{1}{2} \leqslant s \leqslant 1. \end{cases}$$

It is easily shown that this operation satisfies the group axioms on the set $[S^1, Y]$ with fixed base point y_0. The group is called the fundamental group of Y relative to y_0, and is written $\pi_1(Y, y_0)$.

More generally, the higher homotopy groups, $\pi_n(Y, y_0)$, are defined by considering the n-cube I^n (the product of n copies of $[0, 1]$), the boundary ∂I^n of I^n, and the homotopy classes of mappings $[I^n, Y]$ that map ∂I^n into fixed base point y_0. For two such mappings, one sets

$$[f]\cdot[g] = \begin{cases} f(2t_1, t_2, \ldots, t_n) & \text{for } 0 \leqslant t_1 \leqslant \frac{1}{2}, \\ g(2t_1 - 1, t_2, \ldots, t_n) & \text{for } \frac{1}{2} \leqslant t_1 \leqslant 1, \end{cases}$$

and again verifies that this operation makes the above homotopy classes into a group $\pi_n(Y, y_0)$. The group $\pi_n(Y, y_0)$ can also be described by considering the elements of $[S^n, Y]$ that map a fixed point $s_0 \in S^n$ into y_0. Indeed, if the boundary ∂I^n of I^n is identified with the point s_0, the quotient $I^n/\partial I^n$ is topologically equivalent to S^n; and so the elements of $\pi_n(Y, y_0)$ can be identified with the homotopy classes $[S^n, Y]$ that map a fixed point $s_0 \in S^n$ into y_0. If Y is simply connected, $\pi_n(Y, y_0)$ does not depend on the base point y_0, so that the notation $\pi_n(Y)$ is used.

Of prime importance in homotopy theory is the calculation of $\pi_k(S^n)$, the homotopy groups of spheres. The following results are well known:

(1.6.5) $\qquad \pi_k(S^n) \approx \{0\} \qquad \text{for} \quad k < n,$

(1.6.6) $\qquad \pi_n(S^n) \approx \mathbb{Z}.$

The notion of degree of a mapping introduced in Section 1.6A can be used to refine (1.6.6) by giving an effective method for determining the homotopy class of a given mapping f: $S^n \to S^n$. Let $\tilde{f}$ be any continuous extension of f into $\sigma = \{x \mid |x| < 1\}$. We then observe that the degree of f is $d(\tilde{f}, 0, \sigma)$. The following result of H. Hopf shows that this degree is the only homotopy invariant of mappings between spheres of the same dimension.

(1.6.7) **Theorem** Let f and g be two mappings of S^n into S^n. Then f and g are homotopic if and only if the degrees of f and g are equal.

Next we consider the properties of the homotopy groups $\pi_{n+p}(S^n)$ for $p \geqslant 0$. A fact of importance in Chapter 5 states that as $n \to \infty$, the groups $\pi_{n+p}(S^n)$ stabilize (i.e., $\pi_{n+p}(S^n) \approx \pi_{n+1+p}(S^{n+1})$ for n sufficiently large). In fact, the following result of Freudenthal and Serre holds.

(1.6.8) Let p be a nonnegative integer. Then the homotopy groups $\pi_{n+p}(S^n)$ are canonically isomorphic for any integer $n > p + 1$. Moreover, for $p > 0$, these stable groups are *finite*.

In this connection, the notion of *suspension* of a mapping is crucial. To define this notion, let f be a mapping of $S^r \to S^n$, then the *suspension of* f,

$S(f)$, is a mapping of $S^{r+1} \to S^{n+1}$ that is an extension of f (provided we regard S^r and S^n as equators of S^{r+1} and S^{n+1}, respectively). $S(f)$ is constructed by continuously mapping the northern hemisphere of S^{r+1} into the northern hemisphere of S^{n+1}, and similarly for the respective southern hemispheres. Suspension induces a homomorphism E: $\pi_r(S^n) \to \pi_{r+1}(S^{n+1})$ and, in fact, this homomorphism yields the isomorphism of (1.6.8).

The *unstable homotopy groups*, $\pi_{n+p}(S^n)$ for $1 < n \leqslant p + 1$ exhibit particularly interesting properties, and in fact their determination still poses rather deep topological problems. Hopf showed that $\pi_3(S^2)$ is infinite and is in fact isomorphic to the additive group of integers, while $\pi_4(S^3) \simeq \mathbb{Z}_2$, the Abelian group of two elements. In fact every group of the form $\pi_{2n-1}(S^n)$ for n an even integer is infinite, and consequently we conclude that much information is lost under the suspension operation for $p > 0$. Thus *the utilization of these unstable groups in analysis is an interesting problem, which we shall discuss briefly in Chapter 5.*

The homotopy classification of restricted classes of mappings between infinite-dimensional Banach spaces is of great importance in the sequel. To illustrate this point, consider the problem of solvability for nonlinear operator equations. A natural method of approach consists in deforming a given equation to a simpler one in such a way that the solvability of the simpler equation implies the solvability of the given equation. This problem and its relation to infinite-dimensional homotopy will be discussed in Chapter 5.

1.6C Homology and cohomology invariants

(i) Singular homology groups Let δ_p denote the standard Euclidean simplex in $\mathbb{R}^{p+1}$. Then a singular p-simplex defined on a topological space X is a continuous mapping σ of δ_p into X. A singular p-chain on E over an additive abelian group $\mathcal{G}$ is a formal linear combination $c = \sum g_i \sigma_i$ of singular p-simplices σ_i with coefficients g_i in $\mathcal{G}$. The set of such chains $C_p(X, \mathcal{G})$ forms an additive abelian group. If f: $X \to X'$ is a continuous mapping and $c \in C_p(X, \mathcal{G})$, then we can define an induced homomorphism f_1: $C_p(X, \mathcal{G}) \to C_p(X', \mathcal{G})$ by setting $f_1(\sum g_i \sigma_i) = \sum g_i f(\sigma_i)$.

Now if $\delta_p = [x_1, x_2, \ldots, x_p]$ denotes the standard Euclidean p-simplex, then we define the boundary operator d on the associated singular simplex by setting

$$d(x_0, x_1, \ldots, x_p) = \sum_{i=1}^{p} (-1)^i [x_0, x_1, \ldots, \hat{x}_i, \ldots, x_p],$$

where a circumflex denotes the omission of that vertex. For a general singular p-simplex $\sigma(x_1, \ldots, x_p)$ we set $d\sigma = \sigma_1 d(x_0, x_1, \ldots, x_p)$; while for a general element $\alpha \in C_p(X, \mathcal{G})$, $\alpha = \sum g_i \sigma_i$, we extend d by linearity so that $d\alpha = \sum g_i \, d\sigma_i$. Thus d is a homomorphism of $C_p(X, \mathcal{G}) \to C_{p-1}(X, \mathcal{G})$ and $d^2 = 0$.

The kernel of d: $C_p(X, \mathcal{G}) \to C_{p-1}(X, \mathcal{G})$, denoted $Z_p(X, \mathcal{G})$, is called the *p-dimensional cycle group* of X over $\mathcal{G}$; while the image of d: $C_{p+1}(X, \mathcal{G}) \to C_p(X, \mathcal{G})$, denoted $B_p(X, \mathcal{G})$, is called the p-dimensional boundary group of X over $\mathcal{G}$. Now the associated quotient group, denoted $H_p(X, \mathcal{G})$, is called the p-dimensional homology group of X over $\mathcal{G}$, i.e.,

$$(1.6.9) \qquad H_p(X, \mathcal{G}) \equiv Z_p(X, \mathcal{G}) / B_p(X, \mathcal{G}).$$

This definition can be extended by considering a subspace Y of X. Indeed, the homomorphism d: $C_p(X, \mathcal{G}) \to C_{p-1}(X, \mathcal{G})$ maps the subgroup $C_p(Y, \mathcal{G}) \to C_{p-1}(Y, \mathcal{G})$. Thus d induces a homomorphism

$$d_p'\colon C_p(X, \mathcal{G}) / C_p(Y, \mathcal{G}) \to C_{p-1}(X, \mathcal{G}) / C_{p-1}(Y, \mathcal{G})$$

with $d_{p-1} d_p' = 0$. Denoting the kernel of d_p by $Z_p(X, Y, \mathcal{G})$, and the image of d_p by $B_p(X, Y, \mathcal{G})$, we then define the pth relative homology group as

$$(1.6.10) \qquad H_p(X, Y, \mathcal{G}) = Z_p(X, Y, \mathcal{G}) / B_p(X, Y, \mathcal{G}).$$

Clearly, (1.6.10) coincides with (1.6.9) if $Y = \varnothing$.

The rank of the Abelian group $H_q(X, A)$ is called the qth Betti number, $R_q(X, A)$, of the pair (X, A); and the alternating sum $\chi(X, A) = \sum_{q=0}^{\infty} (-1)^q R_q(X, A)$ is called the Euler–Poincaré characteristic of (X, A).

In the sequel, it will be important to determine the homology of some well-known spaces. Thus if $E^N = \{x \mid |x| \leqslant 1, x \in \mathbb{R}^N\}$ and $S^{N-1} = \partial E^N$, we find:

$$H_q(E^N, S^{N-1}, \mathcal{G}) = \begin{cases} 0 & \text{if } q \neq N, \\ \mathcal{G} & \text{if } q = N; \end{cases}$$

$$H_q(S^{N-1}, \mathcal{G}) = \begin{cases} 0 & \text{if } q \neq 0, \quad N-1, \\ \mathcal{G} & \text{if } q = 0, \quad N-1 \quad (\text{for } N \neq 1); \end{cases}$$

$$H_0(S^0, \mathcal{G}) = \mathcal{G} \oplus \mathcal{G}.$$

(ii) Singular cohomology groups The singular cohomology groups of a topological space $X, H^p(X, \mathcal{G})$ relative to a fixed abelian group $\mathcal{G}$ can be defined from the associated singular homology groups, formally, by duality. Actually, the singular cohomology of X has an additional ring structure by defining a "cup" product between elements of the cohomology groups. In Chapter 6, we shall make use of this structure in estimates of the number of critical points of a functional $\mathcal{G}(u)$ defined on an infinite-dimensional manifold X.

(iii) Finite-dimensional critical point theory of M. Morse The singular homology theory just defined is of fundamental importance in the Morse critical point theory of C^2 real-valued functions defined on a finite-dimensional smooth manifold. This theory begins by classifying the critical points of a C^2 real-valued function $F(x)$ defined on $\mathbb{R}^N$. The simplest critical points are called nondegenerate, and consist of the points x_0 at which $\nabla F(x_0) = 0$ but, the Hessian determinant of F at x_0, det $|H_F(x_0)|, \neq 0$. Such points are isolated and can, in fact, be classified by the dimension q of the vector space on which the quadratic form $F''(x_0)\xi \cdot \xi$ is negative definite for $\xi \in \mathbb{R}^N$. This number q is called the index of the critical point x_0, and a lemma of M. Morse shows that if $x = 0$ is a nondegenerate critical point of index q for a C^2 real-valued function $F(x)$, then there is a local coordinate system $(y_1, \ldots, y_N)$ in a neighborhood U of 0 such that

$$F(y) - F(0) = -\sum_{i=1}^{q} y_i^2 + \sum_{i=q+1}^{N} y_i^2. \tag{1.6.11}$$

More generally, if x_0 is any critical point of the C^2 real-valued function $F(x)$ defined on $\mathbb{R}^N$, then x_0 is called degenerate if det $|H_F(x_0)| = 0$. If x_0 is a degenerate but *isolated* critical point of $F(x)$, we may classify x_0 by its *Morse type numbers*, i.e., by the number of nondegenerate critical points of various indices equivalent to x_0. More precisely, the type number of an isolated critical point x_0 of a real-valued function $F(x) \in C^2(\mathbb{R}^N)$, with $F(x_0) = c_0$, is the sequence of positive integers $(m_0, m_1, m_2, \ldots,)$ defined by setting each

$$m_q = R_q(F^{c_0+\epsilon} \cap O(x_0), F^{c_0-\epsilon} \cap O(x_0); \mathbb{Z})$$

(i.e., the qth Betti number of the pair $(F^{c_0+\epsilon} \cap O(x_0), F^{c_0-\epsilon} \cap O(x_0))$ relative to $\mathbb{Z}$), where $\epsilon > O$ is a sufficiently small number, $F^c = \{x \mid F(x) \leqslant c\}$, and $O(x_0)$ is a sufficiently small neighborhood of x_0. The following facts are known about these type numbers.

(1.6.12) The type numbers $(m_0, m_1, m_2, \ldots,)$ of an isolated critical point x_0 of a real-valued function $F(x) \in C^2(\mathbb{R}^N)$ are finite, and $m_q = 0$ for $q > N$.

(1.6.13) The type numbers of a nondegenerate critical point of index q are

$$m_i = \begin{cases} 0 & \text{for } i \neq q, \\ 1 & \text{for } i = q. \end{cases}$$

(1.6.14) The type numbers of an isolated critical point x_0 of C^2 real-valued function $F(x)$ are lower semicontinuous functions of F in the sense that if G has only nondegenerate critical points on a small neighborhood U of x_0, and G is sufficiently close to F in $C^1(U)$, then G has at least m_q nondegenerate critical points of index q in U for $q = 0, 1, 2 \ldots$.

Furthermore, suppose $\mathfrak{M}$ is a compact smooth N-dimensional manifold. Then the notions of nondegenerate critical point and degenerate critical point, index of nondegenerate critical point, and type number can be defined in terms of our definitions on an open set in $\mathbb{R}^N$ by using local coordinate systems. Indeed, each of these notions is invariant under a local diffeomorphism.

On such a compact smooth manifold $\mathfrak{M}$, it is easily proven (by Sard's theorem) that the family of real-valued functions $F(x)$ that possess only nondegenerate critical points on $\mathfrak{M}$ is open and dense in $C^2(\mathfrak{M})$. Moreover, for any such function defined on $\mathfrak{M}$, the set $\mathfrak{M}^a = \{x \mid F(x) \leqslant a\}$ is a deformation retract of $\mathfrak{M}^b = \{x \mid F(x) \leqslant b\}$, provided $[b, a]$ contains no critical level of $F(x)$. On the other hand, if $F^{-1}[a, b]$ contains a single critical point of index q, $\mathfrak{M}^b \approx \mathfrak{M}^a \cup E^q$, i.e., $\mathfrak{M}^b$ is homeomorphic to the disjoint union of $\mathfrak{M}^a$ and a cell E^q of dimension q.

Finally we mention an interesting and useful relation that holds between the Morse indices of a C^2 real-valued function $F(x)$ defined in the neighborhood of an isolated critical point $x_0 \in \mathbb{R}^N$ and the Brouwer degree of the mapping $f = \operatorname{grad} F$ relative to a sufficiently small sphere σ_ϵ centered at x_0. Indeed, under rather general boundary conditions on $\partial\sigma_\epsilon$, if $M_F(x_0) = (m_0, m_1, \ldots, m_N)$, the following formula holds:

$$(1.6.15)\qquad d(f, x_0, \sigma_\epsilon) = \sum_{i=0}^{N} (-1)^i m_i .$$

NOTES

A Historical note on systematic approaches to nonlinear problems of analysis

Nonlinear problems of analysis arose naturally with the advent of calculus. Explicit and ingenious methods of solution abound in the mathematical literature of the seventeenth and eighteenth centuries. This work led Euler and Lagrange to consider the general theory of the calculus of variations. Moreover, in attempting to put Newton's method of undetermined coefficients on a rigorous basis Cauchy was eventually led to the majorant method for analytic nonlinear problems. The widespread use of this method of proof persists to the present day. Cauchy also used minimization methods (the method of steepest descent) systematically in the study of the zeros of simultaneous algebraic or transcendental equations over the reals.

However, a new dimension to our subject was added by Poincaré beginning with his thesis in the 1870's. Poincaré focused attention on the qualitative aspects of nonlinear poblems and thus opened a whole new variety of questions for mathematical investigation. Motivated by a systematic study of physics and geometry, Poincaré introduced new concepts in such diverse areas as bifurcation theory (a term that Poincaré himself coined), the calculus of variations in the large, application of toplogical methods to the study of periodic solutions of systems of ordinary differential equations, to mention only a few.

Hilbert's well-known lecture at the International Congress of 1900 contained a number of intriguing nonlinear problems for analysis and in particular stimulated research on nonlinear elliptic partial differential equations. This last topic proved to be a decisive one for progress on a more abstract level. In particular, the results obtained by S. Bernstein in the early part of the twentieth century on Hilbert's problems for nonlinear elliptic partial differential equations were sufficiently general to provide a basis for later abstraction and generalization.

Somewhat earlier Picard introduced the idea of successive approximation into nonlinear analysis. This idea was a natural extension of the Cauchy majorant method and was subsequently extended by S. Banach in his thesis of 1920 to the contraction mapping principle. This paper marked the quiet birth of nonlinear functional analysis. Other key results of this period include E. Schmidt's work on nonlinear integral equations and Liapunov's study of bifurcation phenomena associated with rotating figures of equilibrium.

A key paper in the development of nonlinear analysis was the paper "Invariant points in function space" by Birkhoff and Kellogg that appeared in 1922. This paper inspired much research on fixed point theorems in infinite-dimensional spaces, as well as other extensions of algebraic topology to analysis. The most penetrating research was due to J. Schauder, who applied his abstract results systematically to problems in nonlinear elliptic partial differential equations. This development reached a high point in 1934 when the paper "Topologie et equations fonctionelles" by Leray and Schauder appeared. Cf. Cacciopoli (1931).

As a final key development in the early study of nonlinear problems we mention the advances in the calculus of variations in the large made by Marston Morse beginning in 1922 and later by Liusternik and Schnirelmann.

The Second World War virtually destroyed the Polish school of functional analysis, and the joint book on nonlinear problems of analysis planned by Banach and Schauder never appeared in print.

B Sources of nonlinear problems in mathematical economics

As an example of the type of nonlinear problems that arise in economics we mention the integrability problem that arises in the theory of consumer behavior. The empirical situation, assumed given, consists of a consumer with fixed income M acting in a commodity space of dimension $n+1$, where each commodity sells at a given price p_j (assumed to be strictly positive), and demand functions

$$x_j = f_j(p_1, p_2, \ldots, p_{n+1}, M) \qquad (j = 1, 2, \ldots, n+1);$$

these functions uniquely determine the amount of the jth commodity selected x_j as a function of prices and income. The integrability problem consists in determining conditions on the functions f_j that ensure that the consumer acts in such a way that he maximizes a "utility function" subject to budget constraints. This problem is thus a simple example of an "inverse problem in the calculus of variations." Moreover, after appropriate normalizations and reductions the problems can be studied via the solution of the nonlinear partial differential equation

$$\frac{\partial M}{\partial p_j} = f_j(p_1, {}_2, \ldots, p_{n+1}, M) \qquad (j = 1, \ldots, n+1), \qquad M(p_0) = M_0.$$

A particularly interesting situation occurs when the demand functions are assumed Lipschitz continuous but not differentiable, a case that has useful applications in mathematical economics. For further discussion of the interesting history and recent developments in this area, the interested reader is referred to the recent book "Preference, Utility and Demand" edited by L. Hurwicz, J. Chipman, and others, and to the article by Berger and Meyers (1971) in that volume.

C Dimensional analysis and integral inequalities

Many integral inequalities of Section 1.4 relating L_p and Sobolev norms hold with an absolute constant independent of the associated domain Ω of $\mathbb{R}^n$. To find additional information about this case a simple device known as dimensional analysis is useful. The device consists in noting that if an inequality holds for a given function $u(x)$, it must also hold for $u(cx)$ where c is a constant that can take any positive value. For example, suppose an inequality of the form

$$\|u\|_{L_p(\Omega)} \leqslant K\|\nabla u\|_{L_2(\Omega)}, \qquad u \in \mathring{W}_{1,2}(\Omega)$$

holds, where K is a absolute constant *independent of the size of* Ω and p is a positive number to be determined. The interested reader will easily show by dimensional analysis that the only value of p for which such an inequality can hold is $p = 2n/(n-2)$, a result totally consistent with (1.4.5).

D Weighted norms and the Kondrachov compactness theorems for unbounded domains

The Kondrachov compactness theorem (1.4.7) fails for general unbounded domains (e.g., $\mathbb{R}^N$), and as mentioned in the text, this loss of compactness is crucial for many interesting nonlinear problems. Thus, it is interesting to note that the Kondrachov compactness theorem can be extended to general unbounded domains provided appropriate weights (decaying at infinity) are introduced into the Sobolev norms. As a simple example, we mention the following result which will prove useful later in Chapter 6: Let Ω be any domain in $\mathbb{R}^n$, then if $\{u_k\}$ is a sequence of functions with uniformly bounded $W_{m,p}\Omega)$ norms, then $\{u_k/|x|^\alpha\}$ has a convergent subsequence in $L_q(\Omega)$ provided $q \geqslant p > 1$, α is a number satisfying $(\alpha - n)/q < s - n/q$, and m, p, and q are related as in Kondrachov's theorem. For further results in this direction the reader is referred to the paper of Berger and Schechter (1972) in the References.

E The Korteweg–Devries equation

The interesting Korteweg–Devries equation

$$u_t + uu_x + u_{xxx} = 0 \qquad (x, t) \in \mathbb{R}^2$$

first arose in the approximate theory of water waves. It possesses traveling wave solutions of the form $u(x, t) = s(x - ct)$ for positive number c. Here $s(x) = 3c\ \mathrm{sech}^2(x\sqrt{c}/2)$ and is called a "solitary wave" or "soliton." Moreover, it has been observed that any one solution of this equation vanishing as $|x| \to \infty$ can be regarded asymptotically as $|t| \to \infty$ as a superposition of a finite number of solitons. This equation also possesses an infinite number of integrals of motion, and in a certain sense is "integrable." See the papers by Lax (1968) and Zakharov and Faddeev (1971) for further discussion.

F Bibliographic notes

Section 1.1: Early uses of qualitative nonlinear analysis in studying closed geodesics on manifolds can be found in Poincaré (1905), Birkhoff (1927), and Morse (1934). Poincaré's conjecture concerning the existence of at least three closed simple geodesics on an ovaloid has provided motivations for many deep researches on nonlinear problems beginning with Ljusternik and Schnirelmann (1930).

The reader is referred to Courant (1950) for a historical discussion of Plateau's problem. The generalization of this problem to higher dimensions has proved a remarkable achievement of recent years requiring the study of geometric measure theory, as in Federer (1969). This topic is however outside the scope of the present book. (See Nitsche, 1974.)

The uniformization of algebraic curves via nonlinear partial differential equations, as described here, is discussed in Poincaré (1890) (cf. Berger, 1969). A good discussion of the equation (1.1.6) concerning conformal metrics of prescribed Gaussian curvature can be found in Kazdan and Warner (1974). The paper of Yamabe (1960) has provided much of the impetus for modern research on the existence of metrics with prescribed curvature properties and the smooth solutions of nonlinear partial differential equations.

The nonlinear aspect of the deformation problem for complex structures on higher dimensional complex manifolds is mentioned in Nirenberg (1964).

A fine survey of nonlinearity in classical mathematical physics can be found in von Kármán (1940). It is remarkable that the abstract structures underlying these nonlinear problems have been so little studied. For further references concerning these topics we recommend the following recent books: Szebehely (1967) for celestial mechanics, Volmir (1967) for nonlinear plates and shells, and Batchelor (1967) for fluid mechanics. The latter book contains many fascinating pictures of vortex rings.

Unfortunately the literature of contemporary mathematical physics is so diverse that unified discussions of the nonlinear phenomena appearing there are singularly hard to find. Our discussion of (1.1.21) is based on Wightman (1974). Einstein's monograph (1955) clearly states the importance for relativity theory of new methods in finding singularity free solutions of nonlinear partial differential equations. Interesting discussions of phase transitions can be found in Titza (1960), Landau (1937), and Brout (1967). Moreover see Rosen (1969) where the result for equation (1.2.6) is obtained.

Section 1.2: Intrinsic properties of nonlinear systems are well discussed in Heissenberg (1967). See the articles of Landau (1944) and Ruelle-Takens (1973) for the relation between turbulence and the critical dependence of nonlinear systems on parameters. Interesting remarks concerning dimension and nonlinear growth can be found in von Neumann (1957).

Historically speaking, studying nonlinear problems with methods developed for linear ones has proved the chief source of nonintrinsic difficulties. Thus the persistence of normal modes under nonlinear Hamiltonian perturbations as defined by equation (1.2.15) can be studied by the majorant method as in Liapunov (1892). However the results obtained are rather weak. A deeper investigation requires a combination of analytic and topological techniques, as in Berger (1970) and Weinstein (1974).

Section 1.3: The material discussed in this section is relatively standard. General references for proofs include the following texts: Smirnov (1964), Riesz and Nagy (1952), Schechter (1971), Dunford and Schwartz (1958, 1963), and Yosida (1965).

Section 1.4: Sobolev's theorem was originally proved in Sobolev (1938). Its sharpening (1.4.1) is due to Nirenberg (1959) and Trudinger (1967). Proofs of much of the material of this section including the Calderon extension theorem can be found in Agmon (1965).

Section 1.5: The reader will find an elementary and informative discussion of weak solutions of elliptic boundary value problems in Sobolev (1950). The use of the bootstrapping procedure to obtain regularity for solutions of semilinear elliptic boundary value problems has been known for some time. The result (1.5.7) of the text was obtained in Berger (1965).

Section 1.6: Proofs of the elementary results of differential topology discussed here can be found in Milnor (1963, 1965). For the results from algebraic topology, the reader is referred to the books Spanier (1966), Hilton (1953), and Wallace (1970). Finite-dimensional Morse theory and the Morse type numbers are well discussed in Morse (1934), Seifert and Threfall (1938), Milnor (1963), and Pitcher (1958).

G Einstein metrics on compact complex manifolds

An important problem going beyond our discussion of Section 1.1 (iv) consists in smoothly deforming a given Kähler metric g on a compact complex manifold M to a Kähler metric $\tilde{g}$ of constant scalar curvature. For higher dimensional complex manifolds, conformal deformations do not accomplish this task. Such a deformation can be achieved by deforming the given metric g to an Einstein–Kähler metric $\tilde{g}$. An Einstein metric $\tilde{g}$ is characterized by the fact that its Ricci tensor $\tilde{R}$ is proportional to $\tilde{g}$. Such a deformation was successfully established recently by S. T. Yau [*Commun. Pure and Appl. Math.*, 1978] by restricting attention to Kähler manifolds (M,g) with ample nonnegative Chern class c_1 (i.e., Kähler manifolds whose Ricci curvature tensor is sufficiently negative). Completing work of E. Calabi and T. Aubin, Yau found the desired deformation to an Einstein metric by proving the existence of a smooth solution to a single nonlinear elliptic partial differential equation globally defined over M. The result also has interesting applications in algebraic geometry, especially to the uniformization problem described in Section 1.1(iii). The nonlinear equation involved was a higher dimensional Monge–Ampere equation, and this kind of nonlinear problem represents an important direction in our subject.

H Instantons for Euclidean Yang–Mills fields

A fascinating problem connecting modern Euclidean quantum field theory and global differential geometry consists in determining all smooth (finite action) fields A defined on $\mathbf{R}^4$ that are critical points of the Euclidean Yang–Mills action functional $S(A)$ relative to the non-Abelian gauge group G. In differential geometric terms, these fields are the smooth critical points A of a functional obtained by integrating the square of the curvature $F(A)$ of a fiber bundle with fixed non-Abelian group G and connection A over $\mathbf{R}^4$. The finite action smooth absolute minima of $S(A)$ are called instantons. For a large class of Lie groups G, these instantons have been determined quite explicitly and (from varying points of view) in recent work of Atiyah, Hitchin, and Singer following pioneering work by the Russian investigators Polyakov, Belavin, Svarc, and Tyupkin. Instantons can be classified by a topologically invariant integer k, their Pontrjagin index. Instantons of a fixed Pontrjagin index k (relative to a non-Abelian simple Lie group G) can be determined by finding smooth solutions of a nonlinear elliptic system S_k defined over $\mathbf{R}^4$ and that tend to zero at infinity. The elliptic system expresses the so-called "self-duality " condition. This nonlinear system S_k can be linearized about any field A of its domain. This linearized system turns out to determine a linear Fredholm operator T_k of generally large positive index $i(k)$ acting between appropriate Banach spaces. In fact, using elementary functional analysis and the Atiyah–Singer index theorem for linear elliptic systems, $i(k)$ can be generally explicitly determined. (See Note H at the end of Chapter 4.)

At this time it is not clear whether the Yang–Mills functional $S(A)$ admits saddle point critical points, or whether smooth solutions of infinite action will play a significant role in the theory. Critical points of the Yang–Mills functional $S(A)$, smooth, apart from prescribed singularities, are known as "merons." Merons seem to be natural both from a differential geometric and physical point of view since on the one hand they seem important for the physical problem of "quark confinement," while on the other hand they constitute a nonlinear Riemann–Roch type analogue for determining smooth metrics on manifolds apart from prescribed singularities at prescribed places.

CHAPTER 2

NONLINEAR OPERATORS

This chapter is divided into seven sections. In the first two sections we develop a calculus to deal with abstract nonlinear operators and show how concretely given operators can be reformulated in this abstract context. The next five sections take up the definition and properties of special classes of nonlinear operators. Each of these special classes will prove useful in the sequel. Two key ideas are used in the definition of these classes: first, defining a class of smooth nonlinear operators by use of the Fréchet derivative (i.e., by linearization); and secondly, defining a class by extending notions of mappings between finite-dimensional spaces.

2.1 Elementary Calculus

Many results of elementary calculus apply equally well to mappings between *infinite-dimensional* spaces. We now explore this important fact. We begin by fixing notation as follows: X and Y denote Banach spaces, and f denotes a given mapping from X to Y, and we write $f \in M(X, Y)$. We shall discuss the following desirable properties for f: boundedness, continuity (with respect to various types of covergence), integrability, differentiability, and smoothness.

2.1A Boundedness and continuity

The map $f \in M(X, Y)$ is called continuous (with respect to convergence in norm) if $x_n \to x$ in X always implies $f(x_n) \to f(x)$ in Y. f is said to be bounded if it maps bounded sets into bounded sets. f is called locally bounded if each point in the domain of f has a bounded neighborhood N such that $f(N)$ is bounded. In the case when f is linear, the two concepts of continuity and boundedness are equivalent; but this is not true in general.

Since continuous maps of a finite-dimensional Banach space X into a Banach space Y are necessarily bounded, one naturally seeks to extend this result to infinite-dimensional spaces. To accomplish this, we introduce the notion of uniform continuity of the mapping f.

(2.1.1) A mapping f is uniformly continuous if for every $\epsilon > 0$ there exists a $\delta(\epsilon) > 0$ such that $\|x - y\| < \delta$ implies $\|f(x) - f(y)\| < \epsilon$.

Clearly a uniformly continuous mapping is continuous. In fact, we have

(2.1.2) A uniformly continuous mapping is bounded.

Proof: It suffices to show that f maps any sphere $S_r = \{x \mid \|x\| \leqslant r\}$ into a bounded set. For any $\epsilon > 0$, by the uniform continuity of f, there is a $\delta > 0$ such that $\|x - y\| < \delta$ implies $\|f(x) - f(y)\| < \epsilon$ for $x,y \in S_r$. Choose n to be any positive integer satisfylng $n\delta > 2r$. Then if $a,b \in S_r$, there are n points $x_i \in S_r$ with $\|x_i - x_{i-1}\| < \delta$ and $x_0 = a, x_{n-1} = b$. Hence

$$\|f(a) - f(b)\| \leqslant \sum_{i=1}^{n-1} \|f(x_i) - f(x_{i-1})\| < (n-1)\epsilon,$$

a number independent of the choice of a and b; from which the result follows.

Actually (besides continuity with respect to convergence in norm) there are three distinct and important notions of (sequential) continuity for mappings f between general Banach spaces X and Y. These notions are obtained by considering the possible actions of f on the weak as well as the strong topologies of X and Y. Thus a map $f \in M(X, Y)$ may (i) map strongly convergent sequences in X into weakly convergent sequences in Y, (ii) map weakly convergent sequences in X into weakly convergent sequences in Y, or (iii) map weakly convergent sequences in X into strongly convergent sequences in Y. This last continuity property (iii) is called *complete continuity* since it implies the other two. Property (ii) is called demicontinuity. The alternative notions of continuity are sometimes useful in proving boundedness of a map f independently of uniform continuity assumptions. In fact, we have

(2.1.3) Let X be a reflexive Banach space and $f \in M(X, Y)$. If f maps weakly convergent sequences in X into sequences weakly convergent in Y, then f is bounded.

Proof: We argue by contradiction. Suppose there is a bounded sequence $\{x_n\}$ in X such that $\|f(x_n)\| \to \infty$. By the reflexivity of X, $\{x_n\}$ has a weakly convergent subsequence $\{x_{n_j}\}$ (say). By hypothesis, $\{f(x_{n_j})\}$ is weakly convergent and hence, by (1.3.11): uniformly bounded. But this fact contradicts the fact $\|f(x_n)\| \to \infty$.

In the sequel continuous mappings between Banach spaces X and Y are denoted $C(X, Y)$.

2.1B Integration

Criteria for the integrability of Banach-space-valued functions can be defined by considering an associated one-dimensional integral. Suppose a function $x(t)$ is defined on a measure space $(T, \mu, \sigma(T))$ with range in a Banach space X. Then a definition by duality is as follows.

Definition $x(t)$ is integrable if there is an element $I_E(x) \in X$ for each element E of the σ-ring $\sigma(T)$ such that

$$(2.1.4) \qquad x^*(I_E(x)) = \int_E x^*(x(t))\, d\mu \qquad \text{(in the Lebesgue sense)}$$

for each $x^* \in X^*$. We set $\int_E x(t)\, d\mu = I_E(x)$. Clearly the operator I_E so defined is linear.

The Hahn–Banach theorem then ensures that $\int_E x(t)\, d\mu$ is well defined and that

$$(2.1.5) \qquad \left\| \int_E x(t)\, d\mu \right\| \leqslant \int_E \|x(t)\|\, d\mu.$$

Thus $x(t)$ is μ-integrable if the operator $I(x^*) = \int_E x^*(x(t))\, d\mu$ is in X^{**} and can be identified with the usual imbedding of $X \to X^{**}$. Hence it is difficult to decide exactly what constitutes the class of integrable functions in the sense of this definition because even if $x^*(x(t))$ is measurable and integrable for each x^*, in general $I_E(x) \in X^{**}$. Note that, for reflexive spaces X the definition works well. In general, we have

(2.1.6) **Theorem** Suppose $x(t)$ is a measurable function on a measure space $M = (T, \mu, \sigma(T))$ with values in a Banach space X (in the sense that it is a step function in M or the limit (in norm) μ-a.e. of a sequence of such step functions). Then $x(t)$ is μ-integrable if $\|x(t)\|$ is μ-integrable.

Proof: First we prove the result for countably valued functions. Suppose $x(t) = x_k$ on E_k for $k = 1, 2, \ldots$ and $\int_E \|x(t)\|\, d\mu < \infty$. Set

$$I_E(x) = \sum_{k=1}^{\infty} x_k\, \mu(E_k \cap E),$$

where E_k is a μ-measurable set. This series is absolutely convergent since

$$\sum_{k=1}^{\infty} \|x_k\|\, \mu(E_k \cap E) = \int_E \|x(t)\|\, dt < \infty.$$

Hence for any $x^* \in X^*$,

$$x^*(I_E(x)) = \sum_{k=1}^{\infty} x^*(x_k)\, \mu(E_k \cap E) = \int_E x^*(x(t))\, d\mu.$$

Thus, in this case, $I_E(x) \in X$ and $I_E(x) = \int_E x(t)\, d\mu$.

Now suppose $x(t)$ is an arbitrary measurable function with values in X. Then, by definition, there is a sequence of countably valued functions $\{x_N(t)\}$ such that (*) $\|x_N(t) - x(t)\| \to 0$ a.e. Suppose for the moment that in addition (**) $\int_E \|x_N(t) - x(t)\|\, dt \to 0$. Clearly, $\|x(t) - x_m(t)\|$ is measurable; and since $I_E(x)$ is a linear function for countably valued functions $x(t)$,

$$\begin{aligned} \left\| \int_E x_N(t)\, d\mu - \int_E x_m(t)\, d\mu \right\| &\leqslant \int_E \|x_N(t) - x_m(t)\|\, d\mu \\ &\leqslant \int_E \|x_N(t) - x(t)\|\, d\mu + \int_E \|x_m(t) - x(t)\|\, d\mu \to 0, \end{aligned}$$

$I_E(x_N) = \{\int_E x_N(t)\, d\mu\}$ is a Cauchy sequence in X. This sequence posseses a limit $\bar{x}_E$ which we define to be $I_E(x)$. Clearly $I_E(x) \in X$ and is independent of any approximating sequence $\{x_N\}$ satisfying(*) and (**) above. Now for[1] any $x^* \in X^*$, $(x^*, x_N(t)) \to (x^*, x(t))$ μ – a.e. and also

$$\int_E |(x^*, x_N(t)) - (x^*, x(t))|\, d\mu \leq \|x^*\| \int_E \|x_N(t) - x(t)\|\, d\mu \to 0.$$

Thus

$$\int_E (x^*, x(t))\, d\mu = \lim_{N\to\infty} \int_E (x^*, x_N)\, d\mu = \lim_{N\to\infty} (x^*, I_E(x_N)) = (x^*, I_E(x)).$$

Consequently $x(t)$ is integrable.

Finally, we construct an appropriate sequence $x_N(t)$ satisfying (*) and (**), provided $\int_E \|x(t)\|\, d\mu < \infty$. Since $x(t)$ is the limit of step functions, there is a countable sequence in X, $\zeta_1, \zeta_2, \ldots$ that is dense in the range of $x(t)$. Set $w_N(t) = \zeta_k$, where k is the smallest integer with $|x(t) - \zeta_k| < 1/N$. The function $w_N(t)$ is measurable and $w_N(t) \to x(t)$ for all t. Furthermore, since the composite of a continuous function and a measurable function is again measurable, $\|w_N(t)\|$ and $\|x(t)\|$ are measurable real-valued functions. Consequently, there is a sequence of simple functions such that $0 \leq g_N(t) \leq \|x(t)\|$ and $g_N(t) \to \|x(t)\|$. Let $x_N(t) = (g_N(t)/\|w_N(t)\|)w_N(t)$ for $w_N(t) \neq 0$, and $x_N(t) = 0$ otherwise. Then $x_N(t)$ is countably valued, clearly measurable, and in addition $\|x_N(t)\| \leq \|x(t)\|$ with $x_N(t) \to x(t)$. Since $\|x(t)\|$ is integrable, $\|x_N(t)\|$ is integrable. Also, $\|x(t) - x_N(t)\| \leq 2\|x(t)\|$, so that by the Lebesgue dominated convergence theorem, $\int_E \|x_N(t) - x(t)\|\, d\mu \to 0$. Hence $\{x_N(t)\}$ is the desired sequence satisfying (*) and (**).

(2.1.7) **Corollary** Let $x(t)$ be a continuous function of $[a, b] \to X$. Then $x(t)$ is integrable and (as usual)

$$\int_a^b x(t)\, dt = \lim_{N\to\infty} \sum_{n=0}^{N} x(t_n^*)\{t_{n+1} - t_n\} \quad \text{as } |t_{n+1} - t_n| \to 0$$

with $t_n^* \in [t_n, t_{n+1}]$.

Proof: The function $x(t)$ is uniformly continuous on $[a, b]$, and hence the step functions are dense in the Banach space of continuous functions $C([a, b]; X)$.

For further results concerning integration we refer the reader to Gross (1964).

2.1C Differentiation

The two main notions of differentiability at a point for an operator $f \in M(X, Y)$ are defined as follows:

Definition $f \in M(X, Y)$ is *Fréchet differentiable* at x_0 if there is a linear operator $A \in L(X, Y)$ such that in a neighborhood U of x_0,

$$\|f(x) - f(x_0) - A(x - x_0)\| = o(\|x - x_0\|).$$

In this case we write $A = f'(x_0)$, and $f'(x_0)$ is called the Fréchet derivative

[1] The notations (x^*, y) and $x^*(y)$ for bounded linear functionals on X are used interchangeably in this chapter.

of f at x_0. If the mapping $x \to f'(x)$ of $X \to L(X, Y)$ is continuous at x_0, f is called C^1 at x_0.

Definition $f \in M(X, Y)$ is *Gateaux differentiable* at x_0 if there is an operator $df(x_0, h) \in M(X \times X, Y)$ such that

$$\lim_{t\to 0} \| f(x_0 + th) - f(x_0) - t\, df(x_0, h)\| = 0$$

for $(x_0 + th) \in U$, a neighborhood of x_0. Furthermore, $df(x_0, h)$ is called the Gateaux derivative of f at x_0, and we write

$$\frac{d}{dt} f(x_0 + th)|_{t=0} = df(x_0, h).$$

Some obvious properties are:

(2.1.8) The Fréchet and Gateaux derivatives are unique.

(2.1.9) $df(x_0, \beta h) = \beta\, df(x_0, h)$ for any scalar β.

(2.1.10) The Gateaux derivative commutes with bounded linear functionals, i.e., if $y^* \in Y^*$ and $f \in M(X, Y)$ is Gateaux differentiable at x_0,

$$\frac{d}{dt} (y^*, f(x_0 + th))|_{t=0} = (y^*, df(x_0, h)).$$

(2.1.11) If f is Gateaux differentiable at $x_0 + th$ $(0 \leqslant t \leqslant 1)$,

$$f(x_0 + h) - f(x_0) = \int_0^1 df(x_0 + th, h)\, dt.$$

Indeed, by (2.1.10),

$$\int_0^1 \frac{d}{dt} (y^*, f(x_0 + th))\, dt = \int_0^1 (y^*, df(x_0 + th, h))\, dt$$

so that $(y^*, f(x_0 + th) - f(x_0)) = (y^*, \int_0^1 df(x_0 + th, h))\, dt$ by the definition of integration. Since $y^* \in Y^*$ is arbitrary, the result is established.

The following result can easily be proven by the interested reader.

(2.1.12) The Fréchet differentiability and derivative of a map $f \in M(X, Y)$ are defined independently of equivalent norms in X or Y.

The relationship between Gateaux and Fréchet differentiability is given by

(2.1.13) **Theorem** If $f \in M(X, Y)$ is Fréchet differentiable at x_0, it is Gateaux differentiable at x_0. Conversely, if the Gateaux derivative of f at x_0, $df(x_0, h)$, is linear in h, i.e., $df(x_0, \cdot) \in L(X, Y)$ and is continuous in x as a map from $X \to L(X, Y)$, then f is Fréchet differentiable at x_0. In either case we have the formula $f'(x_0)y = df(x_0, y)$.

Proof: The fact that Fréchet differentiability implies Gateaux differentiability follows immediately from the definitions. To prove the converse, we first note that by the hypothesis and (2.1.9) above we can write $df(x, h) = df(x)h$, where $df(x) \in L\ (X, Y)$ and $\|df(x, h)\| \leqslant \|df(x)\|\ \|h\|$. Thus utilizing (2.1.11) above,

$$\|f(x+h) - f(x) - df(x)h\| = \left\|\int_0^1 \{df(x+th, h) - df(x, h)\}\, dt\right\|$$
$$\leqslant \int_0^1 \|(df(x+th) - df(x), h)\|\, dt$$
$$\leqslant \int_0^1 \|df(x+th) - df(x)\|\ \|h\|\, dt = o(\|h\|).$$

(2.1.14) Maps with a uniformly bounded Fréchet derivative are uniformly continuous, and hence continuous and bounded.

Rules for Fréchet differentiation are similar to the finite-dimensional case:

(2.1.15) **The Chain Rule** Suppose X, Y, Z are Banach spaces and $U \subset X$, $V \subset Y$ are open sets. Then if $f \in C^1(U, Y)$ and $g \in C^1(V, Z)$ with $f^{-1}(V) \subset U$,

$$[gf(x)]' = g'(f(x)) \cdot f'(x).$$

(2.1.16) **Product Rule** Let $U \subset X$ be an open set and $f \in M(U, \mathbb{R}^1)$, $g \in M(U, Y)$ be differentiable. Then $h(x) = f(x) \cdot g(x)$ is differentiable and $h'(x)y = f'(x)y \cdot g(x) + f(x) \cdot g'(x)y$.

(2.1.17) Suppose U is an open set of X and $f\colon U \to Y$, where Y is the product space $Y = \Pi_{i=1}^N Y_i$. Then if $f = (f_1, f_2, \ldots, f_N)$, where $f_i\colon U \to Y_i$ is differentiable, f is differentiable and $f'(x) = (f_1'(x), f_2'(x), \ldots, f_N'(x))$.

Proof of (2.1.15): Suppose $f(x+y) = f(x) + f'(x)y + o(\|y\|)$, then

$$g \cdot f(x+y) = g(f(x) + f'(x)y + o(\|y\|))$$
$$= g(f(x)) + g'(f(x))[f'(x)y + o(\|y\|)] + o(\|y\|)$$
$$= gf(x) + g'(f(x)) \cdot f'(x)y + o(\|y\|).$$

Proof of (2.1.16): Suppose f and g are expanded in the form $f(x+y) = f(x) + f'(x)y + o(\|y\|)$. Then

$$f(x+y)g(x+y) = f(x)g(x) + [f'(x)y]g(x) + [f(x)]g'(x)y + o(\|y\|).$$

The proof of (2.1.17) is obvious, being the same as in the finite-dimensional case.

Partial derivatives of a mapping $f \in C^1(U, Y)$ can be easily defined provided $U = \Pi_{i=1}^N U_i$ and each U_i is an open subset of a Banach space X_i. Indeed, if $x = (x_1, \ldots, x_N) \in U$ with $x_i \in U_i$, the (Fréchet) partial derivative of f with respect to x_i, $D_i f(x)$, is defined by writing

$$f(x_1, x_2, \ldots, x_i + h, \ldots, x_N) - f(x_1, \ldots, x_i, \ldots, x_N) = \beta(h) + o(\|h\|),$$

where $\beta(h) \in L(X_i, Y)$, and setting $D_i f(x)h = \beta(h)$, provided this expansion is valid. Clearly, $D_i f(x) \in L(X_i, Y)$ if it exists. Moreover, just as in standard calculus text books, we can prove that if $D_i f(x) \in L(X_i, Y)$,

$$f'(x)h = \sum_{i=1}^{N} D_i f(x) h_i. \tag{2.1.18}$$

The mean value theorem of elementary calculus has the following analogue for Fréchet differentiation.

(2.1.19) **Theorem** Suppose $f \in M([a, b], X)$ is a Fréchet differentiable mapping and $\|f'(t)\| \leqslant \zeta'(t)$ for $t \in [a, b]$. Then

$$\|f(b) - f(a)\| \leqslant \zeta(b) - \zeta(a), \tag{2.1.20}$$

$$\|f(b) - f(a)\| \leqslant \sup_{\zeta \in [a, b]} \|f'(\zeta)\| \, |b - a|. \tag{2.1.21}$$

Proof: Clearly for $x^* \in X^*$, the real-valued function $(x^*, f(t))$ is differentiable, and by (2.1.11) above

$$(x^*, f(b) - f(a)) = \int_a^b \frac{d}{dt}(x^*, f(t))\, dt = \int_a^b (x^*, f'(t))\, dt.$$

Thus if x^* is a linear functional of norm 1 such that $(x^*, f(b) - f(a)) = \|f(b) - f(a)\|$, we find

$$\|f(b) - f(a)\| \leqslant \int_a^b \|f'(t)\|\, dt \leqslant \zeta(b) - \zeta(a),$$

and consequently (2.1.21) holds.

2.1D Multilinear operators

The relations between boundedness, continuity, and differentiability are well exemplified by means of multilinear operators. An operator $f \in M(X, Y)$ is called multilinear if $X = \prod_{i=1}^{N} X_i$ and for each element $x = (x_1, x_2, \ldots, x_N)$ and each integer $k = 1, 2, \ldots, N$, $f(x_1, x_2, \ldots, x_N)$ is linear in x_k with all other variables held fixed.

Clearly every multilinear map is Gateaux differentiable. However, not every multilinear map is Fréchet differentiable, and in fact we have

(2.1.22) Let $X = \prod_{i=1}^{N} X_i$ and Y be Banach spaces. Then if $f \in M(X, Y)$ is multilinear, the following facts are equivalent:

(i) f is continuous at each point $x \in X$;
(ii) f is continuous at $x = 0$;
(iii) f is bounded and there is an absolute constant $K \geqslant 0$ such that

$$f(x_1, x_2, \ldots, x_N) \leqslant K \|x_1\|_{X_1} \|x_2\|_{X_2} \cdots \|x_N\|_{X_N}; \tag{2.1.23}$$

(iv) f is Fréchet differentiable.

Proof: We show (ii)$\Rightarrow$(iii) and (iii)$\Rightarrow$(iv) since the other implications are obvious.

(ii)$\Rightarrow$(iii): If $f(x)$ is continuous at $x = 0$, there is some ball $\sigma_\delta = \{x \mid \|x\| \leqslant \delta\}$ such that $\|f(x)\| \leqslant 1$. Since $f(Kx) = K^N f(x)$, f is bounded on every sphere with center $x = 0$, so that f is a bounded operator. Thus setting $|||f||| = \sup \|f(x)\|$ over the unit ball Σ_1, $|||f||| < \infty$. Without loss of generality, we may suppose that $\|x\| = \sup_i \|x_i\|_{X_i}$. With this choice of norm on X we demonstrate (2.1.23) with $K = |||f|||$. The inequality is obvious if any $x_i \equiv 0$, so that we may suppose $x_i \neq 0$ $(i = 1, \ldots, N)$. In that case, setting $y_i = x_i/\|x_i\|$ $(i = 1, \ldots, N)$, $\sup \|f(y_1, y_2, \ldots, y_N)\| \leqslant |||f|||$, and (2.1.23) follows with $K = |||f|||$.

(iii)$\Rightarrow$(iv): A simple computation shows that the Gateaux derivative at $x = (x_1, x_2, \ldots, x_N)$ is given by the formula

(†) $$df(x, h) = \sum_j f(x_1, \ldots, x_{j-1}, h_j, x_{j+1}, \ldots, x_N),$$

where $h = (h_1, h_2, \ldots, h_N)$. Clearly, $df(x, h)$ is linear in h; and so to establish (iv) we apply (2.1.23) after demonstrating that (a) $df(x, h) \leqslant c\|h\|$, and (b) $f'(x)h = df(x, h)$. If we suppose that $\|x_i\| \leqslant M$ $(i = 1, \ldots, N)$, (a) follows immediately from (†) by virtue of (iii). Thus, defining $df(x, h) = f'(x)h$, we find by (2.1.23) that after a short computation,

$$\|f(x + h) - f(x) - f'(x)h\| = o(\|h\|).$$

Thus, f is Fréchet differentiable.

Notation: The set of continuous multilinear operators f: $X_1 \times X_2 \times \cdots \times X_N \to Y$ will be denoted by $L(X_1, X_2, \ldots, X_N; Y)$; and we observe that by (2.1.23) and with the norm

$$|||f||| = \sup\{\|f(x_1, \ldots, x_N)\|/\|x_1\|\,\|x_2\| \cdots \|x_N\|\},$$

$L(X_1, \ldots, X_N; Y)$ is a Banach space. If $X_1 = X_2 = \cdots = X_N = X$, we write this space as $L_N(X, Y)$.

A useful result in this connection is

(2.1.24) **Lemma** The Banach spaces $L(X_1, X_2; Y)$ and $L(X_1; L(X_2, Y))$ are identical up to a linear isometry.

Proof: We shall define two bounded linear mappings ζ_1 and ζ_2 that are inverses for each other with ζ_1: $L(X_1, X_2; Y) \to L(X_1; L(X_2, Y))$. To define ζ_1, let $f(x_1, x_2) \in L(X_1, X_2; Y)$ and regard x_1 as a parameter, so that $f(x_1, x_2)$ is a bounded linear mapping $g \in L(X_2, Y)$, and set $\zeta_1(f) = g$. Clearly, ζ_1 is linear, bounded, and of norm $\leqslant 1$ when regarded as a mapping of $L(X_1, X_2; Y) \to L(X_1; L(X_2, Y))$. To define ζ_2, let $g \in L(X_1; L(X_2, Y))$. Then $g(x_1)x_2 = f(x_1, x_2)$ is a bilinear mapping of x_1, x_2 with values in Y. Letting $\zeta_2(g) = f$, the desired result follows since $\zeta_2 = \zeta_1^{-1}$ maps $L(X_1; L(X_2, Y)) \to L(X_1, X_2; Y)$.

A multilinear form $f(h_1, h_2, \ldots, h_N) \in L_N(X, Y)$ is called symmetric if it is invariant under all permutations of the indices $(1. 2, \ldots, N)$, $\sigma(1, 2, \ldots, N)$. Furthermore, with any multilinear form $f(h_1, h_2, \ldots, h_N) \in L_N(X, Y)$ we can associate a symmetric multilinear form $\in L_N(X, Y)$:

(2.1.25) $$\operatorname{Sym} f(h_1, \ldots, h_N) = \frac{1}{N!} \sum_{\sigma(i_1, \ldots, i_N)} f(h_{i_1}, h_{i_2}, \ldots, h_{i_N})$$

such that $\operatorname{Sym} f = f$ if and only if f is symmetric. In addition, for any symmetric form $f(h_1, \ldots, h_N) \in L_N(X, Y)$, the polar form $f(h)$

$= f(h, h, \ldots, h)$ has the property that $f(\sum_{j=1}^{N} t_j h_j)$ can be expanded by the multinomial theorem, so that we can find $f(h_1, \ldots, h_N)$ by polarization according to the identity

$$f(h_1, \ldots, h_N) = \frac{1}{N!} \frac{\partial^N}{\partial t_1 \cdots \partial t_N} f\left(\sum_{i=1}^{N} t_j h_j \right) \Bigg|_{t_1 = t_2 = \cdots = t_N = 0} \tag{2.1.26}$$

Thus if the polar forms of two symmetric multilinear forms $f(h_1, \ldots, h_N)$ and $g(h_1, \ldots, h_N)$ are equal, then the equality of the multilinear forms, $f(h_1, \ldots, h_N) = g(h_1, \ldots, h_N)$, follows.

2.1E Higher derivatives

A (Fréchet) differentiable operator $f \in M(U, Y)$ is twice (Fréchet) differentiable at x if f': $X \to L(X, Y)$ is differentiable (in the Fréchet sense) at X. The derivative of $f'(x)$, $f''(x) \in L(X, L(X, Y)) = L_2(X, Y)$ by (2.1.24). Furthermore $f \in C^2(U, Y)$ if (a) f is twice (Fréchet) differentiable for each $x \in U$, and (b) $f''(x)$: $U \to L(X, X; Y)$ is continuous. If f is twice Fréchet differentiable, the second derivative $f''(x)$ is unique; and we shall show that $f''(x)$ is symmetric, i.e., $f''(x)(h_1, h_2) = f''(x)(h_2, h_1)$. Continuing by induction, we can define Nth-order derivatives as N-linear forms as follows: f is N-times differentiable at x if the operator $f^{(N-1)}(x)$ is differentiable (in the Fréchet sense) at x. The derivative of $f^{(N-1)}(x)$, $f^{(N)}(x) \in L(X, L(X^{N-1}, Y)) = L_N(X, Y)$ and can thus be regarded as a multilinear operator of order N. In addition, $f \in C^N(U, Y)$ if (i) for each $x \in U \subset X$, f is N-times differentiable in the above sense, and (ii) as a function of x, $f^{(N)}(x)$: $U \to L_N(X, Y)$ is continuous. Again if it exists, $f^{(N)}(x)$ is unique, and we shall show that $f^{(N)}(x)$ is a symmetric N-linear form.

Clearly, by substituting the notion of Gateaux differentiability into the definitions above, higher order Gateaux derivatives $d^N f(x, h)$ can be defined. Essentially

$$d^2 f(x, h_1, h_2) = d(df(x, h_1), h_2) = \frac{d}{dt} df(x + th_1, h_2)\Big|_{t=0}$$
$$= \frac{\partial^2}{\partial t_2\, \partial t_1} f(x + t_1 h_1 + t_2 h_2)\Big|_{t_1 = t_2 = 0}.$$

In the same way,

$$d^N f(x, h_1, h_2, \ldots, h_N) = d[d^{N-1} f(x, h_1, \ldots, h_{N-1}), \hbar_N]$$
$$= \frac{d}{dt_N} [d^{N-1} f(x + t_N h_N, h_1, h_2, \ldots, h_{N-1})]\Big|_{t_N = 0}$$
$$= D_N D_{N-1} \cdots D_1 f\left(x + \sum_{i=1}^{N} t_i h_i\right)\Bigg|_{t_1 = t_2 = \cdots t_N = 0}$$

Since the operators $D_i = \partial/\partial t_i$ and $D_j = \partial/\partial t_j$ are commutative, $d^N f(x, h_N, \ldots, h_1)$ is symmetric if it exists.

The relationship between these higher order derivatives is readily established by the following extension of (2.1.13).

(2.1.27) **Theorem** (i) If f is N times Fréchet differentiable in a neighborhood U of x and $f^N(x)(h_1, \ldots, h_N)$ denotes the Nth Fréchet derivative, then f is N-times Gateaux differentiable and

$$d^N f(x, h_1, h_2, \ldots, h_N) = f^N(x)(h_1, h_2, \ldots, h_N).$$

(ii) Conversely, if the Nth Gateaux derivative $d^N f(x, h_1, \ldots, h_N)$ of f exists in a neighborhood U of x, $d^N f(x, h_1, \ldots, h_N) \in L_N(X, Y)$, and as a function of x, $d^N f(x, h_1, \ldots, h_N)$ is continuous from U to $L_N(X, Y)$, then f is N-times Fréchet differentiable and the two derivatives are equal at x.

Proof: (i): Since $f(x)$ is k-times Fréchet differentiable ($k \leqslant N$), there is a multilinear operator $A(h_1, \ldots, h_k) \in L_k(X, Y)$ such that

$$(*) \qquad \|f^{k-1}(x + h_k)(h_1, \ldots, h_{k-1}) - f^{k-1}(x)(h_1, \ldots, h_{k-1}) - A(h_1, \ldots, h_k)\| = o(\|h_k\|)$$

uniformly on bounded sets of $(h_1, \ldots, h_{k-1})$. We proceed by induction; for $k = 1$, (i) is true by virtue of (2.1.13). Assuming the truth of (i) for $n = k - 1$, we prove (i) for $n = k$. Indeed we show $d^k f(x, h_1, \ldots, h_k) = \mathrm{A}(h_1, \ldots, h_k)$. By the induction hypothesis

$$f^{k-1}(x + h_k)(h_1, \ldots, h_{k-1}) = df^{k-1}(x + h_k, h_1, h_2, \ldots, h_{k-1})$$

and

$$f^{k-1}(x)(h_1, \ldots, h_{k-1}) = df^{k-1}(x, h_1, \ldots, h_{k-1}).$$

Thus $(*)$ implies that as $t \to 0$,

$$\|t^{-1}\{d^{k-1}f(x + th_k, h_1, h_2, \ldots, h_{k-1}) - d^{k-1}f(x, h_1, \ldots, h_{k-1})\} - A(h_1, \ldots, h_k)\| = o(1),$$

so that $f(x)$ is k-times Gateaux differentiable and $d^k f(x, h_1, \ldots, h_k) = \mathrm{A}(h_1, \ldots, h_k)$, which proves (i) for $n = k$. This completes the induction argument, and thus $f(x)$ possesses N Gateaux derivatives and $d^N f(x, h_1, \ldots, h_N) = f^N(x)(h_1, \ldots, h_N)$.

(ii): Again we proceed by induction. For $n = 1$, (ii) is true by (2.1.13). Assuming the truth of (ii) for $n = k - 1$, we prove (ii) for $n = k$. We show that $(*)$ holds with $A(h_1, \ldots, h_k) = d^k f(x, h_1, \ldots, h_k)$. By the induction hypothesis

$$f^{k-1}(x + h_k)(h_1, \ldots, h_{k-1}) - f^{k-1}(x)(h_1, \ldots, h_{k-1}) = \int_0^1 \frac{d}{ds} d^{k-1}f(x + sh_k, h_1, \ldots, h_{k-1})\, ds,$$

so that the right-hand side of (*) can be rewritten as

$$\left\| \int_0^1 d^k f(x + sh_k, h_1, \ldots, h_k)\, ds - d^k f(x, h_1, \ldots, h_k) \right\|.$$

By the mean value theorem (2.1.19), this last expression is

$$\begin{aligned} &\leqslant \|d^k f(x + \xi h_k, h_1, \ldots, h_k) - d^k f(x, h_1, \ldots, h_k)\| \\ &= o(\|h_k\|) \qquad \text{(by hypothesis of the theorem).} \end{aligned}$$

This completes our induction argument and proves (ii) for $k = N$.

(2.1.28) **Corollary** Under the hypotheses of Theorem (2.1.27) the Fréchet derivatives $f^k(x)(h_1, \ldots, h_k)$ are symmetric for $k = 2, 3, \ldots, N$.

(2.1.29) **Corollary** Suppose $f \in M(U, Y)$, where U is an open subset of X, and that the line segment $[x, x + h] \subset U$, and also

(2.1.30) $$\left\| f(x + h) - f(x) - \sum_{i=1}^{n} \frac{1}{i!} a_i(x) h^i \right\| = o(\|h\|^n),$$

where $a_i(x)h^i = a_i(x, h, \ldots, h)$ (the h repeated i times) are symmetric multilinear operators in $L_i(X, Y)$ $(i = 1, \ldots, n)$ which are continuous in x from $U \to L_i(X, Y)$. Then $f \in C^n(U, Y)$ and $a_i(x, h_1, \ldots, h_i) = f^i(x)(h_1, \ldots, h_i)$.

Proof: We show that if (2.1.30) holds then

$$a_j(x, h, h, \ldots, h) = (d^j/dt^j) f(x + th)|_{t=0} \qquad (j = 1, \ldots, n).$$

Indeed, let $y^* \in Y^*$ be arbitrary, then $g(t) = (y^*, f(x + th)) \in C^n(0, 1)$. By the one-dimensional Taylor theorem and (2.1.30),

$$g(t) = (y^*, f(x)) + \sum_{k=1}^{n} \frac{t^k}{k!} \left(y^*, \frac{d^k}{dt^k} f(x + th) \Big|_{t=0} \right) + o(|t|^n).$$

By the uniqueness of Taylor series for real-valued functions,

$$(y^*, a_k(x)h^k) = \left(y^*, \frac{d^k}{dt^k} f(x + th) \Big|_{t=0} \right) \qquad \text{for every } y^* \in Y^*.$$

Consequently,

$$a_k(x)h^k = \frac{d^k}{dt^k} f(x + th)|_{t=0}.$$

Thus, for $k = 2, 3, \ldots, n$, $a_k(x, h_1, \ldots, h_k)$ is symmetric. Hence $a_k(x, h_1, \ldots, h_k) = d^k f(x, h_1, \ldots, h_k)$ and (2.1.27) implies that

$$a_k(x, h_1, \ldots, h_k) = f^k(x)(h_1, h_2, \ldots, h_k) \qquad (k = 1, 2, \ldots).$$

Consequently, $f \in C^n(U, Y)$.

(2.1.31) **Taylor's Theorem (weak form)** If f is $(N - 1)$-times Fréchet differentiable in a neighborhood U of x and $f^N(x)$ exists, then

(2.1.32) $$\|f(x+h) - f(x) - f'(x)h - \cdots - \frac{1}{N!} f^N(x)h^N\| = o(\|h\|^N).$$

Proof: (Cartan) For $N = 1$, the result is the definition of Fréchet differentiation. Proceeding by induction, we assume the truth of the theorem for $k = n - 1$, and establish its validity for $k = n$. Consider the function

$$g(h) = f(x+h) - f(x) - f'(x)h - \cdots - \frac{1}{n!} f^{(n)}(x)h^n,$$

and calculate $g'(h)$ by applying formula (t) of (2.1.27) for the derivatives of multilinear operators. Thus,

$$g'(h) = f'(x+h) - f'(x) - \cdots - \frac{1}{(n-1)!} f^n(x)h^{n-1},$$

and by the induction hypothesis, $\|g'(h)\| = o(\|h\|^{n-1})$. Applying the mean value theorem (2.1.19), $\|g(h) - g(0)\| = o(\|h\|^n)$. Since $g(0) = 0$, $\|g(h)\| = o(\|h\|^n)$. Thus the induction argument is complete, and (2.1.32) is proven.

(2.1.33) **Taylor's Theorem** Suppose $f \in C^{n+1}(U, Y)$ and the line segment $[x, x+h] \subset U$, then

(2.1.34) $$f(x+h) = f(x) + f'(x)h + \tfrac{1}{2}f''(x)h^2 + \cdots + \frac{1}{n!} f^n(x)h^n + R_{n+1}(x, h),$$

where

$$R_{n+1}(x, h) = \int_0^1 \frac{(1-s)^n}{n!} f^{(n+1)}(x+sh)h^{n+1}\, ds.$$

Proof: Let $y^* \in Y^*$, then we apply Taylor's theorem for real-valued functions to the C^{n+1} function $g(t) = (y^*, f(x+th))$ so that for $0 \leqslant t \leqslant 1$,

(2.1.35) $$g(t) = g(0) + \sum_{k=1}^{n} \frac{1}{k!} g^{(k)}(0)t^k + R_{n+1}(t),$$

where

$$R_{n+1}(t) = \int_0^1 \frac{(1-s)^n}{n!} g^{(n+1)}(st)t^{n+1}\, ds.$$

Now $g^{(k)}(t) = (y^*, f^{(k)}(x+th)h^k)$ so that $g^k(0) = (y^*, f^{(k)}(x)h^k)$. Thus, setting $t = 1$ and applying (2.1.35),

$$(y^*, f(x+h)) = \left(y^*, f(x) + \sum_{k+1}^{n} f^{(k)}(x)h^k + R_{n+1}(1)\right),$$

where

$$(y^*, R_{n+1}(1)) = \int_0^1 \frac{(1-s)^n}{n!} (y^*, f^{n+1}(x+sh)h^{n+1})\, ds.$$

Thus by the Hahn–Banach theorem, (1.3.8) follows.

2.2 Specific Nonlinear Operators

In this section, we shall point out some concrete nonlinear operators that will be used in the sequel and discuss the representation of these operators in terms of mappings between Banach spaces. Suppose Ω is a bounded domain in $\mathbb{R}^N$, and X and Y denote Banach spaces of functions defined over Ω. Then we begin by considering the following classes:

2.2A Composition operators

Let $f(x, y)$ be a real-valued function defined on $\Omega \times \mathbb{R}^1$ such that $f(x, y)$ is continuous in y for almost all x, and for all y, $f(x, y)$ is measurable with respect to x. Here Ω is a bounded domain in $\mathbb{R}^N$. In this case, $f(x, y)$ is said to satisfy the *Carathéodory continuity conditions*. Then we consider the nonlinear composition operator $\tilde{f}(u(x))$ defined by setting

$$\tilde{f}(u(x)) = f(x, u(x)) \qquad (\text{for } u(x) \text{ Lebesgue measurable}).$$

Clearly, $\tilde{f}(u(x))$ is Lebesgue measurable; and in fact, it can be shown easily that $\tilde{f}(u(x))$ is continuous with respect to convergence in measure. Moreover, when regarded as a mapping of the Banach space $C(\bar{\Omega})$ into itself, the mapping $\tilde{f}(u(x))$ is also continuous and bounded. A deeper result concerning $\tilde{f}$ that will be used frequently in the sequel concerns the properties of $\tilde{f}$ when regarded as a mapping of $L_{p_1}(\Omega) \to L_{p_2}(\Omega)$. In particular, we state:

(2.2.1) The following statements concerning the operator $\tilde{f}(u(x))$ are equivalent under the growth condition[1]

$$(*) \qquad |f(x, y)| \leq \alpha + \beta |y|^{p_1/p_2} \qquad \text{for some constants} \quad \alpha, \beta \geq 0:$$

(i) $\tilde{f}(u(x))$ maps $L_{p_1}(\Omega)$ into $L_{p_2}(\Omega)$, $1 \leq p_1, p_2 < \infty$.
(ii) $\tilde{f}(u(x))$ is a continuous mapping of $L_{p_1}(\Omega)$ into $L_{p_2}(\Omega)$.
(iii) $\tilde{f}(u(x))$ is a bounded mapping of $L_{p_1}(\Omega)$ into $L_{p_2}(\Omega)$.

[1] Actually the growth condition (*) is a consequence of (i). (See Note A at the end of the chapter.)

The proof of these results follows from straightforward measure theoretic considerations and will be sketched in Note A at the end of this chapter.

An easy induction argument shows that under suitable Carathéodory continuity conditions the multivariable composition operator $\tilde{f}(u_1, u_2, \ldots, u_k) = f(x, u_1, \ldots, u_k)$ regarded as a mapping from $\pi_{i=1}^{k} L_{p_i}(\Omega) \to L_p(\Omega)$ is continuous and bounded if and only if the function $f(x, y_1, \ldots, y_k)$ satisfies the growth condition

$$(2.2.2) \qquad |f(x, y_1, \ldots, y_k)| \leqslant \left\{ c_0 + \sum_{i=1}^{k} c_i |y|^{p_i/p} \right\},$$

where the numbers c_i are constants.

As a simple application of (2.2.1), let us complete the proof of (1.5.7). In the special case proven there, it was assumed that $f(x, u) = k(1 + u^\sigma)$ with $\sigma < (N + 2m)/(N - 2m)$. Now, a close examination of the proof shows that the only way this explicit form was used consisted in guaranteeing the L_p boundedness for $f(x, u)$ for u in various L_p classes. Consequently, the result (2.2.1) shows that the growth condition $|f(x, u)| \leqslant k\{1 + |u|^\sigma\}$ is all that is needed to preserve these L_p boundedness properties provided $f(x, u)$ is Lipschitz continuous. Thus, a repetition of the proof given after the statement of (1.5.7) suffices for the general case.

2.2B Differential operators

A general differential operator A of order m defined over a domain $\Omega \subset \mathbb{R}^N$ is written

$$(2.2.3) \qquad Au = f(x, u, Du, \ldots, D^m u).$$

Usually, A is called an *ordinary differential operator* if $N = 1$, and a *partial differential operator* if $N > 1$. The operator A is *quasilinear* if

$$(2.2.4) \qquad A(u, v) = f(x, u, Du, \ldots, D^{m-1}u, D^m v)$$

is a linear function of v when u is held fixed. A quasilinear operator A is *semilinear* if $A(u, v) = A(u, 0) + A(0, v)$, where $A(0, v)$ is a *linear* function of v and is independent of u.

A differential operator Au is said to be written in divergence form if (2.2.3) can be written

$$Au = \sum_{|\alpha|, |\beta| \leqslant m} D^\alpha \{A_\alpha(x, D^\beta u)\}.$$

Clearly such operators are quasilinear. Operators in divergence form arise naturally since operators of this type are generally the Euler–Lagrange equations of some energy functionals of the form $I(x, u, Du, \ldots, D^m u)$ (cf. Section 1.1C).

A classification of general linear differential operators extends immediately to a classification of large classes of nonlinear differential operators. This can be carried out as follows: (i) If $f \in C^1$ and $Au = f(x, u, \ldots, D^m u)$, one associates with A its *first variation at* u, namely

$$A'(u)v = \sum_{|\alpha| \leqslant m} f_\alpha(x, u, \ldots, D^m u)\, D^\alpha v, \qquad \text{where } f_\alpha \equiv \frac{\partial f}{\partial \xi^\alpha}.$$

Then one defines the *type* of A at u by the type of the linear operator $A'(u)$. (ii) If the operator A is quasilinear, this procedure implies that the type of A is the type of the linear operator $A(0, v)$, provided that this type does not depend on a perturbation by lower order terms.

Thus, for example, nonlinear elliptic differential operators can be defined in analogy with linear elliptic differential operators. In Chapter 1, a linear differential operator of order m, $L = \sum_{|\alpha| \leqslant m} a_\alpha(x)\, D^\alpha$, is called elliptic if the characteristic form of L, $Q(x, \xi) = \sum_{|\alpha| = m} a_\alpha(x)\xi^\alpha$ is definite for each $x \in \Omega$. To extend this notion to a general operator $\mathbf{f} = f(x, u, D^\gamma u)$, $|\gamma| \leqslant m$, we suppose that f is a differentiable function of its arguments. Then, the first variation of $\mathbf{f}$ at $u_0 \in C^m(\Omega)$ is the linear differential operator

$$\mathbf{f}'(u_0)v = \sum_{|\alpha| \leqslant m} \frac{\partial f}{\partial \xi^\alpha}(x, u_0, D^\gamma u_0)\, D^\alpha v.$$

If $\mathbf{f}'(u_0)$ is an elliptic operator, $\mathbf{f}$ is called elliptic at u_0. If $\mathbf{f}$ is written in divergence form and $m = 2M$,

$$\mathbf{f}(u) = \sum_{|\alpha| \leqslant M} (-1)^{|\alpha|} D^\alpha A_\alpha(x, u, \ldots, D^M u), \tag{2.2.5}$$

this definition may be somewhat extended by saying that $\mathbf{f}$ is elliptic if the "top order part" of $\mathbf{f}(u)$ satisfies a positivity condition; i.e., for $\xi \neq \eta$,

$$\sum_{\substack{|\alpha| = M \\ |\beta| < M}} \{A_\alpha(x, \xi^\beta, \xi^\alpha) - A_\alpha(x, \eta^\beta, \eta^\alpha)\}\{\xi^\alpha - \eta^\alpha\} > 0. \tag{2.2.6}$$

The result (2.2.1) and its extension have the following interesting consequence for differential operators defined on Sobolev spaces.

(2.2.7) Suppose Ω is a bounded domain in $\mathbb{R}^N$ whose boundary $\partial\Omega$ is regular, and suppose $f(x, y_1, \ldots, y_k)$ satisfies the Carathéodory conditions and the growth conditions

$$|f(x, y_1, \ldots, y_m)| \leqslant c\left\{1 + \sum_{\alpha=1}^{m} |y_\alpha|^{\sigma_\alpha}\right\},$$

where c is an absolute constant and y_m is a vector variable. Then $f(u) = f(x, u, Du, \ldots, D^m u)$ defines a bounded continuous mapping from

$W_{m,p}(\Omega)$ to $L_s(\Omega)$ provided the numbers $\{\sigma_\alpha\}$ satisfy the inequalities

$$(*) \qquad \sigma_\alpha < \frac{1}{s}\left\{\frac{1}{p} - \frac{m-|\alpha|}{N}\right\}^{-1}$$

The result also holds for $\leqslant$, if the limiting exponent is finite.

Proof: By Sobolev's inequality, we note that for $u \in W_{m,p}(\Omega)$, $D^\alpha u \in L_{p(\alpha)}$ for $1/p(\alpha) \geqslant 1/p - (m-|\alpha|)/N$. Consequently, $|D^\alpha u|^{\sigma_\alpha} \in L_s(\Omega)$ provided $\sigma_\alpha s \leqslant p(\alpha)$, i.e., $\sigma_\alpha \leqslant p(\alpha)/s$. Thus the result follows from (2.2.1) and (2.2.2) since these results imply that it suffices to consider each term $|D^\alpha u|^{\sigma_\alpha}$ individually.

2.2C Integral operators

In the sequel we shall often meet with integral operators Au that can be written in the form

$$(2.2.8) \qquad Au(x) = \int_\Omega K(x,y) f(y, u(y))\, dy,$$

where $K(x,y)$ is some kernel function defined on $\Omega \times \Omega$. Such integral operators are often found associated with boundary value problems for semilinear partial differential equations. Thus, as a simple example: any solution u of the Dirichlet problem for a domain $\Omega \subset \mathbb{R}^N$

$$\Delta u = f(x,u), \qquad u|_{\partial\Omega} = 0,$$

can be written in the form

$$u(x) = \int_\Omega G(x,y) f(y, u(y))\, dy,$$

where $G(x,y)$ is Green's function for (Δ, Ω) relative to the null Dirichlet boundary condition. Thus, as a function of x, $G(x,y) = 0$ for $x \in \partial\Omega$ and

$$G(x,y) = \begin{cases} (2\pi)^{-1} \log|x-y| + \beta(x,y) & (\text{for } N = 2), \\ -((N-2)\omega_N)^{-1}|x-y|^{2-N} + \beta(x,y) & (\text{for } N > 2), \end{cases}$$

where for fixed x, $\beta(x,y)$ is a harmonic function of y. The operator Au defined by (2.2.8) when considered as a map of $L_p(\Omega) \to L_s(\Omega)$ can often be factored in the form $Au = L\tilde{f}(u)$, where $\tilde{f}$ is the composition mapping from $L_p(\Omega) \to L_r(\Omega)$ for some r and L is the linear integral operator

$$Lu(x) = \int_\Omega K(x,y) u(y)\, dy$$

regarded as a mapping from $L_r(\Omega) \to L_s(\Omega)$. Clearly, a sufficient condition for L to define such a bounded linear operator is that

$$\int_\Omega \int_\Omega |K(x, y)|^t \, dx \, dy < \infty, \qquad \text{where} \quad t = \max\left(s, \frac{r}{r-1}\right).$$

Another interesting example is the integral operator associated with solutions of the Neumann problem for a domain $\Omega \subset \mathbb{R}^N$,

$$\Delta u = 0, \qquad \left.\frac{\partial u}{\partial n}\right|_{\partial\Omega} = f(x, u).$$

Apart from an arbitrary constant, any solution of such an equation can be written in the form

$$u(x) = \int_{\partial\Omega} N(x, y) f(y, u(y)) \, dy,$$

where $N(x, y)$ is Green's function for the Neumann problem for Δ, and so has a representation analogous to that of $G(x, y)$ given above.

2.2D Representations of differential operators

There are several distinct approaches that can be used to represent general differential operators by abstract nonlinear mappings between Banach spaces. The methods that will be useful in the sequel are summarized as follows. (More detailed discussions are given later.)

(I) Direct composition representations If $\mathcal{A}u = f(x, D^\beta u)$, $|\beta| \leqslant m$, is a differential operator of order m defined on some domain $\Omega \subset \mathbb{R}^N$ and $f = f(x, \xi)$ is smooth in (x, ξ), say of class $C^{s,\alpha}$, then we can consider the composition operator $\mathcal{A}(u) = f(x, D^\beta u)$, $|\beta| \leqslant m$, defined for u on a Hölder space of functions $C^{s,\alpha}(\Omega)$.

Thus for $u \in C^{s,\alpha}(\Omega)$, $\mathcal{A}(u) \in C^{s-m,\alpha}(\Omega)$, provided of course that $s \geqslant m$. Such a mapping is clearly continuous and bounded from $C^{s,\alpha}(\Omega) \to C^{s-m,\alpha}(\Omega)$. In fact the Fréchet derivative of $\mathcal{A}(u)$ at $u_0 \in C^{s,\alpha}(\Omega)$, $A'(u_0)v$, is easily computed to be

$$A'(u_0)v = \sum_{|\alpha| \leqslant m} f_\alpha(x, u_0, D^\beta u_0) \, D^\alpha v, \tag{2.2.9}$$

where $f_\alpha = \partial f / \partial \xi_\alpha$. The expression on the right of (2.2.9) is just the first variation of $\mathcal{A}(u)$ at u_0. In the same way $\mathcal{A}(u)$ can be regarded as a mapping from $W_{s,p}(\Omega) \to W_{s-m,p}(\Omega)$ provided that the formal derivatives $D^\gamma f(x, \ldots, D^\beta u) \in L_p(\Omega)$, $|\gamma| \leqslant s - m$, with $u \in W_{s,p}(\Omega)$. *This last restriction can be verified by placing growth restrictions on f and its derivatives similar to those of* (2.2.7).

A useful example in studying the boundedness of A is the following:

(2.2.10) Suppose $f(x, \xi)$ is a C^∞ function defined on a domain $\Omega \times \mathbb{R}$, all of whose derivatives are bounded. Then for an integer m sufficiently large, and any $u \in \mathring{W}_{m,p}(\Omega)$,

$$\|f(x, u)\|_{m,p} \leqslant \text{const.}\{1 + \|u\|_{m,p}\}.$$

Proof: For simplicity we consider the case in which f is independent of x. Then, by the chain rule one computes that formally, for $u \in C_0^\infty(\Omega)$,

$$D^k f(u) = \sum_{j=1}^{k} c_{jk} f^{(j)}(u) \left\{ \prod_{\Sigma \beta_i = k} D^{\beta_i} u \right\} \qquad (k = 1, 2, \ldots).$$

To estimate $\|f(u)\|_{m,p}$, it suffices to estimate $\|D^m f(u)\|_{0,p}$. Since by hypothesis the $f^{(j)}(u)$ are bounded,

$$\|D^k f(u)\|_{0,p}^p \leqslant \text{const.} \| \prod_{\Sigma \beta_i = k} D^{\beta_i} u\|_{0,p}^p.$$

By Hölder's inequality with $p_i \beta_i = m$,

$$\|D^k f(u)\|_{0,p}^p \leqslant \text{const.} \prod_{\Sigma \beta_i = k} \|D^{\beta_i} u\|_{pp_i}^p.$$

Then by (1.4.17), since

$$\|D^{\beta_i} u\|_{0,pp_i} \leqslant \text{const.} \|D^m u\|_{0,p}^{1/p_i} \|u\|_{L_\infty}^{1-1/p_i},$$

and

$$\|D^k f(u)\|_{0,p} \leqslant \text{const.} \|D^m u\|_{0,p} \left\{ \prod_{\Sigma \beta_i = k} \|u\|_{L_\infty}^{1-1/p_i} \right\}.$$

In the same way we can show that

(2.2.11) $$\|f(x, u, Du, \ldots, D^\gamma u)\|_{m-\gamma,p} \leqslant \text{const.}\{1 + \|u\|_{m,p}\},$$

provided m is sufficiently large.

(II) Operators defined by Schauder inversion An approach used to great effect since the fundamental studies of J. Schauder is based on the inversion of differential operators (possibly supplemented by appropriate boundary conditions). The basic idea used to define abstract nonlinear mappings associated with boundary value problems for *quasilinear* differential operator equations $Au = g$ consists in writing $A(u) = A(u, u)$ in such a way that:

(1) for fixed elements (v, g) in carefully chosen Banach spaces (X, Y), the *linear* equation $A(v, u) = g$ has one and only one solution $u = T_g(v)$ in X; and

(2) $A(v, u)$ depends continuously on v for fixed $u \in X$.

Then the operator T_g is well defined, so that $T_g \in M(X, X)$ and the fixed points of T_g coincide with the solutions of $Au = g$.

In order to assert the existence of such fixed points of T_g, it will be important to establish the continuity and boundedness of $T_g \in M(X, X)$.

One generally establishes these facts by deriving *a priori estimates* for solutions u of the equation $A(v, u) = g$ with v and g fixed and $\|v\|_X \leqslant R$ of the form

(2.2.12) $\quad \|u\|_X \leqslant c(R)\|g\|_Y,$

where $c(R)$ is a finite positive constant independent of v and u but possibly depending on R. For example, since the differential operator $A(u)$ of order m is quasilinear, one can assume, to establish estimates, that the operator $A(v, u)$ is linear in u with u used to denote only derivatives of order m, and v derivatives of lower order. The continuity and boundedness of T_g then follow in many instances from the following:

(2.2.13) If $A(u)$ is quasilinear and the linear operator $A(v, u)$, defined by fixing v and letting u denote the mth-order derivatives in $A(u)$, admits an estimate of the form (2.2.12), then the operator T_g (so defined) is continuous and bounded as a mapping of X into itself.

Proof: Once (2.2.12) is assumed, the boundedness of T_g is immediate since sup $\|T_g(v)\| < \infty$ over $\|v\|_X \leqslant R$. The continuity of T_g is established as follows. Suppose $T_g v = u$ and $T_g \bar{v} = \bar{u}$, then $A(v, u) = g$ and $A(\bar{v}, \bar{u}) = g$. Consequently, since A is linear in its second argument, we find

$$A(\bar{v}, \bar{u} - u) = A(\bar{v}, u) - A(\bar{v}, \bar{u}) = A(\bar{v}, u) - A(v, u).$$

By virtue of the estimate (2.2.12) and the above

$$\|T_g \bar{v} - T_g v\| = \|\bar{u} - u\| \leqslant c(R)\|A(\bar{v}, u) - A(v, u)\|,$$

where $R = \max(\|v\|, \|\bar{v}\|)$. Thus as $\bar{v} \to v$, since $A(v, u)$ is continuous in v for fixed u, $T_g \bar{v} \to T_g v$, as required.

For the study of elliptic boundary value problems of order $2m$, typical pairs (X, Y) for the Schauder inversion procedure are the Hölder spaces $(C^{2m,\alpha}(\Omega), C^{0,\alpha}(\Omega))$ $(0 < \alpha < 1)$ and the Sobolev spaces $(W_{2m,p}(\Omega), L_p(\Omega))$ $(1 < p < \infty)$. Indeed, for such pairs the estimates (1.4.25)–(1.4.28) can be used to verify (2.2.12). Actually the estimates (1.4.25)–(1.4.28) prove that the operator T_g is also compact (see (2.4.7)), a fact that will be important later. For semilinear elliptic equations, the operator T_g can be exhibited quite explicitly by Green's functions. Thus, for example, if $G(x, y)$ is Green's function for the linear differential operator $Lu = \sum_{|\alpha|=m} a_\alpha(x)\, D^\alpha u$ defined on $\Omega \subset \mathbb{R}^N$ and subject to the Dirichlet boundary conditions $D^\alpha u|_{\partial\Omega} = 0$, $|\alpha| \leqslant m-1$, then the operator T_g described above for the nonlinear system

$$Lu + f(x, u, Du, \ldots, D^\beta u) = 0 \quad \text{in } \Omega, \quad |\beta| < m-1,$$
$$D^\alpha u|_{\partial\Omega} = 0 \quad \text{for} \quad |\alpha| \leqslant m-1$$

coincides with the integral operator

$$Tu(x) = \int_\Omega G(x, y) f(y, u, \ldots, D^\beta u(y))\, dy.$$

(III) Operators defined by duality For differential operators A, defined on a domain $\Omega \subset \mathbb{R}^N$, of divergence form

$$Au = \sum_{|\alpha| \leqslant m} (-1)^{|\alpha|} D^\alpha A_\alpha(x, u, \ldots, D^m u),$$

an especially effective abstract integral representative $\mathcal{Q}$, based on the reflexivity of the Sobolev spaces can often be defined. Suppose that for $u \in C_0^\infty(\Omega)$ and $\phi \in C_0^\infty(\Omega)$, we set

$$F(u, \phi) = \sum_{|\alpha| \leq m} \int_\Omega A_\alpha(x, u, \ldots, D^m u)\, D^\alpha \phi,$$

and by virtue of the inequalities of Section 1.4 we can prove the existence of a bounded function $g(r)$ such that for m a positive integer and $1 < p < \infty$,

$$|F(u, \phi)| \leq g(\|u\|_{m,p})\|\phi\|_{m,p}. \tag{2.2.14}$$

Then by virtue of (2.2.14), $F(u, \phi)$ defines a bounded linear functional in ϕ on $\mathring{W}_{m,p}(\Omega)$ (since $C_0^\infty(\Omega)$ is dense in $\mathring{W}_{m,p}(\Omega)$). Consequently, there is a unique element $\mathcal{Q}(u) \in W_{-m,q}(\Omega)$ (where q is the conjugate index of p) such that

$$F(u, \phi) = (\mathcal{Q}(u), \phi)_{m,p}. \tag{2.2.15}$$

Thus $\mathcal{Q}(u) \in M(\mathring{W}_{m,p}(\Omega), W_{-m,q}(\Omega))$ can be regarded as an abstract representation of the concrete differential operator A.

To determine precise conditions for the validity of the inequality (2.2.14) as well as the boundedness of $\mathcal{Q}$, the definition (2.2.15) is particularly effective. To establish (2.2.14) for bounded domains Ω, we note that the result (2.2.7) is particularly effective. Indeed, by Hölder's inequality

$$|F(u, \phi)| \leq \sum_{|\alpha| \leq m} \|A_\alpha(x, u, \ldots, D^m u)\|_{L_{q(\alpha)}} \|D^\alpha u\|_{L_{p(\alpha)}},$$

where $1/p(\alpha) + 1/q(\alpha) = 1$ and $L_{p(\alpha)}$ is chosen as large as possible so that the following Sobolev inequality holds

$$\|D^\alpha u\|_{L_{p(\alpha)}} \leq c\|u\|_{m,p}.$$

In particular, $1/p(\alpha) = 1/(p - (m - |\alpha|)/N)$; and so $1/q(\alpha)$ is determined. Thus to establish (2.2.14), it suffices to prove that the differential operator $A_\alpha(x, u, \ldots, D^m u)$ is a bounded mapping from $\mathring{W}_{m,p}(\Omega)$ into $L_{q(\alpha)}(\Omega)$ in such a way that

$$\|A_\alpha(x, u, \ldots, D^m u)\|_{L_{q(\alpha)}} \leq g_\alpha(\|u\|_{m,p}) \quad \text{for each } |\alpha| \leq m \tag{2.2.16}$$

for some bounded function $g_\alpha(r)$. This last inequality can be easily established by imposing growth restrictions as in (2.2.7) on the functions $A_\alpha(x, \xi_1, \ldots, \xi_m)$. Similarly, to find conditions ensuring the boundedness of $\mathcal{Q}$, we note that for $u \in \mathring{W}_{m,p}(\Omega)$,

$$\begin{aligned}
\|\mathcal{Q}u\|_{-m,q} &= \sup_{\|\phi\|=1}(\mathcal{Q}(u), \phi) \qquad \text{for} \quad \phi \in C_0^\infty(\Omega) \\
&= \sup_{\|\phi\|=1}(F(u), \phi) \\
&= c\left\{\sum_{|\alpha| \leq m} \|A_\alpha(x, u, \ldots, D^m u)\|_{L_{q(\alpha)}(\Omega)}\right\},
\end{aligned}$$

where c is some absolute constant and $q(\alpha)$ is chosen as above. The boundedness then follows once the inequalities (2.2.2) are established. Once these inequalities are proven, the continuity of $\mathcal{Q}$ is an immediate consequence of (2.2.7).

2.3 Analytic Operators

The important notion of analyticity of mappings between finite-dimensional spaces has a significant extension to the case of infinite-dimensional spaces. Here we define an appropriate concept of analyticity for mappings between Banach spaces and study various properties equivalent to analyticity.

2.3A Equivalent definitions

(2.3.1) **Definition** Let X and Y be Banach spaces over the complex numbers, and let U be a connected open set of X. Then $f \in M(U, Y)$ is complex analytic if for each $x \in U$, $h \in X$, $y^* \in Y^*$, $f(x)$ is single-valued and $(y^*, f(x + th))$ is an analytic function of the complex variable t for $|t|$ sufficiently small.

An immediate consequence of this definition is the fact that for $|t|$ sufficiently small and $x \in U$,

$$(y^*, f(x + th)) = \sum_{n=0}^{\infty} a_n(x, h) \frac{t^n}{n!},$$

where

$$a_n = \frac{d^n}{dt^n} (y^*, f(x + th))_{t=0} = \frac{n!}{2\pi i} \int_{|t|=\rho} \frac{(y^*, f(x + th))}{t^{n+1}} \, dt. \tag{2.3.2}$$

The classical Cauchy estimates then imply that $|a_n(x, h)| \leqslant (\sup_{x \in U}|(y^*, f(x))|)n!\rho^{-n}$. We shall show that $(y^*, a_n(x, h)) = (y^*, f^n(x)h^n)$ $(n = 1, 2, \ldots)$, where $f^n(x)$ denotes the nth Fréchet derivative of f at x. In that case the Hahn–Banach theorem implies that for $\|h\|$ sufficiently small,

$$f(x + h) = \sum_{n=0}^{\infty} \frac{1}{n!} f^n(x)h^n.$$

In fact the following result holds.

(2.3.3) **Theorem** Let U be an open subset of X and f be a locally bounded operator in $M(U, Y)$, then the following properties are equivalent:

(i) f is complex analytic in U;
(ii) f is Gateaux differentiable in U;
(iii) f is Fréchet differentiable in U and $f'(x)h = df(x, h)$;
(iv) f has infinitely many Gateaux derivatives $d^N f(x, h_1, h_2, \ldots, h_N)$ and

$$d^N f(x, h_1, \ldots, h_N) = D_N \cdots D_1 f\left(x + \sum_{i=1}^{N} t_i h_i\right)\Big|_{t_i=0};$$

(v) f has infinitely many Fréchet derivatives $f^N(x)h^N$, and for $N = 1, 2, \ldots,$

$$f^N(x)(h_N, h_{N-1}, \ldots, h_1) = d^N f(x, h_1, \ldots, h_N).$$

(vi) $f(x+h) = \sum_{n=0}^{\infty} (n!)^{-1} f^n(x) h^n$, i.e., to each $y \in U$ there is a positive number $r_y > 0$ such that this series converges uniformly for $\|x - y\| \leqslant r_y$ and $\|h\| \leqslant r_y$.

Proof: (i)$\Leftrightarrow$(ii): That Gateaux differentiable implies complex analytic is immediate. So suppose that f is complex analytic. To show that f is Gateaux differentiable we need only show that

$$g(s, t) = \frac{1}{t}\{f(x + th) - f(x)\} - \frac{1}{s}\{f(x + sh) - f(x)\} \to 0$$

as $t, s \to 0$ independently (i.e., $t_N^{-1}\{f(x + t_N h) - f(x)\}$ is a Cauchy sequence in Y as $t_N \to 0$). By the Cauchy integral formula for $y^* \in Y^*$,

$$(y^*, f(x + th)) = \frac{1}{2\pi i} \oint_C \frac{(y^*, f(x + \xi h))}{\xi - t}\, d\xi,$$

where C is a small circle in $\mathbb{C}^1$ with center at 0 and radius r. Hence

$$(y^*, g(s, t)) = \frac{1}{2\pi i} \oint_C (y^*, f(x + \xi h)) \frac{t - s}{(\xi - s)(\xi - t)}\, d\xi.$$

Since $(y^*, f(x + \xi h))$ is continuous, it is bounded on C; so that by the principle of uniform boundedness (1.3.25), $\|f(x + \xi h)\| \leqslant A$ on C. Therefore, for s, t sufficiently small, say at most $\frac{1}{2} r$, $\|(y^*, g(s, t))\| \leqslant 4Ar^{-2} \cdot |t - s|\, \|y^*\|$. Thus $\|g(s, t)\| \to 0$ as $s, t \to 0$.

(i), (ii)$\Leftrightarrow$(iii): Since $f(x)$ is Gateaux differentiable with derivative $df(x, h)$, it suffices by (2.1.13) to prove that (a) $df(x, h) \in L(X, Y)$ for fixed x and (b) as a function of x, $df(x, h)$ is continuous from $U \to L(X, Y)$. To prove these facts, we use Hartog's theorem.[1] First, $df(x, h)$ is clearly homogeneous of degree 1 in h. Thus we show that $df(x, h_1 + h_2) = df(x, h_1) + df(x, h_2)$. To this end, we note that $(y^*, f(x + t_1 h_1 + t_2 h_2))$

[1] This theorem (cf. L. Hormander, 1966, p. 28) states that a complex-valued function, defined on an open set $\Omega \subset \mathbb{C}^N$, and analytic in each variable separately is analytic in all complex variables jointly.

$= g(t_1, t_2)$ is an analytic function of the two complex variables (t_1, t_2) separately since $\partial g/\partial t_1$, $\partial g/\partial t_2$ exist. Thus Hartog's theorem implies that

$$g(t_1, t_2) = (y^*, f(x)) + t_1 g_{t_1}(0, 0) + t_2 g_{t_2}(0, 0) + o(|t|).$$

Consequently,

$$\begin{aligned}(y^*, df(x, h_1 + h_2)) &= \left(y^*, \lim_{t\to 0}\left\{ \frac{1}{t}\,[f(x + t[h_1 + h_2]) - f(x)]\right\}\right)\\ &= \lim_{t\to 0}\left\{ \frac{1}{t}\,[g(t, t) - g(0, 0)]\right\}\\ &= g_{t_1}(0, 0) + g_{t_2}(0, 0)\\ &= (y^*, df(x, h_1) + df(x, h_2)).\end{aligned}$$

Now the Cauchy estimates imply that for $\|h'\| = \beta$, say,

$$\|df(x, h')\| = \sup_{\|y^*\|=1} \frac{d}{dt}\,(y^*, f(x + th'))_{t=0} \leqslant M.$$

Since $df(x, h)$ is homogeneous of degree 1, for arbitrary h,

$$\|df(x, h)\| = \left\|df\left(x, \frac{\|h\|}{\beta}\,h'\right)\right\| \leqslant \frac{\|h\|}{\beta}\,M, \quad \text{where} \quad \|h'\| = \beta. \tag{2.3.4}$$

This proves (a).

To prove (b) we first show that $df(x, h)$ is complex analytic as a function of x. For $y^* \in Y^*$,

$$(y^*, df(x + t_2h_2, h_1)) = \frac{d}{dt_1}\,(y^*, f(x + t_2h_2 + t_1h_1))|_{t_1=0.}$$

Now $g(t_1, t_2) = (y^*, f(x + t_2h_2 + t_1h_1))$ is an analytic function of the complex variables t_1, t_2 since $f(x)$ is Gateaux differentiable, and again by Hartog's theorem $g_{t_1}(0, t_2)$ is analytic in t_2. Consequently, $df(x, h)$ is analytic in x, and also by the above, Gateaux differentiable in x. Thus for $y^* \in Y^*$,

$$\begin{aligned}(y^*, d(f(x + z), h) - d(f(x), h)) &= \left(y^*, \int_0^1 \frac{d}{ds}\,df(x + sz, h)\,ds\right)\\ &= \int_0^1 \frac{d}{ds}\,(y^*, df(x + sz, h)\,ds)\\ &= \int_0^1 \frac{d}{ds}\left[\frac{1}{2\pi i}\int_C \frac{(y^*, f(x + sz + th))}{t^2}\,dt\right] ds\\ &= \int_0^1 \frac{1}{2\pi i}\int_C \frac{(y^*, f'(x + sz + thz))}{t^2}\,dt\,ds\end{aligned}$$

Thus using the Cauchy estimates and (2.3.4), we find

$$\|df(x+z,h)-df(x,h)\| = \sup_{\|y^*\|=1} (y^*, df(x+z)-df(x,h))$$

$$= o(1) \qquad (\text{as } \|z\|\to 0 \text{ uniformly in } \|h\|=1).$$

(i), (ii)$\Rightarrow$(iv): We proceed by induction. The case $k=1$ coincides with (ii). Assuming the result for $k=n-1$, we prove the theorem for $k=n$. The function

$$g(t_1,\ldots,t_n) = \left(y^*, f\left(x+t_nh_n+\sum_{i=1}^{n-1} t_ih_i\right)\right)$$

is an analytic function of the complex variables $t_1,\ldots,t_n$ by Hartog's theorem for any $y^*\in Y^*$. Thus

$$\tilde{g}(t_n) = D_{n-1}D_{n-2}\cdots D_1 g(t_1,\ldots,t_n)|_{t_1=t_2=\cdots=t_{n-1}=0}$$

is analytic in t_n. By the induction hypothesis $\tilde{g}(t_n)=(y^*, df^{n-1}(x+t_nh_n, h_{n-1}, h_{n-2},\ldots,h_1))$. Consequently, $df^{n-1}(x,h_{n-1},\ldots,h_1)$ is analytic in x and thus Gateaux differentiable.

This argument also shows that

$$f^n(x)(h_n,h_{n-1},\ldots,h_1) = D_nD_{n-1}\cdots D_1 f\left(x+\sum_{i=1}^{n} t_ih_i\right)\Bigg|_{t_i=0},$$

and completes the induction argument. The converse statement is immediate.

(iv)$\Leftrightarrow$(v): To prove this fact we employ Theorem (2.1.27). Thus we show that (a) $d^nf(x,h_1,\ldots,h_n)\in L^n(X,Y)$ and (b) as a function of x, $d^nf(x,h_1,\ldots,h_n)$ is continous from $U\to L^n(X,Y)$. The definition of

$$d^nf(x,h_n,h_{n-1},\ldots,h_1) = D_nD_{n-1}\cdots D_1 f\left(x+\sum_{i=1}^{n} t_ih_i\right)\Bigg|_{t_1=\cdots=t_n=0}$$

shows that $d^nf(x,h_n,\ldots,h_1)$ is linear in h_n and symmetric in h_i ($i=1,\ldots,n$). Consequently, $d^nf(x,h_n,\ldots,h_1)$ is linear in each h_i. By (2.1.22), $d^nf(x,h)$ is a homogeneous polynomial of degree n, and Gateaux differentiable in x. In fact, $d^nf(x,h)$ is analytic in x, for the Cauchy estimates show that $d^nf(x,h)$ is locally bounded in x. Consequently, by polarization, $d^nf(x,h_1,\ldots,h_n)$ depends analytically and hence continuously on x. Thus it remains to show $d^nf(x,h)\in L_n(X,Y)$ for fixed x. This follows from the homogeneity of $d^nf(x,h)$ in h, exactly as in (2.3.4). The converse is immediate.

(v)$\Rightarrow$(vi): This fact follows from the equation (2.3.2) and the remarks thereafter.

(vi)$\Rightarrow$(i): If

$$f(x+h) = \sum_{N=0}^{\infty} \frac{1}{N!} f^{(N)}(x)h^N,$$

we set

$$g(x,h) = \sum_{N=1}^{\infty} \frac{1}{N!} N f^{(N)}(x)h^{N-1}.$$

Then exactly as in the one-dimensional case, for fixed $y^* \in Y^*$ we show that $|(y^*, f(x+th) - f(x) - g(x,h))| = o(1)$ as $|t| \to 0$.

2.3B Basic properties

It will be useful to state explicitly some immediate consequences of the above equivalences.

(2.3.5) **Theorem** Suppose $f(x)$ is a locally bounded complex analytic operator in $M(U, Y)$. Then:

(i) (*maximum principle*) $\sup \|f(x)\|$ over U cannot be attained in U unless $\|f(x)\|$ is constant over U.

(ii) (*Cauchy estimates*) If the sphere $\{\|y - x_0\| \leqslant r_0\}$ lies in U, and $\|x - x_0\| \leqslant r_0/2$, then for all h

$$\|f^{(n)}(x)h^n\| \leqslant n! M(x_0, r_0)\left(\frac{2}{r_0}\right)^{+n} \|h\|^n, \tag{2.3.6}$$

where $M(x_0, r_0) = \sup \|f(x)\|$ over the sphere $\{x \mid \|x - x_0\| = r_0\}$.

Proof: (i): Suppose there is a point $x_0 \in U$ with $\|f(x)\| \leqslant \|f(x_0)\| = M$. Now for $|t|$ sufficiently small, and $y^* \in Y^*$, $(y^*, f(x_0 + th))$ is analytic in t and applying (2.3.2) for ρ sufficiently small, $\|f(x)\| \leqslant (2\pi)^{-1}\int_0^{2\pi}\|f(x + \rho e^{i\theta}h)\|\, d\theta$ so that $\|f(x_0 + th)\|$ is subharmonic for $|t|$ sufficiently small. Thus $\|f(x_0 + th)\| = M$ for $|t|$ sufficiently small. Since h is arbitrary in this argument, $\|f(x)\| = M$ for all $x \in U$ that can be joined to x_0 by a polygonal line in U, and therefore for $x \in U$.

(ii): By the results of Theorem (2.3.3), it follows that $f^{(n)}(x)h^n$ is homogeneous of degree n in h and for $\|h\| \leqslant r_0/2$, $\|f^{(n)}(x)h^n\| \leqslant n!\, M(x_0, r_0)$, where $M(x_0, r_0) = \sup \|f(x)\|$ over the sphere $\{x \mid \|x - x_0\| = r_0\}$. Consequently, for all h, as in (2.3.4), we obtain (2.3.6).

2.4 Compact Operators

Once a result is established for mappings acting between finite-dimensional Banach spaces $\mathbb{B}^N$, it is natural to consider its validity for

Banach spaces of arbitrary dimension by letting $N = \dim \mathbb{B}^N \to \infty$. A class of mappings for which such results are often readily established is defined as follows:

(2.4.1) **Definition** Let X and Y be Banach spaces and U be a subset of X. Then $f \in M(U, Y)$ is compact if f is continuous and maps bounded subsets of U into conditionally compact subsets of Y, and we write $f \in K(U, Y)$. (Here we have used the same notation as in Section 1.3F.)

2.4A Equivalent definitions

It is an interesting and important fact that many nonlinear operators arising in mathematical physics and differential geometry are compact in the above sense, once appropriate Banach spaces X and Y are considered.

Clearly, the compact operators in $M(U, Y)$ are closed under addition and subtraction, and under composition with continuous, bounded operators. Furthermore, all continuous, bounded mappings of a subset U of a Banach space X into a *finite-dimensional* Banach space Y are compact (we denote this class by $K_0(U, Y)$). In fact, the set of compact operators consists of just those mappings that can be approximated by the operators in K_0 in the following sense.

(2.4.2) **Theorem** Suppose U is a bounded subset of X and $f \in M(U, Y)$. Then the following facts are equivalent:

(i) f is compact.

(ii) Given $\epsilon > 0$, there is a continuous, bounded mapping $f_\epsilon \in M(U, Z_\epsilon)$, where Z_ϵ is a finite-dimensional subset of Y, such that $\|f(x) - f_\epsilon(x)\| < \epsilon$. Furthermore, the range of f_ϵ is contained in the convex hull, $\overline{\text{co}}\, f(U)$ of $f(U)$.

(iii) f can be represented by a uniformly convergent series $f(x) = \sum_{n=1}^{\infty} g_n(x)$, where the g_n have finite-dimensional range and $\|g_n(x)\| \leqslant \epsilon/2^n$ for each $x \in U$.

Proof: (i)$\Rightarrow$(ii): If f is compact, $\overline{f(U)}$ is a compact set in Y. Thus given $\epsilon > 0$, $f(U)$ can be covered by a finite number of spheres with centers y_i $(i = 1, \ldots, k)$ and radius ϵ. Let Y_k be the finite-dimensional subspace of Y spanned by $(y_1, y_2, \ldots, y_k)$. We construct a partition of unity on U as follows. For every $i = 1, 2, \ldots, k$, set $\mu_i(x) = \max(0, \epsilon - \|f(x) - y_i\|)$ and $\lambda_i = \mu_i(x)\{\sum_{i=1}^{k} \mu_i(x)\}^{-1}$. Then the $\mu_i(x)$ are continuous, real-valued functions defined on U and $\sum_{i=1}^{k} \mu_i(x) \neq 0$ since for each $x \in U$ some $\mu_i(x) > 0$. Each $\lambda_i(x)$ is a continuous real-valued function defined on U, $0 \leqslant \lambda_i(x) \leqslant 1$ and $\sum_{i=1}^{k} \lambda_i(x) = 1$ for $x \in U$. Now we define the function $f_\epsilon(x)$ for $x \in U$ by setting $f_\epsilon(x) = \sum_{i=1}^k \lambda_i(x) y_i$. Since $f(x) = \sum_{i=1}^k \lambda_i(x) f(x)$, we have

$$\|f(x) - f_\epsilon(x)\| = \left\| \sum_i \lambda_i(x)\{f(x) - y_i\} \right\| \leqslant \sum_i \lambda_i(x) \|f(x) - y_i\| \leqslant \epsilon.$$

Thus f_ϵ is a continuous function with domain U and range contained in the convex hull of the points $y_1, y_2, \ldots, y_k$, and thus in Y_k.

(ii)$\Rightarrow$(iii): For $\epsilon_n = \epsilon/2^{n+2}$, by (ii) there exists a mapping h_n with a finite-dimensional range such that $\|h_n(x) - f(x)\| \leqslant \epsilon/2^{n+2}$. Define the sequence $\{g_n(x)\}$ inductively by $g_0(x) = h_0(x)$: $g_n(x) = h_n(x) - h_{n-1}(x)$. Then since $\sum_{i=1}^n g_i(x) = h_n(x)$, and $h_n(x) \to f(x)$, $\sum_{i=1}^n g_i(x) \to f(x)$. Also,

$$\|g_n(x)\| = \|h_n(x) - h_{n-1}(x)\| \leqslant \|h_n(x) - f(x)\| + \|h_{n-1}(x) - f(x)\|$$
$$\leqslant \epsilon/2^{n+2} + \epsilon/2^{n+1} < \epsilon/2^n.$$

Hence, since $\sum_{n=0}^\infty \|g_n(x)\|$ is uniformly convergent, $\sum_{N=0}^\infty g_n(x)$ is uniformly convergent.

(iii)$\Rightarrow$(i): The sequence of mappings $\sum_{k=1}^n g_k(x)$ is uniformly convergent. Thus (i) follows immediately from the fact that if a sequence of compact operators $f_n \in M(U, Y)$ converges uniformly to f on U, then f is compact. Clearly, such an f is continuous since each f_n is. Also, $\overline{f(U)}$ is compact. Indeed, given $\epsilon > 0$ we can find an integer k satisfying $\|f_k(x) - f(x)\| < \epsilon/2$. Hence any finite $\epsilon/2$-net for $\overline{f_k(U)}$ will be a finite ϵ-net for $\overline{f(U)}$. Thus $\overline{f(U)}$ and therefore f are compact.

2.4B Basic properties

To demonstrate just how one uses Theorem (2.4.2) to extend the validity of theorems proved for finite-dimensional mappings to the infinite-dimensional case, we prove the extension of Brouwer's fixed point theorem (1.6.4) due to Schauder.

(2.4.3) **Schauder Fixed Point Theorem** A compact mapping f of a closed bounded convex set K in a Banach space X into itself has a fixed point.

Proof: Using Theorem (2.4.2), we approximate f by a sequence $\{f_n\}$ of continuous bounded maps with finite-dimensional ranges $Y_{\epsilon,n} \subset K$ such that $\|f(x) - f_n(x)\| \leqslant 1/n$ for any $x \in K$. Restricting f_n to $Y_{\epsilon,n}$, the resulting map $\tilde{f}_n$ has a fixed point $x_n \in K$ so that $\|f(x_n) - x_n\| \leqslant 1/n$. Now $\{f(x_n)\}$ has a convergent subsequence $\{f(x_{n_j})\}$ with limit y since f is compact. Since $\|f(x_{n_j}) - x_{n_j}\| \leqslant 1/n_j$, $x_{n_j} \to y$ as $n_j \to \infty$; and by the continuity of f, $f(y) = y$, so that y is the desired fixed point.

Another useful consequence of Theorem (2.4.2) is:

(2.4.4) **Extension Theorem for Compact Mappings** Let $\bar{U}$ be an open set of a Banach space X and let $f \in K(\bar{U}, Y)$. Then given $\delta > 0$, f can be extended to a compact operator $\tilde{f} \in K(X, Y)$ in such a way that for $x \in X$, $d(\tilde{f}(x), \overline{\text{co}}\, f(U)) \leqslant \delta$.

Proof: By Theorem (2.4.2), $f(x) = \sum_{n=0}^{\infty} f_n(x)$. By Tietze's extension theorem, each $f_n(x)$ can be extended to an $\tilde{f}_n$: $X \to Y_{\epsilon_n}$ with preservation of norm. Hence $\tilde{f}_n$ is a compact map. Now consider the mapping $\tilde{f}(x) = \sum_{n=0}^{\infty} \tilde{f}_n(x)$. For $x \in X$,

$$d(f_0(x), \overline{\text{co}}\, f(U)) \leqslant \|f_0 - f\| \leqslant \sum_{n=1}^{\infty} \frac{\epsilon}{2^n} = \epsilon,$$

also $d(\overline{\text{co}}\, f_0(U), \overline{\text{co}}\, f(U)) < \epsilon$ since for $x \in \overline{\text{co}}\, f_0(U)$,

$$x = \sum_{i=1}^{p} t_i f_0(x_i) \qquad \text{with} \quad t_j = 1 \quad \text{and} \quad x_i \in U,$$

so

$$\|x - \sum t_i f(x_i)\| \leqslant \sum t_i \|f_0(x_i) - f(x_i)\| \leqslant \epsilon.$$

Then choosing $3\epsilon = \delta$, f is the desired extension since

$$d(\tilde{f}(x), \overline{\text{co}}\, f(U)) \leqslant \|\tilde{f}(x) - f_0(x)\| + d(f_0(x), \overline{\text{co}}\, f(U))$$

$$\leqslant \sum_{n=1}^{\infty} \frac{\epsilon}{2^n} + \epsilon = 2\epsilon < \delta$$

We now make some remarks concerning the relations among compactness of an operator and some other basic notions already introduced.

(2.4.5) Suppose $f \in M(U, Y)$ is completely continuous, where U is a closed convex subset of a reflexive Banach space X. Then f is compact.

Proof: Suppose $\{x_n\}$ is a bounded sequence of U. Then by the reflexivity of X, $\{x_n\}$ has a weakly convergent subsequence $\{x_{n_j}\}$. The complete continuity of f implies that $\{f(x_{n_j})\}$ is convergent in the norm topology of Y. Thus f is compact.

(2.4.6) Suppose U is an open subset of X and $f \in K(U, Y)$ is Fréchet differentiable in U. Then the Fréchet derivative $f'(x_0) \in L(X, Y)$ is a compact linear operator for fixed $x_0 \in U$.

Proof: Suppose $f'(x_0)$ were not compact. Then, if σ_1 denotes the unit sphere in X, $f'(x_0)(\sigma_1)$ does not have compact closure in Y. Hence there is a sequence $\{h_i\}$ with $\|h_i\| = 1$ and a number $\epsilon > 0$ such that $\|f'(x_0)\{h_i -$

$h_j\}\| \geqslant \epsilon$ $(i \neq j)$. On the other hand, for $\beta > 0$ and sufficiently small, the Fréchet differentiability of f at x_0 implies,

$$\begin{aligned}\|f(x_0+\beta h_i) - f(x_0+\beta h_j)\| &\geqslant \beta\|f'(x_0)h_i - f'(x_0)h_j\| \\ &- \|f(x_0+\beta h_i) - f(x_0) - \beta f'(x_0)h_i\| \\ &- \|f(x_0+\beta h_j) - f(x_0) - \beta f'(x_0)h_j\| \\ &\geqslant \beta\epsilon - o(|\beta|).\end{aligned}$$

Since ϵ is independent of β, this last inequality implies that $\{f(x_0+\beta h_i)\}$ has no convergent subsequence, which is the desired contradiction.

2.4C Compact differential operators

Heuristically speaking, operators that possess some definite "smoothing property" are generally compact. As a very simple example we consider the operator $Tf(x) = \int_0^x f(s)\,ds$ defined on $C[0, 1]$. Clearly Tf is differentiable on $(0, 1)$, and so T possesses a smoothing property in the sense that T maps continuous functions into differentiable ones. On the other hand, T is compact by virtue of the Arzela–Ascoli theorem (1.3.13). This argument can be extended to the more general abstractly defined operators of Section 2.2D by a careful inspection of the estimates of Section 1.4, as follows:

(i) Consider, for example, the class of abstract operators defined for quasilinear elliptic differential operators by means of the Schauder inversion method of Section 2.2. In this connection, we prove the following abstract result.

(2.4.7) **Lemma** Suppose the operator $A(v, u)$ is linear in u and continuous in v for fixed $u \in Z$ and maps $X \times Z \to Y$, where Z is a linear subspace of X compactly imbedded in X. Then, if the linear equation $A(v, u) = g$ has one and only one solution $u = T_g v$ for $(\|v\|_X \leqslant R)$ satisfying the a priori estimate

(2.4.8) $\qquad \|u\|_Z \leqslant c(R)\|g\|_Y,$

where the positive constant $c(R)$ is independent of v, the mapping T_g: $X \to X$ is compact.

Proof: The continuity of T_g follows as in (2.2.13) since $Z \subset X$. To demonstrate the compactness of $\overline{T_g(\sigma)}$ for any bounded set σ in X, suppose $\{v_n\}$ is any sequence of σ. Then if $u_n = T_g(v_n)$, the estimate (2.4.8) shows that

$$\|T_g(v_n)\|_Z \leqslant c(\sigma)\|g\|_Y,$$

where $c(\sigma)$ is a constant depending only on σ. Since Z is compactly imbedded in X, any bounded set in Z is compact in X. Thus $T_g(v_n)$ has a convergent subsequence in X. Consequently, T_g is a compact mapping.

For elliptic differential operators defined on bounded domains Ω with normal homogeneous boundary conditions, the estimates (1.4.26) and (1.4.28) yield (2.4.8) and the compactness necessary to apply Lemma (2.4.7) to the pairs

$$(Z, X) = (C^{2m, \alpha}(\Omega), C^{\alpha}(\Omega)) \quad \text{or} \quad (Z, X) = (W_{2m, p}(\Omega), L_p(\Omega))$$

for $0 < \alpha < 1$, and $1 < p < \infty$.

(ii) Next we consider the compactness of operators defined implicitly by the duality method. Suppose A is a bounded operator from $\mathring{W}_{m,p}(\Omega) \to W_{-m,q}(\Omega)$ defined implicitly by

$$(Au, \varphi) = \sum_{|\alpha| \leqslant m-1} \int_{\Omega} A_\alpha(x, u, Du, \ldots, D^{m-1}u)\, D^\alpha \varphi.$$

(2.4.9) **Theorem** Suppose the continuous functions $A_\alpha(x, u, \ldots, D^{m-1}u)$ satisfy the growth conditions $(*)$ of (2.2.7), then A is a compact operator from $\mathring{W}_{m,p}(\Omega) \to W_{-m,q}(\Omega)$ and in fact maps weakly convergent sequences in $\mathring{W}_{m,p}(\Omega)$ into strongly convergent sequences.

Proof: Let $u_n \to u$ weakly in $\mathring{W}_{m,p}(\Omega)$, then

$$\|Au_n - Au\| = \sup_{\|\varphi\| = 1} (Au_n - Au, \varphi)$$

and

$$(Au_n - Au, \varphi) = \sum_{|\alpha| \leqslant m-1} \int_{\Omega} [A_\alpha(x, u_n, \ldots) - A_\alpha(x, u, \ldots)]\, D^\alpha \varphi$$
$$\leqslant \sum k_\alpha \|A_\alpha(x, u_n, \ldots, D^{m-1}u_n) - A_\alpha(x, u, \ldots, D^{m-1}u)\|_{L_{q_\alpha}}$$

with q_α so chosen that $A_\alpha(x, u, \ldots, D^{m-1}u) \in K(\mathring{W}_{m-1,p^*}, L_{q_\alpha})$, and $p^* < Np/(N-p)$. Now if $u_n \to u$ weakly in $\mathring{W}_{m,p}(\Omega)$, by (1.3.35), $u_n \to u$ strongly in $\mathring{W}_{m-1,p^*}(\Omega)$. On the other hand, by virtue of the hypothesis on the growth conditions and Lemma (2.2.7), $A_\alpha(x, u, \ldots, D^{m-1}u)$ is a continuous function from $\mathring{W}_{m-1,p} \to L_{q_\alpha}$. Consequently, $\|Au_n - Au\| \to 0$ as $n \to \infty$, so that A is a compact mapping.

2.5 Gradient Mappings

Various classes of linear operators acting between Banach spaces can be easily extended to a more general (nonlinear) setting. Consider the extension of the class of self-adjoint operators acting in Hilbert space.

(2.5.1) **Definition** Let $f \in C(U, X^*)$, where U is an open subset of X and X^* denotes the conjugate space of X. Then f is a *gradient operator* if there is a real-valued function $F(x) \in C^1(U, \mathbb{R}^1)$ such that the Fréchet derivative of $F(x)$, $F'(x) = f(x)$ for each $x \in U$. In this case, F is called an antiderivative of f. We also sometimes used the notation $f(x) = \text{grad } F(x)$.

2.5A Equivalent definitions

To show that gradient operators are actually generalizations of self-adjoint operators, we prove

(2.5.2) **Theorem** Suppose U is a convex set of a Banach space X that contains the origin, and $f \in C^1(U, X^*)$. Then the following statements are equivalent:

(i) f is a gradient operator.

(ii) $\int_C f(x(t))\, dx(t)$ is independent of the path C, provided C is a simple, rectifiable curve in U.

(iii) $\int_0^1 (f(sx), x)\, ds - \int_0^1 (f(sy), y)\, ds = \int_0^1 (f(z(s)), x - y)\, ds$, where $z(s) = sx + (1 - s)y$ and $x, y \in U$.

(iv) $f'(x) \in L(X, X^*)$ is self-adjoint for $x \in U$.

Proof: (i)$\Rightarrow$(ii): if $f(x)$ is a gradient operator, let $f(x) = F'(x)$ and suppose C is parametrized by $C = \{x(t) \mid 0 \leqslant t \leqslant 1\}$. Then

$$\text{(2.5.3)} \qquad \begin{aligned} \int_C f(x(t))\, dx(t) &= \int_C F'(x(t))\, dx(t) \\ &= \int_0^1 \frac{d}{dt} F(x(t))\, dt \\ &= F(x(1)) - F(x(0)). \end{aligned}$$

(ii)$\Rightarrow$(iii): Since x, y and the origin $0 \in U$, two paths in U between x and y are: (1) C_1, the straight line segments joining the origin to x and y; and (2) C_2, the straight line segment joining x and y. By (2.5.3), with $x(t) = tx$ and $y(t) = ty$,

$$\begin{aligned} \int_{C_1} f(x(t))\, dx(t) &= \int_0^1 \frac{d}{dt} F(tx)\, dt - \int_0^1 \frac{d}{dt} F(ty)\, dt \\ &= \int_0^1 (f(tx), x)\, dt - \int_0^1 (f(ty), y)\, dt. \end{aligned}$$

In the same way with $z(t) = tx + (1 - t)y$,

$$\int_{C_2} f(z(t))\, dz(t) = \int_0^1 \frac{d}{dt} F(z(t))\, dt = \int_0^1 (f(z(t)), x - y)\, dt.$$

On the other hand, by (ii), $\int_{C_1} f(x(t))\, dx(t) = \int_{C_2} f(z(t))\, dz(t)$. Thus (iii) is proved.

(iii)$\Rightarrow$(iv): We show (iii) implies that $f(x)$ is a gradient map with $F(x) = \int_0^1 (f(sx), x)\, ds$. Indeed, with this definition (iii) shows that $F(x +$

$\epsilon h) - F(x) = \epsilon \int_0^1 (f(x + s\epsilon h), h)\, ds$. Letting $\epsilon \to 0$ we find that the Gateaux derivative $F(u)$ is $dF(x, h) = (f(x), h)$. Since $f(x) \in C^1(U, X^*)$, $F(u) \in C^2(U, \mathbb{R}^1)$, and by (2.1.28) $F''(u)(h_1, h_2) = f'(x)(h_1, h_2)$ is symmetric in h_1 and h_2.

(iv)$\Rightarrow$(i): Define $F(x) = \int_0^1 (f(tx), x)\, dt$, then we show that (iv) implies that $F(x)$ is differentiable and $F'(x) = f(x)$. Indeed,

$$(\dagger) \qquad F(x + h) - F(x) = \int_0^1 (f(tx + th), h)\, dt + \int_0^1 (f(tx + th) - f(tx), x)\, dt.$$

Now

$$\begin{aligned}
\int_0^1 (f(tx + th) - f(tx), x)\, dt &= \int_0^1 \left\{ \int_0^t \frac{d}{ds} (f(tx + sh), x)\, ds \right\} dt \\
&= \int_0^1 \int_0^1 (f'(tx + sh)h, x)\, ds\, dt \\
&= \int_0^1 dt \int_0^t (f'(tx + sh)x, h)\, ds \\
&\qquad \text{(using (iv))} \\
&= \int_0^1 ds \int_s^1 (f'(tx + sh)x, h)\, dt \\
&\qquad \text{(interchanging order of integration)} \\
&= \int_0^1 (f(x + sh) - f(sx + sh), h)\, ds.
\end{aligned}$$

By $(\dagger)$, $F(x + h) - F(x) = \int_0^1 (f(x + sh), h)\, ds$. Hence a short computation shows that $|F(x + h) - F(x) - (f(x), h)| = o(\|h\|)$, and thus $F'(x) = f(x)$.

Remark: If f is assumed to be continuous but not necessarily C^1, (i)–(iii) are easily proven equivalent.

2.5B Basic properties

It is important to determine the operations under which gradient operators are preserved. A few important cases follow.

(2.5.4) If $f(x)$ is a gradient operator defined on an open subset U of a Hilbert space H, with range in H, and A is a linear self-adjoint operator, then AfA is a gradient operator whose antiderivative is $F(A(x))$, where $F'(x) = f(x)$.

Proof: Let $\tilde{F}(x) = F(Ax)$, and compute $(d/dt)\tilde{F}(x + th)|_{t=0}$.

A gradient map $f \in M(U, H)$ is preserved under projection. Thus if $H = X_1 \oplus X_2$ is a direct sum decomposition of H, and π is the projection of $H \to X_1$, then $\pi f(x)$ is a gradient map of X_1 into itself. More generally,

(2.5.5) If $H = Y_1 \oplus Y_2$ is a direct sum decomposition of the Hilbert space H, and there is a differentiable mapping $g \in C(Y_2, Y_1)$, $y_1 = g(y_2)$, then with π: $H \to Y_2$ denoting the projection map, and assuming $(I - \pi)f(y_1 + y_2) = 0$, $f_1(y_2) = \pi f(y_2 + g(y_2))$ is a gradient map if $f(x)$ is.

Proof: Let $F'(x) = f(x)$. Then we show $G(y_2) = F(y_2 + g(y_2))$ is the antiderivative of $f(y_2)$. Indeed, for $h \in Y_2$, using the assumptions stated,

$$(G'(y_2), h) = (f(y_2 + g(y_2)), h + g'(y_2)h)$$
$$= (\pi f(y_2 + g(y_2)), h + g'(y_2)h).$$

But $(\pi f(y_2 + g(y_2), g'(y_2)h) = 0$ since $g' \in L(Y_2, Y_1)$, while Y_1 and Y_2 are complementary subspaces. Consequently,

$$(G'(y_2), h) = (\pi f(y_2 + g(y_2)), h) = (f_1(y_2), h),$$

as required.

There are important relations between a gradient operator $f(x)$ and its antiderivative $F(x)$. For example, the points x at which $f(x) = 0$ are called the critical points of $F(x)$, so that information on the zeros of $f(x)$ may be found by studying the "graph" of the real-valued function $F(x)$. However, due to the fact that $F(x)$ may be defined on an *infinite-dimensional* Banach space, certain difficulties concerning critical points arise. In particular, the infinum of a C^1 functional $F(x)$ defined on X may not be attained even though $F(x)$ is bounded above $-\infty$, and $F^{-1}[a, b]$ for $-\infty < a, b < \infty$ is bounded. This topic will be discussed in detail in Chapter 6. For the time being, we note the following useful result for gradient mappings $f \in M(X, X^*)$.

(2.5.6) Suppose $F(x)$ is the antiderivative of the gradient mapping $f \in M(X, X^*)$ and $F(0) = 0$. Then:

(i) $F(x) = \int_0^1 (f(sx), x)\, ds$.

(ii) If $f(x)$ is completely continuous, $F(x)$ is continuous with respect to weak convergence in X.

(iii) Suppose f is the polarization of a multilinear operator $\tilde{f} \in L_N(X, X^*)$ such that $I(x_1, \ldots, x_{n+1}) = (\tilde{f}(x_1, \ldots, x_n), x_{n+1})$ is symmetric as a multilinear mapping in $L_{N+1}(X, \mathbb{R}^1)$. Then $f(x, x, \ldots, x)$ is a gradient mapping with antiderivative $F(x) = (1/(n+1))(f(x), x)$.

(iv) $F(x)$ is convex if and only if the monotonicity condition $(f(x) - f(y), x - y) \geqslant 0$ holds for each $x, y \in X$.

Proof: (i): By virtue of (2.5.2(iii)), if one sets $\Phi(x) = \int_0^1 (f(sx), x)\, ds$, then $(d/d\epsilon)\Phi(x + \epsilon h)|_{\epsilon=0} = (f(x), h)$. Since this equation is true for each $h \in X$, $\Phi'(x) = f(x)$; and since

$\Phi(0) = 0$, the uniqueness of the Gateaux derivative implies $F(x) \equiv \Phi(x)$.

(ii): If f is a gradient mapping and $x_n \to x$ weakly in X, then by (2.5.2(iii)), $F(x_n) - F(x) = \int_0^1 (x_n - x, f(x_n(s)))\, ds$, where $x_n(s) = x + s(x_n - x)$. Thus: writing

$$(x_n - x, f_n(s)) = (x_n - x, f(x_n(s)) - f(x)) + (x_n - x, f(x)),$$

we note that since $\{\|x_n\|\}$ is uniformly bounded

$$\lim_{n\to\infty} \{F(x_n) - F(x)\} = \lim_{n\to\infty} \int_0^1 (x_n - x, f_n(s))\, ds = 0.$$

Indeed, f is completely continuous, so the limit on the right-hand side is zero because $f(x_n(s)) \to f(x)$ strongly for each $s \in [0, 1]$ (while the integrand on the right-hand side is uniformly bounded).

(iii): Set $\Phi(x) = (1/(n+1))(f(x), x)$, then using the presumption that $\Phi(x_1, \ldots, x_{n+1}) = (f(x_1, \ldots, x_n), x_{n+1})$ is symmetric, we find that for any $h \in X$, $(d/d\epsilon)\Phi(x + \epsilon h)|_{\epsilon=0} = (f(x), h)$. Hence $\Phi(x)$ is the antiderivative of $f(x)$.

(iv): First we suppose that $(f(x) - f(y), x - y) \geqslant 0$ for each $x, y \in X$ (i.e., that f is a monotone operator) and establish the convexity of $F(x)$. Indeed, without loss of generality we may suppose that $F(0) = 0$, so that by (i) above, for any $t \in [0, 1]$,

$$F(tx + (1-t)y) = F(y) + t\int_0^1 (x - y, f(y + st(x - y)))\, ds$$

$$\leqslant F(y) + t\int_0^1 (x - y, f(y + s(x - y)))\, ds$$

(this last inequality following from the monotonicity of f). By (2.5.2(iii)), the right-hand side of the above inequality can be written as $t\{F(x) - F(y)\}$. Thus $F(tx + (1-t)y) \leqslant tF(x) + (1-t)F(y)$. We now prove the converse; namely, if $F(x)$ is convex, then $(f(x) - f(y), x - y) \geqslant 0$ for each $x, y \in X$. To this end, we first show that the convexity of $F(x)$ implies

(*) $\quad F(x) + (f(x), y - x) \leqslant F(y) \quad$ for each $x, y \in X$.

The convexity of $F(x)$ implies that for each $x, y \in X$ and $s \in [0, 1]$,

$$sF(x) + F(x + s(y - x)) - F(x) \leqslant sF(y).$$

Dividing by s and letting $s \to 0$, we find that (*) results. Next we apply (*), interchanging the roles of x and y to find

(**) $\quad F(y) + (f(y), x - y) \leqslant F(x).$

Adding (*) and (**), we find $(f(x) - f(y), x - y) \geqslant 0$, as required.

2.5C Specific gradient mappings

Generally speaking, if a differential operator A is the Euler–Lagrange derivative of a functional (in the sense described in Section 1.1C), A can be represented abstractly as a gradient mapping $\mathscr{A}$. In order that the mapping $\mathscr{A}$ be defined in Sobolev spaces, the terms of A must satisfy *certain growth conditions*. As an interesting nontrivial example, we consider the partial differential equations defining the von Kármán equations given in Section 1.3B. It is known that generally the deformation equations of elasticity are derived as Euler–Lagrange equations, so it is plausible to expect that the associated operator equations involve only gradient operators. In fact we now prove that our calculus guarantees this fact.

(2.5.7) The weak solutions of the equations (1.1.12) are in one-to-one correspondence with the solutions of the operator equation $u + Cu = \lambda Lu$ in the Sobolev space $\mathring{W}_{2,2}(\Omega)$, where L is a self-adjoint mapping of $\mathring{W}_{2,2}(\Omega)$ into itself and Cu is a gradient mapping of $\mathring{W}_{2,2}(\Omega)$ into itself. Moreover, there is a symmetric bilinear mapping $C(u, v)$ of $\mathring{W}_{2,2}(\Omega)$ into itself, defined by (2.5.9′) below, such that $Lu = C(F_0, u)$ for some fixed element $F_0 \in \mathring{W}_{2,2}(\Omega)$ and $C(u, C(u, u)) = Cu$.

Proof: First we note that (without loss of generality) we may set $\epsilon = 1$ in the equations (1.1.12), and defining F_0 as the solution of $\Delta^2 F = 0$, $D^\alpha F|_{\partial\Omega} = \lambda\psi_0$, we may write a solution (u, F) of (1.1.12) in the form $(u, f + \lambda F_0)$, so that the pair (u, f) satisfies the system:

$$
\begin{aligned}
&\Delta^2 f = -\tfrac{1}{2}[u, u],\\
(2.5.8)\qquad &\Delta^2 u = \lambda[F_0, u] + [f, u],\\
&D^\alpha u = D^\alpha f = 0, \qquad |\alpha| \leq 1,
\end{aligned}
$$

where $[f, g] = (f_{yy}g_x - f_{xy}g_y)_x + (f_{xx}g_y - f_{xy}g_x)_y$. Consequently, in accord with the definition of weak solution given in Section 1.5 and by choosing the inner product in $\mathring{W}_{2,2}(\Omega)$ as

$$(u, v)_{2,2} = \int_\Omega (u_{xx}v_{xx} + 2u_{xy}v_{xy} + u_{xx}v_{yy}),$$

the weak solution (u, f) of the system (2.5.8) can be written as any pair (u, f) that satisfies the following two integral identities for all $\varphi, \eta \in C_0^\infty(\Omega)$:

$$
\begin{aligned}
(2.5.9)\qquad &(u, \eta)_{2,2} = \int_\Omega \{(\bar{f}_{xy}u_y - \bar{f}_{yy}u_x)\eta_x + (\bar{f}_{xy}u_x - \bar{f}_{xx}u_y)\eta_y\}\\
&(f, \varphi)_{2,2} = 2\int_\Omega \{u_x u_{yy}\varphi_x - u_x u_{xy}\varphi_y\}
\end{aligned}
$$

since $\tfrac{1}{2}[u, u] = (u_x u_{yy})_x - (u_{xy}u_x)_y$, where $\bar{f} = \lambda F_0 + f$. Now we define the bilinear operator $C(\omega, g)$ by means *of the duality method* (cf. Section 2.2D(iii)) by setting

$$(2.5.9')\qquad (C(\omega, g), \varphi) = \int_\Omega \{(g_{xy}\omega_y - g_{yy}\omega_x)\varphi_x + (g_{xy}\omega_x - g_{xx}\omega_y)\varphi_y\}$$

for $g, \omega, \varphi \in H$. The operator C is easily seen to satisfy the following properties:

(i) $(C(\omega, g), \varphi)$ is a symmetric function of g, ω, φ (this follows by integration by parts);

(ii) $(C(\omega, g), \varphi) \leq K\|g\|_{2,2}\|\omega\|_{1,4}\|\varphi\|_{1,4}$, where K is an absolute constant. This follows by Sobolev's imbedding theorem and Hölder's inequality. Thus the system (2.5.9) can be written in the form (relative to the inner product in H)

$$(u, \eta) = (C(u, \bar{f}), \eta), \qquad (f, \varphi) = -(C(u, u), \varphi).$$

Since η, φ are arbitrary, we may write these as

$$u = C(u, f) + \lambda C(u, F_0), \qquad f = -C(u, u).$$

Thus, setting $C(u) = C(u, C(u, u))$ and $Lu = C(u, F_0)$, we may rewrite these

$$(2.5.10)\qquad \text{(a)}\ \ u + Cu = \lambda Lu, \qquad \text{(b)}\ \ f = -C(u, u),$$

where it is understood that any solution u of (a) determines f from (b) uniquely.

Now the fact that $C(u)$ defined above is a gradient mapping is now an easy consequence of the fact that the form $(C(\omega, g), \varphi)$ is symmetric in ω, g, and φ. Indeed, in accord with (2.5.2) or (2.5.6), a short computation shows that if we set $I(u) = \frac{1}{4}(C(u), u)$, then $d(I(u + \epsilon v))/d\epsilon|_{\epsilon=0} = (Cu, v)$ for all $u, v \in H$. For the same reason, the operator $Lu = C(u, F_0)$ is self-adjoint.

A simpler but nonetheless significant example concerns the semilinear operator $Au = \Delta u + f(x, u)$ defined on a domain $\Omega \subset \mathbb{R}^N$. Such an operator can always be represented as a gradient mapping of $\mathring{W}_{1,2}(\Omega)$ into itself provided the function $f(x, u)$ satisfies the smoothness and suitable growth conditions. Indeed, to use the duality method of Section 2.2, suppose $\tilde{f}(u) = f(x, u)$ defines a bounded operator from $\mathring{W}_{1,2}(\Omega)$ into L_p for $p < (N+2)/(N-2)$. Then the abstract operator $\mathcal{A}u$ defined implicitly by the formula

$$(\mathcal{A}u, v) = \int_\Omega \{\nabla u \cdot \nabla v - f(x, u)v\} \qquad \text{for} \quad v \in C_0^\infty(\Omega)$$

can easily be verified to be a gradient mapping with antiderivative

$$I(u) = \int_\Omega \{\tfrac{1}{2}|\nabla u|^2 - F(x, u)\}\, dV, \qquad \text{where } F_u(x, u) = f(x, u).$$

2.6 Nonlinear Fredholm Operators

A smooth mapping f between Banach spaces X, Y can be studied by properties of its Fréchet derivative $f'(x)$. This approach was adopted in Section 2.3 for complex analytic mappings and for gradient mappings in (2.5). In the same direction, based on the results of Section 1.3F, we consider the following:

2.6A Equivalent definitions

(2.6.1) **Definition** Let X, Y be Banach spaces and U a connected open subset of X. A mapping $f \in C^1(U, Y)$ is called a nonlinear Fredholm operator if the Fréchet derivative of f, $f'(x)$ is a linear Fredholm map $\in L(X, Y)$ for each $x \in U$ (see Section 1.3F). In this case, the index of f, ind f, is defined by setting $\operatorname{ind} f(x) = \operatorname{ind} f'(x) = \dim \operatorname{Ker} f'(x) - \dim \operatorname{coker} f'(x)$ for $x \in U$.

(2.6.2) ind $f(x)$ is independent of $x \in U$. Indeed, since $f'(x)$ is continuous in x, ind f: $U \to \mathbb{Z}$ is continuous; and since U is connected, $x \in U$ implies ind $f(x)$ is constant. Thus, ind $f(x)$ is independent of $x \in U$.

(2.6.3) Examples of Fredholm maps and the computation of their indices are readily obtained.

(a) Any smooth map between finite-dimensional Banach spaces is a Fredholm map.

(b) Any diffeomorphism between Banach spaces is a Fredholm map of index zero.

(c) If $f(x)$ is any Fredholm map and $C(x) \in C^1(U, Y)$ is a compact operator, then $f + C$ is a Fredholm operator and $\operatorname{ind}(f + C) = \operatorname{ind} f$. This result follows from (2.4.6). Indeed, $\operatorname{ind}(f + C) = \operatorname{ind}(f' + C') = \operatorname{ind}(f')$ since C' is compact, and $\operatorname{ind}(f') = \operatorname{ind} f$.

(2.6.4) **Theorem** Let $f \in C^1(U, Y)$, then for U an open subset of a Banach space Y, the following statements are equivalent:

(i) f is a Fredholm operator.

(ii) For each fixed $x \in U$, the following inequalities hold for each $y \in Y$:

(2.6.5) $$\|y\| \leqslant C_1\|f'(x)y\| + |y|_0,$$

(2.6.6) $$\|y\| \leqslant C_2\|f'^*(x)y\| + |y|_1,$$

where the constants C_1 and C_2 are independent of y and $|y|_0$ and $|y|_1$ are compact seminorms defined on Y.

Proof: By virtue of (1.3.37), the inequalities (2.6.5) and (2.6.6) taken together imply that for each $x \in U, f'(x)$ has closed range and that dim Ker f', dim Ker f'^* are finite. Conversely, (1.3.37) also implies that any linear Fredholm mapping $f'(x) \in L(X, Y)$ satisfies inequalities of the form (2.6.5) and (2.6.6) for each $x \in U$. Thus (i) and (ii) are equivalent.

2.6B Basic properties

The theorems of Morse and Sard as mentioned in (1.6.1) have useful extensions for nonlinear Fredholm mappings. These extensions will be taken up in Chapter 3. As a first step in this direction, we define the notion of singular and regular points of differentiable operators.

(2.6.7) **Definition** Let $f \in C^1(U, Y)$, then $x \in U$ is a regular point for f if $f'(x)$ is a surjective linear mapping in $L(X, Y)$. If $x \in U$ is not regular, x is called singular. Similarly, singular and regular values y of f are defined by considering the sets $f^{-1}(y)$. If $f^{-1}(y)$ has a singular point, y is called a singular value, otherwise y is a regular value.

In this connection, as in the finite-dimensional case, we prove

(2.6.8) **Theorem** The singular points of a Fredholm operator $f \in C^1(X, Y)$ are closed.

Proof: Let $S = \{x \mid f'(x) \text{ is not onto}\}$ and suppose $x_n \in S$ is such that $x_n \to \bar{x}$. By the continuity of the index of f, under small perturbations, $\operatorname{index} f'(x_n) = \operatorname{index} f'(\bar{x})$ for n sufficiently large. Also by Theorem (1.3.38) of Chapter 1, if $\|B\|$ is sufficiently small, and A is a Fredholm map, (2.6.2) implies that

(2.6.9) $\qquad \dim \operatorname{coker}(A + B) \leqslant \dim \operatorname{coker} A.$

Hence, for n sufficiently large,

(2.6.10)
$$\dim \operatorname{coker} f'(\bar{x}) \geqslant \dim \operatorname{coker}[f'(\bar{x}) + (f'(x_n) - f'(\bar{x}))] \geqslant \dim \operatorname{coker} f'(x_n) \geqslant 1.$$

Consequently, $f'(\bar{x})$ is not a surjective linear map, and so $\bar{x} \in S$.

Example: **See Notes H of Chapters 1 and 4 to check how "instantons" are related to nonlinear Fredholm operators.**

2.6C Differential Fredholm operators

The class of nonlinear Fredholm operators arises very naturally in the study of differential systems since many differential operators (possibly supplemented by auxiliary boundary conditions) and their adjoints have only finite-dimensional subspaces of solutions.

Now suppose we are given the nonlinear elliptic operator

(2.6.11) $\qquad N(u) = F(x, u, \ldots, D^{2m}u)$

defined on a bounded domain $\Omega \subset \mathbb{R}^N$ subject to the Dirichlet boundary conditions $D^{\alpha}u|_{\partial\Omega} = 0$, $|\alpha| \leqslant m - 1$. Then we can consider N as a mapping of $C^{2m,\alpha}(\Omega) \to C^{0,\alpha}(\Omega)$. Provided the function $F = F(x, \xi^1, \ldots, \xi^{2m})$ is a C' function of $(\xi^1, \ldots, \xi^{2m})$, the Fréchet derivative of $N(u)$ at u_0 is easily computed to be

(2.6.12)
$$N'(u_0)v = \sum_{|\alpha| \leqslant 2m} F_{\alpha}(x, u_0, \ldots, D^{2m}u_0)\, D^{\alpha}v.$$

By virtue of the Schauder estimates (1.4.27) and (1.4.28), there is a constant c possibly depending on u_0 such that

(2.6.13) $\qquad \|v\|_{C^{2m,\alpha}} \leqslant c\{\|N'(u_0)v\|_{C^{0,\alpha}} + \|v\|_{C^{0,\alpha}}\}.$

Thus, $\|v\|_{C^{0,\alpha}}$ is a compact seminorm defined on $C^{2m,\alpha}(\Omega)$. $N'(u_0)$ has closed range in $C^{0,\alpha}$ and finite-dimensional kernel. To show that N is a nonlinear Fredholm operator one must verify that the system

(2.6.14) $\qquad N'(u_0)v = f, \qquad D^{\alpha}v|_{\partial\Omega} = 0, \qquad f \in C^{0,\alpha},$

can be solved apart from a finite-dimensional subspace of $C^{0,\alpha}(\Omega)$. To this end, we note that the operator $L_1(v) = \sum_{|\alpha| = 2m} F_{\alpha}(x, u_0, \ldots, D^{2m}u_0)D^{\alpha}v$ maps $C^{2m,\alpha}(\Omega)$ onto $C^{0,\alpha}(\Omega)$ and is also one-to-one. On the other hand, $L_2(v) = \sum_{|\alpha| \leqslant 2m-1} F_{\alpha}(x, u_0, \ldots, D^{2m}u_0)D^{\alpha}v$ is compact as a map from $C^{2m,\alpha}(\Omega)$ to $C^{0,\alpha}(\Omega)$. Thus $N'(u_0) = L_1 + L_2 = L_1\{I + L_1^{-1}L_2\}$ can be factored as a homeomorphism acting on a compact perturbation of the identity. Applying the theory of compact operators to $I + L_1^{-1}L_2$ we find that the system (2.6.14) is solvable provided f is orthogonal to a finite-dimensional subspace in $C^{0,\alpha}(\Omega)$. Thus N is a nonlinear Fredholm operator of index zero. More generally the Fredholm property can be established by showing that both the linearized boundary value problem and its adjoint satisfy (2.6.13).

2.7 Proper Mappings

2.7A Equivalent definitions

An operator $f \in C(X, Y)$ is said to be proper if the inverse image of any compact set C in $Y, f^{-1}(C)$, is compact in X. The importance of this notion resides in the fact that the properness of an operator f restricts the "size" of the solution set $S_p = \{x \mid x \in X, f(x) = p\}$ for any fixed $p \in Y$. Thus, it is immediate that the only proper linear operators in $L(X, Y)$ are one-to-one and have closed range. More generally, we prove

(2.7.1) **Theorem** Let $f \in C(X, Y)$, then the following statements are equivalent:

(i) f is proper.

(ii) f is a closed mapping and the solution set $S_p = \{x \mid x \in X, f(x) = p\}$ is compact for any fixed $p \in Y$.

(iii) If X and Y are finite dimensional, then f is *coercive* (in the sense that $\|f(x)\| \to \infty$ whenever $\|x\| \to \infty$).

Proof: (i)$\Rightarrow$(ii): Since any point $p \in Y$ is compact, the properness of f implies that S_p is compact. To prove that f is closed, let K be a closed subset of X and suppose $y_n = f(x_n) \to y$, $x_n \in K$. Then, since the closure of $\{y_n\}$, $\overline{\{y_n\}}$, is compact, the properness of f implies that $\sigma = f^{-1}(\overline{\{y_n\}})$ is compact. Consequently (after possibly passing to a subsequence), since $x_n \in \sigma$, $\{x_n\}$ converges to a point $\bar{x}$. Since K is a closed set, $\bar{x} \in K$, and by the continuity of f, $f(\bar{x}) = y$.

(ii)$\Rightarrow$(i): Now suppose that f is closed and for any $p \in Y$, S_p is compact. Then, to show that f is proper, let C be a compact subset of Y and $f^{-1}(C) = D$. Suppose that D is covered with closed sets D_α that have the finite intersection property. We show that $\bigcap_\alpha D_\alpha \neq \varnothing$, implying that D is compact. To this end, let $\{\alpha_1, \ldots, \alpha_k\} = \beta$ be any subset of $\{\alpha\}$. Then $E_\beta = \bigcap_{i=1}^k D_{\alpha_i}$ is closed and nonempty, so that $f(E_\beta)$ is closed and $C = \bigcup_\beta f(E_\beta)$. Furthermore, the closed sets $f(E_\beta)$ have the finite intersection property since for any finite subset $\gamma \in 2^\alpha$,

$$\bigcap_\gamma f(E_\beta) \supset f\Big(\bigcap_\gamma E_\beta\Big) \neq \varnothing.$$

Therefore the compactness of C implies that $\delta = \bigcap_\beta f(E_\beta) \neq \varnothing$. Now let $y \in \delta$ and $D_y = D \cap f^{-1}(y)$ so that $D_y \neq \varnothing$. By hypothesis, $f^{-1}(y)$ is compact, as is the set $D_y = \bigcup_\alpha \{D_\alpha \cap f^{-1}(y)\}$. Thus it suffices to show that $\{D_\alpha \cap f^{-1}(y)\}$ has the finite intersection property since then

$$\bigcap_\alpha D_\alpha \supset \bigcap_\alpha \{D_\alpha \cap f^{-1}(y)\} \neq \varnothing.$$

Finally, for any finite subset $\gamma = \{\alpha_1, \ldots, \alpha_j\}$ *of* $\{\alpha\}$, since $y \in \bigcap_{\beta \in 2^\alpha} f(E_\beta)$,

$$\bigcap_{i=1}^{J} D_{\alpha_i} \cap f^{-1}(y) = E_\gamma \cap f^{-1}(y) \neq \varnothing.$$

(iii)$\Leftrightarrow$(ii): Let X and Y be finite dimensional. Then the properness of f implies that the inverse image of a bounded subset of Y is bounded in x, which is merely a restatement of the coerciveness of f. Conversely, if f is coercive and C is any compact subset of Y, then $f^{-1}(C)$ is bounded and so is relatively compact in X.

For special classes of operators $f \in C(X, Y)$ acting between infinite-dimensional Banach spaces, the coerciveness of f implies the properness of f. More precisely, we prove the following criteria for properness.

(2.7.2) Suppose $f \in C(X, Y)$ and $\|f(x)\| \to \infty$ as $\|x\| \to \infty$. Then f is proper if either

(i) f is a compact perturbation of a proper mapping; or
(ii) X is reflexive, and $x_n \to x$ weakly in X with $\{f(x_n)\}$ strongly convergent, implies that $x_n \to x$ strongly.

Proof: (i): Let $f(x_n) = y_n$ with $y_n \to y$ in Y. Then, if $f(x) = g(x) + C(x)$, where g is proper and C is compact, the coerciveness of f implies that $\{x_n\}$ is bounded. Consequently, after possibly passing to a subsequence, $\{C(x_n)\}$ is convergent. Thus, since the sequence $g(x_n) = y_n - Cx_n$ is convergent while g is proper, $\{x_n\}$ has a convergent subsequence $\{x_{n_j}\}$ with limit $\bar{x}$. The continuity of f then implies that $f(\bar{x}) = y$; and so f is proper.

(ii): If X is reflexive and $f(x_n) \to y$ in Y, then the coerciveness of f implies that $\{x_n\}$ is bounded. Hence (after possibly passing to a subsequence, once more), we may suppose $x_n \to \bar{x}$ weakly in X and hence, by hypothesis strongly, so that $f(\bar{x}) = y$, so that once again f is proper.

2.7B Basic properties

A simple quantitative property of a proper mapping $f \in C(X, Y)$ is the following one expressing the stability of the solution set $S_p(f) = \{x \mid x \in X, f(x) = p\}$ under small perturbations in p or f.

(2.7.3) **Theorem** Let $f \in C(X, Y)$ be proper. Then:

(i) for every $p \in Y$ and every $\epsilon > 0$, there is a $\delta > 0$ such that

(2.7.4) $$\|f(x) - p\| \leqslant \delta \quad \text{implies} \quad \|x - f^{-1}(p)\| \leqslant \epsilon;$$

(ii) if $g \in C(X, Y)$, then $\|f(x) - g(x)\| \leqslant \delta$ for all $x \in X$ implies that $d(S_p(f), S_p(g)) \leqslant \epsilon$.

Proof: It suffices to prove (i) since (ii) is an immediate consequence of (i).

Thus suppose (i) is false. Then there exists an $\epsilon > 0$, a $p \in Y$, and a sequence $\{x_n\} \in X$ such that for all n

(2.7.5) $\qquad \|f(x_n) - p\| \leqslant 1/n \qquad \text{and} \qquad \|x_n - f^{-1}(p)\| \geqslant \epsilon.$

Since f is proper and $f(x_n) \to p$, by passing to a subsequence if necessary we may suppose that $x_n \to x$. Then since $f \in C(X, Y)$, $f(x) = p$ and $x \in f^{-1}(p)$. But this fact contradicts (2.7.5).

In the same direction we prove

(2.7.6) Let X and Y be Banach spaces and $f \in C(X, Y)$. Suppose U and V are open subsets of X and Y, respectively, such that f maps U onto V, is locally invertible, and proper on U. Then the function c_p = the number of points in $S_p(U) = \{x \mid x \in U, f(x) = p\}$ is finite and constant in each component of $f(U)$. (See Fig. 2.1.)

Proof: Clearly, the local invertibility and properness of f imply that $f^{-1}(p)$ is discrete and compact. Consequently, c_p is finite.

The fact that c_p is locally constant follows in the same way from Theorem (2.7.3), just obtained.

More generally, we now consider proper mappings that are not locally invertible. As in (2.6.7), if $f \in C^1(X, Y)$ we say that x is a singular point if $f'(x)$ is not locally invertible at x. Let the set of such singular points for f be called the singular set S. With the same notation and terminology as in (2.6.7), we then prove

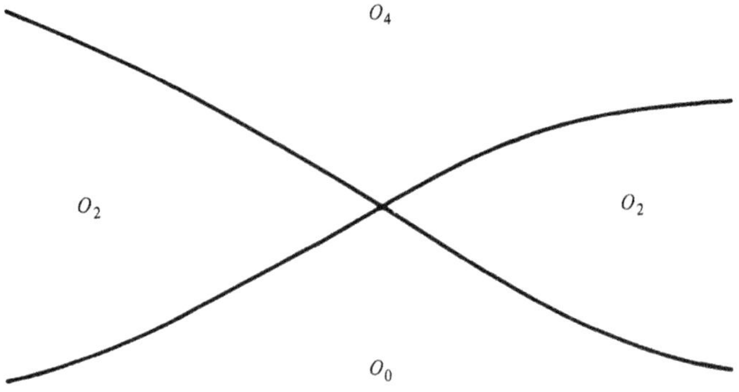

FIG. 2.1 A typical decomposition of the range of a proper Fredholm mapping f by its singular values, into connected components O_i. For $p \in O_i$ the equation $f(x) = p$ has exactly i solutions.

(2.7.7) If $f \in C^1(X, Y)$ is a proper Fredholm operator of index zero and S denotes the singular set of f, then c_y is constant on every (connected) component of $Y - f(S)$. (More generally for proper operators of higher index the sets $f^{-1}(y)$ are homeomorphic.)

Proof: Clearly by (2.6.8) S is closed; and since f is proper, (2.7.1) implies $f(S)$ is closed. Then $U = X - f^{-1}(f(S))$ and $V = Y - f(S)$ are open subsets of X and Y, respectively. Now we can apply (2.7.6) to U and V. Clearly, f maps U into V, is proper, and is locally invertible on U. Thus the result follows since the components of V are arcwise connected.

2.7C Differential operators as proper mappings

Finally, we investigate some criteria for the properness of abstract mappings associated with nonlinear differential operators.

First, consider the concrete operator

$$\mathcal{Q}u = \sum_{|\alpha| \leq m} (-1)^{|\alpha|} D^\alpha \{A_\alpha(x, u, \ldots, D^\alpha u)\}$$

defined on a bounded domain $\Omega \subset \mathbb{R}^N$, and the abstract operator A: $\mathring{W}_{m,p}(\Omega) \to W_{-m,q}(\Omega)$ associated with $\mathcal{Q}$ by the formula (using the duality principle of Section 2.2D)

(2.7.8) $$(Au, \varphi) = \sum_{|\alpha| \leq m} \int_\Omega A_\alpha(x, u, \ldots, D^\alpha u)\, D^\alpha \varphi.$$

We prove the following analogue of (2.7.2):

(2.7.9) **Theorem** Suppose A satisfies the conditions (2.2.7) that ensure that A is a bounded, continuous mapping from $\mathring{W}_{m,p}(\Omega) \to W_{-m,q}(\Omega)$. If (i) $(Au, u)/\|u\| \to \infty$ as $\|u\| \to \infty$, and (ii) $\mathcal{Q}$ is strongly elliptic in the sense that for $p \in (1, \infty)$

$$\sum_{|\alpha| = m} \{A_\alpha(x, y, z) - A_\alpha(x, y, z')\}\{z - z'\} \geq c|z - z'|^p,$$

where c is a constant independent of y, z, z', while the lower order terms satisfy (∗) of (2.2.7), and the mapping A is a proper mapping of $\mathring{W}_{m,p}(\Omega)$ into $W_{-m,q}(\Omega)$.

Proof: First we observe that A is coercive in the sense of (2.7.1) since by (i) $\|Au\| \geq (Au, u)/\|u\| \to \infty$ as $\|u\| \to \infty$. Next, we consider the operator A defined by (2.7.8) and verify the sufficient condition of (2.7.2) for properness. To this end, we write $A = A_1 + A_2$, where

$$(A_1 u, \varphi) = \sum_{|\alpha| = m} \int_\Omega A_\alpha(x, u, \ldots, D^m u)\, D^\alpha \varphi, \quad (A_2 u, \varphi) = \sum_{|\alpha| < m} \int_\Omega A_\alpha(x, u, \ldots, D^\alpha u)\, D^\alpha \varphi.$$

Our hypotheses imply (by (2.4.9)) that A_2 is completely continuous and that A_1 satisfies the inequality

(2.7.10) $$(A_1 u - A_1 v, u - v) \geq k\|u - v\|^p_{m,p}$$

where $k > 0$ is independent of u and v. Consequently, it suffices to prove A_1 is proper. Since the space $\mathring{W}_{m,p}(\Omega)$ is reflexive, it suffices to prove that $u_n \to u$ weakly and $\{A_1 u_n\}$ strongly convergent imply that $u_n \to u$ strongly. But this is an immediate consequence of the inequality (2.7.10).

Finally, we establish the properness of the von Kármán operator defined in (2.5.7), when regarded as a mapping of $\mathring{W}_{2,2}(\Omega)$ into itself. We prove

(2.7.11) The von Kármán operator defined in (2.5.7),

$$\mathscr{A}_\lambda(u) = u + Cu - \lambda Lu \qquad \text{(for fixed } \lambda\text{)},$$

is a proper mapping of $\mathring{W}_{2,2}(\Omega)$ into itself.

Proof: Assuming the complete continuity of the mappings C and L for the moment, it suffices by (2.7.2(ii)) to establish the coerciveness of $\mathscr{A}_\lambda(u)$. To this end, let $\|u_n\| \to \infty$; then for fixed λ,

$$(\mathscr{A}_\lambda(u_n), u_n) = \|u_n\|^2 + (Cu_n, u_n) - \lambda(Lu_n, u_n).$$

By virtue of (2.5.7),

$$(Lu_n, u_n) = (C(F_0, u_n), u_n) = (C(u_n, u_n), F_0),$$

while $(Cu_n, u_n) = \|C(u_n, u_n)\|^2$. Thus for any $\epsilon > 0$,

$$(\mathscr{A}_\lambda(u_n), u_n) \geq \|u_n\|^2 + \|C(u_n, u_n)\|^2 - \epsilon^{-1}\|F_0\|^2 - \epsilon\|C(u_n, u_n)\|^2.$$

Choosing $\epsilon = 1$ in this last equation, we find

$$\|\mathscr{A}_\lambda(u_n)\| \geq (\mathscr{A}_\lambda(u_n), u_n)/\|u_n\| \geq \|u_n\| - \|u_n\|^{-1}\|F_0\|^2.$$

Consequently, $\mathscr{A}_\lambda(u)$ is coercive as an operator from $\mathring{W}_{2,2}(\Omega)$ into itself.

Finally, to determine the complete continuity of the operators L and C, we use the inequality (ii) (following (2.5.9′)) mentioned in the proof of (2.5.7). Indeed, if $u_n \to u$ weakly in $\mathring{W}_{2,2}(\Omega)$, the Sobolev inequality (1.4.18) implies that $u_n \to u$ strongly in $W_{1,4}(\Omega)$, so that as $n \to \infty$, for some absolute constant $K_1 > 0$,

$$\|Lu_n - Lu\| = \sup_{\|\varphi\|=1} (Lu_n - Lu, \varphi) = \sup_{\|\varphi\|=1} (C(F, u_n - u), \varphi) \leq K_1\|F\|_{2,2}\|u_n - u\|_{1,4} \to 0.$$

Similarly, $C(u, u)$ is completely continuous. Thus for any $\varphi \in \mathring{W}_{2,2}(\Omega)$, using (2.5.7) we find

$$\begin{aligned}(C(u_n) - C(u), \phi) &= (C(u_n, u_n), C(u_n, \varphi)) - (C(u, u), C(u, \varphi)) \\ &= (C(u_n, u_n) - C(u, u), C(u, \varphi)) + (C(u_n, u_n), C(u_n, \varphi) - C(u, \varphi)).\end{aligned}$$

Thus for $\|\varphi\|_{2,2} \leq 1$, there is an absolute constant M such that

$$\|C(u_n) - C(u)\| \leq M\{\|C(u_n, u_n) - C(u, u)\| + \|C(u_n, \varphi) - C(u, \varphi)\|\}.$$

Consequently, $C(u_n) \to C(u)$ and the complete continuity of C is established.

Thus we conclude that the map $A_\lambda(u)$ is proper.

In Chapter 6 we study and utilize the consequences of the properness of A_λ (viz. (2.7.6) and (2.7.7)), in conjunction with the calculus of variations in the large, to estimate the number of solutions of the equation $A_\lambda(u) = g$ as g varies.

NOTES

A Mapping properties of the operator $f = f(x, u)$ between L_p spaces

Let $f(x, u)$ be defined on $\Omega \times \mathbb{R}^1$, where Ω is a bounded domain of $\mathbb{R}^n$, with the following continuity properties: f is continuous in u for almost all x and measurable in x for all u relative to Lebesgue measure. Then to prove (2.2.1) proceed as follows:

(i) *f preserves convergence in measure* This is easily demonstrated by first arguing with simple functions, and then successively taking linear combinations, and limits.

(ii) *f is continuous*, given that $f(x, u)$ satisfies the growth condition

$$(*) \qquad |f(x, u)| \leqslant \alpha + \beta |u|^{p_1/p_2} \qquad \text{for some absolute constants } \alpha, \beta > 0.$$

This reduces to proving that the following limit can be taken under the integral sign:

$$\lim_{n\to\infty} \int_\Omega |f(x, u_n) - f(x, u)|^{p_2} \qquad \text{where} \quad \|u_n - u\|_{L_{p_1}} \to 0.$$

The justification for taking the limit under the integral sign is based on Vitali's theorem on absolutely equicontinuous integrals and the given growth condition.

(iii) *f is bounded.* This result follows from the continuity of f at zero, and the continuity of Lebesgue measure.

Although we shall not use this in the text, it can be shown that the growth condition $(*)$ is a consequence of the fact that f maps $L_{p_1}(\Omega)$ into $L_{p_2}(\Omega)$. For this proof, the reader is referred to Krasnoselski (1964).

B Real analytic operators

Our development of complex analytic operators was based on Hartog's theorem on separate analyticity. The analogous situation for real Banach spaces has not been extensively developed. A smooth mapping f defined on an open set D of a real Banach space X with range in another real Banach space Y is called real analytic in D if it possesses Fréchet derivatives of all orders at each point in D, and moreover $f(x)$ can be expanded as a convergent power series in terms of these derivatives as in (2.3.3 (vi)). Certain results on real analytic operators can be deduced from analogous results on complex analytic operators. Indeed, each real Banach space can be imbedded isometrically into a complex Banach space $X + iX$ in a canonical manner; and in fact bounded multilinear, symmetric: mappings of X into Y can be uniquely extended as multilinear symmetrlc bounded mappings of $X + iX$ into $Y + iY$. Thus one shows that a real analytic map can be extended canonically to a complex analytic one. For these results, the reader is referred to Alexicwicz and Orlicz (1954).

C The abstract Navier–Stokes operator

The Navier–Stokes equations (1.1.18)–(1.1.19) can be reformulated via the duality method of Section 2.2.D as an operator equation in a Hilbert space $\mathring{H}$. The condition div $\mathbf{u} = 0$ allows one to restrict attention to solenoidal N-vectors and the Hilbert space $\mathring{H}$ can be chosen to the space of solenoidial N-vectors w obtained by completing each component of $\mathbf{u} \in C_0^\infty(\Omega)$ in the Sobolev space $\mathring{W}_{j,2}(\Omega)$. If we consider the Navier–Stokes equations defined on a bounded domain Ω in $\mathbb{R}^N$, $N = 2, 3$, subject to homogeneous boundary conditions of Dirichlet type we find that these equations can be written in the form

$$f_\lambda(\omega) \equiv w + \lambda N w = \tilde{g} \qquad (\lambda \equiv \text{Reynolds number}).$$

(Note that the pressure term, appearing as a gradient, drops out of these equations and $\tilde{g}$ represents an extended forcing vector.) Using the duality method to justify this fact, given the fact that N is defined implicitly by the formula

$$(**) \qquad (Nw, \varphi)_H = -\int_\Omega \{w \cdot \text{grad } w\}\varphi \equiv \sum_{k=1}^{N} \int_\Omega w_k w_{x_k} \cdot \varphi.$$

It is easily proven that

(i) The operator N so defined is a compact mapping of H into itself.

(ii) For each γ, the associated operator f_γ is a proper mapping. (Note here that $(Nw, w) = 0$ for each $w \in H$, so that the properness follows via (2.7.2).)

(iii) $f_\gamma(w)$ is Fréchet differentiable and consequently $f_\gamma(w)$ is a nonlinear Fredholm operator of index zero.

(iv) For sufficiently small γ, the solution of $f_\gamma(w) = g$ is unique. More generally, off the singular values of f_γ the solutions are finite, in number.

As we shall see in Chapter 5, for a large class of inhomogeneous boundary conditions, the representation (**) remains valid. Moreover, it can be easily shown that the results (i)–(iv) hold in that case as well. See Ladyzhenskaya (1969).

D Bibliographic notes

Section 2.1: The calculus of mappings between infinite-dimensional linear spaces has an interesting history. Early references include Volterra (1930), Hadamard (1903), and Fréchet (1906). More recently the following works have proven interesting: Dieudonné (1960), Nevanlinna (1957), Hille and Phillips (1957), Michal (1958), and Cartan (1970, 1971). Early discussions of the derivative are found in Gateaux (1906) and Fréchet (1925). The book by Ljusternik and Sobolev (1961) contains a more-up-to-date treatment. In his thesis Goldring (1977) has completed the first steps of a Hodge decomposition theorem for nonlinear operators when regarded as differential one forms.

Section 2.2: The books of Krasnoselski (1964) and Vainberg (1964) contain careful discussions of the composition operators of 2.2A. The result (2.2.10) is from Littman (1967). The method of Schauder inversion is a formalization of a technique that recurs in the papers of Schauder mentioned in the bibliography. The duality method for the definition of abstract nonlinear operators has proven to be very effective in many different contexts and is well described in Brézis (1973), Browder (1976), and Lions (1969).

Section 2.3: Our discussion of analytic operators is patterned after Hille (1948). The paper of Taylor (1937) makes interesting reading, while the recent work of Douady (1965) may prove of general value.

Section 2.4: The systematic investigation of compact operators and their connection with algebraic topology is due to Schauder. (See his papers listed in the Bibliography.)

Section 2.5: A useful survey of the basic results on gradient operators can be found in Rothe (1953), and Krasnoselski (1964). Much of this work is based on carrying over the concepts of the calculus of variations to a more formal context. The result (2.5.7) can be found in Berger (1967). Goldring (1977) has extended the motion of gradient mapping by proving various infinite-dimensional versions of the Frobenius integrability theorems.

Section 2.6: Nonlinear Fredholm operators were introduced in Smale (1965). An interesting attempt to extend the index theorem of Atiyah and Singer to a nonlinear context can be found in Palais (1967). There seems little doubt that the concept of nonlinear Fredholm operator will prove important in the future development of our subject.

Section 2.7: A thorough discussion of proper mappings can be found in Bourbaki (1949), while the result (2.7.7) is proved and applied in the paper Ambrosetti and Prodi (1972). The result (2.7.11) is found in Berger (1974).

PART II

LOCAL ANALYSIS

The aim of Part II Here we discuss the local mapping properties of a non-linear operator f restricted to a small neighborhood of a given point of its domain. We then relate these properties to more specific notions of theoretical and concrete importance.

The basic problems to be discussed To fix notation, let f denote an operator defined in a neighborhood $U(x_0)$ of a point x_0 in a Banach space X with range in another Banach space Y. Then we attempt to determine the behavior of $f(x)$ near $f(x_0)$ *in as precise a form as possible* by posing the following questions:

(i) *Linearization problem* If $f(x)$ is differentiable (at x_0), in what sense are the properties of $f(x)$ near $f(x_0)$ reflected by the linear operator $f'(x_0)$?

(ii) *Local solvability problem* If $\|f(x_0 - y\|$ is "small", under what circumstances can we solve $f(x) = y$ for x "near" x_0?

(iii) *Local conjugacy problem* If g is another mapping with domain $U(x_0)$ and range V contained in Y and such that $f - g$ is small (in some sense), in what cases do f and g differ by a local change of coordinates, i.e., there exist local homeomorphisms ("changes of coordinates") h_X: $U(x_0) \to U(x_0)$ and h_Y: $V \to V$ such that $f = h_Y^{-1} g h_X$. In particular, if $f(x) = Lx + O(\|x\|^2)$ near $x = 0$ where L is a linear operator, what properties of L and f ensure that these operators are conjugate near $x = 0$?

(iv) *Stability problem* In what sense are the mapping properties of the operator f near $U(x_0)$ unaffected by a small (but possibly arbitrary) perturbation $f + \epsilon g$ (for ϵ a small real number)? If a given property is destroyed by an "arbitrary" perturbation, can the property be preserved by restricting the class of allowable perturbations?

(v) *Problem concerning local structure of solutions* If $f(x_0) = y_0$, can one give a complete description of the set of solutions $\{x \mid f(x) = y_0, x \in U(x_0)\}$? In particular, are the solutions isolated?

(vi) *Problem of nonlinear effects* What features of the higher order part of the operator f, viz. $f(x) - f'(x_0)\{x - x_0\} - f(x_0)$, are significant in studying the local properties of f near $f(x_0)$?

(vii) *Problem of parameter dependence* If the mapping $f(x) = f(x, \lambda)$ depends continuously (smoothly) on a parameter λ, how do the local properties of f change as λ varies? In particular, describe the behavior of the solutions of $f(x, \lambda) = 0$ near the "bifurcation" set $\Sigma = \{x, \lambda) \mid \operatorname{coker} f_x(x, \lambda) \neq \{0\}, f(x, \lambda) = 0\}$.

(viii) *Problem concerning construction of appropriate solutions* If $f(x) = y$ has a solution

x near x_0 when $f(x_0) - y$ is small, can one construct an explicit approximation x to x_0 such that $\|x - x_0\|$ can be made arbitrarily small?

(ix) *Problem concerning iteration schemes* If a sequence x_n is defined by the rule $x_n = g(x_{n-1}, x_{n-2}, \ldots, x_{n-k})$ (where k is some finite integer independent of n, and g is a continuous map defined on $X \times X \cdots (k \text{ times}) \times X$, under what circumstances does the sequence (or some subsequence) converge to a solution $\bar{x}$ of $x = g(x, x, \ldots, x)$? Furthermore, under the hypothesis of question (viii) above, can an approximate solution be defined by a convergent iteration scheme?

The problems just mentioned arise naturally in connection with the *detailed* study of explicit nonlinear systems. Thus methods for obtaining approximate solutions to a given local problem Π are well known and (generally speaking) readily constructible. For example, Π may be "close" to a problem Π' all of whose solutions are known explicitly, and one assumes (as a first approximation) that the solutions of Π are "close" to those of Π'. Indeed, the well-known techniques of linearization, successive approximation, averaging, undetermined coefficients, and singular perturbation are among the formal constructions for such approximate solutions. Yet the validity of such approximate solutions is often left open by these formal schemes, and in fact, often taken for granted despite much evidence to the contrary. As we shall see in the sequel, the study of local analysis sheds considerable light on such questions. For example, higher order approximations (of arbitrary order N), $x_N(\epsilon) = x_0 + \sum_{n=0}^{N} a_n \epsilon^n$, to the solution of an operator equation $f_\epsilon(x) = 0$ may often be constructed with the property that $f_\epsilon(x_N(\epsilon)) = O(\epsilon^{N+1})$. However, as was mentioned in Section 1.2B, it may happen that $\lim_{N\to\infty} x_N(\epsilon)$ does not exist for any $\epsilon \neq 0$, since the infinite series $\sum a_n \epsilon^n$ diverges. Thus the question of the validity of the approximation $x_N(\epsilon)$ to a true solution $x(\epsilon)$ of $f_\epsilon(x) = 0$ requires further investigation.

Over and above the approaches just mentioned, there is often a close bond connecting local and global approaches to a given problem. Thus there are many problems of local analysis that cannot adequately be studied by power series (or Fourier series) expansions. Such results make their appearance in Section 4.2 in connection with periodic orbits of Hamiltonian systems and the removal of the irrationality conditions from Liapunov's theorem (cf. 1.2 B (v) pages 23-24). The same problem recurs in the "nonlinear desingularization" phenomenon for vortex ring described in Note I of Chapter 4. Indeed in many areas of contemporary science "nonperturbative effects" are playing an increasingly crucial role in understanding.

CHAPTER 3

LOCAL ANALYSIS OF A SINGLE MAPPING

In this chapter we focus attention on a fixed operator f acting between two Banach spaces or (as in Section 3.4) two scales of Banach spaces and discuss the elementary approximation and iteration schemes that relate to the inverse and implicit function theorems. In the first section of the chapter we discuss those results that can be based on the elementary contraction mapping principle. The applications of these results to ordinary differential equations in a Banach space, singularities of mappings, and local properties of extremals of isoperimetric variational problems are then discussed. The classical steepest descent and majorant methods of successive approximation are described in the next two sections. Finally (in Section 3.4) we take up the recent extensions of the iteration schemes associated with the inverse function theorem due to Nash, Moser, Kolomogorov, and Arnold.

3.1 Successive Approximations

The simplest systematic approach to answer the questions of local analysis just raised is based on the method of successive approximations for the solvability of the operator equation $f(x)=0$. In fact all the results of this section are based on this theme. Given $f \in C(\overline{U}, Y)$, the fundamental idea of the method is to define (explicitly) a Cauchy sequence of elements $x_n \in \overline{U}$ such that $f(x_n) \to 0$. Then by the completeness of $\overline{U}$ (assumed to be the closure of an open subset of a Banach space X), x_n converges to some $\bar{x} \in \overline{U}$, and, by the continuity of f, $f(\bar{x})=0$. The simplest case of such a construction is provided by the following.

3.1A The contraction mapping principle

Given a continuous mapping A of a set S into itself, one attempts to locate the fixed points of A by defining a sequence $(x_0, Ax_0, A^2x_0, \ldots, A^nx_0, \ldots)$ for $x_0 \in S$ and by seeking conditions on S and A that ensure the convergence of this sequence. A simple answer is the

(3.1.1) **Contraction Mapping Theorem** Denote by $S(\bar{x}, \rho)$ the sphere of radius ρ and center $\bar{x}$ of a Banach space X. Suppose A maps $S(\bar{x}, \rho)$ into itself and satisfies the condition that for any $x, y \in S(\bar{x}, \rho)$,

(3.1.2) $$\|Ax - Ay\| \leqslant K\|x - y\|,$$

where *K is an absolute constant less than* 1. Then A has one and only one fixed point x_∞ in $S(\bar{x}, \rho)$, and x_∞ is the limit of the sequence $x_n = A^n x_0$ $(n = 0, 1, 2, \ldots)$ for any choice of x_0 in $S(\bar{x}, \rho)$.

Proof: First we show that $x_n = A^n x_0$ is a Cauchy sequence for any $x_0 \in S(\bar{x}, \rho)$. Indeed, for any integers n and p, comparison with the geometric series $K^n + K^{n+1} + \cdots$ yields

$$\|x_{n+p} - x_n\| = \|A^{n+p}x_0 - A^n x_0\| \leqslant \sum_{j=n}^{n+p-1} \|A^{j+1}x_0 - A^j x_0\|$$

$$\leqslant \sum_{j=n}^{n+p-1} K^j \|Ax_0 - x_0\| \leqslant \frac{K^n}{1-K} \|Ax_0 - x_0\|.$$

Hence as $n \to \infty$, $\|x_{n+p} - x_n\| \to 0$ independently of p, so that $\{x_n\}$ is indeed a Cauchy sequence in $S(\bar{x}, \rho)$. Since $S(\bar{x}, \rho)$ is complete, $x_n \to x_\infty$ (say) with $x_\infty \in S(\bar{x}, \rho)$. Hence by the continuity of A

(3.1.3) $$Ax_\infty = \lim_{n\to\infty} Ax_n = \lim_{n\to\infty} x_{n+1} = x_\infty,$$

i.e., x_∞ is a fixed point; and it is unique since if y_∞ were another fixed point, then (3.1.3) would imply

$$\|x_\infty - y_\infty\| = \|Ax_\infty - Ay_\infty\| \leqslant K\|x_\infty - y_\infty\|,$$

which is possible only if $x_\infty = y_\infty$.

Of the many interesting extensions of (3.1.1) the following one is quite useful when the map A depends on a parameter β.

(3.1.4) **Corollary** Suppose $A(x, \beta)$ is a continuous mapping of $S(\bar{x}, \rho) \times B \to S(\bar{x}, \rho)$ for some metric space B, and furthermore that A satisfies (3.1.2) for each $\beta \in B$. Then the mapping $g\colon B \to x_\beta$ (the unique fixed point of $x = A(x, \beta)$) is a continuous mapping of B into X.

Proof: Let $\beta_n \to \beta_\infty$ in B. Then $g(\beta_n) = x_{\beta_n} = A(x_{\beta_n}, \beta_n)$, and similarly for $\beta = \beta_\infty$. Hence

$$\begin{aligned}
\|g(\beta_n) - g(\beta_\infty)\| &= \|A(x_{\beta_n}, \beta_n) - A(x_{\beta_\infty}, \beta_\infty)\| \\
&\leqslant \|A(x_{\beta_n}, \beta_n) - A(x_{\beta_\infty}, \beta_n)\| \\
&\quad + \|A(x_{\beta_\infty}, \beta_n) - A(x_{\beta_\infty}, \beta_\infty)\| \\
&\leqslant K\|x_{\beta_n} - x_{\beta_\infty}\| + \|A(x_{\beta_\infty}, \beta_n) - A(x_{\beta_\infty}, \beta_\infty)\|
\end{aligned}$$

so that

$$\|g(\beta_n) - g(\beta_\infty)\| \leq \frac{1}{1-K} \|A(x_{\beta_\infty}, \beta_n) - A(x_{\beta_\infty}, \beta_\infty)\|.$$

Since A is continuous in β, the right-hand side above tends to 0; and the result follows.

3.1B The inverse and implicit function theorems

We now prove the well-known Banach space analogues of the inverse mapping and implicit function theorems. Both these results are obtained together with constructive iteration schemes. The inverse function theorem gives a first answer to the basic linearizatlon question raised at the beginning of Part II, while the implicit function theorem answers analogous problems involving parameter dependence.

(3.1.5) **Inverse Functlon Theorem** Suppose f is a C^1 mapping defined in a neighborhood of some point x_0 of a Banach space X, with range in a Banach space Y. Then if $f'(x_0)$ is a linear homeomorphism of X onto Y, f is a local homeomorphism of a neighborhood $U(x_0)$ of x_0 to a neighborhood of $f(x_0)$. Furthermore, if $\|y - f(x_0)\|$ is sufficiently small, the sequence

(3.1.6) $$x_{n+1} = x_n + [f'(x_0)]^{-1}[y - f(x_n)]$$

converges to the unique solution of $f(x) = y$ in $U(x_0)$.

Proof: Set $f(x_0) = y_0$. We first attempt to determine ρ so that $f(x_0 + \rho) = y$ provided $\|y - y_0\|$ is sufficiently small, or equivalently

(3.1.7) $$f(x_0 + \rho) - f(x_0) = y - y_0.$$

Since f is C^1 at x_0 and $f'(x_0)$ is invertible, (3.1.7) implies that $f'(x_0)\rho + R(x_0, \rho) = y - y_0$, i.e.,

$$\rho = [f'(x_0)]^{-1}[(y - y_0) - R(x_0, \rho)],$$

where the remainder

$$R(x_0, \rho) = f(x_0 + \rho) - f(x_0) - f'(x_0)\rho = o(\|\rho\|).$$

We show that (3.1.7) has one and only one solution for $\|\rho\|$ sufficiently small, by proving that the operator $A\rho = [f'(x_0)]^{-1}\{y - y_0 - R(x_0, \rho)\}$ is a contraction mapping of a sphere $S(0, \epsilon)$ in X into itself, for some ϵ sufficiently small. Indeed for ρ and $\rho_1 \in S(0,\epsilon)$,

$$\begin{aligned} f'(x_0)\{A\rho - A\rho_1\} &= R(x_0, \rho_1) - R(x_0, \rho) \\ &= f(x_0 + \rho_1) - f(x_0 + \rho) - f'(x_0)(\rho_1 - \rho) \\ &= \int_0^1 \{f'(x_0 + t\rho_1 + (1-t)\rho) \\ &\qquad - f'(x_0)\}(\rho_1 - \rho)\, dt. \end{aligned}$$

Hence

$$\|A\rho - A\rho_1\| = \int_0^1 \|[f'(x_0)]^{-1}\| \, \|f'(x_0 + t\rho_1 + (1-t)\rho) - f'(x_0)\| \, \|\rho_1 - \rho\| dt. \tag{3.1.8}$$

Since f is a C^1 mapping, the middle term of the last intergrand can be made arbitrarily small by choosing $\|\rho\|$, $\|\rho_1\|$ sufficiently small; and hence for some constant $K < 1$ (and independent of $y - y_0$) and sufficiently small $\epsilon > 0$, $\|A\rho - A\rho_1\| \leqslant K\|\rho - \rho_1\|$ for all ρ, ρ_1 in $S(0, \epsilon)$. Furthermore, A maps $S(0, \epsilon)$ into itself. Indeed, $\|A\rho\| = \|A\rho - A(0)\| + \|A(0)\| \leqslant K\|\rho\| + \|A(0)\|$ and $\|A(0)\| = \|[f'(x_0)]^{-1}(y - y_0)\| < (1-K)\epsilon$ provided $\|y - y_0\| < (1-K)\epsilon\|[f'(x_0)]^{-1}\|^{-1}$. Hence A is a contraction map of $S(0, \epsilon)$ into itself, under the last proviso. By the contraction mapping theorem (3.1.1), A has a unique fixed point in $S(0, \delta)$, where $\delta \leqslant \epsilon$ is chosen so small that $f(S(0, \delta)) \subset S(y_0, (1-K)\epsilon\|[f'(x_0)]^{-1}\|^{-1})$. Reversing the steps in the argument, one finds that $f(x_0 + \rho) = y$ has one and only one solution when $\|y - y_0\|$ and $\|\rho\|$ are sufficiently small. That y depends continuously on ρ and hence on $x = x_0 + \rho$ follows immediately from Corollary 3.1.4 and the obvious fact that under the operator $A\rho = [f'(x_0)]^{-1}\{y - y_0 - R(x_0, \rho)\}$ depends continuously on y. Thus $f^{-1}(y) = x$ is a well-defined and continuous mapping from a sphere $S(y_0, \eta)$ in Y to X. Finally, for $\|f(x_0) - y\|$ sufficiently small, $f(x) = y$ has a unique solution $x = x_0 + \rho$, where ρ is the limit of the sequence $\rho_0 = 0$, $\rho_n = A\rho_{n-1}$. Then

$$\begin{aligned} x_n = x_0 + \rho_n &= x_0 + A\rho_{n-1} \\ &= x_0 + [f'(x_0)]^{-1}[y - f(x_0) - R(x_0, \rho_{n-1})] \\ &= x_0 + [f'(x_0)]^{-1}[y + f'(x_0)\rho_{n-1} - f(x_0 + \rho_{n-1})] \\ &= x_{n-1} + [f'(x_0)]^{-1}[y - f(x_{n-1})]. \end{aligned}$$

Hence $x = \lim_{n\to\infty} x_n$ where x_n is defined by the iteration scheme $x_n = x_{n-1} + [f'(x_0)]^{-1}[y - f(x_{n-1})]$.

(3.1.9) **Corollary** Under the hypothesis of Theorem (3.1.5), f^{-1} is differentiable, and $(f^{-1}(y_0))' = (f'(x_0))^{-1}$.

Proof: If $f(x_0) = y_0$ and $f(x_0 + x) = y_0 + h$, then

$$\begin{aligned} f^{-1}(y_0 + h) - f^{-1}(y_0) - f'(x_0)^{-1}h &= f'(x_0)^{-1}\{f'(x_0)x - h\} \\ &= -f'(x_0)^{-1}\{f(x_0 + x) - f(x_0) - f'(x_0)x\} \\ &= o(\|x\|) = o(\|h\|). \end{aligned}$$

Thus f^{-1} is differentiable and $(f^{-1}(y_0))' = (f'(x_0))^{-1}$ at y_0.

Next we find hypotheses so that the equation $f(x, y) = 0$ can be uniquely solved, locally, in the form $y = g(x)$, where the function g is as smooth as f is.

(3.1.10) **Implicit Function Theorem** Let X, Y, and Z be Banach spaces. Suppose $f(x, y)$ is a continuous mapping of a neighborhood U of (x_0, y_0) in $X \times Y$ into Z, $f(x_0, y_0) = 0$, and $f_y(x_0, y_0)$ exists is continuous in x and is a linear homeomorphism of Y onto Z. Then there is a unique continuous mapping g defined in a neighborhood U_1 of x_0, $g\colon U_1 \to Y$, such that $g(x_0) = y_0$ and $f(x, g(x)) = 0$ for $x \in U_1$.

Proof: For fixed x near x_0, we write

$$f(x, y) = f_y(x_0, y_0)(y - y_0) + R(x, y)$$

where $R(x, y) - R(x, y') = o(\|y - y'\|)$ for (x, y) and (x, y') near (x_0, y_0). To solve $f(x, y) = 0$ near (x_0, y_0), we consider the map

$$A_x y = y - [f_y(x_0, y_0)]^{-1} f(x, y) = y_0 - f_y^{-1}(x_0, y_0) R(x, y).$$

The proof of Theorem (3.1.5) ensures that for fixed x (near x_0) A_x is a contraction map of a small sphere centered at y_0 into itself. The unique fixed point $y(x)$ of $A_x(y)$, which exists by (3.1.1), depends continuously on x, by (3.1.4). Furthermore, $y(x_0) = y_0$ and $f(x, y(x)) = 0$. Furthermore, $y(x)$ is the only continuous function with these properties, for any other such function would necessarily be a fixed point of $A_x y$. Thus we need only set $g(x) = y(x)$ to obtain the desired result.

(3.1.11) **Corollary** If, in addition to the hypothesis of the implicit function theorem (3.1.7), $f_x(x, y)$ exists and is continuous for (x, y) near (x_0, y_0), then the function $g(x)$ is continuously differentiable for $x \in U_1$ and

$$g'(x) = -[f_y(x, g(x))]^{-1} f_x(x, g(x)) \tag{3.1.2}$$

Proof: We first establish the Lipschitz continuity of $g(x)$. Under the hypothesis of the corollary, $f(x, y)$ is a C^1 mapping near (x_0, y_0) and in addition for $\|h\|$ sufficiently smooth and (x, y) near (x_0, y_0), $f(x, g(x)) = f(x + h, g(x + h)) = 0$. Hence expanding $f(x + h, g(x + h))$ about $(x, g(x))$ we find

$$\|f_x(x, g(x))h + f_y(x, g(x))[g(x + h) - g(x)]\| = o(\|h\| + \|g(x + h) - g(x)\|).$$

Since $f_y(x, g(x))$ is invertible and continuous in x:

$$\|[f_y(x, g(x))]^{-1} f_x(x, g(x))h + [g(x + h) - g(x)]\| = o(\|h\| + \|g(x + h) - g(x)\|). \tag{3.1.13}$$

Thus there is a constant M independent of h such that $\|g(x + h) - g(x)\| \leqslant M\|h\|$. Now (3.1.13) implies that $g(x)$ is differentiable and that (3.1.12) holds.

Remark: If $f(x, y) \in C^n$ near (x_0, y_0), then the function $g(x)$ is also C^n. This follows immediately from (3.1.12) for $n = 2$, and for general n by induction based on the same formula.

3.1C Newton's method

We now turn to a refinement of the iteration scheme (3.1.5), namely the so-called Newton method, which allows a substantial improvement in the rate of convergence of (3.1.5). This method may be described as follows: Given an initial approximation x_0 to the solution of $f(x) = 0$, we try to find a better approximation $x_1 = x_0 + \rho_1$, where ρ_1 is determined so that $f(x_1) = 0$ is satisfied *apart from higher order terms*. Thus assuming $[f'(x_0)]^{-1}$ exists, $f(x_1) = f(x_0 + \rho_1) = f(x_0) + f'(x_0)\rho_1 + o(\|\rho_1\|)$ so that $\rho_1 = -[f'(x_0)]^{-1}f(x_0)$. Continuing in this manner, at the $(n+1)$th stage an approximate solution $x_{n+1} = x_n + \rho_{n+1}$ is found by setting $\rho_{n+1} = -[f'(x_n)]^{-1}f(x_n)$; so that (provided $[f'(x_n)]^{-1}$ always exists) we find a *formal solution* x_∞ of $f(x)$ in the form $x_\infty = x_0 + \sum_{n=1}^{\infty}\rho_n$. A virtue of the Newton method is the rapid convergence of $x_n \to x_\infty$. That is, instead of an estimate of the form $\|x_{n+1} - x_n\| \leqslant K\|x_n - x_{n-1}\|$ which yields $\|x_\infty - x_N\| = O(K^N)$, one finds exponential convergence: $\|x_{n+1} - x_n\| \leqslant K\|x_n - x_{n-1}\|^2$ for certain absolute constants ϵ_0, K so that $\|x_\infty - x_N\| = O[(\epsilon_0 K)^{2^N}]$, with $\epsilon_0 = \|f'^{-1}(x_0)\|\,\|f(x_0)\|$. Indeed, suppose the sequence $\{x_n\}$ stays in some sphere S in X, f is C^2 and $\|f'(x) - f'(y)\| \leqslant M\|x - y\|$ for $x, y \in S$. Then by Taylor's formula, 2.1.33

$$\|f(x_n) - f(x_{n-1}) - f'(x_{n-1})(x_n - x_{n-1})\| \leqslant M\|x_n - x_{n-1}\|^2. \tag{3.1.14}$$

By definition, $f(x_k) = -f'(x_k)(x_{k+1} - x_k)$ for any k, so that from (3.1.14)

$$\frac{1}{\|f'^{-1}(x_n)\|}\|x_{n+1} - x_n\| \leqslant \|f'(x_n)(x_{n+1} - x_n)\| \leqslant M\|x_n - x_{n-1}\|^2.$$

Thus

$$\|x_{n+1} - x_n\| \leqslant M\|f'^{-1}(x_n)\|\|x_n - x_{n-1}\|^2, \tag{3.1.15}$$

and we can choose $K = M \sup_n \|[f'(x_n)]^{-1}\|$.

(3.1.16) **Theorem** Let f be a C^1 mapping defined on a sphere $S(\bar{x}, \delta)$ of a Banach space X with range in a Banach space Y, and let x_0 be an arbitrary point of $S(\bar{x}, \delta)$. Suppose f is such that for arbitrary $x, y \in S(\bar{x}, \delta)$:

(i) $\|f'(x) - f'(y)\| \leqslant M_1\|x - y\|$;

(ii) $f'(x_0)$ is a linear homeomorphism of $X \to Y$. Then provided $\|f(x_0)\|$ is sufficiently small, the sequence $x_{n+1} = x_n - [f'(x_n)]^{-1}f(x_n)$ is defined and converges to the unique solution x_∞ of $f(x) = 0$. Furthermore, $\|x_N - x_\infty\| = O(\epsilon^{2^N})$ as $N \to \infty$ for some positive number $\epsilon < 1$, with ϵ defined by (3.1.18′) below.

Proof: Proceeding as in the proof of the inverse function theorem, we set $f(x_0) = y_0$ and seek an element ρ such that $f(x_0 + \rho) - f(x_0) = -y_0$. Assuming $(x_0 + \rho) \in S(\bar{x}, \delta)$ and expanding $f(x_0)$ about $x_0 + \rho$ in the form $f(x_0) = f(x_0 + \rho) - f'(x_0 + \rho)\rho + R(x_0 + \rho, \rho)$, we find the following equation for ρ:

(3.1.17) $$-y_0 = f'(x_0 + \rho)\rho + R(x_0 + \rho, \rho)$$

$$\text{where} \qquad \|R(x_0 + \rho, \rho)\| = o(\|\rho\|).$$

Since $f'(x_0)$ is a linear homeomorphism, so is $f'(x_0 + \rho)$ for $\|\rho\|$ sufficiently small, and the operator

$$B\rho = -[f'(x_0 + \rho)]^{-1}\{y_0 + R(x_0 + \rho, \rho)\}$$

is thus well defined for $\|\rho\|$ sufficiently small. We shall show: (i) B defines a contraction mapping of $S(x_0, \delta')$ into itself for δ' sufficiently small, so that the equation (3.1.17) is uniquely solvable; and (ii) the convergent iteration scheme so generated coincides with the one defined in the statement of (3.1.16).

To prove that B is a contraction mapping, we first derive some estimates for the dependence of $[f'(x_0 + \rho)]^{-1}$ on ρ using the notation $[f'(x_0 + \rho)]^{-1} = g(\rho)$. By virtue of (ii) and the identity $L_1^{-1} - L_2^{-1} = L_1^{-1}(L_2 - L_1)L_2^{-1}$,

$$\|g(\rho) - g(\rho')\| \leqslant M_1\|\rho - \rho'\| \, \|g(\rho)\| \, \|g(\rho')\|.$$

Thus setting $\|g(0)\| = C$, we obtain

$$\|g(\rho)\| \leqslant \|g(\rho) - g(0)\| + C \leqslant CM_1\|\rho\| \, \|g(\rho)\| + C$$
$$\leqslant C(1 - CM_1\|\rho\|)^{-1}.$$

Hence for $\|\rho\| \leqslant \delta' < \min\{(2M_1C)^{-1}, \delta\}$, $g(\rho)$ exists and $\|g(\rho)\| \leqslant 2C$. Furthermore, for $\rho, \rho' \in S(x_0, \delta')$,

$$\|B\rho - B\rho'\| \leqslant \|g(\rho)\| \, \|R(x_0 + \rho, \rho) - R(x_0 + \rho', \rho')\|$$
$$+ \|g(\rho) - g(\rho')\| \, \|y_0 + R(x_0 + \rho', \rho')\|.$$

Now

$$R(x_0 + \rho) - R(x_0 + \rho') = f(x_0 + \rho') - f(x_0 + \rho)$$
$$+ f'(x_0 + \rho)\rho - f'(x_0 + \rho')\rho'$$
$$= \int_0^1 \{f'(x_0 + \rho + s(\rho' - \rho))$$
$$- f'(x_0 + \rho)\}(\rho' - \rho)\, ds$$
$$+ [f'(x_0 + \rho') - f'(x_0 + \rho)]\rho.$$

Combining the above results and using the hypotheses of the theorem, we find

$$\begin{aligned}\|B\rho - B\rho'\| &\leqslant 2CM_1\{\|\rho' - \rho\|^2 + \|\rho\|\,\|\rho' - \rho\|\}\\ &\leqslant 4C^2M_1^2(\|y_0\| + \tfrac{1}{2}M_1\|\rho\|^2)\|\rho' - \rho\|\\ &\leqslant \bar{K}\|\rho' - \rho\|,\end{aligned}$$

where $\bar{K} < 1$ is a constant independent of ρ and ρ' provided δ' and $\|y_0\|$ are sufficiently small. In addition $\|B\rho\| \leqslant \|B\rho - B(0)\| + \|B(0)\| \leqslant \bar{K}\|\rho\| + C\|y_0\|$, which implies that $\|B\rho\| \leqslant \delta'$ provided $C\|y_0\| \leqslant (1 - \bar{K})\delta'$. So by choosing $\|y_0\|$ and consequently δ' sufficiently small, B is a contraction mapping of $S(x_0, \delta')$ into itself.

Thus B has a unique fixed point $\bar{\rho} \in S(x_0, \delta)$ defined as the limit of the sequence $\rho_0 = 0$,

$$\begin{aligned}(3.1.18)\qquad \rho_{n+1} = B\rho_n &= -[f'(x_0 + \rho_n)]^{-1}\{y_0 + R(x_0 + \rho_n, \rho_n)\}\\ &= -[f'(x_0 + \rho_n)]^{-1}\{f(x_0 + \rho_n) - f'(x_0 + \rho_n)\rho_n\}\\ &= \rho_n - f'(x_0 + \rho_n)f(x_0 + \rho_n).\end{aligned}$$

Setting $x_n = x_0 + \rho_n$, (3.1.18) becomes the classic Newton iteration scheme $x_{n+1} = x_n - (f'(x_n))^{-1}f(x_n)$ mentioned in the theorem. Setting $x_\infty = \lim_{n\to\infty} x_n$, the estimate for $\|x_\infty - x_N\|$ follows since by the inequality (3.1.15), $\|x_{N+1} - x_N\| = O(\bar{K}^{2^N})$. Indeed,

$$\begin{aligned}\|x_{N+1} - x_N\| &\leqslant \bar{K}\|x_N - x_{N-1}\|^2 \leqslant \cdots \leqslant \bar{K}^{1+2+\cdots+2^{N-1}}\|x_1 - x_0\|^{2^N}\\ &= O\big(\bar{K}^{2^N}\|x_1 - x_0\|^{2^N}\big).\end{aligned}$$

Then if one sets

$$(3.1.18')\qquad \epsilon = \bar{K}\|x_1 - x_0\| = \bar{K}\|[f'(x_0)]^{-1}\|\,\|f(x_0)\|,$$

we find by (3.1.15) and the fact that

$$\|x_\infty - x_N\| \leqslant \sum_{j=0}^{\infty} \|x_{N+j+1} - x_{N+j}\|,$$

that

$$\|x_\infty - x_N\| = O(\|x_{N+1} - x_N\|) = O(\epsilon^{2^N}).$$

3.1D A criterion for local surjectivity

What can be said about the behavior of $f(x)$ near $f(x_0)$ if $f'(x_0)$ is *not invertible*? We shall return to this important question in the next chapter. However, for the present, the methods just developed enable us to study the case in which $f'(x_0)$ is surjective but not (necessarily) injective (a situation often occurring in infinite-dimensional Banach spaces).

(3.1.19) **Theorem** Let X and Y be Banach spaces and f be a C^1 mapping defined on a neighborhood of x_0 in X with range in Y and such that $f'(x_0)$ maps X onto Y. Then f is an open mapping for x near x_0.

Proof: It suffices to show that for y near $y_0 = f(x_0)$, the equation $f(x_0 + \rho) - f(x_0) = y - y_0$ has a solution for small $\|\rho\|$. Since f is C^1 near x_0, this last equation can be rewritten

$$f'(x_0)\rho + R(x_0, \rho) = y - y_0, \qquad \text{where} \qquad \|R(x_0, \rho)\| = o(\|\rho\|).$$

In order to show that this last equation has a solution, we consider the following linear equation for given $\rho \in X$:

$$f'(x_0)\xi = (y - y_0) - R(x_0, \rho).$$

Since $f'(x_0)$ is surjective, (1.3.24) ensures that this equation has a solution $\xi(\rho)$ such that

$$\|\xi(\rho)\| \leqslant M\|(y - y_0) - R(x_0, \rho)\|, \tag{3.1.20}$$

where M is a constant independent of ρ. Thus we can define a sequence $\{\rho_N\}$ by letting ρ_N satisfy:

(i) $f'(x_0)\rho_N = y - y_0 - R(x_0, \rho_{N-1})$;
(ii) $\|\rho_N\| \leqslant M\|y - y_0 - R(x_0, \rho_{N-1})\|$.

Clearly the existence of a limit for $\{\rho_N\}$ in a small sphere about x_0 (for each y near y_0) will prove the theorem. We shall establish this limit by verifying the inequality

$$\|\rho_{N+1} - \rho_N\| \leqslant K\|\rho_N - \rho_{N-1}\|, \qquad (K < 1)$$

of the contraction mapping theorem. Indeed, for $\rho_N \in S(0, \delta_1)$ and $y \in S(y_0, c\delta_1)$, by virtue of (3.1.20) with $\delta_1 < 1$,

$$\begin{aligned}\|\rho_{N+1}\| &\leqslant M_1\{c\delta_1 + \delta_1^2\} \\ &\leqslant \delta_1 \qquad \text{(for } c \text{ sufficiently small).}\end{aligned}$$

Consequently, $\{\rho_N\} \in S(0, \delta_1)$ for all N. On the other hand, for $\rho_{N+1}, \rho_N \in S(0, \delta_1)$

$$\begin{aligned}f'(x_0)(\rho_{N+1} - \rho_N) &= R(x_0, \rho_N) - R(x_0, \rho_{N-1}) \\ &= f(x_0 + \rho_N) - f(x_0 + \rho_{N-1}) - f'(x_0)(\rho_N - \rho_{N-1}) \\ &= \int_0^1 [f'(x_0 + \rho_{N-1} + s(\rho_N - \rho_{N-1})) - f'(x_0)](\rho_N - \rho_{N-1})\, ds.\end{aligned}$$

Thus by the mean value theorem, for some $s_0 \in [0, 1]$,

$$\|\rho_{N+1} - \rho_N\| \leqslant M\|f'(x_0 + \rho_{N-1} + s_0(\rho_N - \rho_{N-1})) - f'(x_0)\|\,\|\rho_N - \rho_{N-1}\|.$$

Since f is C^1, for δ_1 sufficiently small,

$$\|\rho_{N+1} - \rho_N\| \leqslant \tfrac{1}{2}\|\rho_N - \rho_{N-1}\|.$$

Consequently, by the proof of (3.1.1), the sequence $\{\rho_N\}$ is convergent and $\rho_N \to \rho_\infty \in S(0, \delta_1)$. Thus $f(x_0 + \rho_\infty) = y$ so that f is an open mapping for x near x_0.

3.1E Application to ordinary differential equations

An important application of the preceding results concerns the properties of the solutions of the initial value problem

$$\frac{dx}{dt} = f(t, x), \qquad x(0) = x_0, \tag{3.1.21}$$

where we suppose that the function $f(t, x)$ is continuous for fixed x and $t \in [0, \alpha]$ and locally Lipschitz continuous in x for fixed t; i.e., for $x_1, x_2 \in U$ an open subset of a Banach space X with $\|x_i - x_0\| \leqslant R$ $(i = 1, 2)$, there is a positive constant $K(R)$ depending only on R such that

(3.1.22) $$\|f(t, x_1) - f(t, x_2)\| \leqslant K(R)\|x_1 - x_2\|.$$

In fact we prove

(3.1.23) **Theorem** If $K(R)\alpha < 1$ and the above hypotheses on $f(t, x)$ are satisfied, then:

(i) the initial value problem (3.1.21) has one and only one solution $x(t, x_0)$ on the interval $[0, \alpha]$:

(ii) $x(t, x_0)$ is a uniformly continuous function of x_0 for any closed bounded interval on which $x(t, x_0)$ is defined. In fact, if $\|f(t, x) - f(t, y)\| \leqslant K\|x - y\|$, then

(3.1.24) $$\|x(t, x_0) - x(t, y_0)\| \leqslant \|x_0 - y_0\|e^{Kt}.$$

Proof: (i) Clearly (3.1.21) is equivalent to the integral equation

(3.1.25) $$x(t) = x_0 + \int_0^t f(s, x(s))\, ds$$

(i.e., (3.1.21) has a solution $x(t)$ on $[0, \alpha]$ if and only if $x(t)$ satisfies (3.1.25)). Let $Ax(t) = x_0 + \int_0^t f(s, x(s))\, ds$. We shall show that A is a contraction mapping of a sphere of $C\{[0, \alpha], X\}$ into itself for α sufficiently small. Thus by the contraction mapping theorem, (3.1.1) and consequently (3.1.21) will have one and only one solution $x(t, x_0)$. To this end, we recall that $|||x||| = \sup_{[0,\alpha]} \|x(t)\|_X$ is a complete norm defined on $C\{[0, \alpha], X\}$, so that denoting by $\Sigma(x_0, R)$ the sphere of radius R and center x_0 in $C\{[0, \alpha], X\}$,

$$|||Ax - x_0||| \leqslant \int_0^t |||f(s, x(s))|||\, ds \leqslant \{K + |||f(s, x_0)|||\}\alpha;$$

while for $x, y \in \Sigma(x_0, R)$, $|||Ax - Ay||| \leqslant K\alpha|||x - y|||$. Thus A will be a contraction mapping of $\Sigma(x_0, R)$ into itself provided $K\alpha < 1$ and $K\alpha + |||f(s, x_0)|||\alpha \leqslant R$. Clearly these inequalities hold simultaneously if $0 < \alpha < 1/K$ and α is chosen sufficiently small.

(ii): Clearly the mapping A defined in (i) depends continuously on x_0, so that by (3.1.4), the fixed point of A, $x(t, x_0)$, depends continuously on x_0 in the topology of $C\{[0, \alpha], X\}$. To prove the more precise result, we note that if $x(t, x_0)$ and $x(t, y_0)$ denote the solutions of (3.1.21with respect to the initial conditions x_0 and y_0, then

$$\|x(t, x_0) - x(t, y_0)\| \leqslant \|x_0 - y_0\| + \int_0^t \|f(s, x(s, x_0)) - f(s, x(s, y_0))\|\, ds$$
$$\leqslant \|x_0 - y_0\| + K\int_0^t \|x(s, x_0) - x(s, y_0)\|\, ds.$$

Then denoting the left-hand side of the above inequality by $w(t)$, we find that $w(t) \geqslant 0$ and $w(t)$ satisfies the inequality

(3.1.26) $$w(t) \leqslant \|x_0 - y_0\| + K\int_0^t w(s)\, ds.$$

Consequently, for any interval $[0, T]$ on which $x(t)$ exists, $w(t) \leqslant \|x_0 - y_0\|e^{Kt}$. Indeed, multiplying by e^{-Kt}, we find

$$\frac{d}{dt}\left\{e^{-Kt}\int_0^t w(s)\,ds\right\} \leqslant e^{-Kt}\left\{w(t) - K\int_0^t w(s)\,ds\right\} \leqslant \|x_0 - y_0\|e^{-Kt}.$$

Integrating this last inequality from 0 to T, we obtain

$$e^{-KT}\int_0^T w(s)\,ds \leqslant \|x_0 - y_0\|\left\{\frac{1}{K} - \frac{e^{-KT}}{K}\right\}.$$

So from (3.1.26), we obtain (3.1.24).

Next we prove some results on the continuation properties of the solution $x(t, x_0)$ of (3.1.21).

(3.1.27) **Theorem** Suppose $f(t, x)$ is a continuous function defined on $\mathbb{R}^1 \times X$ which is locally Lipschitz in x for fixed t. Then the solution $x(t, x_0)$ of (3.1.21) can be uniquely extended as a solution of (3.1.21) to a maximal interval $[0, A)$. If $x(t, x_0)$ exists on the interval $[0, \beta)$, while $\lim_{t \uparrow \beta} x(t, x_0)$ exists and is finite, $A > \beta$.

Proof: First, we show that any solution $x(t, x_0)$ of (3.1.21) defined on the interval $[0, \gamma)$ is unique. Suppose $x(t, x_0)$ and $y(t, x_0)$ are two solutions of (3.1.21) defined on $[0, \gamma)$, and let $J = \{t | t \in [0, \gamma)$ such that $x(t, x_0) = y(t, x_0)\}$. Certainly by (3.1.23), J is not empty. We shall show that J is open and closed in $[0, \gamma)$, so that since $[0, \gamma)$ is connected, $J = [0, \gamma)$. The set J is certainly closed since both $x(t, x_0)$ and $y(t, x_0)$ are continuous. To show that J is open, suppose $\gamma_0 \in J$, then by the local uniqueness result (3.1.23), there is a $\delta > 0$ such that the system

$$\frac{dx}{dt} = f(x, t), \qquad x(0) = x(\gamma_0, x_0),$$

has the unique solution $x(t, x(\gamma_0, x_0)) = x(t + \gamma_0, x_0)$ for $|t| < \delta$. Consequently, the interval $(\gamma_0 + \delta, \gamma_0 - \delta) \in J$, and J is open.

Now we demonstrate the existence of the maximal interval $[0, \alpha)$ of existence of $x(t, x_0)$ satisfying (3.1.21). Let $\mathfrak{S}$ be the sets of pairs $\{[0, \delta_x), x(t, x_0)\}$ such that $x(t, x_0)$ satisfies (3.1.23) on $[0, \delta_x)$. Clearly by the uniqueness result of the above paragraph, for any two such pairs $\{[0, \delta_{x_1}), x_1(t, x_0)\}$ and $\{[0, \delta_{x_2}), x_2(t, x_0)\}$, $x_1(t, x_0) = x_2(t, x_0)$ for $t \in [0, \min(\delta_{x_1}, \delta_{x_2})$. Let $\alpha = \sup_{x \in \mathfrak{S}} \delta_x$. Then on $[0, \alpha)$, there is precisely one function $x(t, x_0)$ satisfying the initial value problem (3.1.21), and the interval $[0, \alpha)$ is the desired maximal interval.

Finally, if $x(t, x_0)$ exists on $[0, \beta)$ and $\lim_{t \uparrow \beta} x(t, x_0) = \bar{x}$ exists and is of finite norm, we can apply the local existence theorem (3.1.23) to the initial value problem $dx/dt = f(x, t)$, $x(0) = \bar{x}$, and assert the existence and uniqueness of its solution $\tilde{x}(t, \bar{x})$ for some open interval $(-\delta, \delta)$ about $t = 0$. Again by (3.1.23), with $x(\beta, x_0) = \bar{x}$,

$$\tilde{x}(t, \bar{x}) = x(t, x(\beta, x_0)) = x(t + \beta, x_0)$$

for $t \in (-\delta, \delta)$. Thus $x(t, x_0)$ can be uniquely extended to the interval $[0, \beta + \delta)$ as a solution of (3.1.21).

For finite-dimensional Banach spaces, Theorem (3.1.23) can be improved by weakening the assumption of Lipschitz continuity of $f(t, x)$ to just *continuity itself*. A quick proof of this result can be obtained by application of the Schauder fixed point theorem (2.4.3). Consider the initial value problem (3.1.21), where we now suppose that $x(t)$ is an N-vector and $f(t, x)$ is a continuous N-vector function of t and x. Then we prove

(3.1.28) **Peano's Theorem** Under the hypotheses just mentioned, the initial value problem (3.1.21) has at least one solution $x(t, x_0)$ on the interval $[-T, T]$ provided $|T|$ is sufficiently small.

Proof: Without loss of generality we may suppose that $x_0 = 0$. Then a solution of (3.1.21) can be found by solving the integral equation

(3.1.29) $$x(t) = \int_0^t f(s, x(s))\, ds, \qquad x(t) \in \mathbb{R}^N.$$

Denoting the integral on the right-hand side of equation (3.1.29) by $Ax(t)$, and setting

$$\sup_{[0, T]\times[M, M]} |f(s, S)| = K_M,$$

we find that for any continuous N-vector $x(t)$ defined on $[0, T]$ with $|x(t)| \leqslant M$ over $[0, T]$,

(3.1.30) $$|Ax(t)| \leqslant \int_0^t |f(s, x(s))|\, ds \leqslant K_M t \leqslant K_M T.$$

Consequently, with $C_N[0, T]$ denoting the Banach space of continuous N-vector functions defined on $(0, T]$ with sup norm, we find that A is a bounded mapping of $C_N[0, T]$ into itself and that, as before, a solution of (1.3.21) is a fixed point of A in $C_N[0, T]$. We obtain a fixed point of A by requiring that $|T|$ be sufficiently small and invoking Schauder's fixed point theorem (2.4.3). To this end, note that by (3.1.30), A maps the sphere $\Sigma_M = \{x| \; \|x\| \leqslant M, x \in C_N[0, T]\}$ into itself provided $|T| \leqslant M/K_M$. Thus in order to apply Schauder's theorem it suffices to prove that A is a (*continuous and*) *compact mapping* on $\Sigma_M \subset C_N[0, M/K_M]$. The continuity of A follows immediately from the continuity of $f(s, x(s))$. To verify the compactness of A, it suffices by virtue of (3.1.30) and (1.3.13), to prove the equicontinuity of the vectors $A(t)$ for $x(t) \in \Sigma_M$. To this end, we note that

$$|A(t_1) - A(t)| = \left|\int_t^{t_1} f(s, x(s))\, ds\right| \leqslant K_M |t_1 - t|.$$

Hence the requisite equicontinuity property is verified, so that the desired fixed point is obtained and the theorem is established.

3.1F Application to isoperimetric problems

Many questions concerning gradient mappings $G'(x) \in M(H, H)$ (H, a Hilbert space) can be phrased as follows: Find the extremals x of the antiderivative $G(x)$ of $G'(x)$ over a constraint set C. We refer to such problems as abstract isoperimetric problems. Here we shall investigate the operator equations satisfied by such extremals by using the results established so far in this chapter.

As an application of Peano's theorem (3.1.28), we now establish the following result for abstract isoperimetric variational problems.

(3.1.31) Let H be a Hilbert space and suppose u_0 is an extremum of the C^1 functional $G_0(u)$ subject to the constraint $C = \{u | G_i(u) = c_i$

($i = 1, \ldots, N$), where the numbers c_i are constants}. Then there are numbers λ_i (not all zero) such that

$$\sum_{i=0}^{N} \lambda_i G_i'(u_0) = 0, \tag{3.1.32}$$

where $G_i'(x)$ denotes the Fréchet derivative of $G_i(x)$ at x_0.

Proof: We argue by contradiction, by assuming that the vectors $G_i'(u_0)$ ($i = 0, \ldots, N$) are linearly independent. Let the extreme value of $G_0(u)$ on C be c_0, then we show that if (3.1.32) is never satisfied, we can find a curve $u(t) \in C$ for $|t|$ sufficiently small with $u(0) = u_0$ such that $G_0(u(t)) = c_0 + t$. Since t can be positive or negative, this contradicts the facts that u_0 is an extremum for G_0 on C. To this end, let $u(t) = u_0 + \Sigma_{j=0}^{N} a_j(t) w_j$, where the real-valued functions $a_j(t)$ and the vectors w_j are to be determined such that

$$a_j(0) = 0, \qquad G_0(u(t)) = c_0 + t, \qquad G_i(u(t)) = c_i \quad (i = 1, \ldots, N). \tag{3.1.33}$$

Assuming the w_j are given, we can find the functions $a_j(t)$ satisfying (3.1.33) provided that we can solve the initial value problem

$$\frac{d}{dt} G_i\left(u_0 + \sum_{j=0}^{N} a_j(t) w_j\right) = \gamma_i \qquad (i = 0, \ldots, N; \quad a_j(0) = 0), \tag{3.1.34}$$

where $\gamma_0 = 1$ and $\gamma_i = 0$ for $i > 0$. Simplifying (3.1.34) we can rewrite this initial solution to the problem in the vector form

$$\mathcal{Q}(a(t)) \frac{da}{dt} = \gamma, \qquad a(0) = 0, \tag{3.1.35}$$

where $a(t) = (a_0(t), \ldots, a_N(t))$, $\gamma = (1, 0, \ldots, 0)$, and $\mathcal{Q}(a(t) = (a_{ij})$ is the $(N + 1) \times (N + 1)$ matrix with entries $a_{ij} = (G_i'(u_0 + a(t) \cdot w), w_j)$ with $w = (w_0, w_1, \ldots, w_N)$. Now by Peano's theorem (3.1.28), (3.1.35) and consequently (3.1.33) have solutions provided the matrix $\mathcal{Q}(a(t))$ has an inverse, for $|t|$ sufficiently small, that depends continuously on $a(t)$. Clearly this will be the case provided $\det |\mathcal{Q}(a(0))| \neq 0$. Thus we shall choose vectors w_j such that this determinant is different from zero, making use of the fact that the vectors $G_i'(u_0)$ ($i = 0, \ldots, N$) are linearly independent, by assumption.

In fact with $w_j = G_j'(u_0)$, $\det |\mathcal{Q}(a(0))| = \det |(G_i'(u_0), G_j'(u_0))| \neq 0$. If not, the system of linear equations

$$\sum_j \beta_j (G_i'(u_0), G_j'(u_0)) = 0 \qquad (i = 1, \ldots, N)$$

would have a nontrivial solution $\bar{\beta}_j$ (say). Then multiplying the above equation by β_i and summing we find that for $\beta_i = \bar{\beta}_i$ (say),

$$\left\| \sum_{i=0}^{N} \bar{\beta}_i G_i'(u_0) \right\| = 0 \qquad \text{which implies} \qquad \sum_{i=0}^{N} \bar{\beta}_i G_i'(u_0) = 0.$$

Since the vectors $G_i'(u_0)$ are linearly independent, we find that $\bar{\beta}_i = 0$ ($i = 0, \ldots, N$). Thus since $\det |\mathcal{Q}(a(0))| \neq 0$, the curve $u(t) \in C$ exists for $|t|$ sufficiently small, and therefore $G_0(u(t)) = c_0 + t$. This fact is the desired contradiction.

Remark: The result (3.1.31) holds for Banach spaces X since the $\{w_j\}$ elements can be such that $\det |\mathcal{Q}(a(0))| \neq 0$.

A related but somewhat more general result (for a possible infinite number of constraint equations) can be stated as follows:

(3.1.36) **Theorem** Suppose G is a C^1 mapping of a Hilbert space H into a Hilbert space H_1 such that for some x_0, $G'(x_0)$ maps H onto H_1. Then if x_0 is an extremal of a C^1 functional $F(x)$ restricted to the set $M = \{x | G(x) = 0\}$, there is an element $h_1 \in H_1$ such that x_0 is a critical point of the unrestricted functional

$$F(x) - (G(x), h_1).$$

Proof: Let $T = \{x | G'(x_0)x = 0\}$. We first show that for arbitrary $x \in T$, we can write an element $y \in M$ in the form $y = x_0 + x + g$, where $g \in [T]^\perp$ and $\|g\| = o(\|x\|)$ as $\|x\| \to 0$. To this end, we apply the implicit function theorem to the operator equation

$$\mathcal{G}(x, g) \equiv G(x_0 + x + g) = 0.$$

Now (regarded as a linear operator) the partial derivative $\mathcal{G}_g(0, 0) = G'(x_0)$ maps $T^\perp$ injectively onto H_1, and so by Banach's theorem (1.3.20), $G'(x_0)$ (restricted to T) is invertible. Thus the implicit function theorem implies that the equation $\mathcal{G}(x, g) = 0$ has a unique C^1 solution $g = g(x) \in T^\perp$ for $\|x\|$ sufficiently small. To prove that $\|g(x)\| = o(\|x\|)$ as $\|x\| \to 0$, we note for small t and fixed $x \in T$ that

$$G(x_0 + tx + \tilde{g}(t)) = 0,$$

where $\tilde{g}(t) \in T^\perp$ is a C^1 function of t. Differentiating this equation with respect to t and setting $t = 0$,

(3.1.37) $$G'(x_0)x + G'(x_0)\tilde{g}'(0) = 0.$$

Since $x \in T$, $G'(x_0)x = 0$, and (3.1.37) then implies that $\tilde{g}'(0) = 0$ (since $G'(x_0)$ restricted to $T^\perp$ is invertible). Consequently, by (3.1.12), $\|g(x)\| = \|\tilde{g}(1)\| = o(\|x\|)$. Now we observe that for $h \in [T]^\perp$ the expression $f(h) = (F'(x_0), h)$ is a well-defined bounded linear functional on $T^\perp$, and therefore on H_1 since $G'(x_0)$ is a linear homeomorphism of $T^\perp$ onto H_1. Consequently, there is a fixed element $h_1 \in H_1$ such that $(F'(x_0), h) = (y, h_1)$ for each $y \in H_1$. Thus, for each $h \in T^\perp$ with $y = G'(x_0)h$,

(3.1.38) $$(F'(x_0), h) = (G'(x_0)h, h_1).$$

Finally, for arbitrary $h \in M$, $h = n + m$, where $n \in T^\perp$ and $m \in T$. Clearly $G'(x_0)m = 0$. On the other hand, by virtue of the results established in the first paragraph, $h(t) = x_0 + tm + g(t)$, where $\|g(t)\| = o(|t|)$, so that for arbitrary $m \in T$

$$\frac{d}{dt} F(h(t))|_{t=0} = (F'(x_0), m) = 0.$$

Hence (3.1.38) holds not only for $h \in T^\perp$ but for all $h \in H$. Thus x_0 is a critical point on H of the unrestricted functional $F(x) - (G(x), h_1)$.

As a final example, consider the critical points of the C^2 functional $F(x)$ restricted to the hypersurface $\mathfrak{M} = \{x | G(x) = \text{const.}\}$ of a Hilbert space H. If $G'(x) \neq 0$ on $\mathfrak{M}$, then a critical point x_0 of F satisfies the equation

(3.1.39) $$F'(x_0) - \lambda G'(x_0) = 0, \qquad \text{where} \quad \lambda = \frac{(F'(x_0), G'(x_0))}{\|G'(x_0)\|^2}.$$

This second variation denoted $\delta^2 F(x_0, v)$ is a quadratic form defined on tangent vectors to the hypersurface $\mathfrak{M}$, by means of the formula

$$\delta^2 F(x_0, v) = \frac{d^2}{dt^2} F(v(t))|_{t=0}.$$

Here $v(t)$ is a C^1 curve on $\mathfrak{M}$ passing through x_0 such that

$$\frac{d}{dt} v(t)|_{t=0} = v \qquad \text{with} \quad (v, G'(x_0)) = 0.$$

We now compute this second variation of F at x_0 relative to $\mathfrak{M}$. In fact, we have the following simple formula

(3.1.40) The second variation of F restricted to $\mathfrak{M}$ on a Hilbert space H can be written

$$(3.1.41) \qquad \delta^2 F(x_0, v) = ([F''(x_0) - \lambda G''(x_0)]v, v),$$

where $(v, G'(x_0)) = 0$ and λ is given by (3.1.39).

Proof: By arguing as in the proof of (3.1.31), the arc $x(t) = x + tv + a(t)G'(x)$ lies on $\mathfrak{M}$ if x does, where $a(t)$ is defined as the solution of the initial value problem

$$(3.1.42) \qquad a'(t) = -(G'(x(t)), v)/(G'(x(t)), G'(x)), \qquad a(0) = 0.$$

Moreover in deriving (3.1.41) it suffices to consider arcs of the form $x(t)$. Now $(d/dt)F(x(t)) = (F'(x(t), x'(t))$ and

$$(3.1.43) \qquad \frac{d^2}{dt^2} F(x(t)) = (F''(x(t))x'(t), x'(t)) + (F'(x(t)), x''(t)).$$

By choosing $(v, G(x)) = 0$, $a'(0) = 0$ so that $x'(0) = v$. Thus

$$\frac{d^2}{dt^2} F(x(t))|_{t=0} = (F''(x)v, v) + (F'(x), a''(0)G'(x)).$$

Now at a critical point x_0, $F'(x_0) = \lambda G'(x_0)$, implying that

$$(3.1.44) \qquad \delta^2 F(x_0, v) = (F''(x_0)v, v) + \lambda a''(0)\|G'(x_0)\|^2.$$

On the other hand, to compute $a''(0)$, we note that $(G'(x(t)) - G'(x), v) = t\,(G''(x)v, v) + o(t)$. Thus from (3.1.42), since $a''(0) = \lim(a'(t)/t)$ as $t \to 0$, $a''(0) = (G''(x)v, v)/\|G'(x)\|^2$. Finally, from (3.1.44), we obtain

$$\delta^2 F(x_0, v) = ((F''(x_0) - \lambda G''(x_0))v, v) \qquad \text{with} \quad (v, G'(x_0)) = 0.$$

and λ given by (3.1.39).

3.1G Application to singularities of mappings

The notion of singular points and singular values of a C^1 mapping f between Banach spaces was introduced in Section 2.6. This notion represents a direct generalization of the finite-dimensional idea described in Section 1.6. Thus, it is natural to extend the main results summarized in (1.6.1) to an infinite-dimensional context. We begin by proving a useful analogue of Sard's theorem due to Smale (1965).

(3.1.45) Let f be a C^q Fredholm mapping of a separable Banach space X into a separable Banach space Y. Then, if $q > \max(\text{index } f, 0)$, the critical values of f are nowhere dense in Y.

Proof: Since X has a countable basis and nowhere dense sets are closed under countable unions, it suffices to prove the theorem locally. To this end, we first prove that a *Fredholm mapping is locally closed*, i.e., there is a neighborhood $N(x_0)$ of x_0 for any $x_0 \in X$ such that $f|_N$ is closed. Indeed, since $f'(x_0)$ is a linear Fredholm map, we can write X as the direct sum $X = \operatorname{Ker} f'(x_0) \oplus X_1$; and an arbitrary element of X, $x = (z, v)$ with $z \in \operatorname{Ker} f'(x_0)$ $v \in X_1$. Now the partial derivative $f'_v(z, v)$ maps X_1 onto a closed subspace of Y for all $x = (z, v)$ near x_0. Thus by the implicit function theorem, we can find an open neighborhood $D_1 \oplus D_2$ of x_0 in $\operatorname{Ker} F'(x_0) \oplus X_1$ such that $\bar{D}_1$ is compact and f restricted to $z \oplus D_2$ is a differentiable homeomorphism onto its image. Now let $f(x_i) = y_i \to y$ for $x_i = (z_i, v_i) \in D_1 \oplus D_2$. To show that f is locally closed, we show that x_i has a convergent subsequence. Since D_1 is compact, we may assume that $z_i \to \bar{z}$; and since $f(\bar{z}, v_i) \to y$, even that $z_i = \bar{z}$. However, as already mentioned, f restricted to $\bar{z} \times D_2$ is a homeomorphism. Consequently, $v_i \to \bar{v}$, so that $\{x_i\}$ has a convergent subsequence.

By (2.6.8), the critical points of f are closed; and since f is locally closed, it suffices to prove that for any $x_0 \in X$ and any neighborhood $O[f(x_0)]$ of $f(x_0)$ in Y, there is a regular value of f in $O[f(x_0)]$. Indeed, in this case the critical values of f would be nowhere dense in Y. To this end, we use the finite-dimensional result (1.6.1(i)). Since $\dim \operatorname{coker} f'(x_0) < \infty$, $Y = \operatorname{coker} f'(x_0) \oplus Y_1$, and there is a canonical projection $P : Y \to \operatorname{coker} f'(x_0)$. Now $\varphi(z) = Pf(z, v_0)$ is a C^q mapping of $\operatorname{Ker} f'(x_0) \oplus \{v_0\} \to \operatorname{coker} f'(x_0)$, so that Sard's theorem (1.6.1(i)) implies the existence of a regular value z_0 for φ in $P\{O[f(x_0)]\}$. Let $y \in P^{-1}(z_0) \cap O[f(x_0)]$, then y is the desired regular value.

As a useful consequence of this result, we prove a result implying that any problem defined by a Fredholm operator equation of negative index is *not well posed.*

(3.1.46) Let $f : X \to Y$ be any Fredholm map of negative index. Then $f(X)$ contains no interior points; i.e., if $f(x) = y_0$ is solvable in X, then there is a y arbitrarily near y_0 such that $f(x) = y$ is not solvable in X.

Proof: If $f(X)$ contained interior points, by (3.1.45) there would be a y in the range of f such that $f'(x)$ is surjective for some $x \in f^{-1}(y)$. Then index $f'(x) = \dim \operatorname{Ker} f'(x) \geqslant 0$, contradicting the fact that $f(x)$ has negative index.

Additional remarks on nonlinear Fredholm operators: This result demonstrates that nonlinear Fredholm operator equations *of negative index* are not well posed. More precisely, for operator equations, arising naturally in mathematical physics (say), solvability of an equation of the form $f(x) = g$ should not depend on the precise nature of g. Indeed precise knowledge of g is often impossible, due to experimental error or some analogous cause.

Another useful consequence of (3.1.45) for C^q Fredholm maps of positive index r is the fact that if $q > r$ for almost all $g \in Y$, the set of elements $S = \{x | f(x) = g\}$ is a submanifold of X of dimension r or is empty.

We now extend Morse's theorem (1.6.1(ii)) to cover the case of critical values of smooth functionals $F(x)$ defined on a reflexive Banach space X. This case is not covered by (3.1.45) since when regarded as a mapping from $X \to \mathbb{R}^1$, $F(x)$ may not be Fredholm. Indeed, if $F'(x) = 0$, the set $S = \{h|(F'(x), h) = 0\}$ is sometimes infinite dimensional. We shall prove

(3.1.47) Suppose $F(x)$ is a real-valued functional of class C^m defined on a real, separable, reflexive Banach space X, such that $F'(x)$ is a (nonlinear) Fredholm operator from $X \to X^*$ (the conjugate space of X). Then the critical values of $F(x)$ have zero Lebesgue measure (on $\mathbb{R}^1$) provided $m \geqslant \max(\dim \operatorname{Ker} F''(x), 2)$.

Proof: Let x_0 be a critical point of $F(x)$. We shall show that $(*)$ *the critical points in an open neighborhood O_{x_0} of x_0 coincide with the critical points of a C^m real-valued function defined on an open neighborhood of a point in $\mathbb{R}^m$*, $m \geqslant \max(\operatorname{Ker} F''(x_0), 2)$. Then applying the finite-dimensional result (1.6.1(ii)), we find that the critical values associated with critical points of $F(x)$ near x_0 have Lebesgue measure zero. Let C denote the set of critical points of $F(x)$ in X. C can be covered by neighborhoods of the form $O_{x_0} \cap C$, on each of which $F(O_{x_0} \cap C)$ has measure zero. Since X is separable, the covering $\bigcup_{x \in C}\{O_{x_0} \cap C\}$ has a countable subcovering. Thus $F(C)$ has measure zero since the countable union of sets of measure zero also has measure zero.

Thus it remains to prove $(*)$. To this end, since $F'(x)$ is Fredholm, we can write $X = \operatorname{Ker} F''(x_0) \oplus X_2$, so that $x \in X$ can be uniquely written as $x = z + x_2$, where $z \in \operatorname{Ker} F''(x_0)$ and $x_2 \in X_2$. In the same way, we can write $X^* = \operatorname{coker} F''(x_0) \oplus X_2^*$. Also $F'(z, x_2) = (P, F', P_2F')$ $(f_1(z, x_2), f_2(z, x_2))$ where f_1 and f_2 denote the partial derivative operators relative to this decomposition and P_1 and P_2 denote the canonical projections of $X^* \to \operatorname{coker} F''(x_0)$ and $X^* \to X_2^*$, respectively. Then $L = P_2F_2'(x)$ restricted to X_2 is one-to-one and onto, and so invertible by (1.3.20). With this notation, we define a C^1 homeomorphism $h : (z, x_2) \to (\tilde{z}, \tilde{x}_2)$ near x_0 by setting $h(z, x_2) = (\tilde{z}, L^{-1}f_2(z, x_2))$. Then since the critical points of $\varphi(\tilde{z}, \tilde{x}_2) = F(h^{-1}(\tilde{z}, \tilde{x}_2))$ near $h(x_0)$ are solutions of $\varphi'(\tilde{z}, \tilde{x}_2) = F'(h^{-1}(\tilde{z}, \tilde{x}_2))h'^{-1} = 0$, the critical points near $h(x_0)$ of φ will be in one-to-one correspondence with the critical points of F near x_0.

The fact that h is a diffeomorphism is easily justified by the implicit function theorem.

Next, we note that the critical points of $\varphi(\tilde{z}, \tilde{x}_2)$ all belong to the subspace $\operatorname{Ker} F''(x_0)$ so that the critical points of $F(x)$ are in one-to-one correspondence with those of $\zeta(\tilde{z}, 0)$. Indeed, if $(\bar{z}, \bar{x}_2)$ is a critical point of φ, and $h(z, x_2) = (\bar{z}, \bar{x}_2)$, then $f_1(z, x_2) = f_2(z, x_2) = 0$. Thus $\bar{x}_2 = 0$, by the definition of h given above.

Finally, we note that the functional $\varphi(\tilde{z}, 0)$ is of class C^m if $F(x)$ is. This follows from the fact that if $F(x)$ is C^m $(m \geqslant 2)$, then h is of class C^{m-1}. Indeed $\varphi'(\tilde{z}, 0) = (f_1(h^{-1}(\tilde{z}, 0)) + f_2(h^{-1}(\tilde{z}, 0))h'^{-1}$. Since $F_2(h^{-1}(\tilde{z}, 0) = 0$ by definition, $\varphi'(\tilde{z}, 0) = F_1(h^{-1}(\tilde{z}, 0))h'^{-1}$ belongs to C^{m-1}, which implies that $\varphi(\tilde{z}, 0)$ is of class C^m. Thus $(*)$ is established with $\zeta(\tilde{z}, 0)$ as the desired real-valued function, and so the theorem is proven.

3.2 The Steepest Descent Method for Gradient Mappings

For gradient mappings $f(x) = \operatorname{grad} F(x)$ of a Hilbert space H into itself, the method of successive approximations just discussed can be supplemented with alternative techniques. Thus, for example, there are iteration schemes for the solution of $f(x) = 0$ (say) not involving the explicit computation $[f'(x)]^{-1}$ at any point. Perhaps the best known of these is the method of steepest descent, due to Cauchy.

This method consists in solving the initial value problem

$$\frac{dx}{dt} = -f(x), \qquad x(0) = x_0 \qquad \text{with} \quad \operatorname{grad} F = f. \tag{3.2.1}$$

One easily shows that (along a solution $x(t)$ of (3.2.1)) $F(x(t))$ decreases as $t \to \infty$. Provided the solution of (3.2.1) exists for all t, one attempts to show that $\lim_{t\to\infty} x(t) = \bar{x}$ exists and is a solution of $f(x) = 0$. The convergence of the method is in question, and we now take up this problem.

3.2A Continuous descent for local minima

If $F(x)$ possesses a strict relative minimum at some point x_∞, then the method of steepest descent is quite easily justified by the following

(3.2.2) **Theorem** Suppose $F(x)$ is a C^2 real-valued functional defined on a sphere $S(x_0, r)$ of a Hilbert space H, and suppose that for some absolute constant $A > 0$

$$(F''(x)y, y) \geqslant A\|y\|^2 \tag{3.2.3}$$

for $x \in S(x_0, r)$ and $y \in H$. Then provided $\|F'(x_0)\|/A \leqslant r$, the initial value problem (3.2.1) has a unique solution defined for all t, $\lim_{t\to\infty} x(t) = x_\infty$ exists and is the unique minimum of $F(x)$ in $S(x_0, r)$ as well as the unique solution of $f(x) = 0$ in $S(x_0, r)$. Furthermore, we have the following estimate for the rate of convergence of $x(t) \to x_\infty$,

$$\|x(t) - x_\infty\| = O(e^{-At}). \tag{3.2.4}$$

Proof: First we note that the initial value problem (3.2.1) has one and only one solution for small t by virtue of (3.1.23). To ensure that $x(t)$ stays in $S(x_0, r)$ and that $\|dx/dt\| \to 0$ as $t \to \infty$ so that $\|f(x(t))\| \to 0$, we argue as follows. Along a solution $x(t)$ of (3.2.1),

$$\frac{d}{dt} F(x(t)) = (f(x(t)), x'(t)) = -\|x'(t)\|^2. \tag{3.2.5}$$

Hence $F(x(t))$ decreases as t increases. Also,

$$\frac{d^2}{dt^2} F(x(t)) = -2(x''(t), x'(t)) = 2(F''(x(t))x'(t), x'(t))$$
$$\geqslant 2A\|x'(t)\|^2 = -2A \frac{d}{dt} F(x(t)).$$

Hence $F(x(t)) = g$ satisfies the differential inequality $g'' + 2Ag' \geqslant 0$. Consequently, by (3.2.5), $(d/dt)F(x(t)) \geqslant -\|f(x_0)\|^2 e^{-2At}$ and so

$$\|x'(t)\| \leqslant \|f(x_0)\| e^{-At}. \tag{3.2.6}$$

On integrating, we find $\|x(t)-x_0\| \leqslant \|f(x_0)\|/A$ so that $x(t) \in S(x_0, r)$ for all t, so that the solution of (3.2.1) exists for all t. In the same way for $0<t\leqslant t_1$, $\|x(t_1)-x(t)\| \leqslant \|f(x_0)\|A^{-1}e^{-At}$, so that for any sequence $t_n \to \infty$, $x(t_n)$ is a Cauchy sequence and consequently $x = \lim_{t_n\to\infty} x(t_n)$ exists in $S(x_0, r)$ and x_∞ is independent of the sequence t_n chosen since clearly

$$(*) \qquad \|x(t)-x_\infty\| \leqslant \|f(x_0)\|A^{-1}e^{-At}.$$

Also (3.2.6) implies that $\|f(x(t))\| = \|x'(t)\| \to 0$, and hence $f(x_\infty)=0$. The uniqueness of x_∞ follows from $(*)$ since if f vanishes at $x, y \in S(x_0, r)$, and $x(t)$ is the line joining x and y, by (3.2.3)

$$0 = (f(x)-f(y), x-y) = \int_0^1 (f'(x(t))(x-y), x-y) \geqslant A\|x-y\|^2.$$

The fact that $F(x_\infty) = \min_{S(x_0, r)} F(x)$ is unique follows similarly since for $x \in S(x_0, r)$, $f(x_\infty)=0$ and

$$\begin{aligned} F(x)-F(x_\infty) &= \int_0^1 (f(x_\infty + s(x-x_\infty)), x-x_\infty)\, ds \\ &= \int_0^1 (f(x_\infty + s(x-x_\infty)) - f(x_\infty), x-x_\infty)\, ds \\ &\geqslant \tfrac{1}{2}A\|x-x_\infty\|^2. \end{aligned}$$

3.2B Steepest descent for isoperimetric variational problems

It is useful to extend the result (3.2.2) to the case of abstract isoperimetric problems as described in the preceding section. To this end we consider a C^2 functional $F(x)$ restricted to the hypersurface C defined by setting $G(x) = \text{const}$. Assuming $G(x)$ is sufficiently smooth and the constraint set C is arcwise connected, we prove

(3.2.7) **Theorem** Suppose $F(x)$ and $G(x)$ are C^2 real-valued functionals such that $G'(x) \neq 0$ on the constraint set $\mathcal{C} = \{x \mid G(x) = \text{const.}\}$, and the "formal" second variation of F at x relative to $\mathcal{C}$ defined by (3.1.41) satisfies the following inequality for an absolute constant $A>0$ and all $x \in \mathcal{C}_A \equiv \mathcal{C} \cap \{x \mid \|x-x_0\| \leqslant \|f(x_0)\|/A\}$

$$(3.2.8) \qquad \delta^2 F(x, v) \geqslant A\|v\|^2 \quad \text{for} \quad (v, G'(x)) = 0,$$

then the solution of the initial value problem

$$(3.2.9) \qquad x'(t) = -F'(x) + \lambda(x)G'(x), \qquad \lambda(x) = (F'(x), G'(x))/\|G'(x)\|^2, \qquad x(0) = x_0,$$

exists for all $t \geqslant 0$, $\lim_{t\to\infty} x(t) = x_\infty$ exists and is the unique minimum of $F(x)$ in $\mathcal{C}_A$.

Proof: Let $x(t)$ denote the solution of (3.2.9), which certainly exists for sufficiently small t by (3.1.27). We show that along $x(t)$, $F(x(t))$ decreases while $G(x(t)) = \text{const.}$, and moreover $x(t) \in C_A$. This fact implies $x(t)$ exists as a solution of (3.2.9) for all t. Indeed, if $x(t)$ existed only for the maximal time interval $[0, \beta)$, the last statement together with (3.1.27) implies $\beta = \infty$. Moreover, the following simple computations hold:

$$[F(x(t))]' = (F'(x) - \lambda(x)G'(x), x'(t)) = -\|x'(t)\|^2,$$
$$[F(x(t))]'' = (F''(x)x'(t), x'(t)) + (F'(x), x''(t)),$$
$$[G(x(t))]' = 0 \qquad \text{(by virtue of the definition of } \lambda(x)).$$

Consequently, $F(x(t))$ is decreasing along $x(t)$, and

$$[F(x(t))]'' = \delta^2 F(x, x'(t)) - (x'(t), x''(t))$$
$$= \delta^2 F(x, x'(t)) + (\tfrac{1}{2}[F(x(t))]''),$$

so that $[F(x(t))]'' = 2\delta^2 F(x, x'(t))$. Thus (3.2.8) implies that $g(t) = F(x(t))$ satisfies the differential inequality $g''(t) + 2Ag' \geqslant 0$. Consequently, as in (3.2.3) $\|x(t) - x_0\| \leqslant \|f(x_0)\|/A$ so that $x(t) \in C_A$.

Now repeating the argument in (3.2.3) we find that $x_\infty = \lim_{t\to\infty} x(t)$ exists; $F'(x_\infty) - \lambda(x_\infty)G'(x_\infty) = 0$, so that x_∞ is the desired local minimum of $F(x)$ restricted to C_A.

To prove x_∞ is the desired *unique* minimum we prove for any $x_0 \in C_A$,

$$F(x_0) \geqslant F(x_\infty) + \eta(\|x_0 - x_\infty\|),$$

where $\eta(t) > 0$ for $t > 0$. To this end we let $x(0, \mu)$ be any C^1 curve in C_A joining x_0 and x_∞, and let $x(t, \mu)$ be the solution of the initial value problem (3.2.9) with initial value $x(0, \mu)$. Then using (3.2.8) we find, setting $x_\mu = (\partial/\partial\mu)x(t, \mu)$,

$$(d/dt)\|x_\mu\|^2 = 2(x_\mu, x_{\mu t}) = -2(F''(x) - \lambda(x)G''(x)x_\mu, x_\mu)$$
$$\leqslant -2A\|x_\mu\|^2.$$

Thus $\|x_\mu\|e^{At}$ is a decreasing function of t, and

$$\|x(t, 1) - x(t, 0)\| \leqslant \int_\mu^1 \|x_\mu(t, \mu)\|\, d\mu \leqslant e^{-At}\int_\mu^1 \|x_\mu(0, \mu)\|\, d\mu.$$

Consequently, $x_\infty = \lim_{t\to\infty} x(t, 1) = \lim_{t\to\infty} x(t, 0)$. Applying Taylor's theorem (2.1.33) we find for $0 < \tau < t$, suppressing the subscript μ,

$$F(x_0) = F(x(t)) + t[F(x(t))]' + \tfrac{1}{2}t^2[F(x(\tau))]''. \tag{3.2.10}$$

But $F(x(t)) \geqslant F(x_\infty)$, $[F(x(t))]' = -\|x'(t)\|^2$, while $[F(x(\tau))]'' \geqslant 2A\|x'(t)\|^2$, and thus from (3.2.10) $F(x_0) \geqslant F(x_\infty) + (At^2 + t)\|x'(t)\|^2$. Thus x_∞ is the unique minimum, as desired. So applying the analogue of (3.2.6) we find the desired inequality by noting that

$$\|x'(t)\| \geqslant A\|x(t) - x_\infty\| \geqslant A\{\|x_0 - x_\infty\| - \|x_0 - x(t)\|\},$$

so that $\|x_0 - x(t)\|$ can be made small by choosing t sufficiently small since then $\|x_0 - x(t)\| \leqslant \alpha t$ for an absolute constant $\alpha > 0$.

3.2C Results for general critical points

If $F(x)$ has a critical point $\bar{x}$ that is not a relative minimum, then the method of steepest descent generally does not converge to $\bar{x}$ no matter how small $\|x_0 - \bar{x}\| > 0$ may be. Consider, for example, the function $F(x, y) = x^2 - y^2 + \frac{1}{2}y^4$ defined on $\mathbb{R}^2$. The associated initial value problem $dx/dt = -2x$, $dy/dt = 2(y - y^3)$, $(x(0), y(0)) = (x_0, y_0)$ always converges to only one of the three critical points of $F(x, y)$: (0, 0) and (0, $\pm$ 1). The point (0, 0) is a saddle point for $F(x, y)$, and one easily checks that no matter how close (x_0, y_0) is to (0, 0), provided $y_0 \neq 0$, $(x(t), y(t)) = (x_0 e^{-2t}, g(y_0, t))$ always converges to either of the absolute minima $(0, \pm 1)$. Thus there arises the general problem: Does the method of steepest descent lead one to some critical point of $F(x)$ (not necessarily near the initial guess x_0), as in the example just given?

First we prove the following simple result.

(3.2.11) **Theorem** Suppose $F(x)$ is a C^1 real-valued function defined on a Hilbert space H, such that $F'(x)$ is Lipschitz continuous, with the properties:

(i) $F(x)$ is bounded from below on H.

(ii) $F^{-1}(B)$ is bounded for any bounded set B in $\mathbb{R}^1$.

(iii) If $x_n \to x$ weakly and $\{\text{grad } F(x_n)\} \to v$ strongly in H, then $v = \text{grad } F(x)$.

Then the solution of the initial value problem (3.2.1) exists for all t and (weak) $\lim_{t\to\infty} x(t) = \bar{x}$ exists and is a critical point of $F(x)$.

Proof: As in the proof of Theorem (3.2.3), $f(x)$ is locally Lipschitz continuous, so that $x(t)$ the solution of (3.2.1) exists for small t, and along $x(t)$, $F(x(t))$ is decreasing. Suppose that $x(t)$ exists for $t \in [0, t_*)$ but not at $t = t_* < \infty$. Then for $0 < t_1, t_2 < t_*$,

$$\|x(t_2) - x(t_1)\| = \left\|\int_{t_1}^{t_2} \frac{dx}{ds}\, ds\right\| \leqslant \int_{t_1}^{t_2} \|\text{grad } F(x(s))\|\, ds$$
$$\leqslant \left\{\int_{t_1}^{t_2} \|\text{grad } F(x(s))\|^2\, ds\right\}^{1/2} (t_2 - t_1)^{1/2}. \tag{3.2.12}$$

On the other hand, since $F(x(t))$ is bounded from below for $t \in [0, t_*)$,

$$F(x(t_2)) - F(x(t_1)) = \int_{t_1}^{t_2} \frac{d}{ds} F(x(s))\, ds = -\int_{t_1}^{t_2} \|\text{grad } F(x(s))\|^2\, ds. \tag{3.2.13}$$

Combining (3.2.12) and (3.2.13), we find that if $t \to t_*$, $\{x(t)\}$ is a Cauchy sequence in X. Thus $\lim_{t\to t_*} x(t)$ exists and is finite. Consequently, applying the local existence theorem (3.1.23) at $t = t_*$, we obtain that $x(t)$ can be continuously extended for $t > t_*$ satisfying (3.2.1), contradicting the maximality of t_*. Thus $t_* = \infty$.

Next we show that a subsequence of $\{x(t)\}$ converges to a critical point of $F(x)$. Since $F(x)$ is bounded from below, (3.2.13) implies that $\int_0^\infty \|\text{grad } F(x(s))\|^2\, ds < \infty$, so that $\|\text{grad } F(x(t))\| \to 0$ as $t \to \infty$. On the other hand, by (ii), the set $\{x(t)\}$ is bounded since $\{x(t)\} \subset F^{-1}\{\inf_H F(x), F(x(0))\}$. Therefore, $\{x(t)\}$ has a weakly convergent subsequence $\{x(t_i)\}$ as $t_i \to \infty$ with weak limit $\bar{x}$ (say), while grad $F(x(t_i)) \to 0$ strongly. Thus by (iii), grad $F(\bar{x}) = 0$, and $\bar{x}$ is the desired critical point.

A useful result, stronger than (3.2.11), can be obtained if we assume that all the critical points of $F(x)$ are isolated.

(3.2.14) Suppose $F(x)$ is a C^1 real-valued functional such that $F'(x)$ is Lipschitz continuous, defined on a Hilbert space H, bounded from below on H, and satisfying the conditions:

(i) all the critical points of $F(x)$ are isolated;
(ii) any sequence $\{x_n\} \in H$ such that $|F(x_n)|$ is bounded and $F'(x_n) \to 0$ has a convergent subsequence.

Then the solution $x(t)$ of the differential equation (3.2.1) exists for all t, $\lim_{t\to\infty} x(t)$ exists and is a critical point of $F(x)$.

Proof: By repeating the argument of (3.2.11), and using the stronger hypothesis (ii) above, we may assume that $x(t_i) \to x_\infty$ strongly for some subsequence $t_i \to \infty$. Thus it suffices to prove only that $\lim_{t\to\infty} x(t) = x_\infty$ exists.

Indeed, otherwise there would be two spherical neighborhoods O_1 and O_2 centered at x_∞ such that $O_2 \subset O_1$ and $\bar{O}_1 - O_2$ contains no critical points of $F(x)$; while for an infinite sequence of disjoint intervals $[t_i, t_{i+1}]$ there is a number $c > 0$ such that $x(t) \subset \bar{O}_1 - O_2$ for $t \in [t_i, t_{i+1}]$ and $\|x(t_{i+1}) - x(t_i)\| \geqslant c$. By hypothesis (ii), there is a positive constant d with $\|\text{grad } F(x)\| \geqslant d$ for $x \in \bar{O}_1 - O_2$ since otherwise there would be a critical point $x \in \bar{O}_1 - O_2$. Therefore, as $t \to \infty$,

$$\begin{aligned}\lim_{t\to\infty} F(x(t)) &= F(x(0)) - \int_0^\infty \|\text{grad } F(x(s))\|^2 \, ds \\ &\leqslant F(x(0)) - \sum_{i=1}^{\infty} \int_{t_i}^{t_{i+1}} \|\text{grad } F(x(s))\|^2 \, ds \\ &\leqslant F(x(0)) - \sum_{i=1}^{\infty} d \int_{t_i}^{t_{i+1}} \|\text{grad } F(x(s))\| \, ds.\end{aligned}$$

Thus by the above facts,

$$\begin{aligned}\lim_{t\to\infty} F(x(t)) &\leqslant F(x(0)) - \sum_{i=1}^{\infty} d\|x(t_{i+1}) - x(t_i)\| \\ &\leqslant F(x(0)) - \sum_{i=1}^{\infty} \{cd\} = -\infty,\end{aligned}$$

which contradicts the fact that $F(x)$ is bounded from below.

3.2D Steepest descent for general smooth mappings

We end this section with a brief discussion of the applicability of the notion of steepest descent for general mappings. In his early research, Cauchy showed that the techniques just discussed can apply to the study of solvability for smooth mappings. To treat the infinite-dimensional case, assume, for simplicity, that f is a C^q (nonlinear) Fredholm operator of nonnegative index $r < q$, mapping a real Hilbert space H into itself. Then we prove:

(3.2.15) For generic $p \in H$, the solutions of $f(x) = p$ coincide with the critical points of the functional $F(x) = \|f(x) - p\|^2$.

Proof: A simple computation shows that the critical points of $F(x)$ coincide with the solution of the operator equation

(3.2.16) $\qquad [f'(x)]^*\{f(x) - p\} = 0,$

where $[f'(x)]^*$ denotes the adjoint of $f'(x)$. Now by the infinite-dimensional version of Sard's theorem (3.1.45), the singular values $f(S)$ of f form a residual nowhere dense set. Thus for (generic) $p \not\in f(S)$

$$\dim \operatorname{Ker} f'^*(x) = \dim \operatorname{coker} f'(x) = 0.$$

Thus if $\bar{x}$ is a solution of (3.2.16), for such generic p, $f(\bar{x}) = p$. Thus the desired result is established.

Thus assuming the point $0 \notin f(S)$, we can apply the previous results of this section to study the solutions of $f(x) = 0$ by considering the critical points of the functional $F(x) = \|f(x)\|^2$ in each of the results of Sections 3.2A–C.

3.3 Analytic Operators and the Majorant Method

For complex analytic mappings, additional methods are often available, allowing a more complete discussion of the problems raised at the beginning of Part II.

3.3A Heuristics

To illustrate this point, we begin by considering the formal method of undetermined coefficients, and its justification by "Cauchy majorants." This procedure forms the main classical approach to the study of nonlinear problems and still retains an important place in nonlinear analysis. Suppose, for example, that one wishes to solve the (analytic) operator equation $f(x, \lambda) = 0$ defined on a Banach space x for small values of the parameter λ, given that $f(x, 0) = 0$ has the solution x_0. Then to apply the method of undetermined coefficients, we postulate a solution of the form

$$x(\lambda) = x_0 + \sum_{n=1}^{\infty} x_n \lambda^n, \qquad \text{where} \quad x_n \in X.$$

Assuming that such a solution exists with a positive radius of convergence, one expands $f(x(\lambda), \lambda)$ in the form

$$f(x(\lambda), \lambda) = \sum_{n=0}^{\infty} f_n(x_1, \ldots, x_n)\lambda^n,$$

solves the resulting implicit system $f_n(x_1, \ldots, x_n) = 0$ $(n = 1, 2, \ldots)$, and attempts to show that in a small sphere about x_0 that this system has one and only one solution $(\bar{x}_1, \ldots, \bar{x}_n)$. Finally, one justifies the assumption that such a solution $x(\lambda) = x_0 + \sum_{n=1}^{\infty} \bar{x}_n \lambda^n$ has a positive radius of convergence by finding a *majorant series* $x^*(\lambda)$ for $x(\lambda)$. More precisely,

(3.3.1) **Definition** $x^*(\lambda)$ majorizes $x(\lambda)$ (and we write $x(\lambda) \ll x^*(\lambda)$) if $x^*(\lambda) = \sum_{n=0}^{\infty} x_n^* \lambda^n$ with x_n^* positive real numbers such that $\|x_n\| \leqslant x_n^*$.

Clearly if $x^*(\lambda)$ has a positive radius of convergence in the sense that

$$\limsup_{n\to\infty} \|x_n^*\|^{1/n} < \infty,$$

then by the results of Section 2.3, $x(\lambda)$ also has a positive radius of convergence.

3.3B An analytic implicit function theorem

As a simple but typical result of the majorant method, we prove:

(3.3.2) **Analytic Implicit Function Theorem** Suppose the operator $F(x, y)$ is analytic in a neighborhood of (x_0, y_0) of a complex Banach space $X \times Y$ with range contained in a complex Banach space Z. Furthermore, suppose $F(x_0, y_0) = 0$, and that the linear operator $F_y(x_0, y_0)$ is invertible. Then there is one and only one solution $y = f(x)$ of $F(x, y) = 0$ near (x_0, y_0) that is an analytic function of x near x_0. Moreover, the radius of convergence of the associated power series can be estimated by the majorant (3.3.8) below.

Proof: Without loss of generality we may suppose that $(x_0, y_0) = (0, 0)$ and that $F(x, y)$ expanded about $(0, 0)$ is written

$$(3.3.3) \qquad F(x, y) = F_{10}(0, 0)x + F_{01}(0, 0)y + \sum_{i+j=2}^{\infty} F_{ij}(0, 0)(x^i, y^j),$$

where by Cauchy's formulas

$$F_{ij}(0, 0)(x^i, y^j) = \frac{1}{(2\pi i)^2} \int_{|\xi| = r_1} \int_{|\eta| = r_2} \frac{F(\xi x, \eta y)}{\xi^{i+1} \eta^{j+1}} \, d\xi \, d\eta.$$

We now seek a convergent series

$$(3.3.4) \qquad y(x) = \sum_{n=1}^{\infty} a_n(x^n)$$

satisfying the equation $F(x, y) = 0$ for $\|x\|$ sufficiently small. Clearly, since $F_{01}(0, 0) = -L$ is invertible, y must satisfy

$$(3.3.5) \qquad y = L^{-1}F_{10}(0, 0)x + \sum_{i+j=2}^{\infty} L^{-1}F_{ij}(0, 0)(x^i, y^j) = H(x, y).$$

This last equation shows that formally, at least, the coefficients are uniquely determined if x and y satisfy (3.3.3). Indeed substituting (3.3.4) formally into (3.3.5), one finds that the operators a_n must satisfy the system

$$(3.3.6) \qquad a_{n+1}x^{n+1} = f_{n+1}(a_1, \ldots, a_n, x) \qquad (n = 0, 1, 2, \ldots),$$

where the expression f_{n+1} is a multilinear operator of degree $n + 1$ in x involving the forms $a_1, a_2, \ldots$ up to order n and the expressions F_{ij} for $i + j = n + 1$. Thus for $n = 0, 1, 2, \ldots, a_{n+1}x^{n+1}$ can be expressed as a linear combination of multilinear operators of degree $n + 1$ in x and F_{ij} $(i + j = n + 1)$ with positive coefficients. Hence, if we write $a_{n+1}x^{n+1} = Q_{n+1}(x, F_{ij})$, the functions are defined *independent of the particular form of the function $F(x, y)$*. To show that the series (3.3.5) has a positive radius of convergence, we note that by the Cauchy estimates (2.3.5), if $S_r = \{(x, y) \mid \|x\|, \|y\| \leqslant r\}$,

$$(3.3.7) \qquad \|F_{ij}(x^i, y^j)\| \leqslant \frac{M}{r^{i+j}} \|x\|^i \|y\|^j,$$

where $M = \sup_{S_r} \|f(x, y)\|$. Now we consider the notion of majorant appropriate for the proof in this case.

Now by (3.3.7), an immediate majorant $\Phi(\|x\|, \|y\|)$ for the right-hand side of (3.3.5) on S_r is

$$\|L^{-1}\| \left\{ \sum_{i+j=1}^{\infty} \left(\frac{M}{r^{i+j}} \right) \|x\|^i \|y\|^j - \frac{M}{r} \|y\| \right\}$$

$$\leqslant \|L^{-1}\| \left\{ \frac{M}{(1 - \|x\|/r)(1 - \|y\|/r)} - M - \frac{M}{r} \|y\| \right\}.$$

Now comes the *critical step* of the majorant method, namely, the observation that if

$$\Phi(\|x\|, \tilde{y}) \gg H(x, y) = \sum H_{ij}(x^i, y^j)$$

and the equation $\tilde{y} = \Phi(\|x\|, \tilde{y})$ has a convergent analytic solution $\tilde{y} = \sum \tilde{a}_n \|x\|^n$, then the formal solution $y = \sum_{n=0}^{\infty} a_n(x^n)$ of the equation $y = H(x, y)$ is majorized by $\tilde{y}$ and is consequently convergent for $\|x\| < R$, where $R \leqslant \tilde{R}$, and $\tilde{R}$ denotes the radius of convergence of $\tilde{y}$. Indeed if $\tilde{y} = \sum \tilde{a}_n \|x\|^n$, and

$$\Phi(\|x\|, \tilde{y}) = \sum \alpha_{ij} \|x\|^i \tilde{y}^j,$$

then $\tilde{a}_n \leqslant Q_{n+1}(\|x\|, \alpha_{ij})$. As already mentioned, the functions Q_{n+1} have the property that $a_n = Q_{n+1}(x, H_{ij})$ and $\|a_n\| = Q_{n+1}(\|x\|, \alpha_{ij})$ since $\|H_{ij}\| \leqslant \alpha_{ij}$. Thus $y \ll \tilde{y}$. Thus it remains only to compute the radius of convergence for the solution of the equation

$$(3.3.8) \qquad \tilde{y} = c \left\{ \frac{1}{(1 - \|x\|/r)(1 - \tilde{y}/r)} - 1 - \frac{\tilde{y}}{r} \right\},$$

$$\text{where} \quad c = \|L^{-1}\| M.$$

A simple computation given directly below shows that if this equation is regarded as a quadratic equation in $\tilde{y}$, then the root that vanishes for $\|x\| = 0$ is

$$\tilde{y} = \frac{r^2}{2(r + c)} \left[1 - \left(1 - \frac{\|x\|}{r} \right)^{-1/2} \left(1 - \frac{\|x\|}{\alpha} \right)^{1/2} \right],$$

where $\alpha = r(r/r + 2c)^2$ and $c = \|L^{-1}\| M$. Consequently, $\tilde{y}$ is analytic in $\|x\|$ for $\|x\| < \alpha$.

This computation proceeds as follows: regarded as a quadratic in $\tilde{y} = \sigma$ (3.3.8) can be rewritten

$$\sigma^2 - \left\{ \frac{r^2}{r+c} \right\}\sigma - \left\{ \frac{cr^2}{r+c} \right\} \frac{\|x\|}{r} \Big/ \left(1 - \frac{\|x\|}{r}\right) = 0.$$

The smallest root of this equation for fixed x, r (i.e., the one that vanishes when $x = 0$) is obtained by choosing the appropriate negative factor in the usual formula for the roots of a quadratic equation.

Remark on the analytic implicit function theorem: The first part of the conclusion in (3.3.2) can be obtained by noting that the proof of the implicit function theorem given in (3.1.10) is based on the iteration scheme associated with the contraction mapping theorem. The solution $y = y(x)$ is thus the uniform limit of iterates each of which is analytic. Consequently, $y(x)$ is itself analytic. However, the estimate for the radius of convergence is generally sharper in the majorant method.

3.3C Local behavior of complex analytic Fredholm operators

As a simple application of (3.3.2) consider the local structure of the solutions of the equation $f(z) = y$ near a given solution z_0, where f is a complex analytic Fredholm mapping acting between complex Banach spaces.

(3.3.9) **Theorem** If f is a complex analytic Fredholm mapping defined in the neighborhood of a point z_0 of a complex Banach space X with range contained in a complex Banach space Y, then the solutions of $y = f(z)$ near a given solution z_0 form the one-to-one analytic image of a finite-dimensional analytic variety. If the solution z_0 is not isolated, then there is a nonconstant convergent power series,

$$z(\lambda) = z_0 + \sum_{n=1}^{\infty} z_n \lambda^n,$$

satisfying $y = f(z)$.

Proof: Without loss of generality, suppose $z_0 = 0$ and $\|y\|$ is small, so that the equation $y = f(z)$ can be written

(3.3.10) $\quad y = f'(0)z + R(z) \quad$ where $\|R(z)\| = O(\|z\|^2)$

with $R(z)$ complex analytic. Let $X = \operatorname{Ker}(f'(0)) \oplus X_2$, $Y = \operatorname{Range} f'(0) \oplus Y_2$, and for $z \in S(0, \delta) = \{z \mid \|z\| < \delta\}$ let $z = z_1 + z_2$ be the corresponding direct sum decomposition of z. Then we try to solve the equation

(3.3.11) $\quad y = f'(0)z_2 + R(z_1 + z_2).$

By (1.3.38), there is a bounded linear map A_0: $Y \to X$ that is one-to-one for $y \in \text{Range}\, f'(0)$ and has range X_2 such that

$$A_0 f'(0) = I - P,$$

where P is a projection onto $\text{Ker}\, f'(0)$. Premultiplying (3.3.11) by A_0, we find

$$A_0 y = z_2 + A_0 R(z_1 + z_2).$$

Then, by the analytic implicit function theorem (3.3.2), $z_2 = h(z_1)$, where $h(w) = O(\|w\|^2)$ is a complex analytic map. Now $\text{Ker}(A_0) = Y_2$, and A_0 is one-to-one on $\text{Range}(f'(0))$. Hence, since $\dim \text{Ker}(A_0) < \infty$, there is a bounded projection P_2 of Y onto Y_2. Thus (3.3.10) is equivalent to the system

$$0 = P_2[y - f'(0)h(z_1) + R(z_1 + h(z_1))].$$

Choosing a finite basis for Y_2 and $\text{Ker}(f'(0))$, this last system is complex analytically equivalent to a finite system of complex analytic equations in a finite number of complex variables. The results then follow by the theory of analytic varieties of (1.6.2).

3.4 Generalized Inverse Function Theorems

3.4A Heuristics

Here we extend the inverse function theorem (3.1.5) to the case in which the linear operator $f'(x)$ does not possess a bounded inverse. We shall however suppose that for x near x_0, $f'(x)$ possesses an "approximate" inverse. More precisely, let f be a C^1 mapping defined on $S(x, r)$ of a Banach space X into a Banach space Y. Given an approximation x_0 to a solution of $f(x) = 0$, we attempt to construct a sequence of "better" approximations $\{x_N\}$ by writing $x_{N+1} = x_N + \rho_N$, where ρ_N is an "approximate" solution of the equation $f(x_N + \rho_N) = 0$ (linearized about x_N). Newton's method described in Theorem (3.1.16) shows that both the smoothness and the invertibility of f near x_0 ensure that this process converges (and in fact converges quadratically) with $\rho_N = -(f'(x_N))^{-1}f(x_N)$. If $f'(x)$ is not (boundedly) invertible for x near x_0, it may still be possible to find an "approximate" solution ρ'_N for each linearized equation such that $\lim x_{N+1} = \lim(x_N + \rho'_N)$ converges to a solution of $f(x) = 0$. (Indeed, if the degree of approximation of each ρ'_N to a true solution ρ of each linearized equation could be *measured precisely*, the *rapid convergence* of the Newton iteration scheme will be used to *compensate* for the approximate nature of each ρ'_N.) In the theory of differential equations many situations of this type arise, in which the linear operator $(f'(x))^{-1}$ does not preserve the smoothness of the functions in its domain. For example, if $f'(x)$ is a bounded linear mapping of a space with p derivatives, X^p, into a space with $p - m$ derivatives, X^{p-m}, (i.e., $f'(x)$ has order m), then for each p, we expect $(f'(x))^{-1}$ to be a bounded linear map from X^p

into X^{p+m}. However, it often happens that $(f'(x))^{-1}$: $X^p \to X^{p-\beta}$ (and so loses β derivatives), so that after a finite number of steps the sequence $\{(f'(x_N))^{-1}\}$ no longer exists. In this case it is sometimes possible to replace $(f'(x_N))^{-1}$ by an approximate linear smoothed inverse operator $T_{\xi_N}(f'(x_N))^{-1} \doteq L_N x_N$ so that L_N actually maps X^p into X^{p+m} boundedly. This enables one to obtain an approximate solution to the linearized equation.

In order *to measure the degree of approximation* of each ρ_N', we imbed the Banach space X into a one-parameter continuous scale of such spaces $X_\alpha \subset X_{\alpha'}$ for $\alpha > \alpha'$. The imbedding of X_α into $X_{\alpha'}$ is continuous for $\alpha > \alpha'$, and such that $f(\alpha) = \log \|\cdot\|_\alpha$ is a convex function of α for $\alpha \geqslant 0$. Hence for $0 \leqslant \rho \leqslant r$ and $v \in X_r$,

$$\|v\|_\rho \leqslant \|v\|_0^{1-\rho/r}\|v\|_r^{\rho/r}.$$

If $(f'(x))^{-1}$: $Y \to X_\alpha$ is not bounded, it may be bounded when X_α is replaced by $X_{\alpha-\delta}$, for some $\delta > 0$. Thus if f: $X_r \to Y$, the sequence of approximations may diverge in X_r but converge in $X_{r'}$ for $r' < r$. More precisely, we say $f'(u)x = g$ is *approximately solvable of order* λ in X_r if for every $\epsilon > 0$ there is an $x_\epsilon \in X_r$ such that $\|f'(u)x_\epsilon - g\| \leqslant c\|g\|\epsilon^\lambda$, while $\|x_\epsilon\|_r \leqslant K_0/\epsilon$ for $\|u\|_r \leqslant K_0$, where c and K_0 are absolute constants. We now prove a result, utilizing this concept, by solving a sequence of linear equations $f'(x_r)\rho = g$ for $\rho = \rho_N$ approximately (of order λ).

3.4B A result of J. Moser

(3.4.1) **Theorem** Let f be a C^1 mapping from a sphere $S(x_0, \bar{r}) \subset X_r$ into a Banach space Y such that:

(i) for $u \in S(x_0, \bar{r})$, $f'(u)$ is approximately solvable of order λ in X_r;

(ii) $\|f'(u)v\| \geqslant c\|v\|_0$ for $u \in S(x_0, \bar{r})$, where c is independent of u and v;

(iii) for u and $v \in S(x_0, \bar{r})$, there are absolute constants $\beta \in [0, 1)$ and $M > 0$ such that

$$\|Q(u, v)\| = \|f(u+v) - f(u) - f'(u)v\| \leqslant M\|v\|_0^{2-\beta}\|v\|_r^\beta;$$

(iv) $\|f(u)\| \leqslant M\bar{K}$ for $\|u\|_r \leqslant \bar{K}$.

Then, provided the numbers $\bar{r}$ and $\|f(x_0)\|$ are sufficiently small, the sequence of approximations $x_{N+1} = x_N + \rho_N'$, of order λ as constructed above, converges in $X_{r'}$ to $\bar{x}$ for $r' < (\lambda/(\lambda+1))r$. Thus if f: $X_{r'} \to Y$ is continuous, $f(\bar{x}) = 0$.

Proof: (Without loss of generality, we suppose the constant c of (ii) is unity.) Our argument is based on proving three inductive hypotheses:

(1N) $\quad \|f(x_N)\| \leqslant K^{-\mu s^N}$,

(2N) $\quad \|x_N - x_{N-1}\|_0 \leqslant 2K^{-\mu s^{N-1}}$,

(3N) $\quad \|x_N - x_{N-1}\|_r \leqslant \frac{1}{2}K^{s^N}$,

where K, μ, and s are positive numbers to be determined. If these hypotheses are true for all N, the sequence $\{x_N\}$ is a Cauchy sequence in $X_{r'}$ for $r' < r\{\mu/(s+\mu)\}$. Indeed, since

$$\|v\|_{\rho'} \leqslant \|v\|_0^{1-\rho'/r}\|v\|_r^{\rho'/r},$$

by virtue of (2N) and (3N),

$$\begin{aligned}\|x_N - x_M\|_{\rho'} &\leqslant \sum_{j=M+1}^{N} \|x_j - x_{j-1}\|_{\rho'}\\ &\leqslant \sum_{j=M+1}^{N} \|x_j - x_{j-1}\|_0^{1-\rho'/r}\|x_j - x_{j-1}\|_r^{\rho'/r}\\ &\leqslant R\sum_{j=M+1}^{N} K^{s^{j-1}[s\rho'/r-\mu(1-\rho'/r)]}.\end{aligned}$$

Now this series tends to 0 as $N, M \to \infty$, for

$$s\frac{\rho'}{r} - \mu\left(1 - \frac{\rho'}{r}\right) < 0, \qquad \text{i.e.,} \qquad \rho' < r\left\{\frac{\mu}{s+\mu}\right\}.$$

Consequently, if $\bar{x} = \lim x_N$ as $N \to \infty$ and f is a continuous map of $X_{\rho'} \to Y$, then $f(\bar{x}) = 0$. Thus it remains to verify (1N), (2N), and (3N). For $N = 0$, (1N) is a hypothesis, while (2N) and (3N) are vacuous since u_{-1} is not defined. Assuming (1N)–(3N) for $N = 0, \ldots, n$, we demonstrate successively the inequalities $2(n+1)$, $3(n+1)$, and $1(n+1)$.

Verification of $2(n+1)$: First we note that

$$\begin{aligned}\|x_n\|_r &\leqslant \|x_0\|_r + \sum_{j=0}^{n-1}\|x_{j+1} - x_j\|_r\\ &\leqslant K + \sum_{j=0}^{n-1}\tfrac{1}{2}K^{s^{j+1}}, \qquad \text{by } (3n)\\ &\leqslant K^{s^n} \qquad \text{for } K \text{ sufficiently large.}\end{aligned}$$

We determine $x_{n+1} = x_n + \rho_n$ by approximately solving the linearized equation $f'(x_n)\rho_n + f(x_n) = 0$. Setting $\rho_n = \rho_\epsilon$ (where ϵ will be chosen later), we obtain the following estimates:

(a) $\quad \|f'(x_n)\rho_\epsilon + f(x_n)\| \leqslant MK^{s^n}\epsilon^{\lambda}$

(b) $\quad \|\rho_\epsilon\|_r \leqslant K^{s^n}\epsilon^{-1}$.

Hence by (1n),

$$\|f(x_n)\| \leqslant K^{-\mu s^n},$$

and

$$\begin{aligned}\|x_{n+1} - x_n\|_0 = \|\rho_\epsilon\|_0 &\leqslant \|f'(x_n)\rho_\epsilon\| \\ &\leqslant \|f(x_n)\| + \|f'(x_n)\rho_\epsilon - f(x_n)\| \\ &\leqslant K^{-\mu s^n} + M\epsilon^\lambda K^{s^n}.\end{aligned}$$

Now suppose $\epsilon > 0$ is chosen so that

(3.4.2) $\quad M\epsilon^\lambda K^{s^n} \leqslant K^{-\mu s^n}.$

Then $\|x_{n+1} - x_n\|_0 \leqslant 2K^{-\mu s^n}$.

Verification of 3(n + 1): By virtue of (b) above,

$$\|x_{n+1} - x_n\|_r = \|\rho_\epsilon\|_r \leqslant K^{s^n}\epsilon^{-1} \leqslant \tfrac{1}{2}K^{s^{n+1}},$$

provided

(3.4.3) $\quad K^{s^n}\epsilon^{-1} \leqslant \tfrac{1}{2}K^{s^{n+1}}.$

Verification of 1(n + 1): Again by virtue of (a) and (b) above,

$$\begin{aligned}\|f(x_{n+1})\| &\leqslant \|f(x_n) + f'(x_n)\rho_\epsilon + Q(x_n, \rho_\epsilon)\| \\ &\leqslant \|f(x_n) + f'(x_n)\rho_\epsilon\| + \|Q(x_n, \rho_\epsilon)\| \\ &\leqslant MK^{s^n}\epsilon^\lambda + M\|\rho_\epsilon\|_0^{2-\beta}\|\rho_\epsilon\|_r^\beta \\ &\leqslant MK^{s^n}\epsilon^\lambda + M\{2K^{-\mu s^n}\}^{2-\beta}\{\epsilon^{-1}K^{s^n}\}^\beta.\end{aligned}$$

Now we attempt to choose ϵ and μ so that the right-hand side is at most $K^{-\mu s^{n+1}}$. Since (3.4.2) is automatically satisfied if (3.4.5) below is, it suffices to choose ϵ and μ so that

(3.4.4) $$K^{s^n}\epsilon^{-1} = \tfrac{1}{2}K^{s^{n+1}}$$

(3.4.5) $$MK^{s^n}\epsilon^\lambda \leqslant \tfrac{1}{2}K^{-\mu s^{n+1}}$$

(3.4.6) $\quad M(2K^{-\mu s^n})^{2-\beta}(\epsilon^{-1}K^{s^n})^\beta < \tfrac{1}{2}K^{-\mu s^{n+1}}.$

Since (3.4.4) implies that ϵ^{-1} is of the order $K^{s^n(s-1)}$, for K sufficiently large, (3.4.5) and (3.4.6) will be satisfied if

(3.4.7) $\quad s\mu + 1 < \lambda(s-1) \quad$ and $\quad s(\mu + \beta) < \mu(2 - \beta).$

A short computation now shows that if $s > 1$, both relations of (3.4.7) will be satisfied if

(3.4.8) $$0 < \frac{\lambda + 1}{\lambda - \mu} < \left(1 - \frac{\mu + 1}{\lambda + 1}\right)^{-1} < s < \frac{\mu(2-\beta)}{\mu + \beta} \quad (\text{so } 1 < s < 2).$$

This then completes the proof of $1(n+1)$, and consequently the stated result (3.4.1) is established.

3.4C Smoothing operators

In order to apply Theorem (3.4.1), we turn now to the examination of methods that can be used to ensure the approximate solvability of linear operator equations of the form $f'(u)\rho + f(u) = 0$.

Smoothing operators We assume a one-parameter family of linear operators T_ξ: $X_\alpha \to X_{\alpha+\beta}$, for $\xi > 0$ and $\alpha, \beta > 0$, has been constructed with the properties:

(a) For $v \in X_\alpha$, $\|T_\xi v\|_{\alpha+\beta} \leqslant C\xi^\beta \|v\|_\alpha$, where C is an absolute constant;

(b) For $v \in X_r$, $\|(I - T_\xi)v\|_{r-\delta} \leqslant C\xi^{-\delta}\|v\|_r$. We also suppose that the operator $f'(u)$ has a one-sided inverse $L(u)$ with the property that if $u \in X_\alpha$ and $f'(u)v = h$, then there is an absolute constant C_0 such that $L(u)h = v$.

(c) $C_0^{-1}\|L(u)h\|_{\alpha-\sigma} \leqslant \|h\| \leqslant C_0\|L(u)h\|_{\alpha+\sigma_1}$, where the numbers σ, σ_1 represent the "loss of smoothness" by the operator $L(u)$;

(d) $L(u)f(u) \in X_{\alpha+\beta_0}$, $\beta_0 > \sigma_1$ and $\|L(u)f(u)\|_{\alpha+\beta_0} \leqslant C_2\|u\|_{X_\alpha}$.

Thus if $f'(u)$: $X_\alpha \to Y$ is a bounded operator with $\|f'(u)\| \leqslant C_2$, we choose as an approximate solution for the equation $f'(u)\rho + f(u) = 0$, $\rho_\xi = -T_\xi L(u)f(u)$ (which we interpret as a "smoothing" of the element $L(u)f(u)$). Then to ascertain the approximate solvability of the linear equation with this choice, we note that for $u \in X_\alpha$,

$$\begin{aligned}\|f'(u)\rho_\xi + f(u)\| &= \|f'(u)\{(I - T_\xi) - I\}L(u)f(u) + f(u)\| \\ &= \|f'(u)(I - T_\xi)L(u)f(u)\| \\ &\leqslant \|f'(u)\|\ \|(I - T_\xi)L(u)f(u)\|_{\alpha+\sigma_1} \\ &\leqslant \|f'(u)\|\{C\xi^{-\beta}\}\|L(u)f(u)\|_{\alpha+\sigma_1+\beta} \\ &\leqslant C_2C_1\xi^{-\beta}C_0\|u\|_\alpha \qquad \text{for} \quad \beta = \beta_0 - \sigma_1.\end{aligned}$$

On the other hand,

$$\begin{aligned}\|\rho_\epsilon\|_\alpha = \|T_\xi L(u)f(u)\|_\alpha &\leqslant C\xi^{+\sigma}\|L(u)f(u)\|_{\alpha-\sigma} \\ &\leqslant C\xi^{+\sigma}C_0\|f(u)\|.\end{aligned}$$

Consequently, in accord with our notion of approximate solvability $\|\rho\xi\|_\alpha \sim 1/\epsilon$, so setting $\xi^{-\sigma} = \epsilon$, $\xi^{-\beta} = \epsilon^{\beta/\sigma}$, we find that the smoothing operator with properties (a)–(d) defines the approximate solvability of $f'(u)\rho + f(u) = 0$ of degree $(\beta_0 - \sigma_1)\sigma^{-1}$.

Typical smoothing operators are:

A. *Truncation Fourier series* in d variables, $x = (x_1, \ldots, x_d)$. Indeed, if X_r is the space $\tilde{C}^r$ of 2π-periodic functions $v(x)$ of class C^r, over $D = \{x \mid |x_i| \leqslant 2\pi \ (i = 1, \ldots, d), x = (x_1, \ldots, x_d)\}$ with

$$\|v\|_{C^r} = \max_{0 \leqslant \rho \leqslant r} \sup_{x \in D} |D^\rho v|,$$

then each $v \in \tilde{C}^r$ can be represented by the Fourier series

$$v = \sum_{|k| \leqslant \rho} v_k \, e^{i(k \cdot x)}, \qquad k = (k_1, \ldots, k_d).$$

We then define the truncation operator

$$T_N v = \sum_{|k| \leqslant N} v_k \, e^{i(k \cdot x)},$$

and note that the following inequalities hold for $N > 0$:

(i) $\|T_N v\|_{\tilde{C}^{r+s}} \leqslant cN^{s+d+1} \|v\|_{\tilde{C}^r}$,
(ii) $\|(I - T_N)v\|_{\tilde{C}^r} \leqslant cN^{-s+d+1} \|v\|_{\tilde{C}^{r+s}}$.

B. *Convolution operators acting on functions with compact support.* Let $\psi(x)$ be a function whose Fourier transform $\hat{\psi}(\xi)$ is a C^∞ function defined on $\mathbb{R}^d$ and vanishing outside $|\xi| < 1$ and equal to 1 for $|\xi| < \frac{1}{2}$. Then the convolution operator

$$T_\xi u = \xi^d \int \psi(\xi(x - y)) u(y) \, dy$$

satisfies the inequalities (a) and (b) above, with $X_\alpha = C_0^\alpha(\Omega)$, where Ω is a bounded set in $\mathbb{R}^d$. Indeed, (a) follows since differentiation commutes with convolution. Whereas the second inequality follows for the same reason if we verify it with $r = 0$. In this case, since

$$u(x) - \int \psi(z) u\left(x - \frac{z}{\xi}\right) dz = \int \psi(z) \left\{ u(x) - u\left(x - \frac{z}{\xi}\right)\right\} dz,$$

expanding $u(x) - u(x - z/\xi)$ in a Taylor series,

$$\left| u(x) - u\left(x - \frac{z}{\xi}\right) - \sum_{|k| < s} a_k(x) z^k \right| \leqslant \frac{z^s}{\xi^s} C \|u\|_{C^s}.$$

3.4D Inverse function theorems for local conjugacy problems

In Sections 3.4A–C, we discussed an extended version of the Newton iteration scheme $x_{n+1} = x_n - [f'(x_n)]^{-1} f(x_n)$ for a solution of $f(x) = 0$ near a given first approximation x_0. Here we give a similar extension for the iteration scheme $x_{n+1} = x_n - [f'(x_0)]^{-1} f(x_n)$ associated with the inverse function theorem. This last scheme has the advantage that the inverse of

$f'(x)$ need be computed only at x_0. The need for such an extension arises in the study of conjugacy problems for mappings such as those mentioned in the beginning of Part II. Suppose, for example, that f and $f + a$ are C^1 mappings of a solid sphere of radius r and center x_0, $S(x_0, r) \subset X$ (a Banach space) into itself. Moreover, suppose $\|a\|$ small relative to f in $S(x_0, r)$. Then we ask if there is a nonsingular "change of coordinates" u defined on $S(x_0, \mathrm{r})$ (i.e., u is a diffeomorphism of $S(x_0, r)$ into itself) such that for $r > 0$ sufficiently small

$$u^{-1}(f + a)u = f. \tag{3.4.9}$$

In a finite-dimensional Banach space, the rank theorem of advanced calculus implies that f and $f + a$ are conjugate locally in the above sense if their derivatives $f'(x)$ and $f'(x) + a'(x)$ have the same rank in a sufficiently small sphere $S(x_0, r')$. The corresponding Banach space theorem is unfortunately more difficult, although see Note E at the end of this chapter. A formal construction for the homeomorphism u can be given in certain cases by the following iteration scheme: Write u as a perturbation of the identity $u = I + y$. Then we define

$$u_0 \equiv I, \qquad u_{N+1} = u_N \circ \{I + y_{N+1}\}, \tag{3.4.10}$$

where y_{N+1} is determined by solving (in some "approximate" sense)

$$u_{N+1}^{-1} \circ \{f + a\} \circ u_{N+1} = f. \tag{3.4.11}$$

More precisely, defining the operator $F(f, u) = u^{-1}fu$ for homeomorphisms $u \in C(S(x_0, r), S(x_0, r))$, note that the following "semigroup" property holds:

$$F(f, u \circ v) = F(F(f, u), v). \tag{3.4.12}$$

Thus the left-hand side of (3.4.11) can be rewritten

$$F(f + a, u_N \circ (I + y_{N+1})) + F(f_N, I + y_{N+1}),$$

where $f_N = F(f + a, u_N)$. Assuming $F(f, u)$ is a C^1 function of its arguments, noting that $F(f, I) = f$ and $F_f(f, I) = I$, and expanding $F(f, u)$ about (f, I) we find by (2.1.31)

$$F(f_N, I + y_{N+1}) = f + (f_N - f) + F_u(f, I)y_{N+1} + o(\|f_N - f\| + \|y_{N+1}\|).$$

Consequently, a "reasonable" approximation y_{N+1} for a solution of (3.4.11) is a solution of

$$F_u(f, I)y + (f_N - f) = 0. \tag{3.4.13}$$

If $F_u(f, I)$ is invertible, then y_{N+1} is uniquely determined and a formal solution u for (3.4.9) can be written

$$\begin{aligned} u &= \lim_{N\to\infty} u_0 \circ u_1 \circ u_2 \circ \cdots \circ u_N \\ &= \lim_{N\to\infty} (I + y_1) \circ (I + y_2) \circ \cdots \circ (I + y_N). \end{aligned} \tag{3.4.14}$$

Now the convergence of this formal construction is in question, and we shall discuss this question briefly here.

A problem similar to that posed by (3.4.9) occurs in the transformation theory of ordinary differential equations. Consider the following system of N ordinary differential equations near $x = 0$

$$(3.4.15) \qquad \frac{dx}{dt} = f(x) + a(x).$$

Suppose the solution near $x = 0$ of the reduced system $dx/dt = f(x)$ is well known, whereas $a(x)$ is a small perturbation of $f(x)$ (for $|x|$ small). Then it is natural to seek a diffeomorphism near $x = 0$ (i.e., a local coordinate transformation) $x = U(y)$ leaving the origin fixed and transforming the perturbed system (3.4.15) into the known one $dy/dt = f(y)$. In this case, for example, periodic solutions of the perturbed system near $x = 0$ could be found from periodic solutions of the known ones near $y = 0$. Indeed, closed curves near $x = 0$ in the x coordinates correspond to closed curves in the y coordinates near $y = 0$. Now if $f(x)$ is a linear map of $\mathbb{R}^N \to \mathbb{R}^N$ and $a(x) = o(|x|)$, the problem discussed above coincides with the linearization problem mentioned at the beginning of Part II. In this case, letting $F(f, U) = U'^{-1}fU$ for y a differentiable homeomorphism of $S(x_0, r)$ into itself, we again find $F(f, y_1 \circ y_2) = F(F(f, y_1), y_2)$, so that the same formalism for the solution of (3.4.9) can be used to solve the problem posed by (3.4.15). Indeed (3.4.15) can be reduced to $dy/dt = f(y)$ if we find y satisfying

$$(3.4.16) \qquad U'^{-1}(f + a)U = f, \text{ i.e., } F(f + a, U) = f.$$

We shall now reconsider the iteration scheme (3.4.11). To this end, we rewrite (3.4.9) as

$$(3.4.17) \qquad G(x, a) \equiv (I + x)f - (f + a)(I + x) = 0.$$

We seek a solution x for (3.4.17). By analogy with the implicit function theorem (3.1.10), we would expect a solution x to exist provided $\|G(0, a)\|$ is sufficiently small and $G_x^{-1}(0, 0)$ exists. Indeed, one can demonstrate that such an analogy is true even if $[G_x(0, 0)]^{-1}$ is mildly singular, by essentially studying the convergence of the iteration scheme (3.4.11).

We begin by rewriting the iteration scheme (3.4.11) in terms of a solution of $G(x, a) = 0$ near $x = 0$. If $u_{N+1} = I + x_{N+1} = u_N \circ \{I + y_{N+1}\} = (I + x_N) \circ (I + y_{N+1})$, then

$$(3.4.18) \qquad \begin{aligned} x_{N+1} &= (I + x_N)(I + y_{N+1}) - I \\ &= (I + y_1)(I + y_2) \cdots (I + y_{N+1}) - I, \end{aligned}$$

where y_{N+1} is a solution of (3.4.13). Thus in terms of G, (3.4.13) can be rewritten

$$(3.4.19) \qquad G_x(0, 0)y_{N+1} + a_{N+1} = 0,$$

where

$$a_{N+1} = f_N - f = F(f+a, I+x_N) - f \quad (\text{since } f_N = F(f+a, u_N))$$
$$= \big((I+x_N)^{-1}\big)(f+a)(I+x_N) - f$$
$$= [(I+x_{N-1})(I+y_N)]^{-1}(f+a)(I+x_{N-1})(I+y_N) - f$$
$$= \big((I+y_N)^{-1}\big)f_{N-1}(I+y_N) - f,$$

that is,

$$(3.4.19') \qquad a_{N+1} = \big((I+y_N)^{-1}\big)(f+a_N)(I+y_N) - f.$$

The convergence of the scheme (3.4.19–3.4.19′) and its applications to physics and geometry are the subjects of recent books by Sternberg (1969) and Moser (1973b), and so the interested reader is referred to these for further information.

NOTES

A Existence of local solutions of nonlinear elliptic systems

Suppose the following system of k equations in k unknowns is *elliptic* on some domain $\Omega \subset \mathbb{R}^n$

$$(*) \qquad \mathbf{F}(x, \mathbf{u}, \ldots, D^m\mathbf{u}) = 0,$$

where $\mathbf{F}$ and $\mathbf{u}$ denote vector-valued functions, with $\mathbf{F}$ a smooth function of its arguments. Moreover suppose $\mathbf{u}_0(x)$ is a known smooth solution of $(*)$. Then the contraction mapping theorem (3.1.1) can be used to prove the following result: *In a sufficiently smooth ϵ-neighborhood O_{x_0} of any point $x_0 \in \Omega$, there is a smooth solution $\mathbf{u}$ of $(*)$ such that* $\sup|D^\alpha u - D^\alpha u_0| \leqslant C\epsilon^{m-|\alpha|+\sigma}$ *for $|\alpha| \leqslant m$ where $\sigma \in (0, 1)$ and σ and C are absolute constants.* The idea of the proof is to set up the desired solution of $(*)$ as a fixed point of a contraction mapping of the Hölder space $C^{m,\sigma}(O_{x_0})$ into itself. To this end we suppose without loss of generality that $x_0 = 0$ and $u_0(x) \equiv 0$ and introduce a small parameter $\epsilon > 0$ into the problem by setting $x = \epsilon y$ and $u(\epsilon x) = v(y)$ so that by rewriting $(*)$ in the form

$$(**) \qquad Lv = Lv - \epsilon^m F(\epsilon y, v, \ldots, \epsilon^{-m} D_y^m v),$$

where L is the linear elliptic operator with constant coefficients in which only the highest derivatives occur obtained by linearizing $(*)$ about $x_0 = 0$ and $u_0(x) \equiv 0$. Then L is invertible and the equation $(**)$ can be written in the form $v = v - A(\epsilon, v)$, where the right-hand side has $C^{m,\sigma}(O_{x_0})$ norm that is $O(\epsilon)$ as $\epsilon \to 0$. See Nirenberg (1973).

B Isometric imbedding problem for Riemannian manifolds

Let $(\mathfrak{M}^k, g)$ be a given Riemannian metric with metric tensor $g = (g_{ij})$. Then we attempt to imbed $\mathfrak{M}^k$ as a submanifold in some Euclidean space $\mathbb{R}^N$ in such a way that the imbedding is isometric, i.e., the metric induced on $\mathfrak{M}^k$ by this embedding is g. Thus, let $z_1, \ldots, z_N$ be Cartesian coordinates in $\mathbb{R}^N$ and $\mathfrak{M}^k$ be smoothly (isometrically) imbedded in $\mathbb{R}^N$. Then referred to a set of local coordinates $(x_1, \ldots, x_k)$ in $\mathfrak{M}^k$, we require that the functions $z = (z_1, \ldots, z_N)$ satisfy the nonlinear differential system

$$g_{ij}(x) = \sum_{m=1}^{N} \frac{\partial z_m}{\partial x_i} \frac{\partial z_m}{\partial x_j}.$$

Nash (1956) proved that this system could also be solved provided N (dimension of $\mathbb{R}^N$) is made sufficiently large. He proved this result in stages, a key element of which was a forerunner of the implicit function theorem described in (3.4.1). Since this topic is well treated in several monographs, we shall not take up this topic here, but rather refer the reader to the monographs by Sternberg (1969) and Schwartz (1968).

C The center problem

As a direct application of Section 3.4D, we consider the problem of showing that the analytic function

$$f(z) = \lambda z + \sum_{n=2}^{\infty} a_n z^n = f_1(z) + f_2(z)$$

is conjugate to the linear function $f_1(z) = \lambda z$ near $z = 0$ by means of an analytic change of variables $u(z)$. Thus we seek a conformal mapping $u(z) = \sum_{n=1}^{\infty} b_n z^n$ defined near $z = 0$ such that

$$(*) \qquad u(\lambda z) = f(u(z)).$$

If λ is not a root of unity, a formal solution can be found by setting $b_1 = 1$ and then determining b_i recursively from $(*)$ by equating coefficients of like powers of z. Explicitly, for $n > 1$,

$$(\lambda^n - \lambda) b_n = g_n(b_1, \ldots, b_{n-1}).$$

If $|\lambda| \neq 1$, the convergence of this series can be proved by the majorant method described in Section 3.3. However, for $|\lambda| = 1$, the excluded roots of unity are dense; and, in fact, there is a dense set of complex numbers λ at which the formal series diverges. Thus the convergence of the formal series for $|\lambda| = 1$ is in question and a method more refined than the majorant method is required. C. L. Siegel (1942) succeeded in studying the convergence of this formal series by imposing an infinite number of conditions on the number λ of the form

$$(**) \qquad |\lambda^q - 1|^{-1} \leqslant c_0 q^2 \qquad \text{for} \quad q = 1, 2, \ldots .$$

(These conditions ensure that λ is not well approximated by roots of unity.) We shall apply the method of Section 4.3D to prove Siegel's result.

Theorem Suppose $|\lambda| = 1$ and that λ satisfies the inequalities $(**)$. Then $f(z) = \lambda z + f_2(z)$ is conjugate to $f_1(z) = \lambda z$ by an analytic change of variables.

Although the proof of this result can be carried out in various ways, we sketch how it can be carried out on the basis of our discussion of Section 3.4D and refer the interested for the details to the article of Moser (1966) or the monograph of Sternberg (1969). We consider the mapping

$$G(z, a) = (I + z)\{\lambda I\} - \{\lambda I + a\}\{I + z\}$$

with $a(z) = f_2(z)$ defined on a scale of Banach spaces $\{A_n\}$ and with range in another scale $\{B_n\}$, both of which are Banach spaces of holomorphic functions defined on varying neighborhoods of the origin in $\mathbb{C}^1$. A formal calculation of the Fréchet derivative of $G(x, 0)$ with respect to x, evaluated at $(0, 0)$ shows that $G_x(0, 0)v = v(\lambda z) - \lambda v(z)$. Now the linear operator $G(0, 0)$ is invertible as a map between the spaces $\{A_n\}$ and $\{B_n\}$ if properly chosen. However, the "norm" of $G_x(0, 0)$ has a singularity of order 3 in the following sense: The formal solution of the equation $v(\lambda z) - \lambda v(z) = g(z)$ is

$$v(z) = \sum_{k=2}^{\infty} (\lambda^k - \lambda) g_k z^k,$$

where $g(z) = \sum g_n z^n$. If $g(z)$ is analytic on Σ_r, the disk $|z| < r$, then the Cauchy estimates imply that if we set $\|g\|_{A_r} = \sup_{|z|<r} |g(z)|$,

$$|g_k| \leqslant r^{-k} \sup_{\Sigma_r} |g(z)| = r^{-k} \|g\|_{A_r},$$

so that on $\Sigma_{r(1-\epsilon)}$,

$$|v(z)| \leqslant c_0 \left(\sum_{k=2}^{\infty} k^2 \left| \frac{z}{r-\epsilon} \right|^k \right) \|g\|_{A_{r(1-\epsilon)}}$$

$$\leqslant \|g\| c_0 \sum k^2 (1-\theta)^k \leqslant \frac{2c_0}{\epsilon^3} \|g\|_{A_{r(1-\epsilon)}}.$$

This result shows that if we apply the scheme defined in (3.4.19)–(3.4.19′) and take scales of Banach spaces consisting of analytic functions defined on discs whose radii shrink by an amount k^{-n} (for some fixed positive constant k), then convergence can be achieved in some limit domain centered at the origin. This is, in fact, what is proven in the references (mentioned above).

D Integrable almost complex structures

Suppose we are given N first-order differential operators $P_j = \sum_{i=1}^{n} a_{ij} D_i$ defined in a neighborhood Ω of the origin of $\mathbb{R}^N$, where the a_{ij} are complex-valued smooth coefficients. Then if $P_1, P_2, \ldots, P_n$ are their conjugates $\bar{P}_1, \ldots, \bar{P}_n$ are linearly independent, the system $(P_1, \ldots, P_n)$ is called an almost complex structure on Ω. We seek necessary and sufficient conditions that ensure the existence of new local coordinates $\mu_1, \ldots, \mu_{2n}$ so that equations $P_j w = 0$ $(j = 1, 2, \ldots, n)$ are equivalent to the *Cauchy–Riemann equations* $\partial w / \partial \bar{\eta}_j = 0$ $(j = 1, 2, \ldots, n)$ when $\eta_j = \mu_j + i\mu_{j+n}$. In this case we can define the function w to be analytic so that the almost complex structure is actually a "complex" structure, in the sense that analytic functions on Ω can be defined unambiguously. A necessary condition is easily found by noting that if the operators P_j are linear combinations of $\partial/\partial\eta_j$ $(j = 1, \ldots, n)$, then since $\partial_\eta^2 = (\partial/\partial\eta_1, \ldots, \partial/\partial\eta_n)^2 = 0$, $(*)$ *The commutator* $[P_j, P_k]$ *is a linear combination of* $P_1, \ldots, P_n$ *for only* j, k. The condition $(*)$ is called an integrability condition, and since it is independent of coordinate system, makes sense in terms of almost complex structures on manifolds, as well as Ω. Now this condition $(*)$ is also *sufficient* for the desired conjugacy of the system $P_j w = 0$ with the Cauchy Riemann equations, although the proof of this fact is not easy.

This sufficiency problem is a local nonlinear one if we attempt to find the coordinate transformation $\{x\} \to \{\mu\}$. To show this let us choose the new coordinate $\{\eta\}$ so that $P_j \eta = 0$ $(j = 1, \ldots, n)$ and we may suppose that P_j can be written $\bar{\partial}\mu = A(\partial\mu)$, where $A = (a_{ij})$ is a matrix whose entries vanish at the origin to second order and $\bar{\partial} = (\partial/\partial\bar{z}_1, \ldots, \partial/\partial\bar{z}_n)$. Then the condition $(*)$ can be written

$$(\dagger) \qquad P_j a_{ik} = P_k a_{ij}.$$

Since the P_j and P_k involve the entries of A, the system $(\dagger)$ is a nonlinear equation for the entries of A. The sufficiency of $(*)$ was first established in this way by Newlander and Nirenberg (1957). A recent ingenious method for the solution of $(\dagger)$ was given by Malgrange (1969).

E A rank theorem for nonlinear Fredholm operators

The classical rank theorem of advanced calculus is not known for C^1 mappings between Banach spaces. However, for nonlinear Fredholm mappings the following result can be established:

Theorem Let f be a C^1 Fredholm operator of index p defined on a neighborhood U of a point x_0 of a Banach space X with range in a Banach space Y. Then, if $\dim \ker f'(x)$ is constant for $x \in U$, f is conjugate to a linear projection P of $\text{Range}\, f'(x_0) \oplus \text{Ker}\, f'(x_0)$ onto $\text{Range}\, f'(x_0)$ in a sufficiently small neighborhood of U.

For a proof of this result, we refer the reader to the paper by Berger and Plastock (1977).

F Bibliographic notes

Section 3.1: As mentioned earlier, the contraction mapping theorem is due to Banach (1920). It can be regarded as a natural extension of Picard's method of successive approximation. Dieudonné (1960) contains a good discussion of the infinite-dimensional inverse and implicit function theorems, but establishes the rank theorem only in the finite-dimensional case. Newton's method and its classical variants are well discussed in Krasnoselski *et al.* (1972). The result (3.1.19) is due to Graves (1950). Solving the initial value problem for ordinary differential equations in a Banach space is made more difficult by the fact that the direct infinite-dimensional analogue of Peano's theorem (3.1.28) is false (see Dieudonné (1960) for a simple counterexample). The result (3.1.36) can be found in the book of Ljusternik and Sobolev (1961). The infinite-dimensional version of Sard's theorem is due to Smale (1965), while the corresponding generalization of Morse's theorem, (3.1.47), is due to Pohozaev (1968).

Section 3.2: The method of steepest descent dates back to the paper by Cauchy (1847). The results discussed here are based on the paper of Rosenbloom (1956). The result (3.2.14) is due to Browder (1965). This technique of steepest descent is crucial for the more global results of Chapter 6.

Section 3.3: The majorant method for solving local analytic operator equations is also due to Cauchy, and has proven to be a highly flexible tool. Nonetheless, the method is too general for some problems since it often fails to take note of the qualitative features of the operators involved. Excellent articles using this method to obtain infinite-dimensional analogues of the Cauchy–Kowalewski theorem and other related results can be found in Rosenbloom (1961), Treves (1970), and Nirenberg (1972).

Section 3.4: Our discussion of the result (3.4.1) is based on the article by Moser (1966). The principle of using the Newton method to gain rapid convergence can be found in Cartan (1940) (where a problem of factorization of analytic matrices is solved) and a review article of Kolomogorov (1954). The technique of smoothing operators can be found in a paper of Nash (1956). Local conjugacy problems are treated at length in the monographs of Moser (1973b) and Sternberg (1969). An interesting recent article on this topic is Zehnder (1975).

G Decomposition theorems for nonlinear mappings

A useful alternative to studying nonlinear mappings under conjugacy equivalence is obtained by extending the decomposition theorems of Helmholtz, Hodge, and Frobenius to an infinite dimensional context. In such theorems, a nonlinear operator is regarded as a differential 1-form and the notion of gradient mapping is expressed in terms of exterior differentiation. In a 1977 thesis, T. Goldring has recently obtained some important results in this direction, by extending the adjoint operator of exterior differentiation to an infinite dimensional context. Such results promise to have important applications in physics and genetics.

CHAPTER 4

PARAMETER DEPENDENT PERTURBATION PHENOMENA

In this chapter we extend the previous local analysis to discuss certain critical phenomena that occur in the study of a mapping $f(x, \lambda)$, (depending smoothly on a parameter λ), near a singular point (x_0, λ_0) of f. Two specific circumstances are considered: first, bifurcation phenomena in which one assumes $f(x, \lambda)$ is a C^1 Fredholm operator and studies the structure of the zeros (x, λ) of $f(x, \lambda) = 0$ near a given zero (x_0, λ_0) such that, for fixed λ_0, x_0 is a singular point of $f(x, \lambda_0)$; secondly, certain singular perturbation phenomena, in which one attempts to find the sense in which a given formal approximation $x(\epsilon)$ satisfies a C^1 Fredholm operator equation $f(x, \epsilon) = 0$ for $\epsilon > 0$ sufficiently small even though the linear mapping $f'(x(0), 0)$ is not Fredholm. Both cases arise frequently in concrete problems. In bifurcation phenomena, nonuniqueness of solutions is of prime consideration, and it is often found that such considerations, lead directly to relatively sophisticated topological techniques of solution. On the other hand, the singular perturbation problems we study generally require quite sharp analytic estimates on the norm of the linear mapping $f'(x(\epsilon), \epsilon)$ for resolution. Both situations occur naturally in the study of explicit nonlinear eigenvalue problems of the form $Ax = \lambda Bx$. Roughly speaking, letting $\lambda \to \infty$ and setting $\lambda = 1/\epsilon$, it is natural in singular perturbation theory to attempt to prove that there are families of solutions $x(\epsilon)$ near each solution $x(0)$ of $Bx = 0$. On the other hand, assuming that the linear operators $A'(0)$ and $B'(0)$ are both nonzero, bifurcation theory attempts to compare the solutions (x, λ) of $Ax = \lambda Bx$ and the solutions of the linear eigenvalue problem $A'(0)x = \lambda B'(0)x$ for $\|x\|$ sufficiently small.

4.1 Bifurcation Theory—A Constructive Approach

Bifurcation theory is concerned with the structure of the solutions of the equation $f(x, \lambda) = 0$ *as a function of the parameter* λ near a solution (x_0, λ_0) that is also a singular point of the mapping $f(x, \lambda_0)$, (so that $f_x(x_0, \lambda_0)$ is not invertible). Here $f(x, \lambda)$ denotes a C^1 operator mapping a neighborhood of (x_0, λ_0) in the Banach space $X \times Z$ into a Banach space Y. (The parameter space $Z = \{\lambda\}$ will usually be chosen as the real or complex numbers.) Thus at (x_0, λ_0) the linear operator f_x is not invertible, the implicit function theorems of Chapter 3 are not directly applicable, and indeed the behavior of the solutions of $f(x, \lambda) = 0$ near (x_0, λ_0) is indeterminate. As mentioned in Section 1.6 if $X = Y = \mathbb{C}^N$, $Z = \mathbb{R}^1$, and f is a complex analytic mapping, one concludes that the mapping $f(x, \lambda_0)$ is not one-to-one near x_0 and hence $f(x, \lambda_0)$ is not a local homeomorphism there. In this section we shall investigate this "nonuniqueness" phenomenon from a constructive point of view. In Section 4.2, we shall return to the more powerful qualitative methods mentioned above.

It is assumed for historical reasons that the equation $f(x, \lambda) = 0$ *has a known family of solutions* $(x_0(\epsilon), \lambda_0(\epsilon))$ *containing the point* (x_0, λ_0). One of the objects of bifurcation theory is then the assertion that $f(x, \lambda) = 0$ has another family of solutions $(x_1(\epsilon), \lambda_1(\epsilon))$ distinct from the family $(x_0(\epsilon), \lambda(\epsilon))$ near (x_0, λ_0) and such that $(x_1(\epsilon), \lambda_1(\epsilon)) \to (x_0, \lambda_0)$ as $\lambda \to \lambda_0$. (See Fig. 4.1.) In this sense bifurcation theory resembles the spectral theory for a linear operator in which case the known family $(x_0(\epsilon), \lambda_0(\epsilon))$ is the null solution $(0, \lambda)$, whereas the *secondary* families represent linear subspaces of eigenvectors.

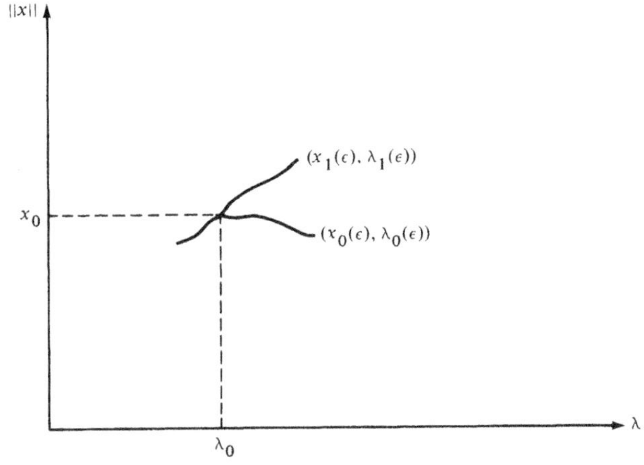

FIG. 4.1 Usual behavior of solutions of $f(x, \lambda) = 0$ near a bifurcation point. The $(\|x\|, \lambda)$ graph of solutions is called a bifurcation diagram.

4.1A Definitions and basic problems

To obtain a better understanding of bifurcation theory, let us examine the origins of the subject. In a paper published in 1885, H. Poincaré attempted to answer the following questions:

(a) Find the possible forms of equilibrium of a homogeneous mass of fluid (subject to gravity) when rotating about a fixed axis with constant angular momentum ω.

(b) Determine the stability or instability of each form.

It was known that: (i) if $\omega = 0$, the only possible form is a sphere; (ii) if ω is small, a family of ellipsoids of revolution M_ω (Maclaurin's ellipsoids) exists and is stable; and (iii) at a certain critical number ω_0 (although continuing to exist) this family becomes "unstable," and a new family of equilibrium forms J_ω (ellipsoids with three unequal axes, otherwise known as Jacobi's ellipsoids) becomes stable. Poincaré found that these ellipsoids (initially deviating slightly from M_ω), in turn become unstable at a higher critical number ω_1. Near ω_1 nonellipsoidal, pear-shaped forms of equilibrium P_ω exist (again initially deviating slightly from J_ω). Poincaré hoped by pursuing this argument to prove that the moon "split off" from the earth by tracing P_ω as ω changes. (See Fig. 4.2.) Unfortunately, it was determined (after much dispute) that P_ω was unstable, and therefore Poincaré's argument for the origin of the moon was abandoned. The mathematical content of Poincaré's ideas, however, had a quite different fate.

Poincaré said that the ellipsoids J_ω *bifurcate* from M_ω at ω_0, and the pear-shaped figures P_ω *bifurcate* from J_ω at ω_1. The families $\{M_\omega\}$, $\{J_\omega\}$, $\{P_\omega\}$ were termed *linear series* or *branches*. The pairs (M_{ω_0}, ω_0) and (J_{ω_1}, ω_1) were called *points of bifurcation*. Stability was determined by showing that the potential energy of a figure (F_ω, say) was a relative minimum. Poincaré termed the transition of stability at $(M\omega_0, \omega_0)$ to (J_{ω_0}, ω_0) *exchange of stability*. In the following sections we shall see how these terms apply to more general circumstances and illustrate their occurrence in many concrete problems.

We now give precision to the above terms relative to the operator equation

(4.1.1) $$f(x, \lambda) = 0 \qquad \text{for } (x, \lambda) \in X \times Z$$

(4.1.2) **Definition** A point (x_0, λ_0) is called a *point of bifurcation* relative to the equation $f(x, \lambda) = 0$ if: (i) (x_0, λ_0) lies on a curve of solutions $(x_0(\epsilon), \lambda_0(\epsilon))$ through (x_0, λ_0); and (ii) every neighborhood of (x_0, λ_0) in $X \times Z$ has a solution of $f(x, \lambda) = 0$ distinct from the family $(x_0(\epsilon), \lambda_0(\epsilon))$.

In case (x_0, λ_0) is a point of bifurcation for the equation $f(x, \lambda) = 0$, the

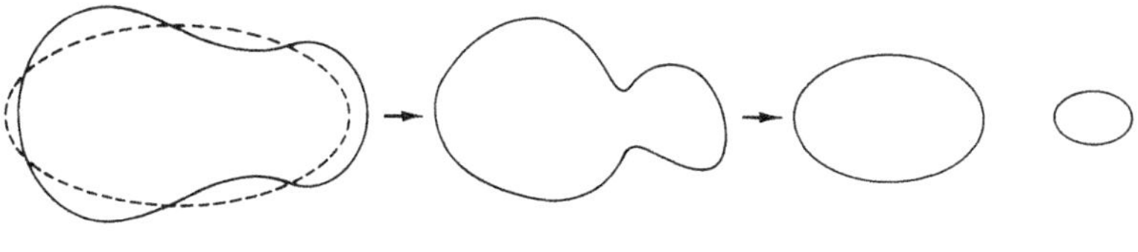

FIG. 4.2 Poincaré's vision of the creation of the moon by fission.

solutions of $f(x, \lambda) = 0$ often consist of distinct continuous curves through (x_0, λ_0) and these curves are called *branches* of solutions. A (global) continuation of such a branch of solutions is a continuous curve $(x(\epsilon), \lambda(\epsilon))$ of solutions for $f(x, \lambda) = 0$ such that $\|x(\epsilon)\| + \|\lambda(\epsilon)\| \to \infty$. A solution $(\tilde{x}, \tilde{\lambda})$ of $f(x, \lambda) = 0$ is called stable if the spectrum of the linear operator $f_x(\tilde{x}, \tilde{\lambda})$ has negative real part; $(\tilde{x}, \tilde{\lambda})$ is unstable if $f_x(\tilde{x}, \tilde{\lambda})$ has a spectral value with positive real part. This implies that the initial value problem (associated with $f(x, \lambda)$ linearized about $(\tilde{x}, \tilde{\lambda})$)

$$\frac{dy}{dt} = f_x(\tilde{x}, \tilde{\lambda})y, \qquad y(0) = y_0,$$

has the property that *all* solutions decay if stable, but not if unstable. This notion is generally called *linear stability theory* since the spectrum of the linear operator $f_x(\tilde{x}, \tilde{\lambda})$ is used as a stability criterion. (A more accurate stability theory would, of course, utilize an initial value problem for the full non-linear operator.)

We can state the basic problems of bifurcation theory in terms of these definitions:

(i) *Existence problem* Determine the points of bifurcation of $f(x, \lambda) = 0$.

(ii) *Structure problem* Determine the complete structure of the solution set of $f(x, \lambda) = 0$ near each point of bifurcation.

(iii) *Continuation problem* Determine circumstances under which branches of solutions admit global continuations. (This problem will be taken up in Part III, due to its nonlocal nature.)

(iv) *Stability problem* Determine stable branches of solutions for the equation $f(x, \lambda) = 0$ near a point of bifurcation.

(v) *Linearization problem* What information concerning the problems of bifurcation theory can be determined from a knowledge of the derivative $f_x(x_0, \lambda_0)$ at a point of bifurcation (x_0, λ_0)?

(vi) *Problem of nonlinear effects* What role does the nonlinearity of the map $f(x, \lambda)$ play in regard to problems (i)–(iv) above?

As a simple example, let $U(x, \lambda)$ be a real analytic function defined on $\mathbb{R}^N \times \mathbb{R}^1$, and consider the solutions of $\nabla U(x, \lambda) = 0$ near a point (x_0, λ_0) at which $\nabla U(x, \lambda) = 0$. If the Hessian determinant, $|U_{x_i x_j}(x_0, \lambda_0)| \neq 0$, the implicit function theorem implies that there is a unique curve $(x(\epsilon), \lambda(\epsilon))$ through (x_0, λ_0). However, if $|U_{x_i x_j}(x_0, \lambda_0)| = 0$, then the point (x_0, λ_0) may be a point of bifurcation, and in particular, there will always be a second curve $(x(\epsilon), \lambda(\epsilon))$ through (x_0, λ_0) (perhaps complex-valued) satisfying the equation $\nabla U(x, \lambda) = 0$. However, since complex-valued solutions are excluded from consideration, the problem of deciding whether or not (x_0, λ_0) is a point of bifurcation requires further investigation.

A related problem (that can often be reduced to a bifuraction question, as posed here) is concerned with the structure of nontrivial solutions of the C^1 Fredholm operator equation $g(x) = 0$ relative to a given curve of solutions $x(t)$ passing through a singular point $x(0) = x_0$ of g. Here the term *nontrivial* referes to solutions of $g(x) = 0$ distinct from $x(t)$. To relate this problem to our discussion we set $f(y, t) = g(x(t) + y)$, then $f(0, t) = 0$ and moreover $f_y(0, 0) = g'(x(0))$ is a linear Fredholm operator.

As an interesting example, consider the problem of finding periodic solutions of the system

$$\ddot{\mathbf{x}} + A\mathbf{x} + \mathbf{f}(\mathbf{x}, \dot{\mathbf{x}}) = 0, \qquad |\mathbf{f}(\mathbf{x}, \mathbf{y})| = o(|\mathbf{x}| + |\mathbf{y}|), \tag{4.1.3}$$

near the singular point $x = 0$. Here $\mathbf{x}(t)$ is an N-vector function of t, A is an $N \times N$ (self-adjoint) nonsingular matrix that possesses k positive eigenvalues $\lambda_1^2 \leqslant \lambda_2^2 \leqslant \ldots \leqslant \lambda_k^2$ (say), and $\mathbf{f}(\mathbf{x}, \mathbf{y})$ is a smooth function of $\mathbf{x}$,$\mathbf{y}$. One generally obtains a first approximation to the periodic solutions of (4.1.3) by a process of linearization. Namely, we seek periodic solutions of (4.1.3) that are near periodic solutions of the linear system

$$\ddot{\mathbf{x}} + A\mathbf{x} = 0 \tag{4.1.4}$$

Since this linear system possesses k distinct "normal mode" periodic families of solutions $\mathbf{z}_1, \mathbf{z}_2, \ldots, \mathbf{z}_k$ (cf. Section 1.2(v)), one attempts to find at least k distinct periodic families for (4.1.3) deviating only slightly from $\mathbf{z}_1, \mathbf{z}_2, \ldots, \mathbf{z}_k$ near $\mathbf{x} = 0$. To relate this problem to our bifurcation theory, it is important to introduce the period of a tentative periodic solution of (4.1.3) explicitly. (The period then plays the role of the parameter λ in the general theory.) This can be accomplished by making the change of variables $t = \lambda s$ in (4.1.3) so that (4.1.3) becomes

$$\mathbf{x}_{ss} + \lambda^2[A\mathbf{x} + \mathbf{f}(\mathbf{x}, \mathbf{x}_s/\lambda)] = 0. \tag{4.1.3$'$}$$

Then solutions of period 1 in s are solutions of period λ in t. An important classic result of Liapunov in this connection is a criterion for any given periodic solution $\mathbf{z}_j$ to be preserved by the higher order perturbation $\mathbf{f}$. The result can be expressed as follows.

Liapunov's criterion Suppose $\mathbf{f}(\mathbf{x}, -\mathbf{y}) = -\mathbf{f}(\mathbf{x}, \mathbf{y})$, then the jth periodic family of (4.1.3) is preserved if $\lambda_i/\lambda_j \neq$ integer for $i \neq j$ ($i = 1, 2, \ldots, k$). If $\mathbf{f}(\mathbf{x}, \mathbf{y})$ is real analytic in $\mathbf{x}$ and $\mathbf{y}$, the family $\mathbf{x}_j(\epsilon)$ and its period $\lambda_j(\epsilon)$ can be written as

$$\mathbf{x}_j(\epsilon) = \sum_{n=1}^{\infty} \boldsymbol{\alpha}_{n,j}(t)\epsilon^n$$

where $\boldsymbol{\alpha}_{1,j}(t) = \mathbf{z}_j(t)$, and $\lambda_j(\epsilon) = 2\pi/\lambda_j + \sum_{n=1}^{\infty} \beta_n \epsilon^n$.

We shall show that this result can be obtained by the general results discussed here. *Indeed, we shall interpret the "unusual" condition* $\lambda_i/\lambda_j \neq$ *integer in terms of the multiplicitly of solutions of the linearized equation,*

(4.1.4). In the classical proofs of this result, this condition was a necessary concommitent of the majorant method for analytic $\mathbf{f}(\mathbf{x}, \mathbf{y})$.

4.1B Reduction to a finite-dimensional problem

(4.1.5) **Theorem** Suppose X and Y are real Banach spaces and $f(x, \lambda)$ is a C^1 map defined in a neighborhood U of a point (x_0, λ_0) with range in Y such that $f(x_0, \lambda_0) = 0$ and $f_x'(x_0, \lambda_0)$ is a linear Fredholm operator. Then all solutions (x, λ) of $f(x, \lambda) = 0$ near (x_0, λ_0) (with λ fixed) are in one-to-one correspondence with the solutions of a finite-dimensional system of N_1 real equations in a finite number N_0 of real variables. Furthermore, $N_0 = \dim \operatorname{Ker} L$ and $N_1 = \dim \operatorname{coker} L$.

Proof: Set $\lambda - \lambda_0 = \delta$ and $x - x_0 = \rho$, then setting $-L = f_x(x_0, \lambda_0)$ we note that the equation $f(x, \lambda) = 0$ can be written in the form

(4.1.6) $$L\rho = R(\rho, \lambda_0, \lambda)$$

$$\text{with} \quad R(\rho, \lambda_0, \lambda) = f(x, \lambda) - f(x_0, \lambda_0) - f_x(x_0, \lambda_0)\rho$$

so that $R(\rho, \lambda_0, \lambda) = O(\|\delta\|) + o(\|\rho\| + \|\delta\|)$. Indeed,

$$\begin{aligned} f(x, \lambda) &= [f(x, \lambda) - f(x_0, \lambda)] + [f(x_0, \lambda) - f(x_0, \lambda_0)] \\ &= f_x(x_0, \lambda_0)\rho + o(\|\rho\| + \|\delta\|) + O(\|\delta\|). \end{aligned}$$

Now we write $\rho = \rho_1 + \rho_2$, where $\rho_1 \in \operatorname{Ker} L$, $\rho_2 \in X_1$ and $X = \operatorname{Ker} L \oplus X_1$, and recall that by (1.3.38), L has a left inverse L_0 with range X_1 and kernel, coker L. Thus premultiplying by L_0, (4.1.6) becomes

(4.1.7) $$\rho_2 = L_0 R(\rho_1 + \rho_2, \lambda_0, \lambda).$$

An application of the implicit function theorem to (4.1.7) now yields a unique solution $\rho_2 = g(\rho_1, \lambda)$ of (4.1.7) provided $\|\rho_1\|, \|\lambda - \lambda_0\|$, and $\|\rho_2\|$ are sufficiently small. Since $Y = \operatorname{Range} L \oplus \operatorname{coker} L$, and L_0 is one-to-one on the range of L, it remains to satisfy (4.1.6) on coker L. Thus if P is the projection of Y on coker L, the solutions of $f(x, \lambda) = 0$ near (x_0, λ_0) are in one-to-one correspondence with the solutions of

(4.1.8) $$PR(\rho_1 + g(\rho_1, \lambda), \lambda_0, \lambda) = 0.$$

By choosing appropriate bases for ker L and coker L the system (4.1.8) is equivalent to N_1 real equations in N_0 real unknowns.

(4.1.9) **Corollary** Under the above hypotheses, suppose $f(x_0, \lambda) = 0$ for λ near λ_0 and $f_x(x_0, \lambda) = I - \lambda L$, where I is the identity operator. If $X = \operatorname{Ker}(I - \lambda_0 L) \oplus X_1$ and (x, λ) is a solution of $f(x, \lambda) = 0$ near (x_0, λ_0) with $x = \rho_1 + g$, $\rho_1 \in \operatorname{Ker}(I - \lambda_0 L)$, $g \in X_1$, then $\|g\| = o(\|\rho_1\|)$.

Proof: Since $f(x, \lambda) = f(x_0, \lambda) + f_x(x_0, \lambda)(x - x_0) + O(\|x - x_0\|^2)$, we may suppose that the equation (4.1.6) can be written as

$$(I - \lambda L)x + T(x, \lambda) = 0,$$

where $T(0, \lambda) = T_x(0, \lambda) = 0$. Consequently, the equation (4.1.8) can be written

(4.1.10) $$h(\rho_1, g) \equiv (I - \lambda L)g - PT(\rho_1 + g, \lambda) = 0.$$

Since $g = g(\rho_1)$ was determined by application of the implicit function theorem, (3.1.11) implies

(4.1.11) $$g_{\rho_1}(\rho_1) = -[h_g(\rho_1, g)]^{-1}[h_{\rho_1}(\rho_1, g)].$$

We show that $\|g_{\rho_1}(\rho_1)\| = o(1)$ as $\|\rho_1\| \to 0$, so that by the mean value theorem (2.1.19), $\|g(\rho_1)\| = o(\|\rho_1\|)$. Indeed $h_g = (I - \lambda L)g - PT_x(\rho_1 + g, \lambda)$, and since $T_x(0, \lambda) = 0$, $h_g(0, 0) = I - \lambda_0 L$, which is invertible on X_1. Consequently, by continuity, $h_g(\rho_1, g)$ is invertible for ρ_1 and g sufficiently small and $\|h_g(\rho_1, g)^{-1}\| \leq \frac{1}{2}\|[h_g(0, 0)]^{-1}\|$. Now

$$\|h_{\rho_1}(\rho_1, g)\| = \|PT_x(\rho_1 + g, \lambda)\| = o(1)$$

as $\|\rho_1\| \to 0$. Here the convergence is uniform for λ near λ_0. Thus by (4.1.11), $\|g_{\rho_1}(\rho_1)\| = o(1)$ as $\|\rho_1\| \to 0$ uniformly for λ near λ_0.

4.1C The case of simple multiplicity

Clearly Theorem 4.1.5 implies that the totality of solutions (x, λ) near (x_0, λ_0) can be completely determined from the finite-dimensional system (4.1.8). This system is referred to as the *bifurcation equations* for $f(x, \lambda) = 0$ at (x_0, λ_0). The most important situation in applications is the case index $f_x(x_0, \lambda_0) = 0$ so that the bifurcation equations consist of N equations in N unknowns with $N = \dim \operatorname{Ker} f_x(x_0, \lambda_0)$. Even in this case, the bifurcation equations do not readily yield general specific answers to the questions raised above unless $N = 1$. To demonstrate the utility of the bifurcation equations, we now discuss this case.

Thus we suppose $f(x, \lambda)$ is a C^2 function of its arguments with $f(0, \lambda) \equiv 0$ in a neighborhood of the point $(0, \lambda_0)$, where $f_x(0, \lambda_0)$ is a linear Fredholm operator of index zero with $\dim \ker f_x(0, \lambda_0) = 1$. Then we prove

(4.1.12) **Bifurcation Theorem for Simple Multiplicity** Suppose the above hypotheses hold and moreover that $f_{\lambda x}(0, \lambda_0)z \not\in \operatorname{Range} f_x(0, \lambda_0)$ for nonzero $z \in \operatorname{Ker} L$. Then $(0, \lambda_0)$ is a point of bifurcation for the equation $f(x, \lambda) = 0$ and there is exactly one continuous curve of nontrivial solutions $(x(\epsilon), \lambda(\epsilon))$ bifurcating from $(0, \lambda_0)$.

Proof: The left-hand side of the bifurcation equations (4.1.8) in the present case can be regarded as a single real-valued function. Indeed, let μ be a nonzero bounded linear functional that vanishes off coker $f_x(0, \lambda_0)$, then the bifurcation equations (4.1.8) can be rewritten

(4.1.13) $$F(\rho, \lambda) \equiv \mu f(\rho + g(\rho, \lambda), \lambda) = 0.$$

By a further translation of origin, if necessary, we may also suppose $\lambda_0 = 0$. Then under the hypotheses of the theorem, we shall show that (0, 0) is a nondegenerate critical point of the C^2 real-valued function $F(\rho, \lambda)$ regarded as a function of the two real variables (ρ, λ). Consequently, by Morse's lemma (1.6.11), after an appropriate change of coordinates $(\rho, \lambda) \to (\tilde{\rho}, \tilde{\lambda})$, the solutions of (4.1.13) near (0, 0) satisfy the equation $\tilde{\rho}^2 - \tilde{\lambda}^2 = 0$, i.e., the solutions (ρ, λ) of (4.1.13) near (0, 0) consist of two curves intersecting at (0, 0). Since one of these curves consists of the $(0, \lambda)$ axis, there is exactly one curve of nontrivial solutions bifurcating from $(0, \lambda_0)$.

To verify that (0, 0) is a critical point of the real-valued function $F(\rho, \lambda)$ we show that $g_\rho(0, 0)\rho = 0$. Indeed, differentiating the expression for F in (4.1.13) we find by the chain rule

$$\mu f_x(0, 0)\{\rho + g_\rho(0, 0)\rho\} = \mu f_x(0, 0)g_\rho(0, 0)\rho = 0.$$

This implies $f_x(0, 0)g_\rho(0, 0)\rho \in Y_1 =$ complement of coker $f_x(0, 0)$. Now $g_\rho(0, 0)\rho \in Y_1$, while $f_x(0, 0)$ is an isomorphism on Y_1. Thus $g_\rho(0, 0)\rho = 0$, as required. Consequently, $F_\rho(0, 0) = 0$ and similarly $F_\lambda(0, 0) = 0$, so that (0, 0) is indeed a critical point of F.

Finally, we verify that (0, 0) is a nondegenerate critical point of F with Morse index one. A simple computation, using the results of the above paragraph, shows that the Hessian matrix $H_F(0, 0)$ of F evaluated at (0, 0) is a 2×2 matrix whose entries are precisely *the second derivatives of* $\mu f(\rho, \lambda)$ *evaluated at* (0, 0) (i.e., the contributions due to the term $g(\rho, \lambda)$ vanish). Since $f(0, \lambda) \equiv 0$ near λ_0, we find $\mu f_{\lambda\lambda}(0, 0) = 0$; while $\mu f_{x\lambda}(0, 0) = \mu f_{\lambda x}(0, 0) \neq 0$ since we have assumed $f_{\lambda x}(0, \lambda_0)\rho \in \operatorname{coker} f_x(0, \lambda_0)$ is nonzero for $\rho \neq 0$. Thus the Hessian matrix $H_F(0, 0)$ is nonsingular and the associated quadratic form is indefinite. Consequently, (0, 0) is a nondegenerate critical point of Morse index one, as desired. Thus completes the proof of the theorem.

Remark on simple multiplicity: To understand the import of the hypotheses of (4.1.12), suppose the equation $f(x, \lambda) = 0$ can be written in the form

(4.1.14) $\qquad (I - \lambda L)x + T(x, \lambda) = 0,$

where $T(x, \lambda)$ is a C^2 function of (x, λ) such that $T(0, \lambda) = T_x(0, \lambda) \equiv 0$. Then since for $z \in \operatorname{Ker}(I - \lambda_0 L)$,

$$f_{\lambda x}(0, \lambda_0)z \equiv -Lz = -\lambda_0^{-1}z \qquad \text{and} \qquad f_x(0, \lambda_0) = I - \lambda_0 L,$$

the condition $f_{\lambda x}(0, \lambda_0)z \notin \operatorname{Range} f_x(0, \lambda_0)$ $\{z \neq 0\}$ implies that the range and the kernel of $I - \lambda_0 L$ intersect only in 0, so that λ_0^{-1} is an eigenvalue of simple multiplicity for L.

Moreover, the result (4.1.12) can be sharpened for the equation (4.1.14) by requiring that $T(x, \lambda)$ be only a C^1 function of its arguments, and also in this case it is easily shown that the curve of nontrivial solutions can be written in the form $(x(\epsilon), \lambda(\epsilon))$ with $x(\epsilon) = \epsilon\rho + o(|\epsilon|)$. This result follows again from the bifurcation equations (4.1.8). This time instead of using

Morse's lemma to resolve them, we use the implicit function theorem to show that the bifurcation equations uniquely determine λ near once λ_0 and $\rho \in \text{Ker}(I - \lambda_0 L)$ are given. (This is demonstrated in (4.1.31) below.)

As an application of these results, we give a

Proof of the Liapunov criterion for periodic solutions of (4.1.3): The main difficulty to overcome is the selection of an appropriate space X for (4.1.3), so chosen that the resulting operator equation has bifurcation points of simple multiplicity. To this end, we shall rewrite (4.1.3) as an operator equation in a closed subspace of the Banach space X_N of C^1 N-vector functions $x(s) = (x_1(s), \ldots, x_N(s))$ defined on $(0, \frac{1}{2})$ that satisfy the boundary conditions $\dot{x}(0) = \dot{x}(\frac{1}{2}) = 0$. The norm of an element $x \in X_N$ is

(4.1.15) $$\|x\| = \sup_{(0, 1/2)} |x(s)| + \sup_{(0, 1/2)} |\dot{x}(s)|.$$

The solutions of this operator equation can then be extended to even periodic solutions of (4.1.3′) by setting $x(-s) = x(s)$ for $s \in [0, \frac{1}{2}]$ and then extending $x(s)$ periodically for all s. Since any element $x(s)$ of X_N can be written uniquely as $x(s) = x_0(s) + x_m$, where x_m is the mean value of $x(s)$ over $(0, \frac{1}{2})$ and $x_0(s)$ has mean value zero over $(0, \frac{1}{2})$, the equation (4.1.3′) can be written as the pair of equations

(4.1.16) $$0 = \ddot{x}_0 + \lambda^2\left[Ax_0 + f(x_0 + x_m, \lambda^{-1}\dot{x}_0) - \tfrac{1}{2}\int_0^{1/2} f(x_0 + x_m, \lambda^{-1}\dot{x}_0)\, ds\right],$$

(4.1.17) $$0 = Ax_m + \tfrac{1}{2}\int_0^{1/2} f(x_0 + x_m, \lambda^{-1}\dot{x}_0)\, ds.$$

Since A is invertible, we can apply the implicit function theorem to (4.1.17) to solve x_m in terms of x_0 and λ, $x_m = g(x_0, \lambda)$; and since f is smooth, g is also a smooth function. Now by the procedure mentioned in Section 2.2D the equation (4.1.3) can be written as the integral equation

(4.1.18) $$x_0(s) = \lambda^2 \int_0^{1/2} G(s, s')\{Ax_0(s') + N(x_0(s'), \lambda)\}\, ds',$$

where $G(s, s')$ is Green's function for $\ddot{x}_0$ over $(0, \frac{1}{2})$ subject to boundary conditions $\dot{x}(0) = \dot{x}(\frac{1}{2}) = 0$, $\int_0^{1/2} x = 0$, and

$$Nx_0(s) = f(x_0 + g(x_0, \lambda), \dot{x}_0/\lambda) - \tfrac{1}{2}\int_0^{1/2} f(x_0 + g(x_0, \lambda), \dot{x}_0/\lambda)\, ds.$$

This integral equation (4.1.18) can of course be written as an operator equation

(∗) $$x_0 = \lambda^2\{Lx_0 + T(x_0, \lambda)\}$$

in the space X_N^0 the closed subspace of X_N consisting of elements of mean value zero over $(0, \frac{1}{2})$. The linear operator L is defined by

$$Lx_0(s) = \int_0^{1/2} G(s, s')Ax_0(s)\, ds$$

is compact, so that $I - \lambda^2 L$ is a Fredholm operator of index zero, while $T(x_0, \lambda)$ is a C^2 operator of higher order since $f(x, y)$ of (4.1.3) is sufficiently smooth. (As we shall see below the space X_N^0 has been chosen to overcome the difficulties of multiplicity in the spectrum of L.)

With these preliminaries out of the way, we can apply the results of (4.1.12) to the equation (∗). To accomplish this, we first calculate the real spectrum of the linear operator $I - \lambda^2 A$ in the Banach space X_N^0; i.e., we find the eigenvalues λ^2 and associated eigenfunctions $x(s)$ of the system $\ddot{x} + \lambda^2 Ax = 0$, satisfying the conditions $\dot{x}(0) = \dot{x}(\frac{1}{2}) = 0$ and $\int_0^{1/2} x(s)\,ds = 0$. Without loss of generality, we may assume that the matrix A is diagonal, so that a short computation shows these eigenvalues are of the form $\{\lambda^2 \mid \lambda^2 = 4\pi^2 N^2/\lambda_i^2;\ N = 1, 2, \ldots;\ i = 1, 2, \ldots, k\}$. We are interested in the behavior near eigenvalues $4\pi^2/\lambda_i^2$ $(i = 1, \ldots, k)$, and we wish to show that (4.1.19) has a branch of nontrivial solutions $(x_0(\epsilon), \lambda_0(\epsilon)) \to (0, 2\pi/\lambda_i)$ as $\epsilon \to 0$. By virtue of (4.1.12), this can be accomplished if the eigenvalue $4\pi^2/\lambda_i^2$ is simple; i.e., if $4\pi^2/\lambda_i^2 \neq N^2 4\pi^2/\lambda_j^2$ $(i \neq j)$ i.e., if $\lambda_j/\lambda_i \neq N$, any integer. This condition is precisely the criterion of Liapunov, which can now be considered justified via (4.1.12). The analyticity of the family $(x_0(\epsilon), \lambda_0(\epsilon))$ is also ensured by virtue of (4.1.15) since one easily shows that the real analyticity of $f(x, y)$ implies that of $T(x, \lambda)$.

4.1D A convergent iteration scheme

We now proceed to construct a convergent iteration scheme for the nontrivial solutions $x(\epsilon, y)$, $\lambda(\epsilon, y)$ of (4.1.14) shown to exist in (4.1.12). We wish to construct approximations to solutions of the equation

(4.1.19) $$F(\beta, x, y) = (\beta I - L)x - T(\beta, x, y) = 0 \qquad \beta = \lambda^{-1},$$

near a simple eigenvector ϵu_0 and eigenvalue β_0 of the linearized equation $(\beta I - L)x = 0$. Assuming ϵ sufficiently small and $\|y\| = O(\epsilon^2)$, we find $\|x(\epsilon, y) - \epsilon u_0\| = O(\epsilon^2)$ and $|\beta(\epsilon, y) - \beta_0| = O(\epsilon)$.

To this end we consider the following iteration scheme (I_N) for a sequence $\{x_N = \epsilon u_0 + v_N, \beta_N\}$: As the initial approximation we take $v_0 = 0$ and $\beta = \beta_0$. Then we compute v_1 and β_1 successively as follows:

(I_1) $$(\beta_0 I - L)v_1 = P^* T(\beta_0, \epsilon u_0, y),$$
$$(\beta_1 I - L)(\epsilon u_0) = PT(\beta_0, \epsilon u_0 + v_1, y).$$

where P^* and P are projection operators. More generally, given v_N and β_N, we compute (v_{N+1}, β_{N+1}) successively by the formulas

(I_{N+1}) $$(\beta_N I - L)v_{N+1} = P^* T(\beta_N, x_N, y),$$
$$(\beta_{N+1} I - L)(\epsilon u_0) = PT(\beta_N, x_{N+1}, y),$$

provided of course that $\beta_{N+1} I - L$ is invertible on Range$(\beta_0 I - L)$.

We shall state and prove a result on the existence and convergence of this scheme. Note that, if the scheme does converge to $\bar{x}(\epsilon) = \epsilon u_0 + \bar{v}$ and $\bar{\beta}$, then $(\bar{x}(\epsilon), \bar{\beta}(\epsilon))$ will satisfy (4.1.19). In addition, if one shows that $\|x_N - \epsilon u_0\| \leqslant K\epsilon^2$ and $\|\beta_N - \beta_0\| \leqslant K\epsilon$, where K is a constant independent of N and ϵ, then this solution will coincide with the solution described in (4.1.12). To this end, we suppose:
$(*)$ the linear operator $\beta_0 I - L$ is Fredholm and dim Ker$(\beta_0 I - L) = 1$.

(4.1.20) **Theorem** Suppose the operator $\beta_0 I - L$ satisfies the condition (*) above, while the operator $F(\beta, x, y)$ satisfies the following estimates for $\|x\|$, $\|y\|$, $|\beta - \beta_0|$ sufficiently small:

$$\text{(a)} \quad \|T(\beta, x, y) - T(\beta, x', y)\| \leqslant M\{\|x\| + \|x'\|\}\|x - x'\|,$$

$$\text{(b)} \quad \|T(\beta, x, y) - T(\beta', x, y)\| \leqslant M\{\|x\|^2 + \|y\|\}|\beta - \beta'|,$$

$$\text{(c)} \quad \|T(\beta_0, 0, y)\| \leqslant M\|y\|,$$

where M is a constant depending only on F. Then for sufficiently small and fixed y, the iteration scheme I_1–I_{N+1} exists for each N and converges to $(\bar{x}, (\epsilon, y), \bar{\beta}(\epsilon, y))$. Moreover, $\bar{x}(\epsilon, y)$ and $\bar{\beta}(\epsilon, y)$ are continuous in ϵ and y, satisfy the equation (4.1.19), and $\|\bar{x}(\epsilon, y) - \epsilon u_0\| \leqslant O(|\epsilon|^2), \|\bar{\beta}(\epsilon, y) - \beta_0\| = O(|\epsilon|)$.

Proof: For fixed ϵ and y, consider the real-valued function $\alpha(\beta, v)$ and the operator $\bar{T}(\beta, v)$ defined successively by the formulas

$$\bar{T}(\beta, v) = (\beta I - L)^{-1} P^* F(\beta, \epsilon u_0 + v, y), \tag{4.1.21}$$

$$\alpha(\beta, v) = \beta_0 + \epsilon^{-1} PF(\beta, \epsilon u_0 + T(\beta, v), y), \tag{4.1.22}$$

where P is the projection of $X \to \{u_0\}$. We shall determine positive numbers K and ϵ_0 such that the mapping $(T(\beta, v), \alpha(\beta, v))$ exists and defines a contraction map of the set $S_{K,\epsilon} = \{(\beta, v) \mid |\beta - \beta_0| \leqslant K|\epsilon|, \|v\| \leqslant K^2\epsilon\}$ into itself for $|\epsilon| \leqslant \epsilon_0$ with respect to the norm on $S_{K,\epsilon}$ defined by $|||(\beta, v)||| = |\beta| + \|v\|$. Then the contraction mapping theorem implies that the sequence $(\beta_{N+1}, v_{N+1}) = (\alpha(\beta_N, v_N), \bar{T}(\beta_N, v_N))$ converges to a unique fixed point in $s_{K,\epsilon}$ which we denote by $(\bar{\beta}(\epsilon, y), \bar{v}(\epsilon, y))$. Clearly this point satisfies (4.1.19) with $\bar{x} = \epsilon u_0 + \bar{v}$. We also prove the convergence is uniform in ϵ for $|\epsilon| \leqslant \epsilon_0$, so that $\bar{\beta}(\epsilon, y)$ and $\bar{x}(\epsilon, y)$ are continuous functions of ϵ since β_N and v_N are.

To carry through the proof, we first note that for any fixed number K and ϵ sufficiently small, $(\beta, v) \epsilon S_{K,\epsilon}$, and the hypotheses of the theorem imply that $\|T(\beta, \epsilon u_0 + v, \epsilon^2 y)\| \leqslant K_1\epsilon^2$, where K_1 is a constant independent of $\beta, v \in S_{K,\epsilon}$ and y. Indeed, setting $x = \epsilon u_0 + v$ throughout,

$$\begin{aligned} \|T(\beta, x, z)\| &\leqslant \|T(\beta, x, y) - T(\beta, 0, y)\| + \|T(\beta, 0, y) - T(\beta_0, 0, y)\| + \|T(\beta_0, 0, y) \\ &\leqslant M(|\epsilon| + \|v\|)^2 + M\|y\|(\beta - \beta_0) + M\|y\| \\ &\leqslant (2M + CM)\epsilon^2. \end{aligned}$$

Furthermore, for any two invertible linear operators $\mathfrak{L}_1$, $\mathfrak{L}_2$ we have $\mathfrak{L}_1^{-1} - \mathfrak{L}_2^{-1} = \mathfrak{L}_2^{-1}(\mathfrak{L}_1 - \mathfrak{L}_2)\mathfrak{L}_1^{-1}$, and since $\mathfrak{L}_1 = \beta_0 I - L$ is invertible on Range$(\beta_0 I - L)$ so is $\mathfrak{L}_2 = \beta I - L$ for $(\beta, v) \in S_{K,\epsilon}$ with ϵ sufficiently small, and in fact we may assume

$$\|(\beta I - L)^{-1}\| \leqslant 2\|(\beta_0 I - L)^{-1}\|.$$

Now we determine the sphere $S_{K,\epsilon}$ such that $\mathcal{G}(\beta, v) = (T(\beta, v), \alpha(\beta, v))$ maps $S_{K,\epsilon} \to S_{K,\epsilon}$. Let $\|P\|$, $\|P^*\|$ be denoted c_P, c_{P^*}, respectively, then for ϵ sufficiently small

$$\begin{aligned} \|\bar{T}(\beta, v)\| &\leqslant \|(\beta I - L)^{-1}\| \, \|P^*\| \, \|F(\beta, \epsilon u_0 + v, y)\| \\ &\leqslant 2\|(\beta_0 I - L)^{-1}\| c_{P^*}(2M + Mc)\epsilon^2 \end{aligned} \tag{4.1.23}$$

(4.1.24) $$|\alpha(\beta, v) - \beta_0| \leqslant \epsilon^{-1}\|P\| \, \|F(\beta, \epsilon n_0 + T(\beta, v), y)\|$$
$$\leqslant \epsilon^{-1}c_P M\big(\epsilon + \|\bar{T}(\beta, v)\|\big)^2 \leqslant 2c_P M\epsilon.$$

Since the estimates (4.1.23) and (4.1.24) are independent of K, we may always choose K sufficiently large and ϵ sufficiently small so that $\mathcal{G}(\beta, v)$ maps $S_{K,\epsilon}$ into itself.

Finally, to demonstrate that $\mathcal{G}(\beta, v)$ is a contraction on $S_{K,\epsilon}$, it suffices to prove that for (β, v) and $(\beta', v') \in S_{K,\epsilon}$,

(4.1.25) $$\|\bar{T}(\beta', v') - \bar{T}(\beta, v)\| \leqslant g_1(\epsilon)\|v - v'\| + g_2(\epsilon)|\beta - \beta'|,$$

(4.1.26) $$|\alpha(\beta, v) - \alpha(\beta', v')| \leqslant g_3(\epsilon)\|v - v'\| + g_4(\epsilon)|\beta - \beta'|,$$

where $g_i(\epsilon) = O(\epsilon)$ as $\epsilon \to 0$ $(i = 1, \ldots, 4)$. To prove (4.1.25), we note that by hypotheses (a)–(c) of the theorem,

(4.1.27) $$\|\bar{T}(\beta, v) - \bar{T}(\beta, v')\| \leqslant \|(\beta I - L)^{-1}\| \, \|P^*\| \, \|F(\beta, \epsilon u_0 + v, \epsilon^2 y)$$
$$- F(\beta, \epsilon u_0 + v', \epsilon^2 y)\|$$
$$\leqslant 2(\beta_0 I - L)^{-1}\|c_{P^*}M(2|\epsilon| + \|v\| + \|v'\|)\|v - v'\|.$$

On the other hand, since $\|(\beta I - L)^{-1} - (\beta_0' I - L)^{-1}\| \leqslant 4|\beta - \beta'| \, \|(\beta_0 I - L)^{-1}\|^2$,

(4.1.28) $$\|\bar{T}(\beta, v') - \bar{T}(\beta', v')\| \leqslant \|(\beta I - L)^{-1}$$
$$- (\beta' I - L)^{-1}\| \, \|P^*\| \, \|T(\beta, \epsilon u_0 + v', \epsilon^2 y)\|$$
$$+ \|(\beta' I - L)^{-1}\| \, \|P^*\| \, \|T(\beta, \epsilon u_0 + v', \epsilon^2 y)$$
$$- T(\beta', \epsilon u_0 + v, \epsilon^2 y)\|$$
$$\leqslant 4K\|(\beta_0 I - L)^{-1}\|^2$$
$$+ 2M\|(\beta_0 I - L)^{-1}\|(1 + K)^2\|y\|c_{P^*}\epsilon^2|\beta - \beta'|.$$

Combining (4.1.27) and (4.1.28) above, we obtain (4.1.25). To prove (4.1.26), we use (4.1.27) and estimate differences exactly as above, so that the details may be omitted.

To complete the proof we show that the sequence $(\beta_{N+1}(\epsilon, y), v_{N+1}(\epsilon, y)) = (\alpha(\beta_N, v_N), \bar{T}(\beta_N, v_N))$ converges uniformly for $|\epsilon|$ sufficiently small, so that $(\bar{\beta}, \bar{x})$ will depend continuously on ϵ. Indeed, setting $(\beta_{N+1}(\epsilon, y), v_{N+1}(\epsilon, y)) = \mathcal{G}(\beta_N, v_N)$, a straightforward induction shows that

$$|||\mathcal{G}(\beta_i, v_i) - \mathcal{G}(\beta_{i-1}, v_{i-1})||| \leqslant (\tfrac{1}{2})^i|||\mathcal{G}(\beta_0, 0) - (\beta_0, 0)||| \leqslant 2K\epsilon_0(\tfrac{1}{2})^i.$$

Hence for any integers n, m,

(4.1.29) $$|||\mathcal{G}(\beta_m, v_m) - \mathcal{G}(\beta_n, v_n)||| \leqslant \sum_{i=n}^{m-1} |||\mathcal{G}(\beta_{i+1}, v_{i+1}) - \mathcal{G}(\beta_i, v_i)|||$$
$$\leqslant 2K\epsilon_0 \sum_{i=n}^{\infty} 2^{-i},$$

so that as $m, n \to \infty$, the above term tends to zero. Hence we conclude that $\{\mathcal{G}(\beta_N, v_N)\}$ is a uniformly convergent sequence of bounded continuous functions on $\mathbb{R}^1 \times X$, and hence the limit function $(\bar{\beta}(\epsilon, y), \bar{v}(\epsilon, y))$ is continuous in ϵ with respect to the appropriate norm. A similar argument yields continuity in y.

4.1E The case of higher multiplicity

We now turn to a more difficult bifurcation problem, in which $\dim \operatorname{Ker} f_x(x_0, \lambda_0) > 1$ and $f_x(x_0, \lambda_0)$ is a linear Fredholm operator of index zero. For simplicity, we suppose that $x_0 = 0$ and

(4.1.30) $$f(x, \lambda) = (I - \lambda L)x + T(x, \lambda) \quad \text{with} \quad T(0, \lambda) = T_x(0, \lambda) = 0.$$

In this case, since $f_x(0, \lambda) = I - \lambda L$, the only possible points of bifurcation $(0, \lambda_0)$ occur at real numbers λ_0 with λ_0^{-1} in the spectrum of L. However, in this case $(0, \lambda_0)$ need not be a point of bifurcation as simple examples show. The hypothesis that $I - \lambda_0 L$ has index zero implies, by (1.3.38), that if λ_0^{-1} is an eigenvalue of multiplicity $N \geq 1$ for L, then

$$\operatorname{Ker}(I - \lambda_0 L) \cap \operatorname{Range}(I - \lambda_0 L) = \varnothing$$

$$\operatorname{Ker}(I - \lambda_0 L) \oplus \operatorname{Range}(I - \lambda_0 L) = X.$$

Since $\dim \operatorname{Ker}(I - \lambda_0 L) = N > 1$ here, for fixed λ, the bifurcation equations involve the determination of real solutions of several equations in as many real variables. Thus these equations may have real nonzero solutions and in fact the solutions of these equations depend essentially on the nature of the higher order term $T(x, \lambda)$. We shall discuss briefly those results that can be proven independently of any assumptions on $T(x, \lambda)$ or the parity of N.

We first show that if (x, λ) is a solution of $f(x, \lambda) = 0$ near $(0, \lambda_0)$, then λ can be uniquely determined in terms of the component of x on $\operatorname{Ker}(I - \lambda_0 L)$. More precisely, we prove

(4.1.31) **Theorem** Let (x, λ) be a solution of $f(x, \lambda) = 0$ sufficiently near $(0, \lambda_0)$. If $x = u + v$ with $u \in \operatorname{Ker}(I - \lambda_0 L)$, $v \in \operatorname{Range}(I - \lambda_0 L)$, then there is a unique function $\lambda = g(u)$ such that:

(i) $f(u + v, g(u)) = 0$,
(ii) $g(u)$ is C^1 in a deleted neighborhood of $u = 0$,
(iii) if $f(u + v', \lambda') = 0$ and $f(u + v, \lambda) = 0$, then $\lambda = \lambda'$ and $v = v'$.

Proof: By virtue of (4.1.8), with $\beta = \lambda^{-1}$ and $u \in \operatorname{Ker}(\beta_0 I - L)$, the bifurcation equations can be written as

(*) $$(\beta I - L)u + PT(u + g(u, \beta), \beta) = 0.$$

Here $g(u, \beta)$ is the function defined in (4.1.5). Let $[u, \bar{u}]$ denote any linear product on $\operatorname{Ker}(\beta_0 I - L)$, then taking the inner product of (*) with u, we find

(**) $$\beta - \beta_0 + [u, u]^{-1}[PT(u + g(u, \beta), \beta), u] = 0.$$

Let the left-hand side of this equation be denoted $F(u, \beta)$. Then we shall show that $F_\beta(u, \beta) \neq 0$ in a small neighborhood of $(0, \beta_0)$, so that by the one-dimensional implicit function theorem, there is a uniquely defined function $\beta = g(u)$ defined near $(0, \beta_0)$ that satisfies (**). Indeed,

$$F_\beta(u, \beta) = 1 + [u, u]^{-1}[PT_x(u + g(u, \beta), \beta)g_\beta + PT_\beta(u + g(u, \beta), \beta), u]$$
$$= 1 + o(1) \quad \text{as} \quad \|u\| + |\beta - \beta_0| \to 0.$$

Thus the conclusions (i) and (ii) of the theorem follow immediately. To prove (iii), note that by the above, if (x, λ) and (x_1, λ_1) satisfy (*) near $(0, \lambda_0)$, with $x = u + f(\lambda, u)$ and $x_1 = u + f(\lambda_1, u)$, then (λ^{-1}, u) and (λ_1^{-1}, u) satisfy (**). Consequently, $\lambda = \lambda_1$, and so $v = v'$ by (4.1.5).

Definite results on bifurcation in the higher dimensional case are important for applications to concrete problems in analysis. Thus our proof of the Liapunov criterion for the periodic solutions of the system (4.1.3) shows that if the "Liapunov irrationality conditions" are violated, bifurcation results on higher multiplicity can be used in studying the preservation of normal modes under nonlinear perturbations. On the other hand, our discussion of the system (1.2.9) given in Chapter 1 illustrates the difficulty of making general statements about the existence of bifurcation points in the case of even multiplicity. Indeed, in the sequel we shall show that *qualitative properties* of the higher order part of $f(x, \lambda)$ often play a crucial role in answering such general bifurcation questions.

As a second application of the bifurcation equations (4.1.8), we consider the problem of determining sub- and supercritical bifurcation of solutions for the operator equation

(4.1.32) $\quad x = \lambda(Lx + Nx)$

defined on a real Hilbert space H, at the point $(0, \lambda_0)$. Here we suppose that $\|Nx\| = o(\|x\|)$ as $\|x\| \to 0$ and that $Nx = Bx + Rx$, where B is a homogeneous operator of order p in x with $(Bx, x) \neq 0$ for $x \neq 0$, while $\|R(x)\| = o(\|x\|^p)$ as $x \to 0$. If $(0, \lambda_0)$ is a point of bifurcation for (4.1.32) and (4.1.32) admits nontrivial solutions (x, λ) in an arbitrarily small neighborhood of λ_0 with $\lambda < \lambda_0$ $(\lambda > \lambda_0)$, then the family (x, λ) is called a *subcritical (supercritical)* family of solutions. (See Figure 4.3) Clearly, it is important to determine sub- or supercritical bifurcation from an inspection of the structure of (4.1.32) but independently of multiplicity considerations. This can be accomplished in a large variety of cases by means of

(4.1.33) **Theorem** If in addition to the above hypotheses, the range of $I - \lambda_0 L$ is closed, $\dim \operatorname{Ker}(I - \lambda_0 L) < \infty$ and $H = \operatorname{Ker}(I - \lambda_0 L) \oplus \operatorname{Range}(I - \lambda_0 L)$, then any family of solutions of (4.1.32) bifurcating from $(0, \lambda_0)$ will be subcritical if $(Bx, x) > 0$ and supercritical if $(Bx, x) < 0$.

Proof: Again the result follows from the bifurcation equation (4.1.8) and the estimate of (4.1.9). The key point is that under the given hypothesis knowledge of the actual solution of the bifurcation equations is unnecessary. Indeed, using the notation of (4.1.5), the bifurcation equations (4.1.8) can be written in the form

(4.1.34) $\quad P[(\lambda - \lambda_0)Lx + \lambda Nx] = 0,$

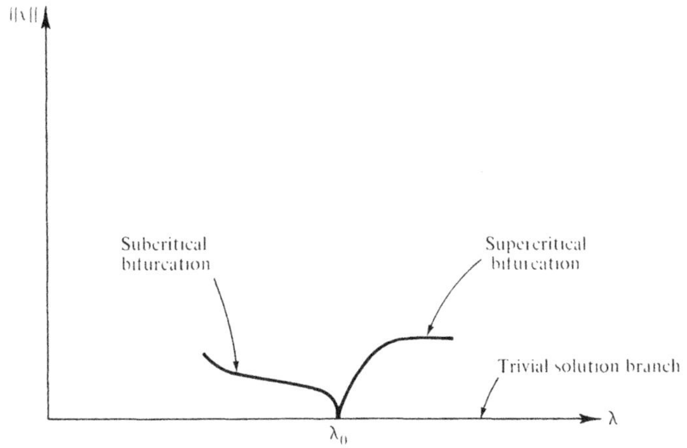

FIG. 4.3 Illustration of subcritical and supercritical bifurcation with a bifurcation diagram.

where P is the projection of $H \to \mathrm{Ker}(I - \lambda_0 L)$ since under the given hypotheses $\mathrm{Ker}(I - \lambda_0 L) \equiv \mathrm{coker}(I - \lambda_0 L)$. Thus, if $x = \rho_1 + g$ with $\rho_1 \in \mathrm{Ker}(I - \lambda_0 L)$, $g \in \mathrm{Range}(I - \lambda_0 L)$, then (4.1.34) becomes

$$[(\lambda_0 - \lambda)/\lambda\lambda_0]\rho_1 = P[B(\rho_1 + g) + R(\rho_1 + g)].$$

Consequently, taking inner products with ρ_1 and using (4.1.9), we find

$$(\lambda_0 - \lambda)/\lambda_0\lambda\|\rho_1\|^2 = (B\rho_1, \rho_1)\{1 + O(\|\rho_1\|)\}.$$

Setting $\rho_1 = |\epsilon|\rho$ with $\|\rho\| = 1$, and suppose $(B\rho, \rho) > 0$ so that $\inf(B\rho, \rho)$ over all $\rho \in \mathrm{Ker}(I - \lambda_0 L)$ with $\|\rho\| = 1$ is greater than some $\alpha > 0$, we find that

$$(\lambda_0 - \lambda)/\lambda\lambda_0 \geqslant |\epsilon|^{p-1}\alpha\{1 + O(|\epsilon|)\}.$$

Thus for λ near λ_0, the result follows immediately from this last equation for the case of subcritical bifurcation. A similar argument if $(B\rho, \rho) < 0$ yields the result on supercritical bifurcation.

4.2 Transcendental Methods in Bifurcation Theory

4.2A Heuristics

It is an interesting fact that significant results in previously unsolved bifurcation problems have been achieved by utilizing results from topology, complex analysis, and critical point theory. This is especially true in the difficult "degenerate" cases (i.e., higher multiplicity) mentioned in Section 4.1E. In this section, we shall explore this topic relative to the equation

(4.2.1) $$f(x, \lambda) \equiv (I - \lambda L)x + T(x, \lambda) = 0,$$

where the operators $I - \lambda L$ and $T(x, \lambda)$ satisfy the hypotheses of Section

4.1, i.e., $I-\lambda L$ is a Fredholm operator of index zero, while $T(0, \lambda) = T_x(0, \lambda) \equiv 0$ so $T(x, \lambda)$ is of higher order in x.

Roughly speaking, the success of these so-called transcendental methods is based on either a qualitative analysis of the higher order term $T(x, \lambda)$ in the equation (4.2.1) or parity considerations of the multiplicity of the derivative $f'(0, \lambda_0)$ at a critical point $(0, \lambda_0)$ of $f(x, \lambda)$. One attempts to distinguish an appropriate "invariant" I_f of a numerical or algebraic nature, for a given operator $f(x, \lambda)$. The invariant I_f is required to have the following properties:

(1) I_f is a measure of the zeros of the operator $f(x, \lambda)$;

(2) I_f is stable under "small" suitably restricted perturbations of $f(x, \lambda)$;

(3) I_f can be approximated by linearization, i.e., by means of the Frechet derivative f_x of f.

We now consider some examples of such invariants and their role in bifurcation theory. We shall use the results of Section 4.1 to reduce the bifurcation problems to finite-dimensional considerations. Thus (apart from Section 4.2C) we shall need only those (topological) invariants I_f for mappings f between finite-dimensional spaces, and we reserve for Part III those results that can be obtained by strictly infinite-dimensional arguments.

More precisely, the reduction to the finite-dimensional problem is based on decomposing (4.2.1) relative to the direct sum decomposition $X = \operatorname{Ker}(I-\lambda_0 L) \oplus \operatorname{Range}(I-\lambda_0 L)$ mentioned in Section 4.1. If P denotes the canonical projection of X onto $\operatorname{Ker}(I-\lambda_0 L)$ and $g(u, \lambda)$ denotes the function $\operatorname{Ker}(I-\lambda_0 I) \times \mathbb{R}^1 \to \operatorname{Range}(I-\lambda_0 L)$ defined by (4.1.8), then as mentioned in (4.1.5), the solution of $f(x, \lambda) = 0$ near a point $(0, \lambda_0)$ at which $f_x(0, \lambda_0)$ is not invertible are in one-to-one correspondence with the solutions of the equation

$$(4.2.2) \qquad (I-\lambda L)u + PT(u + g(u, \lambda), \lambda) = 0, \qquad u \in \operatorname{Ker}(I-\lambda_0 L).$$

4.2B Brouwer degree in bifurcation theory

The degree $d(f, p, D)$ of a continuously differentiable mapping f, with domain a bounded domain $D \subset \mathbb{R}^N$ and range $f(D) = \mathbb{R}^N$, was defined in Section 1.6C and is an invariant of the type we are seeking. Recall that the degree is an integer (positive, negative, or zero) that measures the "algebraic" number of solutions of $f(x) = p$ in D, provided that $f(x) \neq p$ on ∂D.

First, we shall indicate the sense in which the degree $d(f, p, D)$ satisfies properties (1)–(3) of Section 4.2A, and thus qualifies as an integer invariant

I_f. By virtue of its definition in Section 1.6A, $d(f, p, D)$ measures the number of solutions of $f(x) = p$ in D by counting "nondegenerate" solutions of $f(x) = p$ with a + or - depending on the orientation preserving or reversing properties of f at the solution. Here, the solutions of $f(x) = p$ in D are nondegenerate if the Jacobian determinant is not zero at each solution of $f(x) = p$ in D. Furthermore, $d(f, p, D)$ is a homotopy invariant and thus stable under small perturbations in the sense that, if $f(x, t)$ is a continuous map of $\bar{D} \times [0, 1] \to \mathbb{R}^N$ satisfying $f(x, t) \neq p$ on ∂D, then $d(f(x, t), p, D)$ is defined and independent of $t \in [0, 1]$. Finally, in the nondegenerate case,

$$d(f, p, D) = \sum_{f(x)=p} \operatorname{sgn} \det |J_f(x)|,$$

so that the derivative of f determines the "algebraic" count of the solutions of $f(x) = p$ in D.

We now use the fact that $d(f, p, D)$ is an invariant to prove

(4.2.3) **Theorem** Suppose that the operator $f(x, \lambda)$ defined in (4.2.1) satisfies the following hypotheses:

(i) $(I - \lambda_0 L)$ is a Fredholm operator of index zero.
(ii) dim Ker$(I - \lambda_0 L)$ is odd.
(iii) Ker$(I - \lambda_0 L) \cap$ Range$(I - \lambda_0 L) = 0$.
(iv) $T(x, \lambda)$ is a C^1 mapping with $T(0, \lambda) \equiv T_x(0, \lambda) \equiv 0$.

Then $(0, \lambda_0)$ is a point of bifurcation for the equation $f(x, \lambda) = 0$.

Proof: We give a proof by contradiction, by supposing that $(0, \lambda_0)$ is not a point of bifurcation for (4.2.1). Consequently, $(0, \lambda_0)$ is not a point of bifurcation for the equation (4.2.2). Let $h(u, \lambda) = (I - \lambda L)u + PT(u + g(u, \lambda), \lambda)$ and $N = \dim \operatorname{Ker}(I - \lambda_0 L)$. Then for (u, λ) in a small spherical neighborhood D of $(0, \lambda_0)$, $h(u, \lambda) \neq 0$ on ∂D, so $d(h(u, \lambda), 0, D)$ is well defined. In fact, $d(h(u, \lambda), 0, D)$ is necessarily a constant independent of λ, by virtue of the homotopy invariance of the degree function. On the other hand, for λ in a sufficiently small deleted neighborhood of λ_0, since $|h(u, \lambda) - (I - \lambda L)u|$ is small on ∂D, $d(h(u, \lambda), 0, D) = d(I - \lambda L, 0, D)$. A simple calculation of det $|I - \lambda L|$ on Ker$(I - \lambda_0 L)$ now shows that for $\lambda < \lambda_0$,

$$d(I - \lambda L, 0, D) = \operatorname{sgn} \prod_{i=1}^{N} (1 - \lambda\lambda_0^{-1}) > 0;$$

while for $\lambda > \lambda_0$, since dim Ker$(I - \lambda_0 L) = N$ is odd,

$$d(I - \lambda L, 0, D) = \operatorname{sgn} \prod_{i=1}^{N} (1 - \lambda\lambda_0^{-1}) < 0.$$

Thus $d(u, \lambda), 0, D)$ is not constant for λ in any small neighborhood of λ_0;

and we obtain the desired contradiction since the given $\epsilon > 0$, $h[u, t(\lambda_0 - \epsilon) + (1-t)(\lambda_0 + \epsilon)] = 0$ must have a solution on $\{\|u\| = \rho\}$ for every $\rho > 0$ sufficiently small.

Actually a slightly more general result involving the multiplicity of the eigenvalue λ_0 for L instead of dim $\text{Ker}(I - \lambda_0 L)$ can be proved. This result can be stated.

(4.2.3′) The conclusion of Theorem (4.2.3) is valid if hypotheses (ii), (iii) are replaced by the assumption that the multiplicity of λ_0 relative to L is odd.

Proof: The result follows by decomposing the equation $f(x, \lambda) = 0$ defined on X by two equations as in the proof of (4.2.3) but in this case we use the decomposition $X = \bigcup_j \text{Ker}(I - \lambda_0 L)^j \oplus X_1$. By assumption dim $\bigcup_j \text{Ker}(I - \lambda_0 L)^j$ (the multiplicity of λ_0) is odd, while $I - \lambda_0 L$ is invertible when restricted to X_1. Consequently, an argument similar to the proof of (4.2.3) (utilizing the Brouwer degree) yields the desired result.

The results just obtained can be considerably sharpened by placing qualitative restrictions on the higher order term $T(x, \lambda)$. Indeed, suppose that for each λ near λ_0, $T(x, \lambda)$ is a complex analytic mapping of a neighborhood U of $(0, \lambda_0)$ of X in the sense of Section 2.3. In this case we prove the following

(4.2.4) **Theorem** Suppose the operator $f(x, \lambda)$ defined in (4.2.1) is complex analytic for fixed real λ near λ_0, and in addition

(i) $I - \lambda_0 L$ is a Fredholm operator of index zero;

(ii) dim $\text{Ker}(I - \lambda_0 L) > 0$, while $\text{Ker}(I - \lambda_0 L) \cap \text{Range}(I - \lambda_0 L) = \{0\}$.

Then $(0, \lambda_0)$ is a point of bifurcation for the equation $f(x, \lambda) = 0$. In fact, there is an analytic curve

$$x(\epsilon) = \sum_{n=1}^{\infty} \alpha_n \epsilon^n, \qquad \lambda(\epsilon) = \lambda_0 + \sum_{n=1}^{\infty} \beta_n \epsilon^n,$$

of nontrivial solutions of $f(x, \lambda) = 0$ branching from $(0, \lambda_0)$.

Proof: Following the proof of (4.2.3), suppose $(0, \lambda_0)$ is not a point of bifurcation relative to the equation $f(x, \lambda) = 0$. Then $(0, \lambda_0)$ is not a point of bifurcation for the equation $h(u, \lambda) = 0$. Now $\text{Ker}(I - \lambda_0 L)$ has even dimension. Furthermore, the map $h(u, \lambda) = (I - \lambda_0 L)u + PT(u + g(u, \lambda), \lambda)$ is complex analytic in u for fixed λ since both $T(x, \lambda)$ and $g(u, \lambda)$ are. The complex analyticity of $g(u, \lambda)$ follows from 3.3.2 since it was defined by means of the implicit function theorem. Again (by the homotopy invariance of degree (1.6.3)) for λ in a small deleted neighborhood of $[0, \lambda_0]$ and U a small spherical neighborhood of 0 in $\mathrm{Ker}(I - \lambda_0 L)$, $d(h(u, \lambda), 0, U) = d(I - \lambda L, 0, U) = 1$. On the other hand, at $\lambda = \lambda_0$, $d(h(u, \lambda), 0, U)$ is defined since $h(u, \lambda) = 0$ is assumed not to have solutions on the boundary of U. But at $\lambda = \lambda_0$, the Jacobian determinant of $h(0, \lambda_0)$, $\det(I - \lambda_0 L)$, is zero so that the fundamental result on complex analytic mappings mentioned in (1.6.3) implies that $h(u, \lambda_0)$ is not one-to-one in U. Hence $d(h(u, \lambda_0), 0, U) \geqslant 2$. Thus the function $d(h(u, \lambda), 0, U)$ is discontinuous across $\lambda = \lambda_0$, and as a consequence of the homotopy invariance of degree, this implies that $h(u, \lambda) = 0$ has a solution on the boundary of U for λ in some small interval about λ_0. This is the desired contradiction.

To prove the existence of an analytic curve of solutions branching from $(0, \lambda)$, we note that since $(0, \lambda_0)$ is a point of bifurcation from the equation $h(u, \lambda) = 0$, the point $(0, \lambda_0)$ is not an isolated point of the variety $V = \{(u, \lambda) \mid h(u, \lambda) = 0\}$. Since $h(u, \lambda)$ is complex analytic in u and λ, V can be regarded as an analytic set near $(0, \lambda_0)$. Thus V must contain an analytic curve (cf. 3.3.9).

$$u(\epsilon) = \sum_{n=1}^{\infty} a_n \epsilon^n, \qquad \lambda(\epsilon) = \lambda_0 + \sum_{n=1}^{\infty} \beta_n \epsilon^n,$$

Consequently, $x(\epsilon) = u(\epsilon) + g(u(\epsilon), \lambda(\epsilon))$ can also be written as $x(\epsilon) = \sum_{n=1}^{\infty} \alpha_n \epsilon^n$, and so the theorem is proven.

4.2C Elementary critical point theory

Let $f(x) = \operatorname{grad} F(x)$ be a gradient operator defined on the ball $\|x\| \leqslant R$ in a Hilbert space H. Recall from Section 3.2 that a critical point $\bar{x}$ of $F(x)$ restricted to the sphere $\|x\| = R$ satisfies the equation $\lambda_1 x + \lambda_2 f(x) = 0$, where λ_1 and λ_2 are real numbers (with $\lambda_2 \neq 0$). The associated critical value is the real number $c = F(\bar{x})$. It is an interesting fact that in several important instances, certain critical values are numerical invariants of the type mentioned in Section 4.2A.

The simplest example of such an invariant critical value is the supremum of a C^2 weakly sequentially continuous functional $F(x)$, which near the origin has the form $F(x) = \frac{1}{2}(Ax, x) + o(\|x\|^2)$ when restricted to the small sphere $\partial\Sigma_\epsilon = \{x \mid \|x\| = \epsilon\}$. The linear operator A is compact, self-adjoint, and we shall suppose that A has a largest positive eigenvalue λ_1. We now show that the number, $\sup F(x)$ over $\partial\Sigma_\epsilon$, has properties (1)–(3) of an invariant (listed in Section 4.2A). First, if $\alpha = \sup F(x)$ over $\partial\Sigma_\epsilon$ is attained by $\bar{x} \in \partial\Sigma_R$, $\bar{x}$ will be a nontrivial solution of the equation $g(x, \mu) = \mu x - \operatorname{grad} F(x)$ for some real number μ, which we shall see must lie in a small neighborhood of λ_1. This is the sense in which this critical value measures the solutions of the equation $g(x, \lambda) = 0$. In order to show that α is stable under suitably restricted small perturbations, we note that if $F(x)$ is perturbed by the addition of a higher term $G(x) = o(\|x\|^2)$ near $x = 0$, then

$$|\sup_{\partial \Sigma_\epsilon}[F(x) + G(x)] - \alpha| = o(\|x\|^2).$$

In the same way, α can be calculated approximately by linearization since by the variational characterization of λ_1 (mentioned in (1.3.40)), $\epsilon^2\lambda_1 = \sup_{\partial\Sigma_\epsilon}(Ax, x)$ so that

(4.2.5) $$|\alpha - \tfrac{1}{2}\epsilon^2\lambda_1| = |\sup_{\partial \Sigma_\epsilon}\{\tfrac{1}{2}(Ax, x) + o(\|x\|^2)\} - \tfrac{1}{2}\epsilon^2\lambda_1| = o(\epsilon^2).$$

We now indicate the importance of this simple invariant in bifurcation theory. Suppose that the operator equation (4.2.1) defined in the neighborhood of the origin in a Hilbert space H can be written

(4.2.6) $$f(x, \lambda) = x - \lambda\{Lx + T(x)\} = 0.$$

where L is a compact self-adjoint operator and $T(x) = \operatorname{grad} \mathfrak{T}(x)$ is a higher order, completely continuous, operator with $T(x) = o(\|x\|)$. Then we prove

(4.2.7) **Theorem** Let λ_1 be the largest strictly positive eigenvalue of L. Then under the above hypotheses, $(0, 1/\lambda_1)$ is a point of bifurcation for (4.2.6).

Proof: We shall demonstrate that every sufficiently small spherical neighborhood $U_\epsilon = \{x \mid \|x\| \leqslant \epsilon\}$ contains a nontrivial solution $(x(\epsilon), \lambda(\epsilon))$ of (4.2.6) with $\|x(\epsilon)\| = \epsilon$ and $|\lambda(\epsilon) - 1/\lambda_1| = o(1)$. To this end, we make use of the facts concerning $\alpha_\epsilon = \sup\{\tfrac{1}{2}(Lx, x) + \mathfrak{T}(x)\}$ over $\partial\Sigma_\epsilon = \{x \mid \|x\| = \epsilon\}$ just described above. Assuming, for the moment, that α_ϵ is attained by an element $x(\epsilon)$ of $\partial\sigma_\epsilon$, we show that the equation satisfied by $x(\epsilon)$, namely, $\mu_\epsilon x = Lx + T(x)$, is such that $|\mu_\epsilon - \lambda_1| = o(1)$. Once this estimate is obtained

$$|\lambda(\epsilon) - 1/\lambda_1| = |1/\mu_\epsilon - 1/\lambda_1| = o(1) \qquad \text{as} \quad \epsilon \to 0,$$

as required. To this end, taking the inner product of (4.2.6) with x_ϵ we find that, on using (2.5.6),

$$\begin{aligned}
\mu_\epsilon &= \|x_\epsilon\|^{-2}\{(Lx_\epsilon, x_\epsilon) + (Tx_\epsilon, x_\epsilon)\} \\
&= 2\epsilon^{-2}\{(\tfrac{1}{2}(Lx_\epsilon, x_\epsilon) + \mathfrak{T}(x_\epsilon))\} + \epsilon^{-2}\{(Tx_\epsilon, x_\epsilon) - 2\mathfrak{T}(x_\epsilon)\} \\
&= 2\epsilon^{-2}\alpha_\epsilon + \epsilon^{-2}\|x_\epsilon\|\left\{\|Tx_\epsilon\| + 2\sup_{s\in[0,1]}\|T(sx_\epsilon)\|\right\}.
\end{aligned}$$

Thus by (4.2.5) above,

$$|\mu_\epsilon - \lambda_1| = 2\epsilon^{-2}\{\alpha_\epsilon - \tfrac{1}{2}\lambda_1\epsilon^2\} + \epsilon^{-1}o(\epsilon)$$
$$= o(1) \qquad \text{as} \quad \epsilon \to 0.$$

Thus there remains only to show that α_ϵ is attained on $\partial\Sigma_\epsilon$. This fact is immediate from the following argument. Clearly $\alpha_\epsilon < \infty$. Thus if $\{x_n\}$ is a sequence of elements on $\partial\Sigma_\epsilon$ with $F(x_n) = \frac{1}{2}(Lx_n, x_n) + \mathcal{T}(x_n) \to \alpha_\epsilon, \{x_n\}$ possesses a weakly convergent subsequence with weak limit $\bar{x}$. Since $F(x)$ is continuous with respect to weak convergence, $F(\bar{x}) = \alpha_\epsilon$. Now $\bar{x} \in \partial\Sigma_\epsilon$ since otherwise $\|\bar{x}\| < \epsilon$ and for some $t > 1$, $t\bar{x} \in \partial\Sigma_\epsilon$. A simple computation then shows that $F(t\bar{x}) > F(\bar{x})$, contradicting the maximality of $F(\bar{x})$. Thus the proof of the theorem is complete.

In Chapter 6, we shall show that critical values calculated by various minimax principles are numerical invariants satisfying properties (1)–(3) of Section 4.2A. These more subtle invariants also play an important role in bifurcation theory, as will be indicated in Section (6.7A.) These critical values are extensions of the following characterizations of the positive eigenvalues of a compact self-adjoint operator L, arranged in decreasing order and counted by multiplicity

$$\lambda_n = \sup_{[\Sigma]_N} \min_{\Sigma} (Lx, x),$$

where $\Sigma = \{x|\ \|x\| = 1,\ x \in \mathcal{C}_N$, an N-dimensional subspace of $H\}$ and $[\Sigma]_N$ is the class of all such spheres in H. The extension required consists in replacing the sets Σ and $[\Sigma]_N$ by a larger class of "topologically" similar sets.

In order to illustrate the importance of Theorem (4.2.7), we consider the following extension of Liapunov's criterion of Section 4.1A relative to the preservation of normal modes independent of irrationality conditions near a singular point for Hamiltonian systems of the form

(4.2.8) $\qquad \ddot{x} + Ax + \nabla f(x) = 0, \qquad f(x) = o(|x|).$

(4.2.9) **Theorem** Suppose that the $N \times N$ matrix A is a nonsingular self-adjoint matrix with the positive eigenvalues $0 < \lambda_1^2 \leqslant \lambda_2^2 \leqslant \dots \leqslant \lambda_k^2$ $(k \leqslant N)$, and suppose $f(x) = \nabla F(x)$, where $F(x)$ is a smooth function. Then (4.2.8) has a family of (nontrivial) periodic solutions $x(t)$ with minimal period $\tau_\epsilon(t)$ tending to $(0, 2\pi/\lambda_k)$. If $F(x)$ is real analytic, this family depends real analytically on ϵ and can be chosen so that

$$x_\epsilon(t) = \sum_{n=1}^{\infty} \alpha_n(t)\epsilon^n, \qquad \tau_\epsilon(t) = \frac{2\pi}{\lambda_k} + \sum_{n=1}^{\infty} \beta_n\epsilon^n,$$

where $\alpha_1(t)$ is $2\pi/\lambda_k$ periodic and satisfies $\ddot{x} + Ax = 0$ (i.e., the normal mode of smallest period is preserved under a Hamiltonian perturbation).

Idea of the proof: The argument is similar to the one used in (4.1.15) to establish Liapunov's criterion, with the exception that we reformulate (4.2.8) as an operator equation in an appropriate Hilbert space. Then the operators in this equation are gradient maps, so that (4.2.7) is applicable.

Proof: As a first step, we rewrite (4.2.8) as an operator equation in the Hilbert space H of N-vector functions $x(s) = (x_1(s), \ldots, x_N(s))$ such that $x_i(s)$ is absolutely continuous $(0, \frac{1}{2})$ and $\dot{x}_i(s) \in L_2(0, \frac{1}{2})$ for $i = 1, \ldots, N$. For then the solutions of such an equation can be extended to even periodic solutions of (4.2.8). Indeed, the solutions we construct will automatically satisfy the conditions $\dot{x}(0) = \dot{x}(\frac{1}{2}) = 0$. Following the duality method of Section 2.2D, this is easily accomplished by defining the operators $L(x)$, $A(x)$, and $T(x)$ implicitly by the formulas

(4.2.10) $$(Lx, y) = \int_0^{1/2} \dot{x}(s)\cdot\dot{y}(s)\,ds, \qquad (Ax, y) = \int_0^{1/2} Ax(s)\cdot y(s)\,ds$$
$$(T(x), y) = \int_0^{1/2} f(x(s))\cdot y(s)\,ds.$$

Thus the generalized even 1-periodic solutions of (4.2.8) are in one-to-one correspondence with the solutions in H of the operator equation

(4.2.11) $$Lx = \lambda^2[Ax + T(x)].$$

Clearly the linear operators L and A are bounded and self-adjoint, while A and T are compact with $\|T(x)\| = o(\|x\|)$ as $\|x\| \to 0$. Now an element $x(s)$ of H can be written uniquely as $x(s) = x_m + x_0(s)$, where x_m is the mean value of $x(s)$ over $(0, \frac{1}{2})$ and $x_0(s)$ has mean value zero over $(0, \frac{1}{2})$. This corresponds to an orthogonal decomposition of $H = \mathbb{R}^N \oplus H_0$. Since an inner product in H_0 can be chosen to be

$$(x_0, y_0)_{H_0} = \int_0^{1/2} \dot{x}_0(s)\cdot\dot{y}_0(s)\,ds,$$

equation (4.2.8) can be decomposed into the pair of equations

(4.2.12) $$x_0 = \lambda^2[Ax_0 + PT(x_0 + x_m)],$$

(4.2.13) $$0 = Ax_m + (I - P)T(x_0 + x_m),$$

where P is the projection of $H \to H_0$. Since A is invertible, the implicit function theorem applied to (4.2.13) yields the unique, differentiable function $x_m = g(x_0)$. Observe here that g does not depend on λ. Substituting into (4.2.12), this equation becomes

(4.2.14) $$x_0 = \lambda^2[Ax_0 + PT(x_0 + g(x_0))].$$

Now we can apply Theorem (4.2.7) to the operator equation, once we verify the facts that L is a compact self-adjoint operator in H_0, while $PT(x_0 + g(x_0))$ is a higher order *gradient* operator. Both these facts follow immediately from the definitions (4.1.2) and the facts of Section 2.5 concerning gradient operators. In fact, by (2.5.5) if

$$\mathcal{T}(x_0) = \int_0^{1/2} F(x_0(s) + g(x_0(s)))\,ds,$$

then $\mathcal{T}(x_0) = PT(x_0 + g(x_0))$ in H_0. Consequently, by utilizing the calculation of the spectrum of the operator $I - \lambda^2 L$ in Section 1.3, we find that the smallest positive number $\lambda_{\min}$ of the set $\{\lambda^2 \mid \lambda^2 = 4\pi^2N^2/\lambda_i^2\ (N = 1, 2, \ldots, i = 1, 2, \ldots, k)\}$ gives rise to a point of bifurcation $(0, \lambda_{\min})$ for the equation (4.1.4), *irrespective of any irrationality conditions holding among the eigenvalues* $\lambda_1^2, \lambda_2^2, \ldots, \lambda_k^2$. Since $\lambda_{\min} = 2\pi/\lambda_k$, we find a family of periodic solutions of (4.2.11) $x_\epsilon(s)$ with $\|x_\epsilon(s)\|^2_{H_0} = \epsilon$ such that $\lambda(\epsilon) \to 2\pi/\lambda_k$ as $\epsilon \to 0$. In terms of t, we obtain a family of periodic solutions $x_\epsilon(t)$ with period τ_ϵ satisfying $(x_\epsilon(t), \tau_\epsilon) \to (0, 2\pi/\lambda_k)$ as $\epsilon \to 0$. Consequently, if $F(x)$ is real analytic, the operator $T(x)$ will be, and so (by the real analogue of (1.6.2)) the equation (4.2.8) will have an analytic curve of solutions $(x_\epsilon(t))$ with period τ_ϵ with the properties stated in the theorem. Furthermore, τ_ϵ will be the minimal period of x_ϵ since otherwise the family $(x_\epsilon(s), \tau_\epsilon)$ would give rise to a point of bifurcation $(0, \tilde{\lambda})$ for (4.2.8) contradicting the minimality of the eigenvalue $2\pi/\lambda_k$.

4.2D Morse type numbers in bifurcation theory

The Morse type numbers $\{M_F(p)\}$ of an isolated critical point p of a real-valued C^2 function $F(x)$ of a finite number N of real variables were defined in Section 1.6C. $\{M_F(p)\}$ is an $(N+1)$-vector of finite positive integers $(m_0, m_1, \ldots, m_N)$ measuring the "multiplicity" of a (possibly degenerate) critical point p of $F(x)$. We now show that the Morse type numbers are also invariants relative to the equation (4.2.1) if we restrict attention to self-adjoint operators L and gradient operators $T(x, \lambda)$.

Again the discussion at the end of Section 4.2A implies that we need only verify the invariance properties of the Morse type numbers relative to finite-dimensional Hilbert spaces. Here we have made implicit use of a fact (implied by (2.5.5)) that the operator on the left-hand side of (4.2.2) is a gradient operator (for fixed λ and u near 0) since this property is preserved in the passage from (4.2.1) to (4.2.2). If we suppose that $\bar{x}$ is a nondegenerate critical point, then the Morse type numbers of $\bar{x}$ are easily calculated. Indeed, if the Morse index of $\bar{x}$ is k, then $M_F(\bar{x}) = (0, 0, \ldots, 1, 0, \ldots, 0)$, where 1 is the $(k+1)$th coordinate. The Morse index of a nondegenerate critical point $\bar{x}$, can be determined from the linear operator $F''(\bar{x})$; and consequently, at least in the nondegenerate case, the Morse type number can be calculated by linearization. Finally, the Morse type numbers are "stable" under "small perturbations." This fact is immediate if the isolated critical point $\bar{x}$ is nondegenerate. In the more difficult degenerate case, suppose that $\bar{x}$ is an isolated critical point, with Morse type numbers $M_F(\bar{x}) = (m_0, m_1, \ldots, m_N)$. By (1.6.14), if $G(x)$ is a C^2 function sufficiently close to $F(x)$ in the C^1 sense and possessing only nondegenerate critical points near $\bar{x}$, then $G(x)$ possesses at least m_k critical nondegenerate critical points of index k $(k = 0, 1, \ldots, N)$ near $\bar{x}$.

As an application of the Morse type numbers to bifurcation theory, we prove

(4.2.15) **Theorem** Suppose that $I - \lambda_0 L$ is a self-adjoint Fredholm operator that is noninvertible. Then, if $T(x, \lambda)$ is a C^1 higher order gradient operator, $(0, \lambda_0)$ is a point of bifurcation with respect to the equation (4.2.1).

Proof: Suppose $(0, \lambda_0)$ is not a point of bifurcation of (4.2.1). Then $(0, \lambda_0)$ is also not a point of bifurcation relative to (4.2.2). Thus for fixed λ in a small neighborhood of λ_0, $(0, \lambda)$ is an isolated critical point of the real-valued function $H(u, \lambda) = \frac{1}{2}(u - \lambda Lu, u) + \mathcal{T}(u + g(u, \lambda), \lambda)$, where $\mathcal{T}$ is the real-valued function satisfying $\mathcal{T}(0, \lambda) \equiv 0$ and $\mathcal{T}_u(u + g(u, \lambda), \lambda) = PT(u + g(u, \lambda), \lambda)$. Clearly for λ in a small deleted neighborhood of λ_0, $\det |H_{uu}(0, \lambda)| \neq 0$, so that $(0, \lambda)$ is an isolated nondegenerate critical point of $H(u, \lambda)$ for λ fixed. An easy computation of the Hessian $(H_{uu}(0, \lambda))$ shows that the Morse index of $(0, \lambda)$ relative to $H(u, \lambda)$ is 0 for $\lambda < \lambda_0$ and $\dim \operatorname{Ker}(I - \lambda_0 L) = N$ (say) for $\lambda > \lambda_0$. On the other hand, $(0, \lambda_0)$ is an isolated degenerate critical point of $H(u, \lambda_0)$ and suppose at least one of its type numbers is different from zero. By the invariance properties of Morse type numbers mentioned above, since $\sup \|H(u, \lambda) - H(u, \lambda_0)\| \to 0$ in the C^1 sense, as $(u, \lambda) \to (0, \lambda_0)$, the Morse type numbers of $H(u, \lambda)$ at $(0, \lambda)$, $M_\lambda(0)$, satisfy the inequalities $M_\lambda(0) \geqslant M_{\lambda_0}(0)$ in the coordinatewise sense, for each λ in a small neighborhood of λ_0. These inequalities imply that if $M_{\lambda_0}(0) = (m_0, m_1, m_2, \ldots, m_N)$, then for $\lambda < \lambda_0$, $m_0 = 1$ and $m_i = 0$ $(i = 1, \ldots, N)$, while for $\lambda > \lambda_0$, $m_N = 1$ and $m_i = 0$ $(i = 0, \ldots, N-1)$. Thus $m_i = 0$ for $i = 0, 1, 2, \ldots, N$. This is the desired contradiction.. The only remaining case to consider is the possibility that the Morse type numbers $M_{\lambda_0}(0)$ all vanish, but this case is ruled out by the relation (1.6.15) between the Brouwer degree and the Morse type numbers. Indeed, (1.6.15) implies $d(H_u(u, \lambda_0), 0, |u| < \epsilon) = 0$, *but this contradicts the fact that* $d(H_u(u, \lambda_0 - \delta), 0, |u| < \epsilon) = 1$ *for* δ *and* ϵ *sufficiently small since* $u = 0$ *is a relative minimum for* $H(u, \lambda_0 - \delta)$.

Theorem (4.2.15) is clearly a far-reaching extension of the analogous result in (4.2.7). Thus an immediate consequence related to periodic solutions of (4.2.8) near $x = 0$ is the following extension of (4.2.9).

(4.2.16) **Theorem** Under the same hypotheses as (4.2.9), the system (4.2.8) has nontrivial periodic solutions $x_i(t)$ $(i = 1, \ldots, k)$ in any sufficiently small neighborhood of $x = 0$, such that the period τ_i of $x_i(t)$ tends to $2\pi/\lambda_i$ as the diameter of U tends to zero. If the real-valued function $F(x)$ is real analytic, then (4.2.8) will have a real analytic curve of solutions $x_\epsilon(t)$ with periods $\tau_i(\epsilon)$ satisfying

$$x_\epsilon(t) = \sum_{n=1}^{\infty} \alpha_n(t)\epsilon^n \qquad \text{and} \qquad \tau_i(\epsilon) = \frac{2\pi}{\lambda_i} + \sum_{n=1}^{\infty} \beta_n \epsilon^n$$

Proof: The result is immediate by combining the arguments of (4.2.9) with the abstract result (4.2.15) above.

Remark: Weinstein (1974) has recently shown that the periodic solutions referred to in (4.2.16) are all distinct. In Chapter 6 we shall show the Ljusternik–Schnirelmann theory can be used in this connection (cf. Note F at the end of this chapter.)

4.3 Specific Bifurcation Phenomena

The size of certain critical parameters often governs the behavior of many aspects of the natural world. In many such circumstances, bifurcation phenomena play an important role in understanding. In the preceding two sections we illustrated this fact in terms of periodic motion near an equilibrium point of a system of "nonlinearly perturbed" harmonic oscillators. Here we pursue this topic in different areas of mathematical analysis in order to illustrate both the importance of the bifurcation theory developed in the previous sections and the problems involved in applying this theory to specific difficult cases. For simplicity, we have chosen our illustrations from quite well-established disciplines. Yet, despite their classical nature, each topic discussed below is filled with unsolved fundamental problems whose solution requires bifurcation theory.

4.3A Periodic motions near equilibrium points in the restricted three-body problem

The restricted three-body problem can be described as follows: Two particles P_1 and P_2 of mass ratios μ and $1-\mu$ move in circular orbits (under Newtonian attraction) around their center of mass. A third particle P_3 of negligible mass moves in the plane defined by the two revolving bodies. This particle P_3 is subject to the Newtonian attraction of P_1 and P_2, but is assumed not to disturb the circular motion of P_1 and P_2. The problem to be solved is to describe the motion of P_3 under various given initial conditions.

This problem was formulated by Euler in 1772, and has been a central issue in celestial mechanics since the profound studies of Poincaré. Poincaré emphasized the importance of periodic motions in the restricted problem and in fact conjectured that any solution of the restricted problem could be approximated arbitrarily closely for a given time interval by a periodic solution.

Relatively simple differential equations describing the motion of P_3 (in a rotating coordinate system) were found by Jacobi. These equations are autonomous and can be written (in the nondimensional form)

$$x_{tt} - 2y_t = U_x(x, y), \tag{4.3.1}$$

$$y_{tt} + 2x_t = U_y(x, y), \tag{4.3.2}$$

where

$$U(x, y) = \tfrac{1}{2}(x^2 + y^2) + (1-\mu)\{(x-\mu)^2 + y^2\}^{-1/2} + \mu\{(x - 1 + \mu)^2 + y^2\}^{-1/2}. \tag{4.3.3}$$

The system (4.3.1)–(4.3.2) is Hamiltonian and autonomous. Its stationary points (obtained by solving the equations ($U_x = U_y = 0$) are the simplest solutions of this system. It turns out that there are five such stationary points: three of the form $(x_k(\mu), 0)$ $(k = 1, 2, 3)$ usually called L_1, L_2, L_3, and two others L_4 and L_5 either of which forms with the points P_1 and P_2 an equilateral triangle. (See Fig. 4.4.)

Thus the points L_4 and L_5 are called triangular stationary or libration points. Here we shall investigate the possibility of periodic motions in the vicinity of such stationary points, by considerations analogous to those of Sections 4.1 and 4.2. First we observe that it is easy to find *linear approximations* for the desired periodic solutions. Indeed, if (x_0, y_0) are the coordinates of a stationary point of (4.3.1)–(4.3.2), then solutions of the linearized equations about (x_0, y_0) are

$$\xi_{tt} - 2\eta_t = U_{xx}(x_0, y_0)\xi + U_{xy}(x_0, y_0)\eta, \tag{4.3.4}$$

$$\eta_{tt} + 2\xi_t = U_{xy}(x_0, y_0)\xi + U_{yy}(x_0, y_0)\eta. \tag{4.3.5}$$

It turns out that this linearized system has one family of periodic solutions if (x_0, y_0) is chosen to be one of the collinear points L_1, L_2, L_3; while at L_4 and L_5 the system (4.3.4)–(4.3.5) has either two or no distinct periodic solutions depending on whether or not the mass ratio μ is less than or equal to some critical number μ_0. The problem we take up here is the validity of this linear approximation to the periodic solutions of (4.3.1)–(4.3.2). More precisely, *are the periodic solutions of* (4.3.4)–(4.3.5)

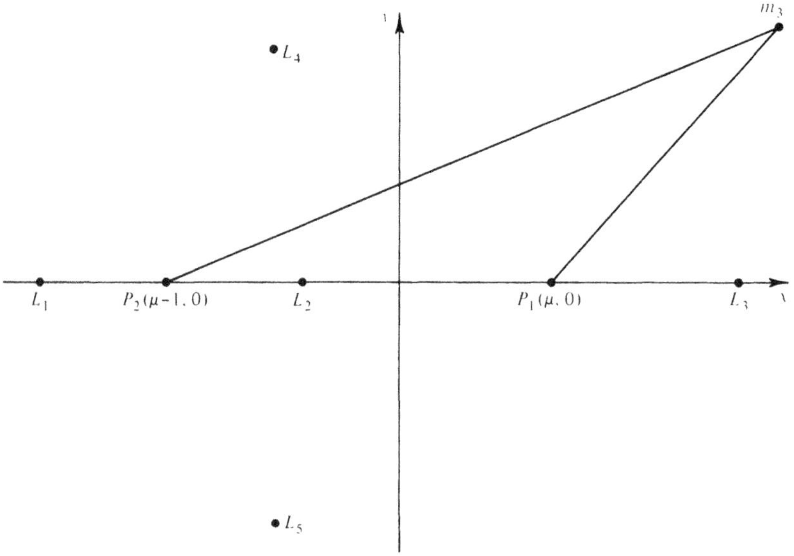

FIG. 4.4 Equilibrium points for the restricted three-body problem.

accurate first approximations to the periodic solutions of (4.3.1)–(4.3.2) *near each of the stationary points* (L_1–L_5)? In Sections 4.1 and 4.2, we partially solved this problem for systems analogous to (4.3.1)–(4.3.2) by bifurcation theory. Indeed, there we considered this problem in terms of the preservation of normal modes under nonlinear perturbations. We intend to show that our methods can be extended to this more difficult case. In fact we prove:

(4.3.6) **Theorem** If the linear system (4.3.4)–(4.3.5) admits a nontrivial periodic solution relative to the points L_i ($i = 1, \ldots, 5$), then in a small neighborhood of L_i the nonlinear system (4.3.1)–(4.3.2) admits at least one nontrivial family of periodic solutions $x_\epsilon(t)$ with period τ_ϵ, both depending analytically on ϵ and such that $(x_\epsilon(t), \tau_\epsilon) \to (0, \tau_i)$, where τ_i is the smallest nonzero period of any nontrivial periodic solution of (4.3.4)–(4.3.5) relative to L_i.

Proof: Our argument is patterned after that used in (4.2.9). The first step consists in setting $t = \lambda s$ in (4.3.1)–(4.3.2) and fixing attention on one-periodic solutions in s of the transformed equations. Following the procedure of (4.2.9), the resulting system can then be written in the form

$$(4.3.7) \qquad (\mathfrak{L} - \lambda B - \lambda^2 L)X - \lambda^2 R(X) = 0$$

in the Hilbert space H consisting of absolutely continuous 2-vectors $X(s) = (x(s), y(s))$ defined on $(0, \frac{1}{2})$ and such that $\dot{X}(s) \in L_2(0, \frac{1}{2})$. Here the operators $\mathfrak{L}$, B, L, and R are defined implicitly by the formulas

$$(\mathfrak{L}X, \varphi) = \int_0^{1/2} X_s \cdot \varphi_s \, ds, \quad (BX, \varphi) = 2\int_0^{1/2} (x_s \varphi_2 - y_s \varphi_1) \, ds,$$

$$(LX, \varphi) = \int_0^{1/2} (H(x_0, y_0)X \cdot \varphi) \, ds, \quad (RX, \varphi) = \int_0^{1/2} \nabla V(X) \cdot \varphi \, ds,$$

where $H(x_0, y_0)$ is the Hessian matrix $(U_{x_i x_j}(x_0, y_0))$ and $V(X) = V(x, y)$ denotes the real-valued function

$$V(x, y) = U(x_0 + x, y_0 + y) - U(x_0, y_0) - \tfrac{1}{2} H(x_0, y_0)\{X \cdot X\}.$$

Thus B and L are self-adjoint compact operators, while R is a completely continuous gradient operator of higher order.

Again, as in (4.2.9), by decomposing a vector $X(s)$ in H into its mean value over $(0, \frac{1}{2})$, X_m, and a vector with mean value zero $X_0(s)$, the equation can be written in the form

$$(4.3.8) \qquad f(X_0, \lambda) = (I - \lambda B - \lambda^2 L)X_0 - \lambda^2 T(X_0) = 0$$

on the closed subspace H_0 of H consisting of those vectors $X_0(s)$ with mean value zero over $(0, \frac{1}{2})$. Now the numbers λ at which the linear operator $f_x(0, \lambda) = I - \lambda B - \lambda^2 L$ is not invertible coincide with the periods of the nontrivial periodic solutions of (4.3.4)–(4.3.5). Since the operators B and L are compact and self-adjoint, $f_x(0, \lambda)$ is a Fredholm operator of index zero for all values of λ. Consequently, the methods of the bifurcation theory developed in Sections 4.1 and 4.2 become applicable. Thus since the Hessian matrix $H(x_0, y_0)$ is nonsingular at each L_i $(i = 1, \ldots, 5)$, by repeating the argument given in (4.1.13), if $\dim \operatorname{Ker}(I - \lambda B - \lambda^2 L) = 1$ for $\lambda = \lambda_0$, then equation (4.3.8) has a unique curve of nontrivial solutions $(x(\epsilon), \lambda(\epsilon))$ depending real analytically on ϵ and bifurcating from $(0, \lambda_0)$. These solutions, in turn, give rise to the desired family of periodic solutions for (4.3.1)–(4.3.2), since the function $V(x, y)$ is real analytic. On the other hand, if $\dim \operatorname{Ker}(I - \lambda B - \lambda^2 L) > 1$ for $\lambda = \lambda_0$, then the argument of Theorems (4.2.15) and (4.2.7) are applicable and we find that the equation (4.3.8) always has a curve of nontrivial solutions $(x(\epsilon), \lambda(\epsilon))$ depending real analytically on ϵ and bifurcating from $(0, \lambda_0)$. These solutions give rise to a nontrivial family of periodic solutions of (4.3.1)–(4.3.2).

(4.3.9) **Corollary** The system (4.3.1)–(4.3.2) has periodic solutions in any small neighborhood of L_1, L_2, and L_3 for any $\mu > 0$. However, there is a critical number $\mu_0 < 1$ so that the system (4.3.1)–(4.3.2) admits periodic solution in any small neighborhood of L_4 and L_5 for $\mu \leqslant \mu_0$, but not for $\mu > \mu_0$. In fact, apart from a countably infinite number of points Σ_μ in $(0, \mu_0)$, (4.3.1)–(4.3.2) possesses two distinct families of periodic solutions near L_4 and L_5.

Proof: By virtue of Theorem (4.3.6), it suffices to determine the numbers λ at which $f_x(0, \lambda) = I - \lambda B - \lambda^2 L$ has a nontrivial kernel, or equivalently, the period λ of the nontrivial solutions of (4.3.4)–(4.3.5). In order to determine the periods of these linear equations, we form the characteristic equations for this system. For L_1, L_2, L_3 this equation can be written as $s^4 + \alpha(\mu)s^2 - \beta^2(\mu) = 0$, where $\alpha(\mu)$ and $\beta(\mu)$ are constants. Since periodic solutions correspond to purely imaginary values of s, we see that there is only one such conjugate pair for any μ since $\beta^2(\mu) > 0$. On the other hand, for L_4 and L_5 the characteristic equation is $s^4 + s^2 + \frac{27}{4}\mu(1 - \mu) = 0$, and this equation will have the desired purely imaginary complex conjugate roots $(\pm is_1, \pm is_2)$ if and only if $1 \geqslant 27\mu(1 - \mu)$. Thus the value μ_0 is the smallest positive solution of the equation $1 = 27\mu(1 - \mu)$. Therefore if $s_2 > s_1$, then by the extended version of Liapunov's criterion applied to the system (4.3.1)–(4.3.2) will have two families of periodic solutions with periods near $2\pi/s_2$, $2\pi/s_1$, respectively, provided $s_2/s_1 \neq N$, an integer. Setting $D = 1 - 27\mu(1 - \mu)$, the excluded values of μ are those values for which $(1 + \sqrt{D})/(1 - \sqrt{D}) \neq N^2$ $(N = 1, 2, \ldots)$. Note that these values have a limit point as $\mu \to 0$.

Remarks on open problems: Clearly unsolved classic problems of importance (and for which our methods apply) in connection with the above results are:

(i) the continuation of the families of periodic orbits for large amplitudes,

(ii) the possibility of the removal of the forbidden values of $\mu \epsilon \Sigma_\mu$ for the existence of two distinct families of periodic solutions near L_4 and L_5. (See Note F at the end of the Chapter.)

4.3B Buckling phenomena in nonlinear elasticity

Many interesting bifurcation phenomena occur in nonlinear elasticity. Perhaps the earliest was the Euler *Elastica* problem mentioned in Chapter 1. Euler's problem consisted in giving a mathematical description of the action of an axial thrust on a uniform elastic rod, and in a paper dated 1744, he reduced this problem to a description of the solutions of the following semilinear boundary value problem:

$$\ddot{w} + Pw[1 - \dot{w}^2]^{3/2} = 0, \qquad w(0) = w(1) = 0.$$

Euler found that the rod deflects out of its plane or "buckles" whenever P, a measure of the axial thrust, exceeds a certain number, the so-called "buckling load," namely the smallest eigenvalue of the associated linear problem:

$$\ddot{w} + Pw = 0, \qquad w(0) = w(1) = 0.$$

He also showed that the quasilinear problem could be explicitly solved in terms of elliptic functions involving the parameter P.

In 1910, Von Kármán proposed a set of two fourth-order quasilinear partial differential equations that can be used to describe an analogous, but more difficult, two-dimensional problem: a mathematical investigation of the buckling of a thin elastic plate subjected to arbitrary forces and stresses along its boundary. In the following years a full treatment of these equations, without extra assumptions on the shape of the plate or symmetry of the buckled plate, proved to be extremely difficult due to the nature of the nonlinearity of the partial differential equations involved. Here we propose to demonstrate the applicability of our bifurcation theory developed in Sections 4.1 and 4.2 to the mathematical investigation of elastic buckling both of plates and the more general thin curved elastic shells.

The formulation of the von Kármán equations is as follows: We consider a thin elastic body B that is flat in its undeformed state subjected to a compressive force (of magnitude λ) acting on the boundary of B. Then the stresses produced in B, as measured by the Airy stress function, $f(x, y) + \lambda F_0(x, y)$ and the displacement of B from its flat state $u(x, y)$ are defined by the following quasilinear elliptic system (cf.(1.1.12))

$$\text{(4.3.10)} \qquad \begin{aligned} \Delta^2 f &= -\tfrac{1}{2}[u, u], \\ \Delta^2 u &= \lambda[F_0, u] + [f, u], \end{aligned}$$

where Δ^2 denotes the biharmonic operator and

$$[f,g] = f_{xx}g_{yy} + f_{yy}g_{xx} - 2f_{xy}g_{xy}.$$

If we represent B as a bounded domain G in $\mathbb{R}^2$ and the boundary of B as ∂G, we may consider the following boundary conditions associated with (4.3.10):

$$(4.3.11)\qquad \begin{aligned} u = u_x = u_y = 0 \\ f = f_x = f_y = 0 \qquad \text{on} \quad \partial G \end{aligned}$$

Here $F_0(x,y)$ is the function obtained by solving an associated inhomogeneous linear problem, and is a measure of the stress produced in the undeflected plate if it were prevented from deflecting. The resulting equilibrium states are called "buckled" states, and the problem is referred to as "elastic buckling."

In order to study the onset of buckling, it is generally supposed that the nonlinear terms $[u,u]$ and $[f,u]$ in the equations (4.3.10) can be neglected. Thus the classical linearized problem of the buckling of plates is known to be described by the following linear eigenvalue problem:

$$(4.3.12)\qquad \Delta^2 w - \lambda[F_0, w] = 0 \qquad \text{in} \quad \Omega,$$

$$(4.3.13)\qquad w = w_x = w_y = 0 \qquad \text{on} \quad \partial\Omega.$$

In terms of our formulation via functional analysis the linearized problem consists in studying the selfadjoint linear eigenvalue problem

$$(4.3.14)\qquad w = \lambda L w \qquad \text{in} \quad \mathring{W}_{2,2}(\Omega).$$

Throughout our work we shall impose the following condition on the function $F_0(x, y)$: The operator $Lw = [F_0, w]$ is a bounded compact operator on $\mathring{W}_{2,2}(\Omega)$. The validity of this assumption will be assured, for example, if all second derivatives of F_0 are uniformly bounded in Ω. Furthermore, we remark that this assumption allows us to consider operators L with positive and negative eigenvalues. This situation corresponds to the physical action of compression on one part of $\partial\Omega$ and tension applied to another part of $\partial\Omega$. Indeed the spectrum of (4.3.14) consists of eigenvalues $\{\lambda_n\}$ forming a sequence of discrete numbers tending to $+\infty$, $-\infty$, or both. The multiplicity of each λ_n is finite, and zero is not an eigenvalue of (4.3.14).

We mention a simple example of linearized problems for clamped plates.

Example *A circular clamped plate subjected to a uniform compressive pressure at its edge* Here the equations (4.3.12)–(4.3.13) reduce to

$$(4.3.15)\qquad \Delta^2 w - \lambda\, \Delta w = 0, \qquad w = w_x = w_y = 0.$$

The study of the solutions of this problem then becomes an analysis of radially symmetric eigenfunctions and nonradially symmetric eigenfunctions.

The radially symmetric solutions $w = w(r)$ of (4.3.15) can be explicitly determined by the zeros of the Bessel function $J_1(r)$. These eigenfunctions are simple and can be characterized as solutions of the second order ordinary differential equation

$$r^2\ddot{\psi} + r\dot{\psi} + (r^2 - 1)\psi = 0$$

which are finite at $r = 0$.

Nonradially symmetric eigenfunctions can be obtained in the form

$$w(r,\theta) = R(r)\begin{cases} \sin\mu\theta \\ \cos\mu\theta \end{cases}.$$

These eigenfunctions are not necessarily simple, but the first eigenfunction is known to be axially symmetric, and simple.

Now we indicate in just what sense (i) the eigenvalues λ_n and eigenfunctions are valid first approximations to the solutions of (4.3.12)–(4.3.13) and (ii) the relevance of eigenvalues λ_n for the understanding of buckling phenomena. To this end, we prove

(4.3.16) **Theorem** (i) Suppose λ_n is an eigenvalue of the linear system (4.3.14). Then $(0, \lambda_n)$ is a point of bifurcation for the nonlinear system (4.3.12)–(4.3.13). Thus for each λ_n there is a one-parameter family of solutions of (4.3.12)–(4.3.13), $(w_\epsilon, f_\epsilon, \lambda_\epsilon)$ depending analytically on ϵ such that

$$w_\epsilon = \sum_{n=1}^{\infty} w_n\epsilon^n, \quad f_\epsilon = \sum_{n=2}^{\infty} f_n\epsilon^n, \quad \lambda_\epsilon = \lambda_n + \sum_{n=1}^{\infty} \beta_n\epsilon^n,$$

and w_1 is a solution of (4.3.13)–(4.3.14).

(ii) The system (4.3.12)–(4.3.13) has no solutions in the interval $(\lambda_{-1}, \lambda_1)$, where λ_1 and λ_{-1} are the smallest positive and negative eigenvalues of (4.3.13)–(4.3.14), respectively.

(iii) Near $w = 0$, $\lambda = \lambda_N$, the nonlinear system (4.3.12)–(4.3.13) has no solutions for $\lambda \leqslant \lambda_N$ $(N = 1, 2, \ldots)$ for $\lambda_N > 0$ and no solutions for $\lambda \geqslant \lambda_N$ for $\lambda_N < 0$.

Proof of (i): In Section 2.5C, we showed that the solutions of the system (4.3.12)–(4.3.13) can be put in one-to-one correspondence with the solutions of the operator equation

(4.3.17) $$w + Cw = \lambda Lw$$

defined on the Hilbert space $H = \mathring{W}_{2,2}(\Omega)$, where the operator $C(w)$ is defined by first defining the bilinear operator $C(w, v)$: $H \times H \to H$ implicitly by

(4.3.18) $$(C(w,v), \varphi) = \int_\Omega [w, v]\varphi, \qquad \varphi \epsilon C_0^\infty(\Omega).$$

Then $C(w) = C(w, C(w, w))$ is a completely continuous gradient operator, homogeneous of degree 3, and such that $(Cw, w) \geqslant 0$. Thus setting $f(w, \lambda) = w + Cw - \lambda Lw$, we note that by virtue of Theorem (4.2.15), all the points $(0, \lambda_N)$ at which $\dim \operatorname{Ker} f_w(0, \lambda_N) > 0$ are points of bifurcation of (4.3.17) and these associated numbers λ_N coincide with the eigenvalues λ_N of (4.3.12)–(4.3.13). Since $f(w, \lambda)$ is real analytic in w and λ, the expansion in (i) follows again as in (4.2.16).

Proof of (II): Suppose (w,f) was a solution of (4.3.12)–(4.3.13) for $\lambda\epsilon[\lambda_{-1}, \lambda_1]$. Then (w, λ) would satisfy (4.3.17). Taking the inner product of (4.3.17) with w, and using the variational characterization of λ_1 and λ_{-1}, we find that $(Cw, w) = 0$. Consequently, $\int_\Omega [w, w]\varphi = 0$ for all $\varphi\epsilon H$ so that w satisfies the system

(4.3.19) $$[w, w] = w_{xx}w_{yy} - w_{xy}^2 = 0 \quad \text{in } \Omega$$

(4.3.20) $$D^\alpha w|_{\partial\Omega} = 0, \qquad |\alpha| \leq 1.$$

Thus the surface $w = w(x, y)$ has zero Gaussian curvature and is therefore developable. This surface is covered by straight lines and by virtue of (4.3.20), $w(x, y) \equiv 0$ in Ω.

Proof of (III): The fact that $(Cw, w) > 0$ for any nontrivial solution of (4.3.12)–(4.3.13) just proved above, and an application of (4.1.33) to the equation (4.3.17) yields the desired result.

Remark on a sharpening of (4.3.16): In Chapter 6 we shall establish an important sharpening of the result (i) above, viz., if λ_n is an eigenvalue of multiplicity k, then, roughly speaking, the nonlinear system (4.3.12)–(4.3.13) has at least k one-parameter families of solutions bifurcating from $(0, 0, \lambda_n)$. (For a more precise result, see Section 6.7C.)

The buckling phenomena associated with thin elastic shells (i.e., elastic structures that are initially curved) are considerably more complex than in the case of elastic plates, despite the similarity in the analogous von Kármán equations. Indeed, it is a well-known experimental fact that *linearization often does not explain the observed deformations.* In order to illustrate this fact, we consider a thin shallow shell S of arbitrary shape whose plane projection is a bounded domain Ω in the xy plane with boundary $\partial\Omega$. Suppose the shell is acted on by an external force $Z(x, y)$ and by forces along $\partial\Omega$. Then, subject to appropriate boundary conditions, the equilibrium states of S will be determined by solving the following von Kármán equations:

(4.3.21) $$\Delta^2 f = -\tfrac{1}{2}[w, w] - (k_1 w_x)_x - (k_2 w_y)_y,$$

(4.3.22) $$\Delta^2 w = [f, w] + (k_1 f_x)_x + (k_2 f_y)_y + Z.$$

Here k_1 and k_2 denote the initial curvatures of the shell in cross sections parallel to the zx and zy planes, respectively. Thus the effect of initial curvature on the von Kármán equations is merely the addition of linear curvature terms in (w, f).

Moreover if $Z = \lambda\Psi_0$ is chosen so that together with the boundary conditions

(4.3.23) $$w = w_x = w_y = 0$$

(4.3.24) $$f_{n\tau} = \lambda\Psi_1 \qquad f_{\tau\tau} = \lambda\Psi_2 \qquad \text{on } \partial\Omega$$

the system (4.3.21)–(4.3.24) has a solution $(w, f) = (0, \lambda F_0)$ for all λ, (i.e., a solution depending linearly on λ, in which the midsurface of the deformed shell is stretched but not bent). Here n and τ represent derivatives in the normal and tangential directions, respectively; Ψ_1 and Ψ_2 represent edge stresses applied on $\partial\Omega$; and λ measures the magnitude of the edge stresses. We note that such a function $Z = \lambda\Psi_0$ depending linearly on λ can always be determined for given smooth Ψ_1, Ψ_2. Indeed if λF_0 is the solution of the Dirichlet problem $\Delta^2 F = 0$ together with the boundary condition (4.3.24) we compute Ψ_0 by the formula $(k_1 F_{0x})_x - (k_2 F_{0y})_y = -\Psi_0$. Hence $w = 0$, $F = \lambda F_0$, satisfies the resulting equations (4.3.21)–(4.3.24).

Now writing a tentative solution of the full system (4.3.21)–(4.3.4) as $w = w, f = F + \lambda F_0$, we find that the following system of equations determines the desired equilibrium states:

(4.3.25) $$\Delta^2 F = -\tfrac{1}{2}[w, w] - (k_1 w_x)_x - (k_2 w_y)_y,$$

(4.3.26) $$\Delta^2 w = [F, w] + \lambda[F_0, w] + (k_1 F)_x + (k_2 F)_y,$$

(4.3.27) $$w = w_x = w_y = 0$$
$$F = F_x = F_y = 0 \quad \text{on} \quad \partial\Omega.$$

The associated linearized system in this case can be written

(4.3.28) $$\Delta^2 F = -(k_1 w_x)_x - (k_2 w_y)_y,$$

(4.3.29) $$\Delta^2 w = \lambda[F_0 w] + (k_1 F)_x + (k_2 F)_y$$

together with the boundary conditions (4.3.27). Clearly (by following the arguments of (2.5.7)) this system can be written as an operator equation in the Hilbert space $H = \mathring{W}_{2,2}(\Omega)$:

(4.3.30) $$w + L_1^2 w = \lambda L w,$$

where Lw is defined as in (2.5.7) and

$$(L_1 w, \varphi) = \int_\Omega (k_1 w_x \varphi_x + k_2 w_y \varphi_y).$$

Under the assumption I made above, the spectrum of (4.3.30) has exactly the same properties as in the plate case.

The relationship of the system (4.3.25)–(4.3.27) and its linearization about $(0, \lambda)$ is expressed in the following

(4.3.31) **Theorem** (i) Let λ_N denote an eigenvalue of the linearized system (4.3.30). Then $(0, \lambda_N)$ is a point of bifurcation of the system (4.3.25)–(4.3.27) and there is a one-parameter family $(w_\epsilon^{(N)}, f_\epsilon^{(N)}, \lambda_\epsilon^{(N)})$ depending analytically on ϵ (for small ϵ) such that

$$w_\epsilon^{(N)} = \epsilon w_N(x) + O(\epsilon^2), \qquad f_\epsilon^{(N)} = O(\epsilon^2)$$
$$\lambda_\epsilon^{(N)} = \lambda_N + O(\epsilon),$$

where w_N is a normalized eigenfunction of (4.3.30).

(ii) the trivial solution $(w, \lambda) = (0, \lambda)$ attains the absolute minimum of the potential energy for $0 < \lambda \leqslant \bar{\lambda}_0$, where $\bar{\lambda}_0$ is the smallest positive eigenvalue of the associated linearized plate equation $w = \lambda L w$. However, for $\lambda \in (\bar{\lambda}_0, \lambda_1)$, the trivial solution $(0, \lambda)$ (although a relative minimum) is (in general) not the solution that attains the absolute minimum of the potential energy.

Proof of (i): Again we can reformulate the system (4.3.25)–(4.3.27) as an operator equation in the Hilbert space H. This operator equation can be written down as in (2.5.7), first as the pair of equations

(4.3.32) $$F = -\tfrac{1}{2} C(w, w) - L_1 w$$

(4.3.33) $$w = C(F, w) + \lambda L w + L_1 F,$$

Then by substituting (4.3.32) into (4.3.33), and setting $Cw\ C(w, C(w_1 w))$ as in (2.5.7)

(4.3.34) $$G(w, \lambda) = w + \tfrac{1}{2} C(w) + C(w, L_1 w) + \tfrac{1}{2} L_1 C(w, w) + L_1^2 w - \lambda L w = 0.$$

Now the points λ_N at which $G_w(0, \lambda)$ is not invertible coincide with the eigenvalues of (4.3.30). To show that each such λ_N is a point of bifurcation *independent of* $\dim \operatorname{Ker}(I + L_1^2 - \lambda_N L)$, we show that $G(w, \lambda)$ is a gradient operator and then we apply Theorem (4.2.15). A simple computation shows that if

$$\mathcal{G}(w, \lambda) = \|w\|^2 + \|\tfrac{1}{2} C(w, w) + L_1 w\|^2 - \lambda(Lw, w),$$

then $\mathcal{G}_w(w, \lambda) = 2G(w, \lambda)$, so that $G(w, \lambda)$ is a gradient operator. Furthermore, $G(w, \lambda)$ can be put in the standard form $G(w, \lambda) = (I - \lambda L)w + T(w)$ by defining a Hilbert space norm on $\mathring{W}_{2,2}(\Omega)$ equivalent to the norm $\|u\|_{2,2} = \int_\Omega |\Delta w|^2$, by setting $\|w\|_{2,2}^2 + \|Lw\|_{2,2}^2$ as new norm.

Proof of (ii): The potential energy of the system can be represented by the functional $\mathcal{G}(w, \lambda)$ defined in the proof of (i). By the variational characterization of $\bar{\lambda}_0$, for $\lambda \in [0, \lambda_0]$, $\mathcal{G}(w, \lambda) \geqslant \| \frac{1}{2} C(w, w) + L_1 w \|^2 \geqslant 0$. Since $\mathcal{G}(0, \lambda) = 0$, the trivial solution $(0, \lambda)$ attains the infimum of $\mathcal{G}(w, \lambda)$ for $\lambda \in [0, \lambda_0]$. On the other hand, for $\lambda \in [\bar{\lambda}_0, \lambda_1]$ and arbitrary $z \in H$, the quadratic form

$$(\mathcal{G}_{ww}(0, \lambda)z, z) = (z, z) + (L_1^2 z, z) - \lambda(Lz, z)$$
$$\geqslant 0 \qquad \text{(by the variational characterization of } \lambda_1).$$

Thus for $\lambda \in [\bar{\lambda}_0, \lambda_1]$, the trivial solution is a relative minimum of the potential energy function $\mathcal{G}(w, \lambda)$. In order to show that (in general) the trivial solution is not an absolute minimum, we note that if $(\bar{w}, \bar{F}, \lambda)$ is a solution of (4.3.24), then

$$\mathcal{G}(\bar{w}, \bar{\lambda}) = -\tfrac{1}{4} \| C(\bar{w}, \bar{w}) \|^2 - \tfrac{1}{2} (C(\bar{w}, \bar{w}), L\bar{w}).$$

Now if $\dim \operatorname{Ker}(I + L^2 - \lambda_1 L_1) = 1$ and $(C(u_1, u_1), Lu_1) \neq 0$ as is the case in general for $u_1 \in \operatorname{Ker}(I - L^2 - \lambda_1 L_1)$, then $\mathcal{G}(\bar{w}, \bar{\lambda})$ can be made negative for $\lambda < \lambda_1$. Thus, in general, the trivial solution is not an absolute miminum for a range of values λ below λ_1.

It is important to study the stability properties of solutions of the system (4.3.10)–(4.3.11). In fact there are various theories to explain the physical principles by which the plate "chooses" a particular solution or "bifurcated state" among the various possibilities present at each value of λ. Here we state one such principle which dates back to Dirichlet.

Principle of Least Potential Energy: At a particular value of λ, the plate selects a state at which its potential energy is least. Conversely, an equilibrium state that is not a relative minimum of its potential energy is unstable.

For the present problem, the potential energy of an equilibrium state defined by $u(x, y)$ is defined, up to a constant factor, by

(4.3.35) $$V(u) = (u, u) + \tfrac{1}{2}(Cu, u) - \lambda(Lu, u).$$

Therefore, the unbuckled state of the plate u_0 has potential energy $V(u_0) = 0$.

(4.3.36) **Theorem** Any buckled state of the plate defined by $u = u(x, y)$ has strictly negative potential energy, i.e., $V(u) < 0$. Consequently, the unbuckled state is unstable for $\lambda > \lambda_1$.

Proof: For any buckled state,

$$(u, u) + (Cu, u) = \lambda(Lu, u).$$

Thus

$$V(u) = -\tfrac{1}{2}(Cu, u) < 0, \qquad \text{since} \quad u \neq 0.$$

Hence, by virtue of the principle of least potential energy, the plate always buckles out of the plane when $\lambda > \lambda_1$; the trivial solution is unstable for $\lambda > \lambda_1$.

4.3C Secondary steady flows for the Navier–Stokes equation

In a large number of cases the structure of the steady solutions of the Navier–Stokes equations governing the motion of a viscous incompressible fluid depends crucially on a single real dimensionless parameter R, called the Reynolds number. For sufficiently small R, there is a unique "laminar" stationary flow that satisfies the Navier–Stokes equations. In fact, in Chapter 5, it will be shown that under very general circumstances the Navier–Stokes equations always admit at least one stationary solution for any positive value of the Reynolds number R. Nonetheless, it is observed in many circumstances that these stationary solutions of the Navier–Stokes equations are unstable for large Reynolds numbers. Indeed, as the Reynolds number increases, unsteady, highly irregular, (turbulent) fluid motions are observed experimentally. The explanation for this phenomenon on the sole basis of the nonlinearity of the Navier–Stokes equations is an outstanding and unsettled problem. *Here we investigate the beginning of this transaction from laminar steady flow, by means of bifurcation theory*. In particular, (in certain cases) we show first that the eigenvalues of certain linear operators can be associated with points of bifurcation of the Navier–Stokes equations; and secondly, that at these points of bifurcation an "exchange of stability" phenomenon takes place.

(i) General problem of secondary stationary flow Let Ω be a bounded domain in $\mathbb{R}^N$ ($N = 2, 3$) with boundary $\partial\Omega$. The Navier–Stokes equations for the motion of a viscous fluid flow under assigned forces f and a are

$$(4.3.37)\qquad \begin{aligned} &\beta\,\Delta v = (v\cdot \operatorname{grad})v + \nabla P + f,\\ &\operatorname{div} v = 0,\\ &v\,|_{\partial\Omega} = a. \end{aligned}$$

Here the vector v and the scalar $P(x)$ are unknown. We suppose that $f = \nu f_0$ depends linearly on a parameter ν and that the system (4.3.37) admits a known solution of the form

$$(4.3.38)\qquad v(\nu) = \nu v_0, \qquad P(\nu) = P_0(x_0,\nu) \qquad \text{for all real } \nu.$$

In this case we seek other solutions of (4.3.37) of the form $v = \nu w + v(\nu)$, $P = cp + P_0$. Then to determine w and p, setting $\lambda = \nu/\beta$, we investigate the nontrivial solutions of the equation

$$(4.3.39)\qquad \begin{aligned} &\Delta w = \lambda\{(w\cdot \operatorname{grad})v + (v\cdot \operatorname{grad})w + (w\cdot \operatorname{grad})w\} + \nabla p,\\ &\operatorname{div} w = 0,\\ &w\,|_{\partial\Omega} = 0. \end{aligned}$$

Clearly, by utilizing the results of Section 2.2D (cf. Note C of Chapter 2), we can reformulate the solutions of (4.3.39) as solutions of an operator equation of the form

$$f(w, \lambda) \equiv w - \lambda\{Lw + Nw\} = 0 \tag{4.3.40}$$

in the *real* Hilbert space $\mathring{H}_1$ of solenoidal N-vectors obtained by completing each component of the solenoidal N-vectors in $C_0^\infty(\Omega)$ in the Sobolev space $\mathring{W}_{1,2}(\Omega)$. The operators L and N are defined implicitly by the formulas (cf. Note C of Chapter 2)

$$(Lw \cdot \varphi)_{\mathring{H}_1} = -\int_\Omega [\{w \cdot \text{grad } v\} + \{v \cdot \text{grad } w\}] \cdot \varphi,$$

$$(Nw, \varphi)_{\mathring{H}_1} = -\int_\Omega \{w \cdot \text{grad } w\} \cdot \varphi.$$

As mentioned in Note 2C, the Sobolev imbedding theorem implies that L and N are compact mappings of $\mathring{H}_1 \to \mathring{H}_1$.

Actually we shall restrict consideration to those problems in which it can be established that dim Ker$(I - \lambda L)$ is odd in $\mathring{H}_1$. Indeed the operator N is neither a gradient map nor complex analytic on the *real* Hilbert space $\mathring{H}_1$, so that the only general results applicable to (4.3.40) are those that place a restriction on the parity of dim Ker$(I - \lambda L)$. In order to analyze Ker$(I - \lambda L)$, it will be necessary to specialize Ω as well as the vectors f and a. Thus our bifurcation analysis of Sections 4.1–4.2 shows that in order to prove that the Navier–Stokes equations admit secondary stationary flows distinct from $(v(\nu), P(\nu))$, it suffices to find points of bifurcation $(0, \lambda)$ of the equation (4.3.40). Consequently, we must determine the real eigenvalues λ of the linear operator $f'(0, \lambda)w = (I - \lambda L)w$ in $\mathring{H}_1$, and determine which of these correspond to points of bifurcation of (4.3.40).

A solution $u(x, t)$ of the time-dependent Navier–Stokes equations (1.1.18)–(1.1.19) is called stable or unstable according to whether any small perturbation of the data defining the solution $u(x, t)$ gives rise to a solution $v(x, t)$ that does or does not remain close to $u(x,t)$ for all t, in an appropriate norm. For stationary states $u(x)$, this stability criterion can sometimes be verified *by linearization* in the following manner. One considers a solution of the equations (1.1.18)–(1.1.19) in the form $v(x, t) = e^{\sigma t}w(x) + u(x)$ neglecting higher order terms in $w(x)$; i.e., we consider the spectrum of the Navier–Stokes operator linearized about $u(x)$. If this operator has an eigenvalue with positive real part, then $u(x)$ is called unstable according to linearized stability theory; while if all eigenvalues of the linear operator have negative real part, $u(x)$ is called stable (in the linearized sense). Thus in order to study linearized stability theory it will be necessary to study the nontrivial solutions (w, σ) of the equation

$$f_w(u, \lambda)w = -\sigma \mathfrak{L} w, \tag{4.3.41}$$

where $\mathfrak{L}$ is the imbedding of $W_{1,2}(\Omega) \to L_2(\Omega)$ in the Hilbert space $\mathring{H}_1$, and $f_w(u, \lambda)$ denotes the Fréchet derivative of the operator $f(w, \lambda)$ (defined in (4.3.40)) evaluated at the stationary solution (w, λ).

As a first example, to illustrate the nonuniqueness phenomena just described, suppose $\overline{\Omega}$ is a smooth surface of revolution that does not contain points on the axis of revolution z. Then for this special geometry (in terms of cylindrical polar coordinates (r, θ, z) we seek secondary, rotationally symmetric, stationary flows $w = w(r, z)$. We suppose that the external force $F = (0, F_0(r), 0)$ is such that (4.3.37) admits a trivial solution $v(r) = (0, \nu v_0(r), 0)$ for $\nu \in (-\infty, \infty)$. From (4.3.40), the resulting Navier–Stokes equations for $w = (w^{(r)}, w^{(\theta)}, w^{(z)})$ in cylindrical coordinates are written in the operator form $f(w, \lambda) = [I - \lambda L]w + \lambda T(w)$ in the closed subspace A_0 of $\mathring{H}_1$ consisting of rotationally symmetric, solenoidal vectors with inner product

$$(w, v)_{A_0} = \int_D (\nabla w \cdot \nabla v) r \, dr \, dz,$$

where D is the axial cross section of Ω. Then, if $w_0 = -v_0/r$ and $g = -(dv_0/dr + v_0/r)$,

$$(4.3.42) \qquad (Lw, \varphi)_{A_0} = \int_D (2 w_0 w^{(\theta)} \varphi^{(r)} + g w^{(r)} \varphi^{(\theta)})$$

In general, L is not a self-adjoint operator; but if $v_0(r) = r^{\beta}$ with $\beta < -1$, it can be made self-adjoint by a slight change in the inner product on A_0. Indeed, we define an equivalent inner product on A_0 by setting

$$[w, \varphi]_{\tilde{A}_0} = (w, \varphi)_{A_0} + \int_D \left(w^{(r)} \varphi^{(r)} - \left(\frac{2}{\beta + 1} \right) w^{(\theta)} \varphi^{(\theta)} + w^{(z)} \varphi^{(z)} \right) r \, dr \, dz.$$

Since $2w(r) = (-2/(\beta + 1)) g(r)$,

$$[Lw, \varphi]_{\tilde{A}_0} = \int_D 2w(r) \{ w^{(\theta)} \varphi^{(r)} + w^{(r)} \varphi^{(\theta)} \} r \, dr \, dz$$

$$= [w, L\varphi]_{\tilde{A}_0},$$

the operator L is both compact and self-adjoint. Consequently, there is a countably infinite sequence of isolated real numbers $0 < \lambda_1 \leqslant \lambda_2 \leqslant \cdots \leqslant \lambda_n \to \infty$ such that $0 < \dim \operatorname{Ker}(I - \lambda_i L) < \infty$ $(i = 1, 2, \ldots)$.

An immediate consequence of these facts is the nonuniqueness of stationary states for (4.3.37).

(4.3.43) **Theorem** Let Ω be described as above. Then there are vectors f and boundary conditions a such that the corresponding rotationally symmetric stationary states of the Navier–Stokes equations (4.3.37) are not unique.

Proof: Let $\lambda = \nu/\beta$ be any eigenvalue λ_i with eigenvector u_i described above. Let $v_1 =$

$\frac{1}{2}(a + u_i)$ and $v_2 = \frac{1}{2}(a - u_i)$, where $a = (0, \lambda r^\beta, 0)$ and $\beta < -1$. Then v_1 and v_2 are solenoidal, and take the same value on $\partial\Omega$. Furthermore in Ω, $(I - \lambda L)v_1 + \lambda T(v_1) = (I - \lambda L)v_2 + \lambda T(v_2)$. Consequently, if the vector f is this common value, both v_1 and v_2 are the desired rotationally symmetric stationary states.

We now prove

(4.3.44) **Theorem** Under the above conditions, if dim Ker$(I - \lambda_i L)$ is odd, then the Navier–Stokes equations (4.3.37) admit a family of secondary rotationally symmetric stationary states $w_i = w_i(r, z, \epsilon)$ for $\lambda(\epsilon)$ near λ_i such that

$$w_i(r, z, \epsilon) = \epsilon u_i(r, \theta) + O(|\epsilon|^2), \qquad \lambda(\epsilon) = \lambda_i + O(|\epsilon|),$$

where u_i is a solution of the linearized equation for $\lambda = \lambda_i$.

Proof: The proof is immediate from (4.2.3) and the fact that in $\tilde{A}_0$ the operator equation $f(w, \lambda) = (I - \lambda L)w + Tw = 0$ is such that L is compact and self-adjoint, while T is a C^1 operator of higher order near $w = 0$.

(4.3.45) **Corollary** Let λ_1 denote the smallest positive number such that dim Ker$(I - \lambda L) > 0$. Then the trivial solution $v_0(\nu)$ is stable (in the linearized sense) for $|\lambda| < \lambda_1$ and unstable (in the linearized sense) for $|\lambda| > \lambda_1$.

Proof: For $0 < \lambda < \lambda_1$, the equation $(I - \lambda L)u = -\sigma \mathfrak{L} u$, $\|u\| = 1$, implies that $(\mathfrak{L}u, u)\sigma = -((I - \lambda Lu), u)$. Thus by the variational characterization of λ_1, for $u \neq 0$, $(\mathfrak{L}u, u)\sigma < 0$. Consequently, $\sigma < 0$ and the stationary state associated with the trivial solution, $(0, \lambda)$, is linearly stable. For $\lambda > \lambda_1$, the smallest eigenvalue σ_1 of $I - \lambda L$ is characterized as

$$(4.3.46) \qquad -\sigma_1 = \inf \frac{((I - \lambda L)u, u)}{(\mathfrak{L}u, u)} \leqslant \frac{((I - \lambda L)u_1, u_1)}{(\mathfrak{L}u_1, u_1)} \leqslant \frac{(\lambda_1 - \lambda)\|u_1\|^2}{\lambda_1(\mathfrak{L}u_1, u_1)} < 0.$$

Consequently, the trivial solution $(0, \lambda)$ is unstable for $\lambda > \lambda_0$, according to the linearized stability theory.

(II) Taylor vortices The flow of a viscous incompressible fluid between two rotating concentric cylinders of infinite length is an excellent (although more complicated) example of the type of secondary stationary flows just mentioned. Suppose the radii of these cylinders are denoted R_1, R_2 with $R_1 < R_2$, and these cylinders rotate with angular velocities ω_1 and ω_2, respectively. Suppose that cylindrical coordinates (r, θ, z) are chosen with the z axis coinciding with the common axis of the cylinder. It can then be easily shown that with $f = (0, \nu f_0, 0)$ and $a = 0$, the Navier–Stokes equations admit a solution $(v(r, \nu), P(\nu))$ with $v(r, \nu) = (0, v_\nu^{(\theta)}(r), 0)$. This solution is called *Couette flow*, and we shall find relations between r_1, r_2, ω_1, and ω_2 such that (4.3.37) admits "periodic" secondary flows of the form $v = v(r, \nu) + w(r, z)$ (i.e., axisymmetric flows) that are periodic in z. (See Fig. 4.5.) Such flows are called Taylor vortices after G. I. Taylor who discovered them experimentally in 1923, and studied them mathematically by linearizing the Navier–Stokes equation about the Couette flows.

Proceeding as above, we seek vectors $w = (w^{(r)}, w^{(\theta)}, w^{(z)})$ that (i) vanish for $r = r_1, r_2$, (ii) are $2\pi/\alpha_0$ periodic in z, where α_0 is to be determined, (iii)

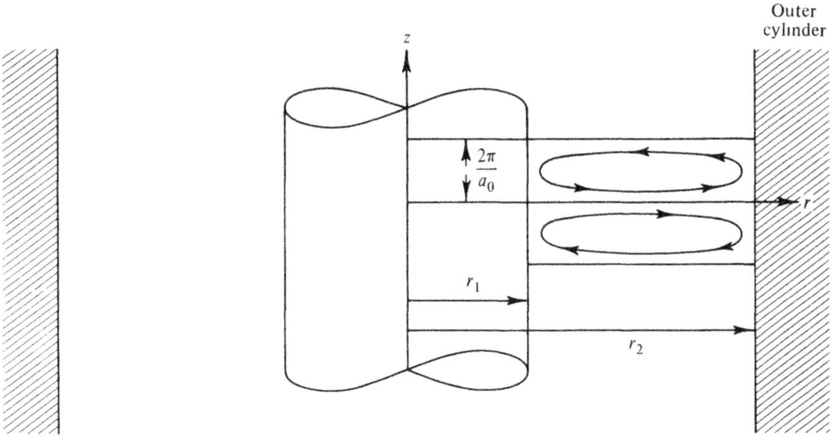

FIG. 4.5 Notation for Taylor vortices.

possess no net mass flow in the z direction, i.e., $\int_{r_1}^{r_2} w^{(z)} r\, dr = 0$, and (iv) $w^{(r)}$, $w^{(z)}$ are even in z, and $w^{(\theta)}$ is odd in z. Such conditions can easily be incorporated into an admissible class of solutions for (4.3.37). Clearly this class $\mathring{K}_1$ is a closed subspace of $\mathring{H}_1$. In order to prove the existence of Taylor vortices, we prove the existence of a point of bifurcation for the operator equation (4.3.40) in $\mathring{K}_1$

Actually one can prove

(4.3.47) **Theorem** If $\omega_1 > 0$ and $\omega_2 \geqslant 0$, then the Navier–Stokes equations (4.3.37) admit secondary Taylor vortices for the configuration just described above, provided $\omega_2 r_2^2 < \omega_1 r_1^2$ (i.e., if the inner cylinder rotates with a sufficiently large angular velocity).

Proof: By what has been mentioned above, it suffices to analyze the spectrum of the linearized operator L in the Hilbert space $\mathring{K}_1$. Thus we wish to analyze the system

(4.3.48)
$$\begin{aligned} \Delta u^{(r)} - u^{(r)}/r^2 - \partial q/\partial r + \lambda\omega(r)u^{(\theta)} &= 0, \\ \Delta u^{(\theta)} - u^{(\theta)}/r^2 + \lambda g(r)u^{(r)} &= 0, \\ \Delta u^{(z)} - \partial q/\partial z &= 0, \\ (1/r)(\partial/\partial r)(ru^{(r)}) + \partial u^{(z)}/\partial z &= 0, \qquad \int_{r_1}^{r_2} u^{(z)} r\, dr = 0 \end{aligned}$$

where $\omega(r) = a + b/r^2$ and $g(r) = -2a$. Thus we seek solutions of this system of the form $u = (u^{(r)}, u^{(\theta)}, u^{(z)}) = (u(r)\cos \alpha z,\ v(r)\cos \alpha z;\ \omega(r)\sin \alpha z)$. Substituting these into (4.3.48) and eliminating q we find the functions $u(r)$, $v(r)$ and eigenvalues λ by solving the system of ordinary differential equations

$$\begin{aligned} &(L - \alpha^2)^2 u = 2\alpha^2\lambda\omega(r)v, \\ &(L - \alpha^2)v = -\lambda g(r)u, \\ &u(r_i) = v(r_i) = u_r(r_i) = 0 \qquad (i = 1, 2), \end{aligned}$$

where $L = d^2/dr^2 + (1/r)\,d/dr - 1/r^2$. Let $G_1(r, r')$ and $G_2(r, r')$ be Green's functions for the differential operators $-r(L - \alpha^2)$ and $r(L - \alpha^2)^2$ subject to the above boundary conditions. Both G_1 and G_2 are continuous and symmetric in r and r'. In fact, finding the solutions of (4.3.48) in $\mathring{K}_1$ is clearly equivalent to finding the nontrivial solutions of the system

(4.3.49) $\quad \mu = uG_2\omega G_1 gu,$

(4.3.50) $\quad v = \mu G_1 g G_2 \omega v, \qquad \mu = 2\alpha^2\lambda^2.$

Now we prove that if a number μ is a simple eigenvalue of (4.3.49), then $\lambda = \mp\sqrt{\mu/2\alpha^2}$ is also a simple eigenvalue of L. Finally, we observe that Green's functions G_1 and G_2 are "oscillatory kernels", (cf. Karlin (1968) and thus, provided $\omega(r)$ and $g(r)$ are both positive, the operator $B = G_2\omega G_1 g$ is also. Therefore, the operator equation (4.3.49) has a sequence of simple eigenvalues $0 < \mu_1(\alpha) < \mu_2(\alpha) < \cdots < \mu_n(\alpha) < \cdots$. Thus the spectrum of $I - \lambda L$ consists of the eigenvalues $\lambda_{ik}(\alpha_0) = \pm[\mu_i(k\alpha_0)/2k^2\alpha_0^2]^{1/2}$. Now to ensure that (4.3.48) has simple eigenvalues, we note that the functions $\mu_k(\alpha)$ are real analytic functions of α. Setting $\Lambda_i(k\alpha) = \lambda_{ik}(\alpha)$, the function $\Lambda_{ikrs}^{(\alpha)} = \Lambda_i(k\alpha) - \Lambda_r(s\alpha)$ is real analytic in α (and is easily shown as not identically zero for $i \neq r$). Thus the zeros of $\Lambda_{ikrs}^{(\alpha)}$ are denumerable and so the set $\Sigma = \{\alpha \,|\, \Lambda_{ikrs}^{(\alpha)} = 0;\ r, s = 1, 2, \ldots;\ r, s \neq i, k\}$ is denumerable. Clearly the positive numbers α in the complement of Σ generate simple eigenvalues of L, so that for such values α, $\dim \operatorname{Ker}(I - \lambda(\alpha)L) = 1$ and $\operatorname{Ker}(I - \lambda L) \cap \operatorname{Range}(I - \lambda L) = \{0\}$. Thus $(0, \lambda(\alpha))$ is a point of bifurcation of (4.3.40), in this case by (4.1.12).

Before proceeding to a discussion of the occurrence of bifurcation phenomena in the theory of complex manifolds hinted at in Section 1.1, we refer the reader to the brief discussion on analysis on complex manifolds in Appendix B.

4.3D Bifurcation of complex structures on compact complex manifolds

Complex manifolds often depend critically on parameters in the sense that the *complex structures* associated with these manifolds change as the associated parameters, defining the structure, vary. Each member of a family of complex structures M_ω depending analytically on complex parameters ω may be complex analytically homeomorphic, in which case we consider the family "trivial" (from the point of view of bifurcation theory). Here we consider the question of finding *nontrivial deformations* of a given complex structure, i.e., a family of complex structures M_ω depending analytically on complex parameters ω, but such that the members of the family M_ω are not complex analytically homeomorphic. The approach to this problem via nonlinear partial differential equations was mentioned briefly in Section 1.1, and here we shall explore certain analytic aspects of this problem in more detail. (For a more complete discussion, we refer the reader to the monograph by Kodaira and Morrow (1971).)

The connection with nonlinear partial differential equations is made as follows. Let $\mathfrak{M}_0$ be a compact complex analytic manifold of complex dimension N. We shall apply the bifurcation result of Section 4.1 to construct a nontrivial family $\mathfrak{M}_t$ of complex structures on $\mathfrak{M}_0$ depending continuously on a finite number of complex parameters t, for $|t|$ sufficiently small, and with $\mathfrak{M}_0$ corresponding to $t=0$. Two complex structures $\mathfrak{M}_0$ and $\tilde{\mathfrak{M}}$ defined on the same manifold $\mathfrak{M}$ are close if in terms of the local holomorphic coordinates $\xi^1, \ldots, \xi^n$ of $\tilde{\mathfrak{M}}$, the form $d\xi^j$ may be expressed in terms of suitable local holomorphic coordinates $z^1, \ldots, z^n$ of $\mathfrak{M}_0$ by setting

$$d\xi^j = dz^j + \sum_{k=1}^{n} \varphi_{j\bar{k}}(z^1, \ldots, z^n) d\bar{z}^k, \tag{4.3.51}$$

where $\varphi_{j\bar{k}}$ is small in some common coordinate patch.

The complex structure of $\tilde{\mathfrak{M}}$ defines a splitting of the complex first order differential forms on $\mathfrak{M}$ into the direct sum of an n-dimensional subspace T and its complex conjugate space; and so defines an almost complex structure on $\mathfrak{M}$. Thus if $\tilde{\mathfrak{M}}$ is an almost complex structure on $\mathfrak{M}$ satisfying (4.3.51), the form $\omega = \sum \omega_k d\bar{z}^k$ is a well-defined vector (0, 1) differential form on $\mathfrak{M}_0$. It is known (by the Newlander–Nirenberg theorem, cf. Note D of Chapter 3) that this almost complex structure defines a complex structure on $\mathfrak{M}$ if and only if the following "integrability condition" is satisfied:

$$\bar{\partial}\omega - [\omega, \omega] = 0, \tag{4.3.52}$$

where for any vector-valued $(0, p)$ and $(0, q)$ forms ω and σ, $[\sigma, \omega]$ is a $(0, p+q)$ form defined whose i-th component is

$$[\sigma, \omega] = \tfrac{1}{2} \sum_J \left(\sigma^J \wedge \partial_j \omega^i + (-1)^{\sigma} \omega^J \wedge \partial_j \sigma^i\right)$$

where $\partial_j = \partial/\partial z_j$ and $\sigma = pq + 1$. So that the bilinear operator $[\sigma, \omega]$ satisfies the following identities

(4.3.53) (i) $[\sigma, \omega] = (-1)^{\sigma}[\omega, \sigma]$

(ii) $\bar{\partial}[\sigma, \omega] = [\bar{\partial}\sigma, \omega] + (-1)^{P}[\sigma, \bar{\partial}\omega]$;

(iii) $[[\sigma, \sigma], \sigma] = 0$.

In this sense the determination of nontrivial families of complex manifolds can be reduced to a study of (4.3.52). Indeed, complex structures on $\mathfrak{M}$ can be constructed by finding solutions of the equation (4.3.52) and, in fact, we shall find a family of solutions of (4.3.52) near $\omega = 0$. More precisely, we prove (following Kuranishi (1965))

(4.3.54) **Theorem** Let m be the dimension of the vector space $H^1(\mathfrak{M}, \Theta)$, the first cohomology group on the compact complex analytic manifold $\mathfrak{M}$ with coefficients in the sheaf of germs of holomorphic vector fields (Θ). Then the equation (4.3.52) has a family of solutions near $\omega = 0$ depending on m complex parameters and lying on a complex analytic set near the origin in C^m. Moreover, if the cohomology group $H^2(\mathfrak{M}, \Theta)$ vanishes, there is a nontrivial family of deformations for each element of $H^1(\mathfrak{M}, \Theta)$.

Proof: The result is obtained in a sequence of steps by considering the solution of the enlarged system of partial differential equations

(4.3.55) $\quad$ (i) $\bar{\partial}\omega = [\omega, \omega]$; $\qquad$ (ii) $\bar{\partial}^{\mathrm{T}}\omega = 0$,

together with its linearization near $\omega = 0$, (following Kuranishi)

(4.3.56) $\quad$ (i) $\bar{\partial}\omega = 0$; $\qquad$ (ii) $\bar{\partial}^{\mathrm{T}}\omega = 0$.

Here the operator $\bar{\partial}^{\mathrm{T}}$ is the L_2 adjoint of $\bar{\partial}$.

Now, as described in Appendix B, the solutions of (4.3.56) coincide exactly with solutions of the vector Laplace equation

(4.3.57) $\quad \square\omega = 0, \qquad$ where $\square = \bar{\partial}^T\bar{\partial} + \bar{\partial}\bar{\partial}^T$.

Thus the solutions of (4.3.56) coincide with (complex) the vector-valued harmonic (0, 1) forms $H_{0,1}(\mathfrak{M}, \Theta)$ defined on the complex analytic manifold $\mathfrak{M}$, and these forms, in turn, correspond to the elements of $H^1(\mathfrak{M}, \Theta)$. This last conclusion follows from Hodge theory.

Step 1. Reformulations: In order to apply the bifurcation theory of Section 4.1 to this problem, we first rewrite the system (4.3.55) by means of the Hodge–Kodaira decomposition theorem (of Appendix B). Using the fixed Hermitian metric on $\mathfrak{M}_0$, we introduce an L_2 scalar product on $\wedge_{0,p}$, denoted for $\omega, \sigma \in \wedge_{0,p}$ by $\langle\omega, \sigma\rangle$. Then by definition, $\langle\bar{\partial}\sigma, \omega\rangle = \langle\sigma, \bar{\partial}^{\mathrm{T}}\omega\rangle$. Let H be the projection of $\wedge_{0,p} \to H_{0,p}(\mathfrak{M})$, and $G\square$ be the projection of $\wedge_{0,p} \to [H_{0,p}(\mathfrak{M})]^{\perp}$. Then G commutes with $\bar{\partial}$ and $\bar{\partial}^{\mathrm{T}}$, while if one sets $Q = \bar{\partial}^{\mathrm{T}}G$, one finds that for $\omega \in \wedge_{0,p}$ (since H and $G\square$ are complementary projections)

(4.3.58) $\quad \omega = H\omega + \bar{\partial}Q\omega + Q\bar{\partial}\omega \qquad$ or $\qquad \omega = H\omega + \square G\omega$

Consequently, (4.3.55) implies that ω satisfies

(4.3.59) $\quad \omega = H\omega + Q[\omega, \omega]$.

Here we have used the fact that, since $\bar{\partial}^{\mathrm{T}}\omega = 0$, so that $\bar{\partial}Q\omega = \bar{\partial}G\bar{\partial}^{\mathrm{T}}\omega = 0$. Conversely, we shall show that

(*) if ω satisfies (4.3.59) (and is sufficiently small), then ω will also satisfy the system (4.3.55) on $\mathfrak{M}$, provided $H[\omega, \omega] = 0$, where H is the projection of $\wedge_{0,2}$ onto the vector-valued (0, 2) harmonic forms $H_{0,2}(\mathfrak{M}, \Theta)$.

As a preliminary step in this direction we note first that if ω satisfies (4.3.59), $\bar{\partial}^{\mathrm{T}}\omega = 0$ since $\bar{\partial}^{\mathrm{T}}Q$ and $\bar{\partial}^{\mathrm{T}}H$ are both identically zero; and secondly, by operating on (4.3.59) with $\bar{\partial}$, and then using (4.3.58),

$$(4.3.60) \qquad \bar{\partial}\omega = \bar{\partial}Q[\omega, \omega] = \Box G[\omega, \omega] - \bar{\partial}^{\mathrm{T}}\bar{\partial}G[\omega, \omega]$$

$$\bar{\partial}\omega - [\omega, \omega] = -H[\omega, \omega] - Q\bar{\partial}[\omega, \omega].$$

Moreover, we shall show later that $H[\omega, \omega] = 0$ implies that $Q\bar{\partial}[\omega, \omega] = 0$.

To investigate the smoothness of the solutions of (4.3.59), we consider the semilinear elliptic equation

$$(4.3.61) \qquad \Box\omega - \partial^{\mathrm{T}}[\omega, \omega] = 0$$

defined on $\mathfrak{M}_0$. We begin by observing that the smooth solutions of this equation include the smooth solutions of (4.3.59) so that the regularity properties of these solutions can be deduced from the properties of (4.3.61). Indeed, if ω satisfies (4.3.59), then

$$(4.3.62) \qquad \Box\omega = \Box Q[\omega, \omega] = \Box G\bar{\partial}^{\mathrm{T}}[\omega, \omega] = \bar{\partial}^{\mathrm{T}}[\omega, \omega].$$

Thus we conclude that if ω is any known solution of (4.3.59) in the Sobolev space W_k on k-times differentiable (0, 1) forms, the smoothness properties of ω can be deduced from the regularity theory for the nonlinear strongly elliptic second order systems, provided k is chosen sufficiently large. In fact, ω can be considered as a C^∞ function. (For details, see Kodaira and Morrow (1971).)

The equation (4.3.59) can now be reformulated as the following operator equation in W_k:

$$(4.3.63) \qquad L\omega + B(\omega) = 0 \quad \text{with} \quad L\omega = \omega - H\omega, \quad B(\omega) = Q[\omega, \omega],$$

where the bounded bilinear operator B satisfies the following estimates for arbitrary $\omega, \bar{\omega} \in H$ and sufficiently large k:

$$\|B(\omega) - B(\bar{\omega})\|_k \leqslant c(\|\|\omega_k, \|\bar{\omega}\|_k)\|\omega - \bar{\omega}\|_k,$$

where $c(x, y) \to 0$ as $|x| + |y| \to 0$. This fact is a consequence of the following routine but lengthy estimates:

$$\|[\omega, \sigma]\|_{k-1} \leqslant C\|\omega\|_k\|\sigma\|_k; \qquad \|Q\omega\|_{k+1} \leqslant C\|\omega\|_k,$$

where C is an absolute constant and k is chosen sufficiently large so that by Sobolev's inequalities (1.4.12) yields pointwise estimates. Moreover, we note that since W_k is a Hilbert space over the complex numbers, the mapping B is complex analytic.

Step 2. Application of bifurcation theory: Thus applying the theorem (4.1.5) to (4.3.63) we find that the solutions of (4.3.59) near $\omega = 0$ coincide with the solutions of the finite-dimensional system of equations

(4.3.64) $$PB(\omega_0 + g(\omega_0), \omega_0 + g(\omega_0)) = 0,$$

where P is the projection of W_k onto Ker L, ω_0 is an arbitrary element of Ker L, and g is a *complex analytic mapping* of Ker $L \to [\text{Ker } L]^{\perp}$ in W_k. Clearly, Ker L coincides with the harmonic (0, 1) forms so that $P = H$, and consequently the equation (4.3.64) is *automatically satisfied* since $PB \equiv HQ \equiv QH \equiv 0$. Consequently, (4.3.60) shows that (4.3.52) will hold precisely if:

(4.3.65a) $$H[\omega_0 + g(\omega_0), \omega_0 + g(\omega_0)] = 0,$$

(4.3.65b) $$Q\bar{\partial}[\omega_0 + g(\omega_0), \omega_0 + g(\omega_0)] = 0.$$

Step 3. Recapitulation: Finally, gathering our results together by showing that (4.3.65a) implies (4.3.65b) provided ω_0 is sufficiently small, (so that the equations (4.3.65a) are the only obstructions to solving (4.3.59) near $\omega = 0$ with $\omega = \omega_0 + g(\omega_0)$). To prove this fact, suppose ω satisfies (4.3.59) and (4.3.65a), then by (4.3.53)

$$\begin{aligned} Q\bar{\partial}[\omega, \omega] &= 2Q[\bar{\partial}\omega, \omega] = 2Q\big[\bar{\partial}Q[\omega, \omega], \omega\big] \\ &= 2Q\big\{[[\omega, \omega], \omega] - \big[Q\bar{\partial}[\omega, \omega], \omega\big]\big\} \qquad \text{(by (4.3.58))} \\ &= -2Q\big[Q\bar{\partial}[\omega, \omega], \omega\big]. \end{aligned}$$

Thus setting $\sigma = Q\bar{\partial}[\omega, \omega]$, we find $\sigma = -2Q[\sigma, \omega]$. Therefore, by virtue of (4.3.63) and the fact that, for some absolute constant $\bar{c} > 0$,

$$\|\sigma\|_H \leqslant \bar{c}\|\sigma\|_H\|\omega\|_H.$$

Hence, for $\|\omega\|_H$ sufficiently small, $\|\sigma\|_H = 0$, so that for ω sufficiently small, (4.3.59) and (4.3.64) imply (4.3.65b). (This completes the proof of (*) of step 1.)

Thus we find that if $H^2(\mathfrak{M}, \Theta) = 0$, the projection mapping $H \equiv 0$ (in (4.3.65a)), and so the equation (4.3.52) has a family of solutions near $\omega = 0$ for each $\omega_0 \in H^1(\mathfrak{M}, \Theta)$. More generally, if $H^2(\mathfrak{M}, \Theta) \neq 0$, the equation (4.3.52) is solvable for fixed ω_0 if and only if the system (4.3.65a) holds. This latter system can be interpreted as an "analytic set" depending on $m_1 = \dim H^1(\mathfrak{M}, \Theta)$ complex parameters since fixing a basis ω_i for Ker L and setting $\omega_0 = \sum t_i u_i$, we find (4.3.65a) depends analytically on the complex variable $t = (t_1, \ldots, t_{m_1})$. Moreover, it can be shown that the deformations $\mathfrak{M}_t$ just shown to exist are not complex analytically homeomorphic and "locally complete" in the sense that any other deformation near $\mathfrak{M}_0$ is equivalent to some $\mathfrak{M}_t$ constructed as above. For the details of this argument, we refer the reader once more to the monograph of Kodaira and Morrow (1971).

4.4 Asymptotic Expansions and Singular Perturbations

4.4A Heuristics

Let $f_\epsilon(x)$ be a C^1 mapping of a Banach space X into a Banach space Y depending continuously on a small real parameter ϵ. The following abstract situation is commonly encountered in physical problems associated with the solutions of the operator equation $f_\epsilon(x) = 0$. There is a sequence $x_n(\epsilon) \in X$ $(n = 0, 1, 2, \ldots, N)$ such that:

(i) $\|f_\epsilon(x_n(\epsilon))\| = O(\epsilon^{n+1})$ for n fixed, as $\epsilon \to 0$;

(ii) for small nonzero ϵ, there is a solution $\bar{x}(\epsilon)$ of $f_\epsilon(x) = 0$ such that $\|\bar{x}(\epsilon) - x_n(\epsilon)\| = O(\epsilon^{n+1})$ for n fixed, as $\epsilon \to 0$.

Under these circumstances we say that $x_n(\epsilon)$ is an *asymptotic* approximation to the solution $\bar{x}(\epsilon)$. Clearly, for ϵ sufficiently small and fixed n, an asymptotic $x_n(\epsilon)$ provides as accurate an approximation to the solution $\bar{x}(\epsilon)$ as desired, even though in many important applications the sequence $\{\|x_n(\epsilon)\|\}$ *diverges* as $n \to \infty$, for fixed $\epsilon \neq 0$ (cf. the discussion of (1.2.12) in Section 1.2B). Clearly, one must determine circumstances ensuring property (ii) of asymptotic approximation of a given sequence $\{x_n(\epsilon)\}$ satisfying (i) since numerical schemes for satisfying this latter property are well known.

As a concrete problem in this connection, consider the semilinear Dirichlet problem

$$(\Pi_\epsilon) \qquad \epsilon^2 \Delta u + u - g^2(x)u^3 = 0, \qquad u|_{\partial\Omega} = 0,$$

defined on a domain Ω in $\mathbb{R}^N$, where $g(x)$ is a smooth strictly positive function of Ω. We wish to investigate the solutions of this system for small ϵ. In particular, we wish to justify the heuristic idea that the one-signed solutions obtained by setting $\epsilon = 0$ in (Π_ϵ) (namely, $u_0(x) = \pm g(x)$) should be "zero-order" approximations to solutions of (Π_ϵ) for small ϵ if properly modified near $\partial\Omega$ so as to satisfy the Dirichlet boundary condition on $\partial\Omega$ Such a justification is relatively easy if $N = 1$, $g(x) = 1$, and $\Omega = (0, 1)$ since in that case an explicit solution in terms of Jacobian elliptic function[1] $\operatorname{sn}(x, k)$ is

$$u_1 = \left(\frac{2k^2}{1 + k^2} \right)^{1/2} \operatorname{sn}(K(k)x, k),$$

where $1/\epsilon = 2(1 + k^2)K(k) > \pi$. Now, since

$$\xi = \int_0^v \frac{dw}{(1 - w^2)^{1/2}(1 - k^2 w^2)^{1/2}} \leqslant \int_0^v \frac{dw}{1 - w^2} = \operatorname{arctanh} v,$$

$$\tanh \xi \leqslant \operatorname{sn}(\xi, k) \leqslant 1 \qquad \text{for} \quad 0 \leqslant \xi \leqslant K(k).$$

[1] Here the Jacobian elliptic function $\operatorname{sn}(\xi - \xi_0, k)$ is the solution of the differential equation $v_\xi^2 = (1 - v^2)(1 - k^2 v^2)$. The function $\operatorname{sn}(\xi, k)$ is a periodic function of ξ with quarter-period $K(k)$, with the same symmetry as $\sin \xi$, and maximum value 1 at $\xi = K(k)$. Furthermore, $\operatorname{sn}(\xi, 0) = \sin \xi$, $K(0) = \pi/2$, and as $k \uparrow 1$ with ξ fixed,

$$\operatorname{sn}(\xi, k) \sim \tanh \xi, \qquad K(k) \sim \ln\{4(1 - k^2)^{-1/2}\},$$

and $K(k)$ is an increasing function on $[0, 1)$.

Set $\delta(\epsilon) = (1 - k^2)^{1/2}$. As $\epsilon \downarrow 0$, $k \uparrow 1$, $\delta \sim 4 \exp(-1/2^{3/2}\epsilon)$ then

$$\tanh \frac{x}{\epsilon(2-\delta^2)^{1/2}} \leqslant \left(1 + \frac{\delta^2}{2(1-\delta^2)}\right)u_1 \leqslant 1 \qquad \text{for} \quad 0 \leqslant x \leqslant \tfrac{1}{2}.$$

Thus as $\epsilon \downarrow 0$ and x is bounded away from 0, 1, the function u_1 differs from 1 by terms exponentially small in ϵ and such that $u_1 < 1$. Near $x = 0, 1$ there is a *boundary layer* of width $O(\epsilon)$ in which $u_1 \sim \tanh(x/\sqrt{2}\,\epsilon)$.

However, for $N > 1$, the explicit justification analogous to that just described is impossible and so a qualitative discussion is required. To this end we prove a general result in the next subsection and apply it to (Π_ϵ) thereafter.

4.4B The validity of formal asymptotic expansions

The above example is clearly quite distinct from those described in the bifurcation theory earlier in the chapter. It is an example of a class of singular perturbation problems that can be posed abstractly as follows: Let $f_\epsilon(x)$ be a C^1 mapping of the type just described. Suppose there is a sequence $\{x_n(\epsilon)\}$ $(n = 0, 1, 2, \ldots, N)$ of elements X such that (i) $\|f_\epsilon(x_n(\epsilon))\| = O(\epsilon^{n+1})$ for fixed n as $\epsilon \to 0$, but (ii) the linear operator $f'_\epsilon(x_n(0))$ is not necessarily invertible for any n. Under what circumstances is $x_n(\epsilon)$ an *asymptotic approximation* to a solution $x(\epsilon)$ of $f_\epsilon(x) = 0$ (in the sense defined in Section 4.4A)?

Generally speaking, the asymptotic approximations $x_n(\epsilon)$ that are commonly used can be written as a power series in ϵ, $x_n(\epsilon) = \sum_{i=0}^{n} \alpha_i \epsilon^i$, where α_i may depend on ϵ but $\|\alpha_i\|$ is bounded independently of ϵ. Such truncated power series are called asymptotic expansions. In this case we shall assume hypotheses (I)–(III) below in order to establish an answer to the singular perturbation problem just raised.

(I) There is a constant M independent of ϵ and $x, y, \rho \in X$ such that:

(a) $\|f_\epsilon(x+y) - f_\epsilon(x) - f'_\epsilon(x)y\| \leqslant M\|y\|^2$ for all x, y with $\|x\|$, $\|y\| \leqslant R$; and

(b) $\|(f'_\epsilon(x) - f'_\epsilon(y))\rho\| \leqslant M\|x - y\|\,\|\rho\|$ for all $\|\rho\| \leqslant R$.

(Note that this condition is automatically satisfied if the map $f_\epsilon(x)$ has a uniformly bounded second derivative.)

(II) There is an element $x_n(\epsilon) \in X$ for all integers $n \leqslant N$ (a given integer) such that $x_n(\epsilon) = \sum_{i=0}^{n} \alpha_i \epsilon^i$ with $\|\alpha_i\| \leqslant A_i < \infty$, where A_i (but not necessarily α_i) is independent of ϵ and $\|f_\epsilon(x_n(\epsilon))\| = O(|\epsilon|^{n+1})$.

(III) There are constants $c, p > 0$ independent of ϵ and ρ such that for $\epsilon > 0$, $f'_\epsilon(x_i)$ is invertible and

$$\|f'_\epsilon(x_i)\rho\| \geqslant c\epsilon^p\|\rho\| \qquad \text{for some } i \geqslant p.$$

(*Notational remark:* For convenience, the symbol $\|\cdot\|$ is used to denote appropriate norms in either X or Y.)

We now state:

(4.4.1) **Theorem** Suppose $f_\epsilon(x)$ is a one-parameter family of continuously differentiable maps of X into Y for small nonnegative ϵ, satisfying hypotheses (I)–(III). Then for $N \geqslant 2p$ and $n \leqslant N-p$, there is a solution $\bar{x}_\epsilon = x_n(\epsilon) + \rho_n(\epsilon) \in X$ of $f_\epsilon(x) = 0$ with $\|\rho_n(\epsilon)\| = O(\epsilon^{n+1})$ for each n and each ϵ in some interval $(0, \epsilon_0)$, where ϵ_0 is some small positive number. This solution $\bar{x}_\epsilon$ is independent of n, and is the unique solution of $f_\epsilon(x) = 0$ such that $\|x - x_p\| = O(\epsilon^{p+1})$.

Idea of the proof: The result is demonstrated in four stages:

(i) the solution ρ_n of the equation $f_\epsilon(x_n + \rho_n) = 0$ is rewritten in the form $\rho_n = T_\epsilon \rho_n$ where T_ϵ is a bounded mapping of X into itself;

(ii) for any integer $n \geqslant N-p$ (in particular for $n = N$) an application of the contraction mapping theorem to the equation $\rho = T_\epsilon \rho$ yields a unique solution ρ_n with $\|\rho_n\| = O(\epsilon^{n-p+1})$;

(iii) for any integer $n \leqslant N-p$, it is shown that (ii) implies that $\rho_n = \rho_N + \sum_{j=n+1}^{N} \alpha_j \epsilon^j$ satisfies $\rho = T_\epsilon \rho$ with $\|\rho_n\| = O(\epsilon^{n+1})$, as required;

(iv) for any two solutions ρ_n and ρ_n' of $\rho = T_\epsilon \rho$ both of order $O(\epsilon^{n+1})$ with $n \geqslant p$, their difference $\delta_n = \rho_n - \rho_n'$ satisfies the equation $\delta = T_\epsilon(\rho_n + \delta) - T_\epsilon \rho_n$. This fact is shown to imply that $\delta_n = 0$.

Note that once (i)–(iii) are established $\bar{x}_\epsilon = x_n(\epsilon) + \rho_n$ is independent of n for $0 \leqslant n \leqslant N-p$ since if $\bar{x}_\epsilon = x_n(\epsilon) + \rho_n$ and $\bar{x}_\epsilon' = x_m(\epsilon) + \rho_m$ for $0 \leqslant m, n \leqslant N-p$, $\bar{x}_\epsilon = \bar{x}_\epsilon'$ follows immediately from the definition of ρ_n and ρ_m given in (iii).

Proof: (i): First we show that by virtue of the hypotheses of the theorem

(4.4.2) $$\|f_\epsilon'(x_n)\rho\| \geqslant \tfrac{1}{2} c\epsilon p \|\rho\|$$

provided $n \geqslant i$ and ϵ is sufficiently small. Indeed

$$\begin{aligned} \|f_\epsilon'(x_n)\rho\| &\geqslant \|f_\epsilon'(x_i)\rho\| - \|(f_\epsilon'(x_n) - f_\epsilon'(x_i))\rho\| \\ &\geqslant c\epsilon^p \|\rho\| - M\|x_n - x_i\|\,\|\rho\| \qquad \text{(by (I), (iii))} \\ &\geqslant \left(c\epsilon^p - \sum_{j=i+1}^{n} A_j \epsilon^j\right)\|\rho\| \qquad \text{(by (II))} \\ &\geqslant \tfrac{1}{2} c\epsilon^p \|\rho\| \qquad \text{(for sufficiently small } \epsilon\text{) since } i \geqslant p. \end{aligned}$$

Hence $f_\epsilon'(x_n)$ is an invertible linear map for $n \geqslant i$:

(4.4.3) $$\|f_\epsilon'^{-1}(x_n)\| \leqslant (2/c)\epsilon^{-p}$$

Now let a tentative solution be denoted $\bar{x}_\epsilon = x_n + \rho_n$ and $f_\epsilon(x_n) = -\epsilon^{n+1} g_n(x, \epsilon)$, where $\|g_n\| \leqslant \tilde{c}_n$, a constant independent of ϵ. Hence we wish to determine ρ_n such that $f_\epsilon(x_n + \rho_n) = 0$; i.e., we wish to solve

(4.4.4) $$f_\epsilon(x_n + \rho_n) - f_\epsilon(x_n) = \epsilon^{n+1} g_n.$$

Now by virtue of hypothesis (I), the equation (4.4.4) can be rewritten

(4.4.5) $$f'_\epsilon(x_n)\rho_n + R_\epsilon(x_n, \rho_n) = \epsilon^{n+1}g_n,$$

where

(4.4.6) $$\|R_\epsilon(x_n, \rho_n)\| \leqslant M\|\rho_n\|^2.$$

By virtue of (4.4.3), (4.4.5) may be rewritten

(4.4.7) $$\rho_n = f_\epsilon'^{-1}(x_n)\{\epsilon^{n+1}g_n - R_\epsilon(x_n, \rho_n)\}.$$

(ii): To obtain a solution ρ_n of (4.4.7) we apply the contraction mapping theorem (3.1.1) to the map

$$T_\epsilon\rho = f_\epsilon'^{-1}(x_n)\{\epsilon^{n+1}g_n - R_\epsilon(x_n, \rho)\}$$

acting on the sphere $S(\delta, s) = \{\rho \mid \rho \in X, \|\rho\| \leqslant \delta\epsilon^s\}$, where the numbers δ and s are to be determined so that T_ϵ maps $S(\delta, s)$ into itself. To this end, we note the following estimates for $\rho \in S(\delta, s)$:

$$\|\epsilon^{n+1}f_\epsilon'^{-1}(x_n)g_n\| = O(\epsilon^{n+1-p}) \qquad \text{(by (4.4.3))},$$

$$\|f_\epsilon'^{-1}(x_n)R_\epsilon(x_n, \rho)\| = O(\epsilon^{2s-p}) \qquad \text{(by (4.4.6))}.$$

Thus if we choose s such that (a) $n+1-p \geqslant s$ and (b) $2s - p > s$, T_ϵ will map $S(\delta, s)$ into itself for δ sufficiently large (in fact $\delta > 2\tilde{c}_n/c$) and ϵ sufficiently small. Together (a) and (b) imply that s may be chosen so that $n+1-p \geqslant s > p$. Thus we set $s = n-p+1$ for any $n \geqslant 2p$. (Note that such an n exists since $N \geqslant 2p$.) On the other hand, for any $\rho, \rho' \in S(\delta, n-p+1)$

$$\|T_\epsilon\rho - T_\epsilon\rho'\| \leqslant \|f_\epsilon'^{-1}(x_n)\|\,\|R_\epsilon(x_n, \rho) - R_\epsilon(x_n, \rho')\|.$$

Since

(4.4.8) $$\begin{aligned} R_\epsilon(x_n, \rho) - R_\epsilon(x_n, \rho') &= f_\epsilon(x_n + \rho) - f_\epsilon(x_n + \rho') - f'_\epsilon(x_n)(\rho - \rho') \\ &= \int_0^1 [f'_\epsilon(x_n + t\rho + (1-t)\rho') - f'_\epsilon(x_n)][\rho - \rho']\, dt, \end{aligned}$$

applying the mean value theorem and hypothesis (i), we obtain

$$\|T_\epsilon\rho - T_\epsilon\rho'\| \leqslant \|f_\epsilon'^{-1}(x_n)\|\{M(t_0\|\rho\| + (1-t_0)\|\rho'\|)\|\rho - \rho'\|\}$$

for some $t_0 \in (0, 1)$. Thus since $\|\rho\|, \|\rho'\| \leqslant \delta\epsilon^{n-p+1}$ and by virtue of (4.4.3),

$$\begin{aligned} \|T_\epsilon\rho - T_\epsilon\rho'\| &\leqslant \left(\tfrac{1}{2}c\epsilon^p\right)^{-1}(M\delta\epsilon^{n-p+1})\|\rho - \rho'\| \\ &\leqslant \frac{2}{c}M\delta\epsilon^{n-2p+1}\|\rho - \rho'\| \qquad \text{with} \quad n \geqslant 2p. \end{aligned}$$

Thus by choosing $\epsilon_0 < c/2M\delta$, T_ϵ is a contraction map of $S(\delta, n-p+1)$ into itself and so has a unique fixed point ρ_n with $\|\rho_n\| = O(\epsilon^{n-p+1})$. In particular for $n = N$ we have

(4.4.9) $$\|\rho_N\| = O(\epsilon^{N-p+1}).$$

(iii): Now suppose $n \leqslant N - p$, we find a ρ_n such that $f(x_n + \rho_n) = 0$ and $\|\rho_n\| = O(\epsilon^{n+1})$. Indeed let

$$\rho_n = \sum_{i=n+1}^{N} \alpha_i \epsilon^i + \rho_N,$$

then

$$f(x_n + \rho_n) = f\left(x_n + \sum_{i=n+1}^{N} \alpha_i \epsilon^i + \rho_N\right) = f(x_N + \rho_N) = 0.$$

Furthermore,

$$\|\rho_n\| \leqslant \|\sum_{i=n+1}^{N} \alpha_i \epsilon^i\| + \|\rho_N\| \leqslant \sum_{i=n+1}^{N} A_i \epsilon^i + O(\epsilon^{N-p+1}) \qquad \text{(by (4.4.9))}.$$

Hence $\|\rho_n\| = O(\epsilon^{n+1}) + O(\epsilon^{N-p+1})$ and provided $n \leqslant N - p$,

$$\text{(4.4.10)} \qquad \|\rho_n\| = O(\epsilon^{n+1}).$$

Thus for $n \leqslant N - p$, we have found the desired solution ρ_n to the equation (4.4.4) and furthermore the desired estimate (4.4.10).

(iv): Finally, we demonstrate the uniqueness of ρ_n for $n \geqslant p$ with $\|\rho_n\| = O(\epsilon^{n+1})$. For $n \geqslant 2p$, this fact is an immediate consequence of the fact that T_ϵ is a contraction mapping. Furthermore, for $n \geqslant p$, suppose there are two solutions ρ_n and $\rho_n + \delta_n$ of (4.4.5) satisfying the estimate (4.4.10). Then from (4.4.7) and (4.4.8), ρ_n is a solution of the equation

$$\text{(4.4.11)} \qquad v = -f_\epsilon'^{-1}(x_n)\int_0^1 [f_\epsilon'(x_n + \rho_n + tv) - f_\epsilon'(x_n)]v\, dt.$$

The uniqueness of ρ_n is a consequence of the following

(4.4.12) **Lemma** Equation (4.4.11) has the unique solution $v = 0$ for $\|v\| = O(\epsilon^{p+1})$ for ϵ sufficiently small.

Proof: Set $J(v) = \int_0^1 [f_\epsilon'(x_n + \rho_n + tv) - f_\epsilon'(x_n)]v\, dt$. Suppose v and $v + \delta$ are solutions of (4.4.11), then $f_\epsilon'(x_n)\delta = J(v) - J(v + \delta)$, so that by (III)

$$\text{(4.4.13)} \qquad \|J(v + \delta) - J(v)\| = \|f_\epsilon'(x_n)\delta\| \geqslant \tfrac{1}{2} c\epsilon^p \|\delta\|.$$

On the other hand, after suitable rearrangement, setting $y_n = x_n + \rho_n$,

$$J(v + \delta) - J(v) = \int_0^1 \{[f_\epsilon'(y_n + t(v + \delta)) - f_\epsilon'(y_n + tv)]v + [f_\epsilon'(y_n + t(v + \delta)) - f_\epsilon'(x_n)]\delta\}\, dt.$$

So by hypothesis (i)

$$\|J(v + \delta) - J(v)\| \leqslant M\|\delta\|\,\|v\| + M\{\|v\| + \|\delta\| + \|\rho_n\|\}\|\delta\|.$$

Assuming $\|\delta\| \neq 0$, and combining (4.4.13) and the above, we obtain

$\tfrac{1}{2} c\epsilon^p \leqslant M\|v\| + M\{\|v\| + \|\delta\| + \|\rho_n\|\}$ with $\|\rho_n\| = O(\epsilon^{n+1})$, which contradicts the fact that $\|v\| = \|\delta\| = O(\epsilon^{p+1})$. Hence $\delta = 0$ and the lemma is proved.

Remarks: In case $X = Y$ is a Hilbert space, the following condition can be substituted for hypothesis (III):

(III′) there is a Banach space $X' \supset X$ such that

$$(4.4.14)\qquad \text{(a)}\quad (f_\epsilon'(x_i)\rho, \rho) \geq c_1\|\rho\|_{X'}^2,$$

$$\text{(b)}\quad (f_\epsilon'(x_i)\rho, \rho) \geq c_2\epsilon^p\|\rho\|_X^2 - K\|\rho\|_{X'}^2,$$

where c_1, c_2, and K are positive constants independent of ϵ.

Indeed multiplying (a) by K, (b) by c_1, and adding, one obtains

$$(c_1 + K)(f_\epsilon'(x_i)\rho, \rho) \geq c_2c_1\epsilon^p\|\rho\|_X^2,$$

so that an application of the Cauchy–Schwarz inequality yields

$$\|f_\epsilon'(x_i)\rho\|_X \geq \left(\frac{c_2c_1\epsilon^p}{c_1 + K}\right)\|\rho\|_X.$$

The invertibility of $f_\epsilon'(x_i)$ follows from the Lax–Milgram theorem (1.3.21).

Special consideration is required for the important case of the asymptotic behavior of the approximation x_0, given a first order approximation $x_1(\epsilon) = x_0 + \alpha_1\epsilon$. More generally, in case $N = 2p - 1$ $(p \geq 1)$ an analogue of Theorem (4.4.1) holds. In fact, the interested reader will easily prove:

(4.4.15) **Theorem** Suppose in addition to the hypotheses of Theorem (4.4.1) that $c^2 > 4M\|f_\epsilon(x_N(\epsilon))\|/\epsilon^{N+1}$ for all positive ϵ sufficiently small. Then the conclusions of Theorem (4.4.1) are valid for $N = 2p - 1$ and $n \leq p - 1$, except (possibly) the uniqueness statement for $n = p - 1$.

Example (a sequence $\{x_n(\epsilon)\}$ such that $\epsilon^{-n}f_\epsilon(x_n) \to 0$ as $\epsilon \to 0$ for any fixed n, that is not an approximation to a solution of $f_\epsilon(x) = 0$ for any $\epsilon \neq 0$) Consider the problem

$$(4.4.16)\qquad \epsilon^2k\,\frac{d^2x}{dt^2} + x = g(t), \qquad x \to 0 \quad \text{as} \quad t \to \pm\infty,$$

where $k = \pm 1$ and g is a given function that is $N + 2$ times continuously differentiable on $(-\infty, \infty)$ with $g^{(r)}(t) = O(|t|^{-1/2-\delta})$, $\delta > 0$, for $t \to \infty$ and $r = 0, 1, \ldots, N + 2$.

For both values of k there exist functions $x_n(t, \epsilon)$ that appear to be approximate solutions of the problem. In fact, an obvious iterative scheme starts with $x_0(t) = g(t)$ and yields

$$(4.4.17)\qquad x_{2m}(t, \epsilon) = g(t) - \epsilon^2kg''(t) + \cdots + \epsilon^{2m}(-k)^mg^{(2m)}(t),$$

$(0 < 2m = N)$, and an elementary calculation shows that

$$(4.4.18)\qquad \epsilon^2k\,\frac{d^2x_{2m}}{dt^2} + x_{2m} = g(t) + O(\epsilon^{2m+2}) \quad \text{uniformly in } t,$$

$$x_{2m} \to 0 \qquad \text{as} \quad t \to \pm\infty.$$

However, for $k = 1$ this result is misleading. The general solution of the differential equation can be written explicitly, and on this basis it is straightforward to show that:

(a) *for $k = 1$, $\epsilon \neq 0$ and $g = e^{-t^2}$, say, the problem* (4.4.16) *has no solution*;

(b) *for $k = -1$, the problem has a solution $\bar{x}(t, \epsilon)$ and x_{2m} is an asymptotic approximation to $\bar{x}$.*

We shall now show that both these conclusions are in accord with the (sufficient) conditions of Theorem (4.4.1). Choose X and Y to be the Sobolev space $W_{1,2}(-\infty, \infty)$, which is a real Hilbert space with inner product

$$(x, y) = \int_{-\infty}^{\infty} \left\{ \frac{dx}{dt} \frac{dy}{dt} + xy \right\} dt.$$

Then $f_\epsilon(x)$ and $f_\epsilon(x)$ (which will be written f'_ϵ in the present case, because here it is independent of x) are implicitly defined by

$$(f_\epsilon(x), \varphi) = \int_{-\infty}^{\infty} \left\{ -\epsilon^2 k \frac{dx}{dt} \frac{d\varphi}{dt} + x\varphi - g\varphi \right\} dt \qquad \text{for all } \varphi \in X,$$

$$(f'_\epsilon \rho, \varphi) = \int_{-\infty}^{\infty} \left\{ -\epsilon^2 k \frac{d\rho}{dt} \frac{d\varphi}{dt} + \rho\varphi \right\} dt \qquad \text{for all } \varphi \in X,$$

where we have used the Riesz representation theorem for linear functionals in Hilbert space and the fact that, for fixed x and ρ, the integrals are bounded linear functionals defined for all $\varphi \in X$.

It follows at once that condition (I) in satisfied, with $M = 0$. Condition (II) is also satisfied (we can define $x_{2m+1} = x_{2m}$): Using the differentiability of the x_{2m}, the Schwarz inequality, and (4.4.18) in which the O-term is actually $\epsilon^{2m+2}(-k)^m g^{(2m+2)}(t)$, we obtain

(4.4.19) $$\|f_\epsilon(x_{2m})\| = \sup_{\|\varphi\|=1} \int_{-\infty}^{\infty} \left(\epsilon^2 k \frac{d^2 x2m}{dt^2} + x_{2m} - g \right) \varphi \, dt$$

$$\leqslant \left\{ \int_{-\infty}^{\infty} \left(\epsilon^2 k \frac{d^2 x_{2m}}{dt^2} + x_{2m} - g \right)^2 dt \right\}^{1/2} = O(\epsilon^{2m+2}).$$

Therefore it remains to show that *condition* (III) *is violated for* $k = 1$ *and holds for* $k = -1$.

To establish the former, consider the function

(4.4.20) $$z(t, \epsilon, \mu) = \zeta(\mu t) \cos(t/\epsilon),$$

where μ is a small positive number and $\zeta(s)$ is a mollifier as follows: $\zeta \in C^\infty(-\infty, \infty)$, $\zeta(s) = 1$ for $|s| \leqslant 1$, $\zeta(s) = 0$ for $|s| \geqslant 2$, and $0 \leqslant \zeta(s) \leqslant 1$ for all s. Clearly, $z \in X$, and if we can show that

(4.4.21) $$\frac{\|f'_\epsilon z\|}{\|z\|} \to 0 \quad \text{as} \quad \mu \to 0 \quad (\text{with } \epsilon \text{ fixed, and } k = 1),$$

then no inequality of the form (III) is possible. To prove (4.4.21), we have (cf. the derivation of (4.4.19))

$$\|f'_\epsilon z\| \leqslant \left\{ \int_{-\infty}^{\infty} \left(\epsilon^2 \frac{d^2 z}{dt^2} + z \right)^2 dt \right\}^{1/2} = O(\mu^{1/2}),$$

upon substitution of (4.4.20). Also

$$\|z\|^2 > \int_{-1/\mu}^{1/\mu} \left\{ \left(\frac{dz}{dt} \right)^2 + z^2 \right\} dt$$

$$= \int_{-1/\mu}^{1/\mu} \left\{ \left(\frac{1}{\epsilon} \sin \frac{t}{\epsilon} \right)^2 + \cos^2 \frac{t}{\epsilon} \right\} dt \geqslant \frac{2}{\mu} \qquad \text{for } \epsilon \leqslant 1,$$

and so (4.4.21) is proved.

To show that (III) holds for $k = -1$, we refer to the alternative form (III′), with $X' = L_2(-\infty, \infty)$; for all $\rho \in X$ we have

$$(f'_\epsilon \rho, \rho) \geqslant \|\rho\|^2_{X'} \quad \text{and} \quad (f'_\epsilon \rho, \rho) \geqslant \epsilon^2 \|\rho\|^2_X - \|\rho\|^2_{X'}.$$

4.4C Application to the semilinear Dirichlet problem (Π_ϵ)

In order to illustrate the applicability of Theorem (4.4.1), we consider the semilinear Dirichlet problem (Π_ϵ) mentioned earlier. The following fact can be established:

(A) *For sufficiently small values of* ϵ, *the problem* (Π_ϵ) *has a unique smooth positive solution* $u(x, \epsilon)$ *that tends to* $1/g(x)$ *as* $\epsilon \to 0$, *outside a narrow "boundary layer" of width* $O(\epsilon)$ *concentrated near* $\partial\Omega$.

This result is proven by (B) constructing an approximate solution $U_M(x, \epsilon) = \sum_{m=0}^{M} \epsilon^m u_m(x, \epsilon)$ such that $u(x, \epsilon) - U_M(x, \epsilon)$ is $O(\epsilon^{M+1})$ uniformly on $\bar{\Omega}$ (for any integer M). Then we show that for $M = 0$ this expansion is indeed asymptotic to a true solution of (Π_ϵ) for ϵ sufficiently small, and consider the behavior of $U_0(x, \epsilon)$ for small ϵ. In order to apply (4.4.1), we construct approximations $U_m(x, \epsilon)$ for $m \geqslant 1$ satisfying

(4.4.22) $$K_\epsilon(U_M) = \epsilon^2 \Delta U_m + U_M - g^2(x) U_M^3 = O(\epsilon^{M+1}) \quad \text{uniformly in } \Omega,$$

$$U_M|_{\partial\Omega} = 0.$$

To display the principal ideas in the construction of such approximations, we describe the steps leading to the lowest approximation $U_0(x, \epsilon)$.

(a) By neglecting $\epsilon^2 \Delta u$ in the differential equation $K_\epsilon u = 0$, we find the approximation $v_0(x) = 1/g(x)$, which is expected to differ from the positive solution u by $O(\epsilon^2)$ for $\epsilon \downarrow 0$ with x bounded away from $\partial\Omega$.

(b) To describe u near the boundary $\partial\Omega$, we first attach labels (s, t) to the points in a fixed neighborhood $\Omega_* = \{x \mid 0 < t < t_*\}$ of $\partial\Omega$; here $s = x_0 \in \partial\Omega$ labels the normal to $\partial\Omega$ through the point x_0, while t measures distance from $\partial\Omega$, as in Fig. 4.6. For explicit calculations, s is replaced by an $(N-1)$-dimensional surface coordinate σ. Making the *stretching transformation* $t = \epsilon\tau$, we then seek a function $w_0(s, \tau)$ which is to approximate u in the "boundary layer"; more precisely, $u - w_0$ is to be $O(\epsilon)$ for $\epsilon \downarrow 0$ with τ fixed (so that $t \downarrow 0$). Writing the operator K_ϵ in terms of s, τ, and ϵ and keeping only the dominant terms of this form of K_ϵ, we obtain the boundary layer problem

(4.4.23a) $$\frac{\partial^2 w_0}{\partial \tau^2} + w_0 - g_*^2(s, 0) w_0^3 = 0,$$

(4.4.23b) $$w_0|_{\tau=0} = 0, \quad w_0 \to \frac{1}{g_*(s, 0)} \quad \text{as} \quad \tau \uparrow \infty,$$

where $g_*(s, t) = g(x)$, and where the condition for $\tau \uparrow \infty$ is a "matching" condition, suggested by the fact $v_0(x) \to 1/g_*(s, 0)$ as $t \downarrow 0$. The solution of

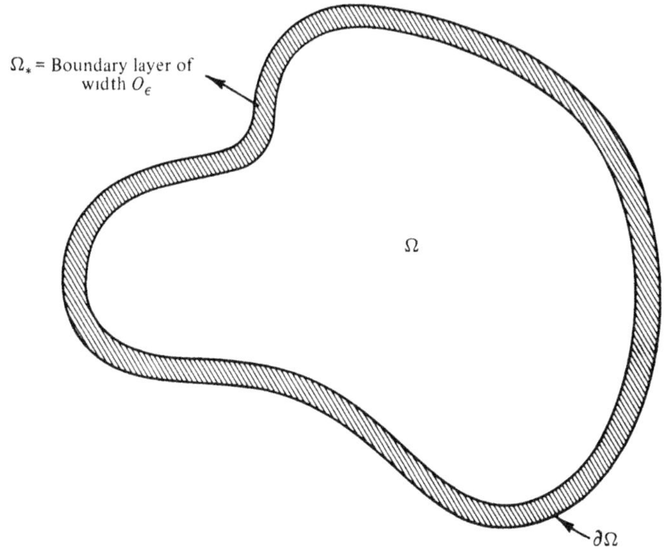

FIG. 4.6 Illustration of the boundary layer phenomena for (Π_ϵ).

the problem (4.4.23) in which s has the role of parameter is

$$w_0 = \frac{1}{g_*(s, 0)} \tanh \frac{\tau}{2^{1/2}} .$$

(c) Because $\lim_{t\downarrow 0} v_0(x) = \lim_{\tau\uparrow\infty} w_0(s, \tau)$ (a particular case of what is often called the "asymptotic matching principle"), we can now use the prescription for forming the leading term of the "composite expansion." We define

$$U_0^*(x, \epsilon) = v_0(x) + w_0\left(s, \frac{t}{\epsilon}\right) - \lim_{\tau\uparrow\infty} w_0 \tag{4.4.24}$$

$$= \frac{1}{g(x)} + \frac{1}{g_*(s, 0)}\left(\tanh \frac{t}{\epsilon 2^{1/2}} - 1\right)$$

on $\overline{\Omega}_*$: $0 \leqslant t \leqslant t_*$. To overcome the trivial difficulty that t is not uniquely defined on $\Omega - \Omega_*$, we note that $\tanh(t/\epsilon 2^{1/2}) - 1$ is transcendentally small for $\epsilon\downarrow 0$ with t fixed and positive, and introduce a mollifier $\zeta(x) \in C^\infty(\overline{\Omega})$ such that $\zeta(x) = 1$ on $0 \leqslant t \leqslant 1/2t_*$, $\zeta(x) = 0$ on $\Omega - \Omega_*$, and $0 \leqslant \zeta \leqslant 1$ on $\overline{\Omega}$. We can then define

$$U_0(x, \epsilon) = \frac{1}{g(x)} + \frac{\zeta(x)}{g_*(s, 0)}\left(\tanh \frac{t}{\epsilon 2^{1/2}} - 1\right), \tag{4.4.25}$$

and it is readily verified that this satisfies (4.4.22) with $M = 0$. This is because of the dual role of the final term in (4.4.24) which to the lowest order cancels v_0 in the boundary layer, cancels w_0 at points bounded away from $\partial\Omega$, and makes U_0 more accurate than either v_0 or w_0 in the intermediate region defined by $t = \theta(\epsilon)$, where θ is any function such that $t = \theta(\epsilon)\downarrow 0$ and $\tau = \epsilon^{-1}\theta(\epsilon)\uparrow\infty$ as $\epsilon\downarrow 0$.

For higher approximations U_M, $M \geqslant 1$, the constituent functions in (4.4.24) are replaced by finite series

$$\sum_{m=0}^{M} \epsilon^m v_m(x), \qquad \sum_{\mu=0}^{M} \epsilon^\mu w_\mu(s,\tau), \qquad \text{and} \qquad \sum_{\mu=0}^{M} \epsilon^\mu \sum_{n=0}^{\mu} w_{\mu,n}(s)\tau^n,$$

respectively. The proof of (4.4.22) is then complicated by the fact that the coefficient functions v_m, w_μ, and $w_{\mu,n}$ are defined by elaborate recurrence relations.

Now assuming the approximations $U_M(x,\epsilon)$ have been constructed, we must reformulate the Dirichlet problem as an operator equation $f(x,\epsilon) = 0$, in appropriately defined Banach spaces X and Y, in such a way that we can verify the hypotheses (I)–(III) of Theorem (4.4.1). The duality procedure described in (Section 2.2D) shows that (at least for $N \leqslant 3$) the solutions of (Π_ϵ) can be regarded as generalized solutions of an operator equation of the form

$$L_\epsilon u + Nu = 0,$$

where L_ϵ and N are bounded mappings of $\mathring{W}_{1,2}(\Omega)$ into itself defined implicitly by the formulas:

$$(L_\epsilon u, v)_{1,2} = \int_\Omega (\epsilon^2\, \nabla u \cdot \nabla\phi - u\cdot\phi),$$

$$(Nu, v)_{1,2} = \int_\Omega g^2(x) u^3 \phi.$$

Now if one defines $f_\epsilon(u) = L_\epsilon u + Nu$, then hypotheses (I) and (II) are easily verified since the second derivative of $f_\epsilon(x)$ is uniformly bounded in any sphere $\|u\|_{1,2} \leqslant R$, and by virtue of (4.4.22),

$$\begin{aligned} \|f_\epsilon(U_M(\epsilon))\|_{1,2} &= \sup_{\|v\|_{1,2}\leqslant 1} (L_\epsilon U_M + NU_M, v) \\ &= \sup_{\|v\|\leqslant 1} \left|\int_\Omega K_\epsilon(U_M)v\right| \\ &\leqslant O(\epsilon^{M+1}). \end{aligned}$$

However, the verification of hypothesis (III) is more subtle and, in fact, *motivates our selection of X as $\mathring{W}_{1,2}(\Omega)$* since, in this case, integral estimates are easier to obtain then pointwise ones. We now assume that, as in the case $M = 0$, the approximations

$$(4.4.26) \qquad U_M(x,\epsilon) = 1/g(x)\left\{\zeta(x)\tanh\frac{\tau}{2} + [1-\zeta(x)]\right\}\{1 + O(\epsilon)\}.$$

We set $f_\epsilon'(U_M) = \mathfrak{L}_{M,\epsilon}$ and prove the estimate contained in

(4.4.27) **Lemma** There exist positive numbers $\nu(M)$ and $\epsilon_0(M)$ such that, for any $\rho \in \mathring{W}_{1,2}(\Omega)$ and for $0 < \epsilon \leqslant \epsilon_0$,

(4.4.28) $$\langle \mathfrak{L}_{M,\epsilon}\rho, \rho\rangle = \int_\Omega \{\epsilon^2(\nabla\rho)^2 + (3g^2U_M^2 - 1)\rho^2\} \geqslant \epsilon^2\nu^2 \int_\Omega (\nabla\rho)^2,$$

where ν is independent of ϵ and ρ.

Proof: The primary steps in the proof are, first, a sharp form for the Poincaré inequality, and secondly, a simple approximation to the function U_M.

(i): Writing ρ as a line integral along the inward normal to $\partial\Omega$, and using the Schwarz inequality, we have

$$\rho^2 = \left\{\int_0^t \rho_{t'}(s, t')\,dt'\right\}^2 \leqslant t\int_0^t \rho_{t'}^2\,dt'.$$

Integrating over the subset of $\overline{\Omega}_*$ determined by $0 \leqslant t \leqslant l$, we obtain

(4.4.29) $$\int_{t<l} \rho^2 \leqslant H(l)\,\frac{l^2}{2}\int_{t<l} \rho_t^2,$$

where $H(l)$ is a continuous function resulting from the curvature(s) of $\partial\Omega$, such that $H(l) \to 1$ as $l \to 0$.

(ii): We define

$$h_M(x, \epsilon) = 3 - 3g^2U_M^2.$$

Then one shows that (in terms of the mollifier $\zeta(x)$ defined before (4.4.25))

$$\frac{1}{3}h_M - \zeta(x)\operatorname{sech}^2\frac{\tau}{2^{1/2}} = \zeta(1-\zeta)\left(1 - \tanh\frac{\tau}{2^{1/2}}\right)^2 + O(\epsilon),$$

and at those points $(t_*/2 < t < t_*)$ where $\zeta(1-\zeta) \neq 0$, the function $1 - \tanh(\tau/2^{1/2})$ is exponentially small for $\epsilon \downarrow 0$. Accordingly

(4.4.30) $$h_M(x, \epsilon) = 3\zeta(x)\operatorname{sech}^2\frac{\tau}{2^{1/2}} + r_M, \qquad |r_M| \leqslant k\epsilon,$$

where $k = k(M)$ is a constant independent of x and ϵ.

(iii): We can now estimate the functional

$$\langle \mathfrak{L}_{M,\epsilon}\rho, \rho\rangle = \int_\Omega \{\epsilon^2(\nabla\rho)^2 + (2 - h_M)\rho^2\}.$$

By (4.4.29) with $l = a\epsilon$ (where a is to be determined),

$$\int_\Omega \epsilon^2(\nabla\rho)^2 \geqslant \frac{2}{a^2H(a\epsilon)}\int_{t\leqslant a\epsilon} \rho^2,$$

and by the approximation (4.4.30)

$$-\int_\Omega h_M\rho^2 \geqslant -3\int_{t\leqslant a\epsilon}\rho^2 - 3\operatorname{sech}^2\frac{a}{2^{1/2}}\int_{t\geqslant a\epsilon}\rho^2 - k\epsilon\int_\Omega \rho^2,$$

where $t \geqslant a\epsilon$ refers to the complement of $t < a\epsilon$ in the whole domain Ω (even though t is not uniquely defined on $\Omega - \Omega_*$). Then

$$\langle \mathfrak{L}_{M,\epsilon}\rho, \rho\rangle \geqslant \left(\frac{2}{a^2H(a\epsilon)} - 1 - k\epsilon\right)\int_{t\leqslant a\epsilon}\rho^2 + \left(2 - 3\operatorname{sech}^2\frac{a}{2^{1/2}} - k\epsilon\right)\int_{t\geqslant a\epsilon}\rho^2.$$

Now $(2/a^2H(0)) - 1 > 0$ for $a < 2^{1/2}$, and $2 - 3\,\mathrm{sech}^2(a/2^{1/2}) > 0$ for $a > \frac{2}{3}2^{1/2}$. Choose $a = \frac{5}{6}2^{1/2} \equiv a_0$, and ϵ_0 so small that, for $0 < \epsilon \leqslant \epsilon_0$,

$$\min\left\{ \frac{2}{a_0^2 H(a_0\epsilon)} - 1,\ 2 - 3\,\mathrm{sech}^2\frac{a_0}{2^{1/2}} \right\} - k\epsilon_0 \geqslant \mu^2 > 0,$$

where $\mu(M)$ is independent of ϵ. Then

$$\langle \mathcal{L}_{M,\epsilon}\rho, \rho\rangle \geqslant \mu^2 \int_\Omega \rho^2 \quad \text{for} \quad 0 < \epsilon \leqslant \epsilon_0. \tag{4.4.31}$$

(iv): Since (4.4.30) implies that $2 - h_M \geqslant -1 - k\epsilon$, we also have

$$\langle \mathcal{L}_{M,\epsilon}\rho, \rho\rangle = \int_\Omega \{\epsilon^2(\nabla\rho)^2 + (2 - h_M)\rho^2\} \geqslant \epsilon^2 \int_\Omega (\nabla\rho)^2 - (1 + k\epsilon_0)\int_\Omega \rho^2,$$

for $0 < \epsilon \leqslant \epsilon_0$. Multiplying this by $\mu^2/(1 + k\epsilon_0)$, adding (4.4.31), and defining $\nu^2 = \mu^2/(1 + k\epsilon_0 + \mu^2)$, we obtain the result (4.4.27).

Thus the hypotheses of Theorem (4.4.1) are completely verified so that since the approximations $U_M(x, \epsilon)$ can be computed for $M = 2$, $U_0(x, \epsilon)$ is an asymptotic approximation to a true solution in the $\mathring{W}_{1,2}(\Omega)$ norm. However, there remains the problem of verifying (A) in the pointwise sense. The statement (A) follows immediately from the form of $U_0(x, \epsilon)$ discussed in (4.4.25), provided we show that $U_0(x, \epsilon)$ is an asymptotic approximation in the $C(\bar{\Omega})$ norm. For $N = 1$, this fact follows from Sobolev's inequality (1.4.1); indeed, $\|U_0(x, \epsilon)\|_C \leqslant \text{const.}\|U_0(x, \epsilon)\|_{1,2}$, where the constant is independent of ϵ. For $N = 2, 3$, the result follows the details of the proof of (4.4.1) and the L_2 regularity theorem for linear elliptic equations of Section 1.5. Indeed, utilizing the notation and results of Theorem (4.4.1),

$$u(x, \epsilon) = U_0 + \sum_{i=1}^{3} U_i\epsilon^i + \rho_4, \tag{4.4.32}$$

where $\|\rho_4\|_{1,2} = O(\epsilon^3)$. The L_2 regularity theory implies that for $N = 3$, ρ_4 is a generalized solution of the linear equation $\epsilon^2\,\Delta u = f(\rho_4)$ with $f \in L_2(\Omega)$ and $\|f\|_{0,2} = O(\epsilon^3)$. Consequently, $\|\rho_4\|_{2,2} = \epsilon^{-2}\|f(\rho_4)\|_{0,2} = O(\epsilon)$. Now the Sobolev imbedding theorem yields the estimate $\|\rho_4\|_{C(\bar{\Omega})} = O(\epsilon)$, so that (4.4.32) implies $|u(x, \epsilon) - U_0| = O(\epsilon)$.

Remark For $N > 3$ and nonlinearities that grow faster than u^3, an exact analogue of the result (A) holds. See the Notes at the end of the chapter for references and the idea of the proof.

4.5 Some Singular Perturbation Problems of Classical Mathematical Physics

Many problems P_ϵ in mathematical physics can be made to depend smoothly on a small real parameter ϵ, in such a way that when $\epsilon = 0$ a solution x_0 to P_0 can be exhibited explicitly. One then attempts to show that *the small simplifying change in passing from* P_ϵ *to* P_0 *has a correspondingly small effect in the solution* $x(\epsilon)$ *of* P_ϵ; i.e., the problem P_ϵ has a solution $x(\epsilon) = x_0 + o(1)$ as $\epsilon \to 0$. What *actually* occurs in a broad class of important problems P_ϵ defined by equations of the type $f_\epsilon(x) = 0$ is that a *formal* solution $x(\epsilon) = \sum_{i=0}^{\infty} \alpha_i \epsilon^i$ for $f_\epsilon(x) = 0$ exists with $x(0) = \alpha_0$ satisfying $f_0(x) = 0$. However, such series often *actually diverge* for $\epsilon \neq 0$ *since the magnitude of the coefficients* α_n *do not tend to zero*. Nonetheless, one hopes that if $x(\epsilon)$ is truncated to $x_n(\epsilon) = \sum_{i=0}^{n} \alpha_i \epsilon^i$, then $x_n(\epsilon)$ is *asymptotic* (in the sense mentioned in Section 4.4A) to a true solution $\tilde{x}(\epsilon)$ of $f_\epsilon(x) = 0$. In this section we point out three problems of this class, in which this asymptotic property of a formal solution can be proven, on the basis of Theorem (4.4.1). In all the cases considered below, we note that (4.4.1) reduces the problem to finding sharp bounds for the norms of certain linear operators.

4.5A Perturbation of an anharmonic oscillator by transient forces

Consider the ordinary differential equation

(4.5.1) $$\ddot{x} + x = f(x) + \epsilon g(t); \quad f(x) = O(|x|^2) \quad \text{at } x = 0,$$

where $g(t)$ is a continuous function on $(-\infty, \infty)$ decaying exponentially at ∞, and ϵ is a small parameter. We are interested in the behavior of the solution $x(t)$ of the Cauchy problem with zero initial conditions for (4.5.1) as $t \to \infty$. This solution does not tend to zero as $t \to \infty$, at least for small $|\epsilon|$. Actually what can be proved is the following.

(4.5.2) *For* ϵ *sufficiently small and* $f(x)$ *real analytic near* $x = 0$, *the solution* $x(t)$ *tends asymptotically as* $t \to \infty$ *to a periodic solution* $u(t)$ *(near* $x = 0$*) of the equation* $\ddot{x} + x = f(x)$.

Instead of giving a proof of this result, it is perhaps somewhat more appropriate to discuss it in the context of Section 4.4A. If, using a majorant method, one attempts to compute the solution $x(t)$ of (4.5.1) as a power series in ϵ, $x(t) = \sum_{n=1}^{\infty} a_n(t) \epsilon^n$, one finds that the coefficients $a_n(t)$ are unbounded functions of t over the interval $[0, \infty]$. A different method that does succeed consists in denoting a possible solution $x(t)$ of (4.5.1) as $x(t) = u(t, \lambda, b) + y(t)$, where $u(t, \lambda, b)$ is the solution of (4.5.1) with period λ and phase b, and such that $y(t) \to 0$ as $t \to \infty$. Next one finds formal asymptotic expansions for the period $\lambda(\epsilon)$, the phase $b(\epsilon)$, and the remainder $y(t)$. One then justifies these asymptotic expansions on the basis of (4.4.1). In fact this is the approach used in a paper by Ter-Krikorov (1969), where the interested reader will find detailed proofs.

4.5B The membrane approximation in nonlinear elasticity

The partial differential equations governing the equilibrium states of a thin elastic plate naturally contain a small parameter ϵ^2, a measure of the plate's thickness (cf. (1.1.12)). The *membrane approximation* to the problem of determining the equilibrium states under given body forces consists in setting $\epsilon^2 = 0$ in the above-mentioned equations, and finding the solutions of this reduced system. More explicitly, let Ω be a bounded domain in $\mathbb{R}^2$ with boundary $\partial\Omega$, then the partial differential equations defined over Ω can be written in the form

$$(4.5.3) \qquad \Delta^2 F + \tfrac{1}{2}[w, w] = 0, \qquad \epsilon^2 \Delta^2 w - [w, F] = g,$$

where the bilinear form $[f, g] = f_{xx}g_{yy} + f_{yy}g_{xx} - 2f_{xy}g_{xy}$. The physical quantities F, w, and g have been described earlier. For simplicity, we assume that the plate is clamped, so that a given problem is specified by adding the boundary conditions

$$(4.5.4) \qquad D^\alpha w|_{\partial\Omega} = 0 \quad \text{for } |\alpha| \leqslant 1, \qquad F_{\tau\tau}|_{\partial\Omega} = T, \qquad F_{\eta\tau}|_{\partial\Omega} = S.$$

In Section 6.2, we shall show this system always has solutions that minimize the associated potential energy of the physical problem. Now we consider the problem of comparing such solutions (w_ϵ, F_ϵ) with the solutions of the degenerate system (Π_0) obtained by setting $\epsilon = 0$ in (4.5.3), together with the boundary conditions

$$(4.5.5) \qquad w|_{\partial\Omega} = 0, \qquad F_{\tau\tau}|_{\partial\Omega} = T, \qquad \text{and} \quad F_{\eta\tau}|_{\partial\Omega} = S.$$

One of the main difficulties with this problem is the nonuniqueness of the solutions of the system (4.5.3) – (4.5.4). Nonetheless one shall establish circumstances under which a solution of (Π_0) is the leading term of an asymptotic approximation to a special class of solutions of (4.5.3). The asymptotic nature of the membrane approximation is a consequence of the fact that the boundary condition $D^\alpha u|_{\partial\Omega} = 0$ for $|\alpha| = 1$ is omitted. In fact, one expects that as $\epsilon \to 0$ a solution of (4.5.3) – (4.5.4) will tend uniformly to a solution of (4.5.3) and (4.5.5) with $\epsilon = 0$ everywhere except in the vicinity of the boundary $\partial\Omega$, where an *edge effect* (or boundary layer) appears (i.e., *near* $\partial\Omega$ *the gradient of the function* w *or* $D^\alpha w$ *changes rapidly*), just as in the example (Π_ϵ) of Section 4.4C.

In order to justify the membrane approximation, it is known to be necessary to restrict the forces acting on the elastic body. We shall do this by considering only those systems for which the degenerate problem (Π_0) possesses a positive solution (w_0, F_0), i.e., a solution such that $F_{0,xx}$ and $[F_0, F_0] > 0$ in Ω. It is easy to show that such positive solutions are unique (if they exist), and in fact using the techniques of Part III one can show that they do exist in a large class of elastic problems. More importantly, the following result is known.

(4.5.6) **Theorem** Suppose a positive solution (w_0, F_0) for the system (4.5.3) – (4.5.5) exists, then for $\epsilon > 0$ sufficiently small the system (4.5.3)–(4.5.4) has a unique positive solution (w_ϵ, F_ϵ) such that $(w_\epsilon, F_\epsilon) \to (w_0, F_0)$ uniformly apart from a narrow region of Ω near $\partial\Omega$.

The proof of (4.5.6) is parallel to that of (A) of Section 4.4C. One first constructs formal asymptotic expansions for functions $w_m(x, \epsilon)$ and $F_m(x, \epsilon)$ so that they satisfy the boundary conditions (4.5.4) exactly and the equations apart from terms of order $O(\epsilon^{m+1})$. Then one represents the solutions of (4.5.3)–(4.5.4) as solutions $x = (w, F)$ of an operator equation of the form $f_\epsilon(x) = 0$ acting on the Sobolev space $\mathring{W}_{2,2}(\Omega)$ exactly as in (2.5.7). The Sobolev inequalities show that the second derivative $f_{\epsilon xx}(x)$ is uniformly bounded on sets $\|x\| = \{\|w\|_{2,2}^2 + \|F\|_{2,2}^2\}^{1/2} \leqslant R$. Thus to apply Theorem (4.4.1) in this case, it remains to establish a lower bound for $\|f_\epsilon'(x_k)\|$. To this end, we note that in $\mathring{W}_{2,2}(\Omega)$, the operator $f_\epsilon(x)$ can be written as $(F_* + \frac{1}{2}C(w, w), \epsilon^2 w - C(w, F))$, where $F_* = F(x, y) - g$ and g is the unique function satisfying the system $\Delta^2 g = 0$ and the latter two boundary conditions of (4.5.4). Thus, if $y = (\bar{w}, \bar{F})$,

$$f_\epsilon'(x)y = \left(\bar{F} + C(w, \bar{w}), \epsilon^2\bar{w} - C(\bar{w}, F) - C(w, \bar{F})\right).$$

Hence

$$(f'(x_k)y, y) = \int_\Omega \{|\Delta F|^2 + \epsilon^2|\Delta w|^2\} + \int_\Omega \{F_{k,xx}w_y^2 + F_{k,yy}w_x^2 - 2F_{k,xy}w_x w_y\}.$$

By the positivity of x_0, $(f_\epsilon'(x_0)y, y) \geqslant \epsilon^2\|y\|^2$, and in fact there is a positive constant c independent of ϵ such that the last integral above is $\geqslant c\int_\Omega |\nabla w|^2$. Furthermore, by the construction of the approximations F_k, $\sup_\Omega |D^\alpha(F_k - F_0)| = O(|\epsilon|)$ for $|\alpha| = 2$, and $k \geqslant 0$, so that $(f_\epsilon'(x_k)y, y) \geqslant \frac{1}{2}\epsilon^2\|y\|_{2,2}^2$ for ϵ sufficiently small. Therefore, as in the argument following (4.4.14), by the Lax–Milgram theorem $f_\epsilon'(x_k)$ is invertible and $\|f_\epsilon'(x_k)\| \geqslant \frac{1}{2}\epsilon^2$. Thus the hypotheses of (4.4.1) are verified and Theorem (4.5.6) is proven, once one notices that the leading terms in the asymptotic expansions for F_k and w_k are (w_0, F_0). For more details, see Srubshchik (1964).

4.5C Perturbed Jeffrey–Hamel flows of a viscous fluid

The Navier–Stokes equations defining the steady plane radial flow of a viscous incompressible fluid between two inclined planes (meeting at an angle 2α) admits exact solutions. The solutions, known as Jeffrey–Hamel flows $G(\alpha, R)$, exist for all Reynolds numbers R, and for given parameters (α and R) are nonunique (in general). Most of the Jeffrey–Hamel flows exhibit a combination of inflow and outflow along a profile. Thus it is of interest to consider a perturbation of the geometric situation defining these flows with a view to proving that among the special Jeffrey–Hamel flows there is one that is a first approximation to a solution of the perturbed problem. As was observed by L. E. Fraenkel, an interesting perturbed problem consists of a class of two-dimensional symmetric channels C with walls whose radii of curvature are uniformly large relative to the local channel width. Fraenkel proved the existence of a unique Jeffrey–Hamel flow $G_0(\alpha, R)$ depending analytically on α near $\alpha = 0$, and he constructed a formal solution[1] $u(x, \epsilon) = G_0 + \sum_{n=1}^{\infty} \epsilon^n \psi_n$ of the Navier–Stokes equations in C in terms of the small curvature ϵ of the walls of the channel. For appropriate values of the physical parameters R and α this formal solution exhibits a separation phenomenon in which the velocity field of the flow is divided into distinct regions of forward and reversed flow by a zero velocity curve, which is itself separated from the wall of the channel.

An important mathematical problem in this connection, (solvable by using the methods of (4.4.1)), is the justification of the use of the formal solution $u(x, \epsilon)$ as an approximation to a true solution of the physical problem. The formal solution $u(x, \epsilon)$ mentioned above cannot be expected to converge for $\epsilon \neq 0$ since repeated differentiation of certain functions in the construction of $u(x, \epsilon)$ makes the coefficients of ϵ^n roughly comparable with $n!$. Thus it is natural in this situation to show that the truncated formal solution is an asymptotic approximation to a true solution of the Navier–Stokes equations in the sense of Section 4.4A.

To formulate this problem mathematically, we proceed as follows. Suppose the channel C is the image of the strip $\Omega = \mathbb{R}^1 \times (-1, 1)$ under the conformal mapping $z = z(w, \epsilon)$ characterized by the equation $dz/dw = he^{i\theta}$ and the function $\alpha(\epsilon w) = (d/dw)(\log(dz/dw))$. Thus with $z = x + iy$ and $w = u + iv$, $|dz| = h|dw|$, and $\alpha(\epsilon w)$ turns out to be approximately the angle which the upper channel wall makes with the x axis. Then the "vorticity form" of the Navier–Stokes equations becomes

$$\text{(4.5.7)} \qquad \left[\Delta + R\left\{\psi_u \frac{\partial}{\partial v} - \psi_v \frac{\partial}{\partial u}\right\}\right] \frac{1}{h^2}\,\Delta\psi = 0,$$

where ψ is proportional to the stream function and Δ is the Laplacian with respect to (u, v). Setting $h_u = hk$ and $h_v = h\lambda$, and using the fact that

[1] The parameter α is a fixed parameter in the Jeffrey–Hamel problem, but a slowly varying function in the perturbed problem.

$\Delta(\log h) = 0$, we find that (4.5.7) can be rewritten as

(4.5.8) $$f_\epsilon(\psi) \equiv \left\{\Delta - 4\left(k\frac{\partial}{\partial u} + \lambda\frac{\partial}{\partial v}\right) + 4(k^2+\lambda^2)\right\}\Delta\psi - R\left\{\psi_v\left(\frac{\partial}{\partial u} - 2k\right) - \psi_u\left(\frac{\partial}{\partial v} - 2\lambda\right)\right\}\Delta\psi = 0.$$

This equation is supplemented by the boundary conditions

(4.5.9) $$\psi = \pm 1, \qquad \psi_v = 0 \qquad \text{for} \quad v = \pm 1,$$

(4.5.10) $$\psi_u \to 0 \qquad \text{as} \quad |u| \to \infty.$$

If α is a real constant, $k = \alpha$, $\lambda = 0$, and ψ is independent of u, then (4.5.8) becomes

(4.5.11) $$\psi_{vvvv} + (4\alpha^2 + 2R\alpha\psi_v)\psi_{vv} = 0.$$

Solutions of the two-point boundary problem (4.5.9) and (4.5.11) are precisely the Jeffrey–Hamel flows mentioned above. To define the formal solutions $u(x, \epsilon)$ we set $u = \sigma/\epsilon$ in (4.5.8) and suppose $|\psi_u| = O(\epsilon)$ so that $k = \alpha(\sigma) + O(\epsilon^2)$, while $\lambda = O(\epsilon)$. Then (4.5.8) can be written as

$$\psi_{vvvv} + 4\alpha^2\psi_{vv} + 2R\alpha\psi_v\psi_{vv} = O(\epsilon).$$

Now if we assume that $\psi(x, \epsilon) = G_0(v, R, \alpha(\sigma)) + \sum_{n=1}^{N}\epsilon^n\psi_n$, then the ψ_n can be computed iteratively, each being the unique odd solution of

$$L\psi = \psi_{vvvv} + 4\alpha^2\psi_{vv} + 2R\alpha\left(\frac{\partial G_0}{\partial v}\psi_v\right)_v = F_n(G_0, \psi_1, \ldots, \psi_{n-1}),$$

$$\psi = \psi_v = 0 \qquad \text{for} \quad v = \pm 1,$$

where F_n is found from the exact equation (4.5.8). Now we note (without proof) that the truncated formal solution $\bar{\psi}_N = G_0 + \sum_{n=1}^{N}\epsilon^n\psi_n$ constructed as above is an approximate solution of (4.5.8) in the sense that in the notation of Section 4.4A:

(i) $f_\epsilon(\Psi_N) = O(\epsilon^{N+1})$ uniformly on Ω;
(ii) Ψ_N satisfies the boundary conditions (4.5.9), (4.5.10);
(iii) $\|f_\epsilon(\Psi_N)\|_{L_2(\Omega)} \leqslant k_N(R, \alpha)\epsilon^{N+1/2}$, for ϵ sufficiently small.

We can now state

(4.5.12) For ϵ sufficiently small and the quantities R, α suitably restricted, (4.5.7) has a classical solution $\psi = G_0 + \rho_0$. Furthermore, ψ is the only solution such that $\|\psi - G_0\|_{\mathring{W}_{2,2}(\Omega)} = O(\epsilon)$. Here $G_0 = G_0(\alpha, R)$ is the Jeffrey–Hamel flow mentioned at the beginning of Section 4.5C.

Proof: We apply (4.4.1) with $p = 0$, $n = 1$, and $N = 1$ by rewriting (4.5.8) as an operator equation in the closed subspace H of odd functions belonging to $\mathring{W}_{2,2}(\Omega)$. An appropriate norm for H can be chosen to be

$$\|u\|_H^2 = \sum_{|\alpha|=2} \int_\Omega |D^\alpha u|^2.$$

Indeed by replacing ψ in (4.5.8) by $G_0 + \rho$, we may suppose ρ satisfies a null boundary condition. Clearly by arguing as in Section 2.2D, the equation (4.5.8) can be then rewritten as an operator equation *in* H of the form $\tilde{f}_\epsilon(\rho) = L_\epsilon \rho + RN_\epsilon \rho = O(\epsilon)$, where L and N are bounded continuous mappings of H into itself defined formally by the equation

$$f_\epsilon(G_0 + \rho) = f_\epsilon(G_0) + L_\epsilon \rho + N_\epsilon(\rho)$$

so that the linear operator L_ϵ is implicitly defined by

$$(L_\epsilon \psi, \varphi) = \int_\Omega [(\Delta\psi)(\Delta\varphi) + \Delta\psi(4k\varphi_u + 4\lambda\varphi_v) + 4(k^2 + \lambda^2)\varphi\, \Delta\psi] + O(R).$$

Clearly $\|N_\epsilon \psi - N_\epsilon \varphi\| \leq k\{\|\psi\| + \|\varphi\|\}\|\psi - \varphi\|$ for some constant k independent of ψ and φ, and N satisfies hypothesis (I) of (4.4.1). Thus by (4.4.1), to prove (4.5.12) it suffices to prove that

(*) L_ϵ is invertible and $\|L_\epsilon \psi\| \geq k\|\psi\|$, where k is a constant independent of ϵ.

By virtue of (1.3.21), (*) will be established once we find a positive constant k_1 independent of ϵ such that $(L_\epsilon \psi, \psi) \geq k_1 \|\psi\|_H^2$. Now a simple computation with $p = 4\alpha + R(\partial G_0/\partial v)$ and $q = 4\alpha^2 + 2R\alpha(G_0/\partial v)$ shows that

$$(L_\epsilon \psi, \psi) = \int_\Omega \{|\Delta\psi|^2 + (p\psi_u + q\psi)\Delta\psi + q_v \psi\psi_v + p_{vv}\psi\psi_u\} + O(\epsilon)\|\psi\|_H^2.$$

Integration by parts using the fact that $\alpha = \alpha(\epsilon u)$ gives

(†) $$(L_\epsilon \psi, \psi) = \int_\Omega \{|\Delta\psi|^2 + R(\partial^2 G_0/\partial v^2)\psi\psi_{uv} - q(\psi_u^2 + \psi_v^2)\} + O(\epsilon)\|\psi\|_H^2.$$

Thus it suffices to find a suitable lower bound for the first quadratic form on the right-hand side of (†). To this end, by suitably restricting R and α, we shall find a constant $\lambda = \lambda(R, \alpha) > 0$ independent of ϵ such that for all odd (in v) functions $\psi \in C_0^\infty(\Omega)$,

(4.5.13) $$Q_1(\psi) = \int_{-1}^{1} \{2\psi_{uv}^2 + \psi_{vv}^2 + RG_{vv}\psi\psi_{uv} - q(\psi_u^2 + \psi_v^2)\}\, dv \geq \lambda \int_{-1}^{1} (2\psi_{uv}^2 + \psi_{vv}^2)\, dv.$$

Once this inequality is proved, (*) is established by adding ψ_{uu}^2 to both sides of (4.5.13) and integrating with respect to u. To prove (4.5.13), we note that the inequality

(4.5.14) $$\int_{-1}^{1} (\varphi_{vv}^2 - q\varphi_v^2)\, dv \geq \mu \int_{-1}^{1} \varphi_v^2\, dv \qquad \text{for } \varphi \in \mathring{W}_{2,2}(-1, 1)$$

can be established by suitably restricting R and α and if φ is odd in v, μ can be increased to μ_1 (say). Applying this inequality to ψ and $\int_{-1}^{v} \psi_u\, dv$, we find that

$$\int (\psi_{vv}^2 - q\psi_v^2) \geq \mu_1 \int \psi_v^2,$$

$$\int (\psi_{uv}^2 - q\psi_u^2) \geq \mu \int \psi_u^2.$$

Then arguing as in (4.4.14), these two inequalities imply the existence of positive constants $\mu_2 < \mu_3$ such that

$$\int_{-1}^{1} \{\psi_{uv}^2 + \psi_{vv}^2 - q(\psi_u^2 + \psi_v^2)\}\, dv \geq \int_{-1}^{1} (\mu_2\psi_{uv}^2 + \mu_3\psi_{vv}^2)\, dv.$$

From (4.5.13) and the fact that $\int_{-1}^{1}\psi^2 \leq c\int_{-1}^{1}\psi_{vv}^2$ for any $\delta > 0$,

$$Q_1(\psi) \geq \int_{-1}^{1} \{\psi_{uv}^2 + \mu_2\psi_{uv}^2 + \mu_3\psi_{vv}^2\}\, dv - \beta\left\{\delta \int \psi_{uv}^2\, dv + \frac{1}{\delta}\int \psi_{vv}^2\right\}$$

$$\geq (1 + \mu_2 - \beta\delta)\int_{-1}^{1} \psi_{uv}^2\, dv + (\mu_3 - \beta/\delta)\int_{-1}^{1} \psi_{vv}^2\, dv.$$

It remains to choose δ so that $1 + \mu_2 - \beta\delta > 0$ and $\mu_3 - \beta/\delta > 0$. Clearly this can be accomplished by restricting R and α so that $\beta^2 < (1 + \mu_2)\mu_3$.

Thus it remains to investigate the validity of (4.5.14) for some fixed $\mu > 0$. To this end, it is necessary to investigate the lowest eigenvalue of the linear eigenvalue problem

$$w^{(\text{iv})} + (qw')' + \lambda w'' = 0, \qquad w = w' = 0, \quad v = \pm 1.$$

A continuity argument shows that $\mu > 0$ for all $R \geq 0$ provided α lies in the interval where the formal approximation $G_0(\alpha)$ is the unique Jeffrey–Hamel flow depending analytically on α. For details see Fraenkel (1973).

NOTES

A Linear stability of bifurcated branches of solutions

As was mentioned in Section 4.1 there is often an "exchange of stability" in physical systems after a bifurcation. In the elasticity result of (4.3.36) this was demonstrated by energy considerations. In general for non-Hamiltonian systems the less precise linear stability criterion, mentioned in Section 4.1 is useful. This criterion is based on information concerning the spectrum of the linear operator $f_x(\tilde{x}, \tilde{\lambda})$ at a solution $(\tilde{x}, \tilde{\lambda})$ of $f(x, \lambda) = 0$. The Leray–Schauder degree theory is often useful in this connection. Thus, for example, in finite-dimensional systems linear stability is determined by proving that the real parts of the eigenvalues of $f_x(\tilde{x}, \tilde{\lambda})$ are negative, while instability results from any eigenvalue with positive real part. Thus if we compute the Brouwer degree of $f(x, \lambda)$ at 0 as a function of λ, as λ crosses a bifurcation point λ we shall be able to ascertain facts about the spectrum of f_x at any solution $(\tilde{x}, \tilde{\lambda})$ of $f(\tilde{x}, \tilde{\lambda}) = 0$ since the Brouwer degree can be computed additively by linearization at nonsingular solutions. A similar results holds for the infinite-dimensional case. For further information the interested reader is referred to the paper by Sattinger (1971).

B Bifurcation for general operator equations at eigenvalues of odd multiplicity

Let $f(x, \lambda)$ be a C^1 mapping of a neighborhood of (0, 0) of the Banach space $X \times R$ into Y such that (i) $f(0, \lambda) \equiv 0$ for all λ near 0, (ii) $f_{\lambda x}(x, \lambda)$ is a C^1 function of t, (iii) the linear operator $f_x(0, 0)$ is a linear Fredholm map of index zero and dim Ker $f_x(0, 0)$ is odd, and (ii). For $x \in \operatorname{Ker} f_x(0, 0)$, $f_{x\lambda}(0, 0)x \notin \operatorname{Range} f_x(0, 0)$. Then, as a generalization of the results (4.1.12) and (4.2.3), one can prove that (0, 0) is a point of bifurcation relative to the operator equation $f(x, \lambda) = 0$. This result is obtained by decomposing the Banach space as in (4.1.12) and applying the properties of the Brouwer degree to the associated bifurcation equations. The full details are given in Westreich (1973).

C Reduction of the bifurcation equations under symmetry assumptions

In many bifurcation problems of the multiplicity m of an eigenvalue λ of an associated linear problem is caused by invariance properties of the associated operator equation under a group of isometries. In many cases this symmetry allows a reduction in both the number of equations in the bifurcation equations and in the number of unknowns in this equation. Thus in the secondary steady flows of the Navier–Stokes equation associated with convection in a horizontal fluid heated from below (the so-called Bénard problem), the secondary solutions observed have a hexagonal cellular pattern. This result can be obtained from the bifurcation theory developed in Section 4.1 by finding solution of the associated nonlinear boundary value problem in a Banach space of vector-valued functions which themselves possess "hexogonal" symmetry. This was first carried out in Judovitch (1968). A general study of this situation was carried out by Loginov and Tregonin (1972).

D Boundary layer phenomena for semilinear Dirichlet problems

A generalization of the result given in Section 4.4C can be carried out for boundary value problem defined on $\Omega \subset R^N$

(†) $$\epsilon^2 \Delta u + f(x, u) = 0, \qquad u|_{\partial\Omega} = 0,$$

where $f(x, u)$ is a C^∞ function of its arguments with the properties (i) there is a C^∞ positive function $T(x)$ defined on $\bar{\Omega}$ such that $f(x, T(x)) \equiv 0$ on $\bar{\Omega}$ and

(ii) $$f_u(x, T(x)) < 0 \quad \text{on} \quad \bar{\Omega} \qquad \text{with} \quad \int_v^{T(x)} f(s, y)\, dy > 0$$

for fixed $v \in [0, T(x)]$. There $f(x, u)$ *is not subjected to any growth restriction*. Then, in order to justify the approximation $u_0 = T(x)$ for a positive solution $f(x)$ apart from a small boundary layer of width $O(\epsilon)$ near $\partial\Omega$ as $\epsilon \to 0$, one can work in Hölder spaces $C^{m,\mu}(\Omega)$ using the procedure of Section 4.4B. However, in order to obtain the crucial estimate for the form of the linear operator $L = [f_\epsilon'(u_i)]^{-1}$ the Sobolev space context is essential. Once an estimate for L has been obtained, the Sobolev inequalities give a pointwise estimate for L when regarded as mappings between the appropriate Hölder spaces. For the complete details the interested reader is referred to De Villiers (1973). See also Fife (1973).

E Bibliographic notes

Section 4.1: Poincaré's original paper on bifurcation theory can be found in Poincaré (1885). This paper was devoted to determination of equilibrium forms for a rotating ideal fluid was inspired by a number of conjectures in the treatise by Kelvin and Tait (1879). Later treatments of this problem include Liapunov (1906–1914), Lichtenstein (1933), and Appell (1921). A comprehensive modern study is still noticeably absent. As mentioned in earlier notes, Liapunov's criterion (4.1.4) is generally proved by the majorant method, see Siegel and Moser (1971) for a modern treatment. The reduction of bifurcation theory to a finite-dimensional problem as in (4.1.5) is generaly called the Liapunov–Schmidt method due to the fundamental papers Liapunov (1906) and Schmidt (1908). A recent book on this subject is Vainberg and Tregonin (1974). Our treatment of simple multiplicity (4.1.12) is due to Diustermatt, and is intended to suggest the importance of the techniques on the recent theory of singularities in more difficult bifurcation problems. Other recent treatments of bifurcation in the case of simple multiplicity include Crandall and Rabinowitz (1971) and Westreich (1972). Krasnoselski's books (1964) yield very detailed information on this problem. The relationship between bifurcation theory and nonlinear normal modes can be found in Berger (1969). The iteration scheme of Section 4.1D can be found in Berger and Westreich (1974). A survey of constructive methods in the higher multiplicity case can be found in Sather (1973), and in numerous other papers and books. Unfortunately, such methods often are not useful in practice since definite results are obtained by making assumptions that are not easy to verify.

Section 4.2: The use of transcendental methods in bifurcation theory is well described in Krasnoselski (1964), Berger (1970a) and Cronin (1964). In particular the use of the degree of a mapping is due to Krasnoselskii (1964) who also noted the importance of gradient operators to obtain sharp results on bifurcation points that are independent of multiplicity. Prodi (1971) seems to be the first paper utilizing Morse theory in bifurcation problems. See also Berger (1973). The sharp results on bifurcation theory for complex analytic mappings such as (4.2.4) were described in Cronin (1953). Our proof follows the paper Schwartz (1963). The use of the Liusternik Schnirelmann theory of category in bifurcation theory is described in the papers Berger (1969, 1970). Recent interesting results on applying the higher homotopy groups of spheres to bifurcation problems involving more than one parameter can be found in Ize (1975). See Amer. Math. Soc. memoir # 174.

Section 4.3: A good survey of the periodic solutions of the restricted three-body problem near the equilibrium points L_1–L_5 can be found in the article of Deprit and Henrard (1969). The treatment of buckling phenomena in nonlinear elasticity given here follows Berger (1967). Our discussion of the onset of turbulence for viscous fluids as a bifurcation phenomena follows Judovitch (1966, 1967). Other useful results include Kirchgassner and Sorger (1969) and Gortler *et al.* (1968). The Taylor vortices were first described in Taylor (1923), but their mathematical study via the full nonlinear Navier–Stokes equations began with the papers Velte (1966) and Judovitch (1966). Our discussion of the bifurcation of complex structures on higher dimensional complex manifolds follows Nirenberg (1964) and Kuranishi (1965). Recently Kuranishi has used the Nash–Moser implicit function theorem to obtain results on deformations with singularities. An alternative approach to this problem using nonlinear functional analysis can be found in a recent paper by Forster (1975). The purely algebraic approach to this problem suffers from the fact that the formal power series solutions constructed generally diverge.

Section 4.4: The result (4.4D) is adapted from the papers by Berger and Fraenkel (1969, 1970). The method for more general elliptic boundary value problems has been further developed by Fife (1973).

Section 4.5: An excellent survey of singular perturbation problems in mathematical physics is Friedrichs (1955). The result on the perturbation of periodic solutions of an anharmonic oscillator is due to Ter-Krıkorov (1969). Our discussion of the validity of the membrane approximation in nonlinear elasticity is adapted from Schrubshik (1964), while our discussion of perturbed Jeffrey–Hamel flows can be found in papers by Fraenkel (1962, 1973).

F Normal Modes for Nonlinear Hamiltonian Systems

As described in Section 4.2, transcendental methods of bifurcation theory are crucial in establishing general results on the preservation of normal modes of a linear autonomous Hamiltonian system of ordinary differential equations under a nonlinear Hamiltonian perturbation of higher order. This situation is described for second-order systems in the papers of Berger (1969, 1970a, 1971c). Berger's methods have been adapted to first-order systems in a recent unpublished paper of Westreich, although in this case certain restrictions appear due to an example of Siegel, see Siegel and Moser (1971) pp. 109–110. Interesting finite-dimensional approaches to the problem for first-order systems can be found in Weinstein (1974) and Moser (1976) who both use the Ljusternik–Schnirelmann transcendental method analogous to Berger (1970a). See Section 6.7. A possible extension to hyperbolic partial differential equations is described in Berger (1973).

G Applications of Singularity Theory to Bifurcation Phenomena

A new and promising approach to constructive bifurcation theory is provided by the theory of singularities initiated by H. Whitney and continued by R. Thom, J. Mather, and V. Arnold. Despite the overblown claims made about "catastrophe theory," there is little doubt that the careful consideration of specific bifurcation problems involving gradient mappings by the methods of singularity theory will lead to progress. Important in this direction are the recent careful considerations by V. Arnold of the singularities associated with the degenerate critical points of a real-valued function defined in a neighborhood of a point in $\mathbf{R}^N$. This study can be applied to the "case of higher multiplicity" of Section 4.1E.

In applications of singularity theory to infinite-dimensional bifurcation problems, one reduces the problem to the bifurcation equations described in Section 4.1 and considers as equivalent any two bifurcation problems whose bifurcation equations differ by "restricted" coordinate changes. One then obtains a classification of the behavior near a bifurcation point into normal canonical forms. The need to restrict coordinate changes is due to the fact that the "bifurcation parameter λ" must generally be treated with great care in bifurcation phenomena. These ideas are currently being pursued by a number of investigators including M. Golubitsky, D. Schaeffer, J. Hale, S. Chow, and J. Mallet Paret.

H Instantons and Local Analysis

As mentioned in Note H of Chapter 1, instantons are by definition smooth (finite-action) absolute minima of the Yang–Mills action functional $S(A)$ defined on $\mathbf{R}^4$ relative to the gauge group G, and the homotopy class of fields A of Pontrjagin index k. Surprisingly for specific non-Abelian Lie groups G, the number of instantons with fixed k can be determined by the local analysis results of Part II together with results concerning elliptic systems of partial differential equations and the representation theory of non-Abelian Lie groups. Thus, for $G = \mathrm{SU}(2)$, Atiyah, Hitchin, and Singer [*Proc. Nat. Acad. Sci.* 1977 and subsequent papers] have shown that the number of inequivalent instantons of index k depends on $8k-3$ parameters. Similar explicit results also hold for higher gauge groups, although several of these cases present new phenomena.

PART III

ANALYSIS IN THE LARGE

The aim of Part III is the extension of the local results described to a global context in such a way that the specific problems (discussed in Chapter 1) can be successfully treated. In order to accomplish this goal, methods mixing analysis with topology will prove quite useful. Such transcendental methods were quite valuable for the bifurcation theory of Chapter 4. For global problems, however, this mixing is often essential for a proper understanding of nonlinear phenomena. It is this combination of analytic and topological techniques that has lead to the highest achievement in our subject.

In practical terms, the extension from local to global is particularly important for two major reasons. First, a sufficiently accurate first approximation to the solution of a given problem may not be available. Secondly, although such an approximation may exist, many problems require the totality of solutions $\mathbb{S}$ of a given problem to be considered as a whole. Indeed the set $\mathbb{S}$ is often partitioned into distinct classes $\{\mathbb{S}_\alpha\}$ with various classes of prime importance under differing circumstances.

Rather than put forward a very general global theory, we shall be content to describe a realm of ideas midway between classical linear functional analysis and the theory of nonlinear functional analysis on general infinite-dimensional manifolds.

We now summarize (in the abstract):

The problems to be discussed. To fix notation, let f be a bounded mapping defined on an open domain U of a Banach space X with range in another Banach space Y. Then we pose the following questions:

(i) (*Mapping problem*) Under what circumstances is f surjective (i.e., $f(U) = Y$), univalent, or a homeomorphism of U onto $f(U)$?

(ii) (*Linearization problem*) What global properties of f can be deduced from the local behavior of f? In particular, if $f \in C^1(U, Y)$, what global properties can be deduced from $f'(x)$?

(iii) (*Solvability problem*) Determine necessary and sufficient conditions for the solvability of the operator equation $f(x) = y$ for fixed $y \in Y$, analogous to the classic linear Fredholm theory.

(iv) (*Problem concerning global structure of solutions*) Can one obtain a description of the set of solutions of the operator equation $f(x) = y$ for fixed $y \in Y$? In particular, under what circumstances does the equation have at least a given number of solutions?

(v) (*Problem concerning classification of solutions*) Determine a classification of solutions of the operator equation $f(x) = y$ that is invariant under small (suitably restricted) perturbations of the mapping f.

(vi) (*Deformation problem*) Under what circumstances can any of the above problems be answered by smoothly deforming the given mapping $f \in M(U, Y)$ to a simpler mapping $f_1 \in M(U, Y)$?

(vii) (*Problem concerning parameter dependence*) Suppose the given mapping f depends continuously on a parameter λ, then exactly how do the solutions of $f(\lambda) = 0$ depend on the parameter λ?

(viii) (*Approximation problem*) Under what circumstances can the properties of the mapping f acting between infinite-dimensional spaces be deduced from finite-dimensional approximations for the range of f?

(ix) (*Generalized operator theory*) What parts of the theory of linear operators in classical functional analysis extend directly to a nonlinear context?

(x) (*Problem of nonlinear effects*) What qualitative features of the operator f affect the answers to the above problems?

(xi) (*Problem of infinite dimensional effects*) What answers to the above questions are related to the infinite dimensionality of the Banach spaces X and Y?

Typical of the situations we consider are nonlinear eigenvalue problems of the form $Ax = \lambda Bx$, where at least one of the operators A and B is nonlinear. The problem is to find the *totality* of nontrivial solutions (x, λ) of this equation, and to classify these solutions in a manner consistent, if possible, with linear eigenvalue theory. Since the solutions we seek are not local in nature, the methods of Part II must be supplemented by more all-encompassing ones. The interesting nonlinear phenomena for these problems (e.g., "continuous spectra" (see Fig. 6.3) p. 381) thus provide motivation for the development of deeper methods of study.

CHAPTER 5

GLOBAL THEORIES FOR GENERAL NONLINEAR OPERATORS

The study of mapping properties of general nonlinear operators calls for new methods of study quite distinct from the theory of Part II. Here three such general methods are discussed. The first, linearization, is based on piecing together local information about the Fréchet derivative $f'(x)$ of a mapping $f \in C^1(X, Y)$ by considering the geometry of the image $f(X)$. Secondly, we take up methods of approximation for an operator f acting between infinite-dimensional Banach spaces X, Y by a sequence of mappings $\{f_n\}$ acting between finite-dimensional subspaces of X and Y. Finally, we consider infinite-dimensional homotopy theory of a mapping $f \in C(X, Y)$. In this theory one attempts to answer questions concerning the mapping properties of f by homotopy, i.e., by continuously deforming f to a simpler $\tilde{f}$, for which the mapping property can be easily determined. This last theory leads to the association of various numerical topological invariants with f (in many cases). We end the chapter by applying the invariants to a variety of specific problems. In Chapter 6, we take up additional methods that can be used for gradient mappings.

This distinction is well illustrated by the partial differential equations defining fluid flow described in Chapter 1. For problems involving ideal (i.e., inviscid) steady flow, the associated Euler equations generally define gradient mappings between appropriate Banach spaces. Indeed this fact will be utilized in the problem of global vortex rings (to be described in Section 6.4). However for steady viscous flow, the more general Navier–Stokes equations *do not* define a gradient mapping and so to study problems concerning such flows, methods of study utilizing the theory of this Chapter, must be considered. Moreover, in Section 5.5, we show the power of these methods by solving a classic problem of periodic ideal fluid flow.

5.1 Linearization

Let $f(x)$ be a C^1 operator defined on a domain D of a Banach space X with range in a Banach space Y, i.e., $f \in C^1(D, Y)$. In Part II, various local properties of the operator $f(x)$ near a point $x_0 \in D$ were derived from the behavior of the Fréchet derivative $f'(x_0)$. Here it is natural to piece this information together and consider those global properties of the mapping $f(x)$ that can be determined from the behavior of $f'(x)$ at each point $x \in D$.

Simple examples show that even if the Fréchet derivative $f'(x)$ is a linear homeomorphism or surjective (as a linear map from X to Y) for each $x \in D$, $f(x)$ may not share these properties. In this section we clarify this situation by the use of simple topological considerations based on ideas concerning covering spaces in the case of homeomorphisms and connectivity properties of $f(D)$ for surjectivity.

5.1A Global homeomorphisms

Suppose that the Fréchet derivative $f'(x)$ of an operator $f \in C^1(X, Y)$ is a linear homeomorphism from X to Y for each $x \in X$. Then the inverse function theorem of Chapter 3 implies f is a homeomorphism of a neighborhood U_x of each point x onto $f(U_x)$. Thus we inquire: *What are the properties that f must possess in order for f to be a homeomorphism of X onto Y?* A topological approach to this question that proves to be quite natural in this context is the notion of *covering space*. Indeed from the well-known definition and properties that we summarize below, it follows easily that a necessary and sufficient condition for f to be a homeomorphism of X onto Y is that (X, f) be a covering space for Y. This follows from the simple connectivity of the Banach space Y.

Definition Suppose X is a connected, locally connected Hausdorff topological space. (X, f) is a covering space of a connected Hausdorff topological space Y if:

(i) f is a continuous mapping of X onto Y, and
(ii) every $y \in Y$ has an open neighborhood U_y about it such that $f^{-1}(U_y)$ is the disjoint union of open sets O_i in X each of which is mapped homeomorphically onto U by f.

The following properties of covering spaces will be useful in the sequel. Indeed it is these properties that are crucial for applications to analysis.

Suppose (X, f) covers Y, then the covering map f has the properties:

(i) *Unique path lifting* Suppose $f(x_0) = y_0$ and $L(t)$ is a continuous path in Y with $L(0) = y_0$, then there is one and only one continuous path $p(t)$ in X with $p(0) = x_0$ and $fp(t) = L(t)$ for $t \in [0, 1]$.

(ii) *Covering homotopy property* Suppose $L_1(t)$ and $L_2(t)$ are continuous paths in Y with fixed base point that are homotopic, then these paths can be lifted to continuous paths $\mathfrak{L}_1(t)$ and $\mathfrak{L}_2(t)$ in X that are also homotopic with fixed base point.

(iii) For each $y \in Y$, the number of points in $f^{-1}(y)$ is constant.

(iv) Suppose Y is simply connected, then f is a homeomorphism.

For the proofs of these results we refer the reader to the book by Hu (1959). See also Spanier (1966).

Thus to answer the homeomorphism problem we shall determine necessary and sufficient conditions for a homeomorphism to be a covering map. To this end we prove

(5.1.1) Suppose D is a domain of X and $f \in C(D, Y)$, then necessary and sufficient conditions for the pair (D, f) to cover $f(D)$ are:

(i) f is a local homeomorphism; and

(ii) f lifts line segments, i.e., for any finite line segment $L(t) \in f(D)$ joining $y_0 = f(x_0)$ and $y_1 \in f(D)$ for any $x_0 \in D$, there is a curve $x(t)$ such that $f(x(t)) = L(t)$ with $x(0) = x_0$. (See Fig. 5.1.)

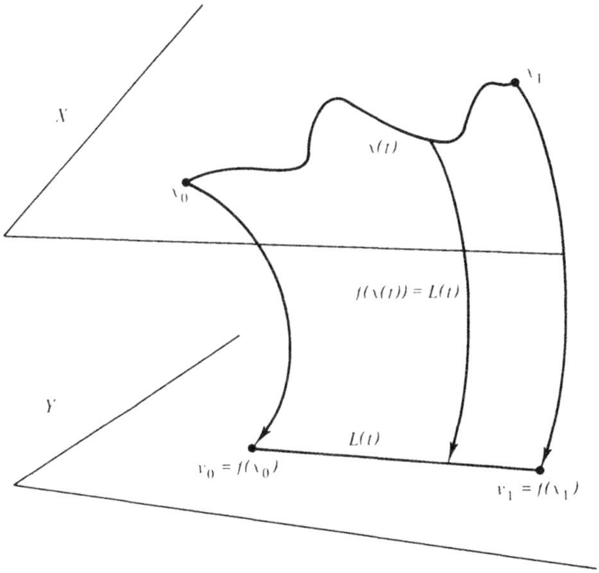

FIG. 5.1 Finding a curve $x(t)$, lifting the line segment $\{L(t), 0 \leqslant t \leqslant 1\}$.

Proof: (R. Plastock) The necessity of (i) and (ii) follows from the results mentioned in the above discussion of covering spaces.

To prove the sufficiency of (i) and (ii), we proceed as follows. Since f is a local homeomorphism, f is an open mapping and possesses the unique lifting property for line segments. Hence if $y \in f(D)$ and $\{x_\alpha\} = f^{-1}(y)$, then there is an open ball $B(y, r) = \{z \mid \|z - y\| < r, z \in Y\}$ contained in $f(D)$ and the set of curves emanating from x_α,

$$O_{x_\alpha} = \{x_\alpha(t) \mid t \in [0, 1), \quad x_\alpha(t) \in X, \quad x_\alpha(0) = x_\alpha, \quad f(x_\alpha(t)) = y + tr\bar{z}\}$$

is well defined for any $\bar{z}$ of norm 1. (The curve $x_\alpha(t)$ exists by hypothesis (ii).) Now we show that the sets O_{x_α} are disjoint open sets mapped homeomorphically onto $B(y, r)$, and $f^{-1}(B(y, r)) = \bigcup_\alpha O_{x_\alpha}$. Clearly once these facts are established, we have shown that (D, f) covers $f(D)$. By construction, it will also follow that $\bigcup_\alpha O_{x_\alpha} = f^{-1}(B(y, r))$.

Before proceeding further with the sufficiency half of the proof, we make the following observations, assuming f satisfies (i) and (ii).

(a) Distinct paths in O_x are mapped onto distinct radii in $B(y, r)$.

(b) If $P_1(t)$ and $P_2(t)$ are any two paths that intersect,then either P_1 and P_2 are mapped onto the same radius, or the point of intersection can only occur at $t = 0$.

(c) If $f : D \to f(D)$ is a local homeomorphism with M_1 and M_2 open subsets of D with nonempty intersection on each of which f is a homeomorphism, then f is a homeomorphism of $M_1 \cup M_2$ onto $f(M_2)$ provided $f(M_1) \cap f(M_2)$ is connected; thus if $f(M_1)$ and $f(M_2)$ are balls, $f(M_1) \cap f(M_2)$ is convex and thus connected.

From the construction of O_x and (a) above, we can conclude that each O_x is mapped onto $B(y, r)$ in a one-to-one way, and thus homeomorphically since f itself is an open map. It remains to show that the O_x are disjoint open sets whose union is $f^{-1}(B(y, r))$. *The disjointness follows from* (b). For if $P_1(t_1) = P_2(t_2) = \bar{x} \in O_{x_1} \cap O_{x_2}$ $(x_1 \neq x_2)$, then showing that $P_1(0) = P_2(0)$ will be the desired contradiction. However, if P_1 and P_2 are mapped onto the same radius, then necessarily $t_1 = t_2$. Thus $s = \{t \mid P_1(t) = P_2(t),\ 0 \leqslant t \leqslant 1\}$ is nonempty, open, and closed. Hence $s = [0, 1]$ and in particular, $P_1(0) = P_2(0)$.

Next we show that *each O_x is open*. Let $u \in O_x$, $f(u) = v$, $f(p(t)) = (1 - t)y + tv$ and $p(0) = x$. By compactness we cover $p(t)$ with a finite number of open sets D_j each of whose image under f is a ball. Then by (c) above, $\Delta = \bigcup_j D_j$ is mapped homeomorphically onto $f(\Delta)$. Now we claim that there is some number $\epsilon > 0$ such that if $\|v - w\| < \epsilon$ the line joining y and $w \in f(\Delta)$. Indeed otherwise there are sequences $w_n \to v$ such that $y(t_n) = (1 - t_n)y + t_n w_n \not\subseteq f(\Delta)$, so that $y(t_n) \to (1 - \bar{t})y + \bar{t}v \in f(\Delta)$. But this is a contradiction since $f(\Delta)$ is an open set. Consequently, by restricting f to Δ, we observe that $f^{-1}|_\Delta (w \mid \|w - v\| < \epsilon)$ is an open set in O_x containing u.

Finally, we show that $f^{-1}(B(y, r)) = \bigcup_{x \in f^{-1}(y)} O_x$. Since it suffices to show the inclusion of the right-hand side in the left, we suppose $x \in f^{-1}(B(y, r))$. Let $L(t) = (1 - t)f(x) + ty \subseteq B(y, r)$. By hypothesis, there is a path $P(t)$ such that $P(0) = x$ and $f(P(t)) = L(t)$. In particular, $P(1) \in f^{-1}(y)$. If we let $\tilde{L}(t) = L(1 - t)$ and $\tilde{P}(t) = P(1 - t)$, then $f(\tilde{P}(t)) = \tilde{L}(t)$, $\tilde{P}(0) \in f^{-1}(y)$, and $\tilde{P}(1) = x$. Thus $x \in O_{P(1)}$.

Thus the proof of (5.1.1) is complete.

Thus to show that the C^1 mapping $f : X \to Y$ is a global homeomorphism, we need only prove that f is a local homeomorphism and for any $y \in Y$ and $x_0 \in X$ there is a curve $x(t) \in X$ such that

$$(5.1.2) \qquad f(x(t)) = ty + (1-t)y_0, \qquad x(0) = x_0, \qquad \text{for} \quad t \in [0, 1].$$

(This requires that we know the surjectivity of f a priori.) In this sense then the homeomorphism question is reduced to a simpler one-dimensional problem.

A useful explicit method for constructing the curve $x(t)$ satisfying (5.1.2) is based on the theory of ordinary differential equations in Banach spaces. Indeed, by differentiating the relation (5.1.2) with respect to t, we find that the curve $x(t)$ satisfies the initial value problem

$$(5.1.3) \qquad dx/dt = [f'(x)]^{-1}(y - y_0), \qquad x(0) = x_0.$$

Conversely, if (5.1.3) has a solution $x(t)$ that exists for $t \in [0, 1]$, the curve $x(t)$ will satisfy (5.1.2). This idea is useful for finite-dimensional problems due to Peano's theorem (3.1.28). For infinite-dimensional problems, the study of (5.1.3) is less useful.

The argument based on ordinary differential equations however, can easily be abstracted. This is clearly seen in the proof of the following result of Banach and Mazur.

(5.1.4) **Theorem** (i) Let $f \in C(X, Y)$. Then f is a homeomorphism of X onto Y if and only if f is a local homeomorphism and a proper mapping. (ii) Provided $f \in C^1(X, Y)$, f is a diffeomorphism if and only if f is proper and $f'(x)$ is a linear homeomorphism for each $x \in X$.

Proof: (i) If $f \in C(X, Y)$, the necessity of the conditions stated are immediate.

To demonstrate the sufficiency, we first note that f maps X onto Y. Indeed, since f is a local homeomorphism, $f(X)$ is an open set; while the fact that f is proper implies that $f(X)$ is closed. Consequently, the connectedness of Y implies that $f(X) = Y$. Thus by (5.1.1), we need only demonstrate the existence of a curve $x(t)$ satisfying (5.1.2). The fact that f is a local homeomorphism, implies that for some small $\epsilon > 0$ and $t \in [0, \epsilon)$, there is a curve $x(t)$ satisfying $f(x(t)) = ty + (1-t)y_0$. Let $\beta > 0$ be the largest number for which $x(t)$ can be continuously extended to satisfy $f(x(t)) = ty + (1-t)y_0$ for $0 \leqslant t < \beta$, and suppose $t_i \to \beta$. Since $L = \{y(t) \mid y(t) = ty + (1-t)y_0,\ t \in [0, 1]\}$ is compact and f is proper, $f^{-1}(L)$ is compact, so that $x(t_i)$ has a convergent subsequence $x(t_{i_n}) \to \bar{x}$, as $t_{i_n} \to \beta$. By continuity, $f(\bar{x}) = \beta y + (1-\beta)y_0$. By virtue of the fact that f is a local homeomorphism, $x(t)$ can be continuously extended for $t > \beta$. This contradicts the maximality of β, and we conclude that $x(t)$ exists for $t \in [0, 1]$ independently of $x_0 \in X$ and $y \in Y$.

(ii): If $f \in C^1(X, Y)$ is a diffeomorphism with inverse g, then $f'(x)$ is certainly a linear homeomorphism for each $x \in X$ since the relations $fg(y) = y$, $gf(x) = x$ can be differentiated. Thus the necessity of the conditions stated is clear. To demonstrate the sufficiency of the conditions, we use the inverse function theorem and (i) above to prove that f is a homeomorphism, with inverse g (say). Then since $fg(y) = y$ for each $y \in Y$, the differentiability of f implies the differentiability of g.

The following quantitative criterion is due to Hadamard, for the case of finite dimensions.

(5.1.5) (Hadamard) Suppose $f \in C^1(X, Y)$ is a local homeomorphism and $\zeta(R) = \inf_{\|x\| \leqslant R} (1/\|[f'(x)]^{-1}\|)$. Then if $\int^\infty \zeta(R)\, dR = \infty$, f is a homeomorphism of X onto Y. In particular, if $\|[f'(x)]^{-1}\| \leqslant M$ for all $x \in X$, f is a homeomorphism of X onto Y.

Proof: We shall prove that (X, f) covers $f(X)$, and in addition that $f(X) = Y$. By (5.1.1), (X, f) will cover $f(X)$ if and only if f lifts line segments. To establish this fact we argue as in the proof of (5.1.5). Let $x_0 \in f^{-1}(y_0)$, and $y \in Y$, then we seek a curve $x(t)$ such that $f(x(t)) = ty + (1-t)y_0$. Since $f(x)$ is a local homeomorphism, $x(t)$ exists for small t. Let β be the largest number such that $x(t)$ can be continuously extended to $0 \leqslant t < \beta$ satisfying $f(x(t)) = ty + (1-t)y_0$. We shall prove that $\lim_{t\to\beta} x(t)$ exists and is finite. Assuming this fact for the moment, then as in the proof of (5.1.4), $x(t)$ exists for $t \in [0, 1]$, so that f lifts line segments. Since $y \in Y$ is arbitrary, this argument shows that $f(X) = Y$.

Consequently, we need only use the hypotheses of the theorem to prove that $\lim_{t\to\beta} x(t)$ exists and is finite. If $\|[f'(x)]^{-1}\| \leqslant M$, this fact can be easily established. Indeed, if $t < \beta$, $x(t)$ satisfies the equation

$$(*) \qquad x'(t) = [f'(x(t))]^{-1}(y - y_0),$$

so that for any $t_1, t_2 < \beta$,

$$\|x(t_2) - x(t_1)\| \leqslant \left\| \int_{t_1}^{t_2} [f'(x(t))]^{-1}(y - y_0) \right\| \leqslant M\|y - y_0\|\, |t_2 - t_1|.$$

Thus $x(t)$ satisfies a Lipschitz condition for $t < \beta$ and since X is complete, $\lim_{t\to\beta} x(t)$ axists and is finite.

More generally, if $\int^\infty \zeta(t) = \infty$, the argument just given can be modified as follows. We can define the length of $x(t)$ for $0 \leqslant t < \beta$ with respect to the weight $g(x) = 1/\|[f'(x)]^{-1}\|$, as

$$L_g(x(t), [0, \beta)) = \int_0^\beta g(x(t))\|x'(t)\|\, dt.$$

Now we reach the desired conclusion by proving that if $\int^\infty \zeta(t) = \infty$, the metric defined on X as above is complete in the sense that if $L_g(x(t),$

$[0, \beta)) < \infty$, then $\lim_{t \to \beta} x(t)$ exists and is finite. Let $0 < s < \beta$. By our above estimates, since $d\|x(t)\|$ is of bounded variation,

$$\infty > \int_0^s g(x(t))\|x'(t)\|\, dt \geqslant \int_0^s g(x(t))\, d\|x(t)\|$$

$$\geqslant \int_0^s \zeta(\|x(t)\|)\, d\|x(t)\| \geqslant \int_{\|x(0)\|}^{\|x(s)\|} \zeta(\tau)\, d\tau.$$

Consequently, since $\int^\infty \zeta(t)\, dt = \infty$, $\|x(t)\|$ is uniformly bounded for $t \in [0, \beta)$. On the other hand, since $\zeta(t)$ is nonincreasing and $\int_0^\infty \zeta(t)\, dt = \infty$, $\sup\{t \mid \zeta(t) > 0\} = \infty$, so that $g(x)$ is bounded from below on any bounded set. In particular, $g(x(t))$ is bounded from below for $0 \leqslant t < \beta$, say $\| g(x(t))\| \geqslant G$. Now let $t_i \uparrow \beta$, then

$$(**) \qquad \sum_{i=1}^{n} \|x(t_{i+1}) - x(t_i)\| \leqslant \sum_{i=1}^{n} \sup_{t \in [t_i, t_{i+1}]} \|x'(t)\|(t_{i+1} - t_i)$$

$$\leqslant \int_0^\beta \|x'(t)\|\, dt \leqslant \frac{1}{G} \int_0^\beta g(x(t))\|x'(t)\|\, dt < \infty.$$

Thus $(**)$ implies that $\sum_{i=1}^{n}\|x(t_{i+1}) - x(t_i)\| < \infty$, so that $\{x(t_i)\}$ is a Cauchy sequence in X. Consequently $\lim_{t_i \to \beta} x(t)$ exists and is finite.

As an application of (5.1.4), we prove

(5.1.6) Suppose that $f(x)$ is a continuous mapping from a reflexive Banach space X into its conjugate space X^* with the property:

$$(\dagger) \qquad (f(x) - f(y), x - y) \geqslant \eta(\|x - y\|)\|x - y\|,$$

where $\eta(r)$ is a positive function satisfying $\eta(0) = 0$; and $\eta(r) \to \infty$ as $r \to \infty$. Then f is a homeomorphism of X onto X^*.

Proof: We prove that f is proper and a local homeomorphism. To show that f is proper, we note that $(\dagger)$ implies that whenever $x_n \to x$ weakly and $f(x_n) \to z$, then $f(x) = z$ and $x_n \to x$ strongly. Consequently, suppose $\{f(x_n)\}$ is convergent in X^*, then by $(\dagger)$ and the reflexivity of X, $\{x_n\}$ is bounded, and after possibly passing to a subsequence, we may suppose that $x_n \to \bar{x}$ weakly in X. So $\{x_n\}$ is strongly convergent, by property $(\dagger)$.

To prove that f is a local homeomorphism, we see by virtue of $(\dagger)$ that f maps X injectively onto its range. In addition, if $x_1 = f^{-1}(z_1)$ and $x_2 = f^{-1}(z_2)$, then $(\dagger)$ also implies that $\|z_1 - z_2\| \geqslant \eta(\|f^{-1}(z_1) - f^{-1}(z_2)\|)$, so that f^{-1} is continuous. Thus by (5.1.4), f is a global homeomorphism of X onto Y since it can be shown that the range of F is open.

In practice, the application of Hadamard's theorem often requires an initial decomposition of the appropriate Banach spaces X, Y as well as the utilization of special properties of the mapping f relative to this decomposition. Such an example occurs in the differential geometric problem of finding a Riemannian metric of constant Gaussian curvature -1 on a smooth two-dimensional Riemannian manifold $(\mathfrak{M}, g)$ of negative Euler–Poincaré characteristic $\chi(\mathfrak{M})$. This problem was mentioned in Section 1.1A in connection with the uniformization problem for algebraic curves. By virtue of our discussion there (cf. the equation (1.1.5)), the existence of such a metric can be ensured by finding a smooth solution defined on $\mathfrak{M}$ for the following partial differential equation

(5.1.7) $$\Delta u - K(x) - e^{2u} = 0.$$

In this connection we shall use Hadamard's theorem to prove

(5.1.8) **Theorem** A necessary and sufficient condition for the solvability of (5.1.7) is that $\int_{\mathfrak{M}} K(x)\, dV_g < 0$. Consequently if $K(x)$ denotes the Gaussian curvature of $(\mathfrak{M}, g)$ and $\chi(\mathfrak{M}) < 0$, (5.1.7) is always solvable, by virtue of the Gauss–Bonnet theorem.

Proof: The proof requires three steps. First, we obtain a reformulation of the problem into an appropriate operator equation defined on the Sobolev space $W_{1,2}(\mathfrak{M}, g)$ carried out in such a way that any (generalized) solution u to the operator equation is automatically a smooth solution of (5.1.7). Secondly, in order to apply Hadamard's theorem we must decompose the resulting abstract operator equation relative to the facts that the Laplace operator Δ has a kernel on $(\mathfrak{M}, g)$ consisting of the constant functions; and in fact the desired conclusion itself asserts that the associated mapping f, unless modified, certainly will not be a global homeomorphism. The final step consists in estimating the size of the Fréchet derivative of the appropriately modified f in such a way as to satisfy the hypotheses of Hadamard's theorem (5.1.5).

Step 1: The desired reformulation of (5.1.7) is easy in this case, provided we use the duality method of Section 2.2D. Indeed defining the operators L and N implicitly by the formulas

$$(Lu, v) = \int_{\mathfrak{M}} \nabla u \cdot \nabla v, \qquad (Nu, v) = \int_{\mathfrak{M}} e^{2u} v,$$

we see that L is a bounded self-adjoint mapping of the Sobolev space $W_{1,2}(\mathfrak{M}, g)$ into itself, while by virtue of the estimate (1.4.6) N is a C^1 mapping acting between the same spaces. Consequently the partial differential equation (5.1.7) can be written in the form

(*) $$Lu + Nu = f \quad \text{where} \quad (f, v) = -\int K(x) v.$$

The verification that solutions of this operator equation in $W_{1,2}(\mathfrak{M}, g)$ are automatically smooth enough (possibly redefined on a set of Lebesgue measure zero) to satisfy (5.1.7) pointwise is a consequence of the L_p regularity theory described in Section 1.5B and the estimate (1.4.6).

Step 2: We now decompose the operator equation (*) by setting $W_{1,2}(\mathfrak{M}, g) = H$ and writing $H = \text{Ker } L \oplus H_1$ and denote by P the canonical projection of H onto H_1. Then if a tentative solution $u = w + c$ of (*) relative to this decomposition is given, (*) is equivalent to the pair

$$(**) \qquad Lw + PN(c + w) = Pf; \qquad e^{2c}\int_{\mathfrak{M}} e^{2w} = -\int_{\mathfrak{M}} K(x)\, dV_g.$$

This result follows since H_1 is the subspace of functions in H of mean value zero over $(\mathfrak{M}, g)$. Now the second equation shows that the constant c is determined as a function of w if and only if $\int_{\mathfrak{M}} K < 0$. Using this value $c = c(w)$ in the first equation of the pair (**), we see that our problem reduces to showing the operator on the left-hand side of the first equation of (**) is a global homeomorphism of H_1 onto itself.

Step 3: We conclude the proof by showing that when regarded as a C^1 mapping of H_1 into itself the operator $\tilde{f}(w) = L + PN(c(w) + w)$ has a Fréchet derivative $\tilde{f}'(w)$ that is an invertible linear operator for all $w \in H_1$ and in fact the Fréchet derivative $[\tilde{f}'(w)]^{-1}$ is uniformly bounded. Then Hadamard's theorem (5.1.5) implies $\tilde{f}$ is a global homeomorphism of H_1 onto itself as required. To this end we must compute and estimate $\tilde{f}'(w)$. This is most easily done, in the present case, by returning to the implicit definitions of L and N and estimating the following quadratic form (defined on H_1), as follows with $v \in H$.

$$\left(\tilde{f}'(w)v, v\right) = \int_{\mathfrak{M}} |\nabla v|^2 + 2[e^{2c(w)}e^{2w}]v^2 \geq \int_{\mathfrak{M}} |\nabla v|^2 = \|v\|^2_{H_1}.$$

Note that on H_1, the norm reduces to the Dirichlet integral. Consequently, the Lax–Milgram lemma (1.3.21) implies $\tilde{f}'(w) \in L(H_1, H_1)$ is an invertible linear operator and $\|[\tilde{f}'(w)]^{-1}\| \leq 1$, which is the desired uniform bound. Thus our result is established.

5.1B Mappings with singular values

Our results on global homeomorphisms, just obtained, may still be useful even though a C^1 operator f has singular values. Indeed, f may be a global homeomorphism when certain sets of finite codimension are excised from its range and domain. As we shall see this reduces the study of the mapping properties of f to a finite-dimensional problem.

We illustrate this idea in the simplest case where the sets in question are linear subspaces. First, we present an abstract result and then apply it to determine the exact structure of the range of a specific semilinear elliptic differential operator (supplemented with Dirichlet boundary conditions). (Additional applications of this fundamental abstract idea will be given in Section 5.3.) The abstract result we have in mind is embodied in

(5.1.9) **Reduction Lemma** Let X and Y be Banach spaces, with $L \in L(X, Y)$ a Fredholm operator of nonnegative index p and $N \in C^1(X, Y)$. Moreover, suppose for some fixed number $\epsilon > 0$, the following result holds *off* some finite-dimensional subspace $W = \text{Ker } L \oplus V$ (i.e., for $x \in W^{\perp}$):

(5.1.10) $\quad Lx + PN'(u)x$ is invertible and $\|Lx + PN'(u)x\| \geqslant \epsilon\|x\|$,

where P is the canonical projection of Y onto $L(W^{\perp})$. Then $f \in \text{Range}(L + N)$ if and only if a certain system of $\dim W - p$ equations in $\dim W$ unknowns, defined by (5.1.12) below, is solvable and the solutions of $Lx + Nx = f$ correspond precisely to the solutions of (5.1.12). Moreover if $Lx + Nx = f$ with $x = w_0 + w_1$, $w_0 \in W$, $w_1 \in W^{\perp}$, the following estimate holds

(5.1.11) $\quad \|w_1\| \leqslant c_1\|w_0\| + c_2 \qquad (c_1, c_2$ absolute constants)

provided $N'(x)$ is uniformly bounded.

Proof: We decompose X into the direct sum $X = W \oplus W^{\perp}$ and $Y = L(W^{\perp}) \oplus Y_0$ with P_0 the canonical projection of Y onto Y_0. Then for $x = w_0 + w_1$ and $f \in Y$, the equation $Lx + Nx = f$ can be rewritten as the system

(5.1.12) $\quad$ (i) $Lw_0 + P_0Nx = P_0f$,

$\quad$ (ii) $Lw_1 + PNx = Pf$.

Now regarded as a C^1 mapping of $W^{\perp}$ into $L(W^{\perp})$, $Aw_1 = Lw_1 + PNx$ has (by virtue of the inequalities (5.1.10)) a Fréchet derivative $A'(u)w_1$ with lower bound

$$\|A'(u)w_1\| = \|Lw_1 + PN'(u)w_1\| \geqslant \epsilon\|w_1\|.$$

Moreover, by Banach's theorem (1.3.20), since $L \in L(W^{\perp}, L(W^{\perp}))$ is invertible, so is $A'(u)$. Consequently, Hadamard's theorem (5.1.5) implies that (5.1.12 (ii)) can be uniquely solved by an element $w_1 = w_1(w_0, Pf)$ that depends smoothly on w_0 and Pf. Moreover regarding w_1 parametrized by w_0 (for Pf fixed) we find by differentiating (5.1.12(ii)) with respect to w_0 in the direction v

$$Lw_1'(w_0[v]) + PN'(x)\{v + w_1'(w_0[v])\} = 0$$

Consequently, a simple estimate using (5.1.10) and the uniform boundedness of $N'(x)$ shows that $\|w_1'(w_0)[v]\|/\|v\|$ is uniformly bounded for all v and so

$$\epsilon\|w_1\| \leqslant K_0\|w_0\| + K_1,$$

where K_1 and K_0 are absolute constants. Now, this estimate implies the bound (5.1.11).

Finally, we observe the system (5.1.12 (i)) can now be regarded as involving dim Y_0 equations in dim W unknowns since x can be written $x = w_0 + w_1(w_0, Pf)$. Since L is a Fredholm operator of index p, we note that dim $Y_0 =$ dim $W - p$, so the lemma is established.

(5.1.11′) Moreover, if the operator $N(x)$ is uniformly bounded, the function $w = w(x_1, Pf)$ is also uniformly bounded as w_0 varies.

Proof: We now suppose N is uniformly bounded. Then (5.1.12) implies

$$(*) \qquad \|Lw\| \leqslant \|Pf\| + \|PN(w_0 + w)\| \leqslant K,$$

where K is a positive absolute constant. Since L is invertible as a linear mapping between $W^{\perp}$ and $L(W^{\perp})$. Thus, $(*)$ implies that $\|w\|$ is uniformly bounded.

Turning to a simple but informative application of the reduction lemma (in which an infinite-dimensional problem reduces to a one-dimensional one), we consider the mapping A defined by the explicit semilinear elliptic partial differential operator

$$(5.1.13) \qquad Au \equiv \Delta u + f(u), \qquad u\,|_{\partial\Omega} = 0$$

defined on a bounded domain $\Omega \in \mathbf{R}^N$ (augmented by null Dirichlet boundary conditions on Ω). Here Δ denotes the Laplace operator relative to Ω with eigenvalues $\lambda_1 < \lambda_2 \leqslant \lambda_3$ with $\lambda_1 > 0$, and f is a C^2 strictly convex function satisfying $f(0) = 0$ and the asymptotic relations

$$(5.1.13') \qquad 0 < \lim_{t \to -\infty} f'(t) < \lambda_1; \qquad \lambda_1 < \lim_{t \to +\infty} f'(t) < \lambda_2.$$

Under these conditions we prove the following analogue for solving quadratic equations over the real numbers.

(5.1.14) Regarded as a mapping of $X = \mathring{W}_{1,2}(\Omega)$ into $Y = W_{-1,2}(\Omega)$, the range of the mapping A has the following properties:

(a) The singular values of A form a connected manifold M of codim 1 in Y such that $Y - M$ has exactly two components O_0, O_2. (See Fig. 5.2.)

(b) For $g \in O_j$, the equation $Au = g$ has exactly j solutions ($j = 0, 2$), while for $g \in M$, the equation $Au = g$ has exactly one solution.

(c) For given $g \in L_2$, the number of solutions of $Au = g$ is completely determined by the size of the projection of g on the one-dimensional eigenspace S_1 of Δ associated with λ_1. Thus if $u_1 \in S_1$ denotes the positive eigenfunction u_1 (of L_2 norm 1) and $g = \alpha u_1 + g_1$ with $u_1 \perp g_1$ (in the L_2 sense), there is a continuous real-valued function of g_1, $\alpha(g_1)$, such that if $\alpha = \alpha(g_1) g \in M$, while $g \in O_0$ (or O_2) depending on whether $\alpha < \alpha(g_1)$ or $\alpha > \alpha(g_1)$, respectively.

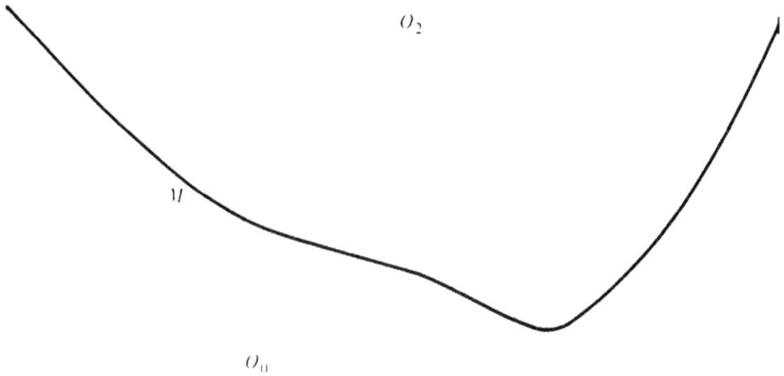

FIG. 5.2 Illustration of the solvability result for (5.1.13).

(d) All the singular values of A can be written $g = \alpha(g_1)u_1 + g_1$.

Remark: Moreover, if the convexity of $f(t)$ is removed, note that the solvability of (5.1.13) (but not the number of solutions) is still determined by (c) of (5.1.14). Finally see note F at the end of the chapter for the "complete integrability" of this problem.

Proof: The result is established in three steps: first, a reformulation of the problem as an abstract operator equation; next, an application of the reduction lemma to reduce the problem to the study of the solutions of *one equation in one unknown*; and finally, a resolution of the one-dimensional problem defined on S_1 by a simple graphical consideration.

Step 1: (*Reformulation*) We first use the duality method (Section 2.2 (iii)) to represent the equation $Au = g$ in the form $Lu - Nu = -g$, where $g \in H$ (a Hilbert space). Indeed we define L and N implicitly by the formulas

$$(Lu, v) = \int_\Omega (\nabla u \cdot \nabla v - \lambda_1 uv), \qquad (Nu, v) = \int_\Omega [f(u) - \lambda_1 u]v.$$

Moreover over $v \in C_0^\infty(\Omega)$, we observe that the weak solutions of (5.1.13) satisfy (5.1.12) provided we set $H = W_{1,2}(\Omega)$ and define g by the relation $\int_\Omega gv = (g, v)$. Clearly L defined by the above formula is a self-adjoint Fredholm mapping of $\mathring{W}_{1,2}(\Omega)$ into itself such that dim Ker $L = 1$ and $(Lu, u) \geq 0$ for all $u \in H$. Moreover N is a bounded mapping of H into itself with a uniformly bounded Fréchet derivative $N'(u)$ defined implicitly by the relation $(N'(u)w, v) = \int_\Omega (f'(u) - \lambda_1)wv$. The verification of these properties is routine. Actually, N defined as above is a C^1 mapping (and this fact will be useful in the sequel). To verify this fact we observe that the asymptotic properties (5.1.13′) imply that $f''(t) \to 0$ as $|t| \to \infty$, so that $f''(t)$ is uniformly bounded over $(-\infty, \infty)$. Moreover if $u_n \to u$ in $H, f'(u_n) \to f'(u)$ in measure over Ω. Consequently an application of the Lebesgue dominated convergence theorem shows that $N'(u)$ calculated directly from the implicit definition of N exists and is continuous in u.

Step 2: (Reduction to a one-dimensional problem): We apply the reduction lemma to the operator $f = L - N$. In the present case $p = 0$ since L is self-adjoint. Moreover, in the present case the subspace W in the statement of the reduction lemma coincides with the one-dimensional subspace Ker $L = \{v \mid \Delta v + \lambda_1 v = 0, v \in H\}$. (Here we have used the fact that the lowest eigenvalue of Δ on Ω is always simple.) Indeed, for $v \perp \operatorname{Ker} L$, by (5.1.13) and the properties of f

$$(Lv - PN'(u)v, v) = \int_\Omega [|\nabla v|^2 - f'(u)v^2] \geq [1 - (\lambda_2 - \epsilon_+)/\lambda_2] \int_\Omega |\nabla v|^2.$$

Thus, defining the norm in H by setting $\|u\|_H^2 = \int_\Omega |\nabla u|^2$ and by applying (1.3.20), we find that W can be chosen to be Ker L. Consequently, according to the reduction lemma it suffices to study the one-dimensional problem $-P_0 N(tu_1 + w(t_1 g_1)) = -\int_\Omega g u_1$, where u_1 is the positive normalized eigenfunction of Δ associated with Ω. Denoting the negatives of either sides of the above by $h(t)$ and K respectively, we need only study the solutions of the equation

$$h(t) \equiv -\lambda_1 t + \int_\Omega f[tu_1 + w(t, g_1)]u_1 = K. \tag{5.1.15}$$

Moreover from (5.1.11) the following estimate holds for $\|w(t, g_1)\| \leq c_1 t + c_2$.

Step 3: We now establish the results (a)–(d) mentioned in the statement of (5.1.14), by proving the following two properties of the function $h(t)$ defined by (5.1.15): (α) $h(t) \to \infty$ as $|t| \to \infty$ and (β) $\inf_t h(t) > -\infty$ is the only critical point of $h(t)$ and is attained exactly once at $t = t_1$ (say). Moreover t_1 is uniquely determined by g_1 and is a continuous function of g_1. Assuming (α) and (β) for the time being, we observe that (a)–(d) as stated in (5.1.14) are immediate consequences upon setting $\alpha(g_1) = h(t_1)$. Thus to demonstrate that singular values $A(S)$ of A are precisely the points of the form $g = h(t_1)u_1 + g_1$ as g varies over $(\operatorname{Ker} L)^\perp$, we first differentiate the equation $A(tu_1 + w(t, g_1)) = h(t)u_1 + g_1$ with respect to t and set $t = t_1$, obtaining $A'(t_1 u_1 + w(t_1, g_1))\,(u_1 + w'(t_1)) = 0$ so $t_1 u_1 + w(t_1, g_1 0) \in S$. Conversely, suppose $u \in S$ and $A(u) = c_1 u_1 + g_1 \in A(S)$. So, by our results so far, $u = tu_1 + w(t_1 g_1)$ and for some $v = \alpha u_1 + w_1$, $A'(u)v = 0$. Thus if P denotes the projection of H onto $(\operatorname{Ker} L)^\perp = H_1$, $PA'(u)(\alpha u_1) = -PA'(u)w_1$; and since $-PA'(u)$ is invertible on H_1,

$$w_1 = -(PA'(u))^{-1}PA'(u)u_1 = w'(t_1, g_1).$$

Thus $v = \alpha u_1 + \alpha w'(t, g_1)$ and since $(I - P)A'(u)v = 0$ implies $h'(t) = 0$, we find $t = t_1$ by (β); so $u = t_1 u_1 + w(t_1, g_1)$ as required.

Finally, we prove (α) and (β). The basic problem is the justification of the simple idea that both facts are easily obtained if the contributions to $h(t)$ due to $w(t_1, g_1)$ are neglected. Given a subset $\{t\}$ with $|t| \to \infty$, we note that it suffices to show $h(t) \to +\infty$ for any two sequences $t = t_n \to +\infty$ and $t = s_n \to -\infty$. This will follow by finding an absolute constant $c_1 > 0$ such that

$$(*) \qquad \lim_{t_n \to \infty} \frac{h(t_n)}{t_n} \geq c_1 \qquad \text{and} \qquad (**) \qquad \lim_{s_n \to -\infty} \frac{h(s_n)}{s_n} \leq -c_1.$$

We demonstrate $(*)$ here since $(**)$ is completely analogous. To this end, we note that the a priori estimate mentioned at the end of step 2 implies that for t sufficiently large $\{w(t)/t\}$ is uniformly bounded and so can be assumed to possess a weakly convergent subsequence $\{w(t_n)/t_n\}$ that without loss of generality can be assumed to be strongly convergent in $L_2(\Omega)$ and pointwise convergent (a.e.) to an element $\bar{w} \in H$.

Next partitioning Ω into three sets Ω_+, Ω_-, and Ω_0 according as $u_1(x) + w(x)$ is positive, negative, or zero, we find from (5.1.15) that setting $\lim_{t \to \infty} f'(t) = \lambda_2 - \epsilon_+$ and $\lim_{t \to +\infty} f'(t) = \lambda_1 - \epsilon_-$,

$$\lim_{n \to \infty} \frac{h(t_n)}{t_n} = -\lambda_1 + (\lambda_2 - \epsilon_+)\int_{\Omega_+}(u_1 + w)u_1 + (\lambda_1 - \epsilon_-)\int_{\Omega_-}(u_1 + \epsilon)u_1. \tag{5.1.16}$$

Here we have used the Lebesgue bounded convergence theorem and the observation that the integral of $f(t_n u_1 + w(t_n))/t_n$ over Ω_0 also tends to zero as $n \to \infty$. Moreover, $\int_\Omega w u_1 = 0$ since this result is preserved under weak convergence in H. Consequently

$$\int_{\Omega_+}(u_1 + w)u_1 + \int_{\Omega_-}(u_1 + w)u_1 = \int_\Omega (u_1 + w)u_1 = \int u_1^2 = 1.$$

Thus we can estimate the right-hand side of (5.1.16) from below by changing the coefficient of the negative integral from $\lambda_1 - \epsilon_-$ to $\lambda_2 - \epsilon_+$. Thus combining (5.1.16) with the equality displayed above we find that, for $c_1 = \lambda_2 - \lambda_1 - \epsilon_+ > 0$, since $w \perp u_1$ in the L_2 sense,

$$\lim_{t_n \to \infty} \frac{h(t_n)}{t_n} = -\lambda_1 + (\lambda_2 - \epsilon_+)\int (u_1 + w)u_1 = c_1.$$

Thus the desired result (α) is established by *making strong use of the asymptotic linearity of f to invoke the orthogonality of u_1 and $w(t)$.*

The proof of (β) is somewhat more intricate utilizing, as we shall see, the convexity of f strongly since the *asymptotic linearity* of f defined by the relations (5.1.13) cannot be used. We show below that at a critical point t_0 the function $h(t)$ has the property

$$(*) \qquad h''(t_0) = \int_\Omega f''(u(t_0))\{u_1 + w'(t_0)\}^3$$

in case the function $h(t)$ is twice differentiable. Then we use the positivity of the eigenfunction u_1 to prove $\operatorname{sgn}[u_1 + w'(t_0)] > 0$, so that $h''(t_0) > 0$. To this end we rewrite (5.1.15) in the form $h(t) = PA(u(t))$, where P is the projection of H onto Ker L. Then a formal calculation, assuming $h'(t_0) = 0$, yields the formula $h''(t_0) = (A''(u(t_0)\ (u'(t_0), u'(t_0)), u'(t_0))$ and consequently the representation (*). Moreover $u'(t) = u_1 + w'(t)$ satisfies the linear equation $A'(u(t))u'(t) = 0$ which in the present case implies $v = u'(t)$ is an eigenfunction of the equation

$$(5.1.17) \qquad \Delta v + f'(u(t))v = 0, \qquad v|_{\partial\Omega} = 0$$

corresponding to the eigenvalue $\lambda = 1$. Now the asymptotic relations satisfied by $f'(s)$ and the extremal characterization of eigenvalues imply $\lambda = 1$ is the smallest eigenvalue of (5.1.17) and moreover $\operatorname{sgn} u'(t) = \operatorname{sgn}(u_1 + w'(t))$ is constant in Ω. Now since $\int_\Omega u_1 w'(t) = 0$, $w'(t) > 0$ for some open subset Ω' of Ω. Thus $\operatorname{sgn}(u_1 + w'(t)) > 0$ over Ω, as desired; and (β) is proven in case the function $h(t)$ is twice differentiable. However, *in general, $h(t)$ is not C^2*, so a modification of the direct idea just given is necessary. To this end we note it suffices that for $|t - t_0|$ sufficiently small, $h(t) > h(t_0)$, or equivalently that $\operatorname{sgn} h'(t) = \operatorname{sgn}(t - t_0)$. To establish this we use the defining relation $A'(u(t))u'(t) = 0$, to find

$$\int_\Omega f'(u(t))u'(t)w'(t_0) = \int_\Omega f'(u(t_0)u'(t)w'(t_0)).$$

Consequently a short computation gives the relation

$$(5.1.18) \qquad h'(t) - h'(t_0) = \int_\Omega \{f'(u(t)) - f'(u(t_0))\}u'(t)u'(t_0).$$

Now instead of computing $h''(t_0)$ directly we divide the above integral on the right into two parts an integral over $\Omega_1 = \{x | u'(t_0) \leqslant 1\}$ and an integral over $\Omega - \Omega_1$. On Ω_1 an easy application of the Lebesgue dominated convergence theorem to (5.1.18) shows that since f'' is uniformly bounded, $\lim_{t \to t_0}(h'(t) - h'(t_0))/(t - t_0)$ exists and coincides with the right-hand side of (*), and thus is positive. Consequently it remains to discuss (5.1.18) on $\Omega - \Omega_1$. Since $u(t) \to u(t_0)$ in $L_2(\Omega)$, we first note that any sequence $t_n \to t_0$ has a subsequence (which we again relabel t_n) such that for $n \geqslant n_0$ and some $\epsilon > 0$

$$u'(t_n) \geqslant \tfrac{1}{2} \quad \text{(a.e.)} \qquad \text{and} \qquad (u(t_n) - u(t_0))/(t_n - t_0) \geqslant \epsilon.$$

Thus for $n \geqslant n_0$ on $\Omega - \Omega_1$, $f(u(t_n)) \geqslant f(u(t_0))$ and $u'(t_n) > 0$, so the integral in (5.1.18) on

$\Omega - \Omega_1$ is positive for $t > t_0$, and the desired result is established.
(See Note B at the end of the chapter for extensions of (5.1.14).)

5.2 Finite-Dimensional Approximations

Properties of solutions of an operator equation $f(x) = 0$ defined on a Banach space X can be studied by approximating both f and X by sequences of mappings $\{f_n\}$ and spaces $\{X_n\}$. Assuming that the pair (f_n, X_n) converges to (f, X) in a well-defined sense, one attempts to analyze the solutions $\{x_n\}$ of $f_n(x_n) = 0$ on X_n as $n \to \infty$ with a view to showing that appropriate subsequences of $\{x_n\}$ converge to solutions of $f(x) = 0$ on X. Assuming $\dim X = \infty$, we shall discuss this circle of ideas by approximating X by a sequence of finite-dimensional subspaces X_n and f by a sequence $\{f_n\}$ (with finite-dimensional range) obtained by restricting the domain of f to X_n.

5.2A Galerkin approximations

More precisely, let X be a real reflexive separable Banach space (with conjugate space X^*) and let $\{X_n\}$ be a fixed sequence of finite-dimensional subspaces of X such that $X_n \subset X_{n+1}$ and $\bigcup_{n=1}^{\infty} X_n$ is dense in X. To fix notation, let P_n denote the projection of X onto X_n, P_n^* denote the conjugate operator, and $X_n' = P_n^* X^*$. Thus if f is a bounded continuous mapping of X into X^* and $g \in X^*$, then a sequence of finite-dimensional approximations to the equation $f(x) = g$ defined on X can be denoted

$$(5.2.1)_n \qquad P_n^* f(x) = P_n^* g, \qquad x \in X_n.$$

The system $(5.2.1)_n$ is often referred to as the Galerkin approximation for the equation $f(x) = g$. Here we have limited attention to equations defined on reflexive Banach spaces X in order to use the weak compactness of bounded sets in X. Indeed, in order to deduce the solvability of $f(x) = g$ on X from the solvability of $(5.2.1)_n$ for n sufficiently large, we need only:

(i) determine an a priori bound for the solutions $\{x_n\}$ of $(5.2.1)_n$, $\|x_n\|_X \leqslant M$ (say), where M is independent of n, so that (after possibly passing to a subsequence) one can assume that $\{x_n\}$ is weakly convergent with unique weak limit $\bar{x}$;

(ii) use the qualitative properties of f and the Galerkin construction itself to show that $f(\bar{x}) = g$. Thus (in the simplest case) if f is weakly sequentially continuous as a mapping from X to X^*, (i) immediately implies that $f(\bar{x}) = g$.

Actually a closer study of the approximation procedure $(5.2.1)_n$ allows this weak continuity assumption just mentioned to be considerably reduced. To this end, we observe that two characteristic qualitative properties of the solutions $\{x_n\}$ of the Galerkin approximations can be expressed by the two sequences of equations

$$(5.2.2)\qquad \text{(a)}\quad (f(x_n), x_n) = (g, x_n);$$
$$\text{(b)}\quad (f(x_n), z) = (g, z), \qquad z \in X_n.$$

These results are obtained by taking inner products of $(5.2.1)_n$ with x_n and z, respectively. Thus for convergence purposes, once (i) is established, we can suppose not only that $x_n \to \bar{x}$ weakly, but also that $f(x_n) \to \xi$ weakly (since f is bounded). Furthermore, (5.2.2b) implies that $\xi = g$, while (5.2.2a) yields the fact that $(f(x_n), x_n) \to (g, \bar{x})$. Hence the crucial property of the operator f necessary to establish (ii) from (i) can be expressed as

Condition (G) If $x_n \to x$ weakly in X, $f(x_n) \to y$ weakly in X^*, and $(f(x_n), x_n) \to (y, x)$, then $f(x) = y$.

This condition is satisfied by all continuous mappings between finite-dimensional Banach spaces, all weakly sequentially continuous mappings, and, as we shall see in the sequel, a large class of mappings associated with quasilinear elliptic partial differential equations.

We are now in a position to prove the following:

(5.2.3) **Theorem** Suppose f is a bounded mapping of a real, reflexive separable Banach space X into its conjugate space X^* continuous and satisfying the conditions

(I) $(f(x), x)\|x\|^{-1} \to \infty$ as $\|x\| \to \infty$, and
(II) f satisfies Condition (G).

Then f is a surjective mapping of X onto X^*, and for any $g \in X^*$ a solution of $f(x) = g$ can be obtained as the weak limit of an appropriate subsequence of the solutions of the Galerkin approximations $(5.2.1)_n$.

Proof: The main idea is to use condition (I) to ensure both the existence of a solution x_n of $(5.2.1)_n$ and a uniform a priori bound for the sequence $\{x_n\}$. Then, as mentioned above, (II) is used to show that a weakly convergent subsequence of $\{x_n\}$ converges to a solution of $f(x) = g$ for any $g \in X^*$.

Now suppose that g is an arbitrary element of X^*. Then to show that $(5.2.1)_n$ is solvable for each n, we show that the finite-dimensional mapping $f_n = P_n^* f$, regarded as a mapping from $X_n \to P_n^* X_n$, is surjective. To this end, we observe that for $x \in X_n$, hypothesis (I) implies

$$\|x\|_{X_n}^{-1}(f_n(x), x) = (f(x), x)\|x\|_X^{-1} \to \infty \qquad \text{as} \quad \|x\|_{X_n} \to \infty.$$

Thus the Brouwer degree of f_n at any fixed element $\tilde{g}$ of $P_n^* X_n$ relative to a sufficiently large sphere $\{x \mid \|x\| \leqslant R\}$ is unity, so that $f_n(x) = \tilde{g}$ has a solution in X_n. Consequently, there is an $x_n \in X_n$ satisfying $(5.2.1)_n$. Futhermore, by (5.2.2a) and Schwarz's inequality,

$$(f_n(x_n), x_n) = (f(x_n), x_n) = (g, x_n) \leqslant \|g\| \, \|x_n\|.$$

Thus hypothesis (I) implies that the sequence $\{x_n\}$ is uniformly bounded.

Hence after passing to suitable subsequences, we may suppose that $\{x_n\}$ converges weakly to $\bar{x}$ in X, and, by virtue of the boundedness of f, that $f(x_n) \to g$ weakly in X^*. Furthermore, repeating the argument given after (5.2.2), we may also suppose that $(f(x_n), x_n) \to (g, \bar{x})$. Then, since f satisfies Condition (G), $f(\bar{x}) = g$. Since g was arbitrary, f is surjective and the theorem is established.

The applicability of the result just obtained will now be clarified by (i) determining a large class of mappings f satisfying Condition (G) and (ii) demonstrating the necessity of some hypothesis such as Condition (G) for the convergence of Galerkin approximations.

(5.2.4) **Example** (Theorem (5.2.3) is false if hypothesis (II) is removed.) Let $X = l_2$, the Hilbert space of square summable sequences, and let an element $x \in l_2$ be written as $x = (x_1, x_2, \ldots)$ with $\|x\|^2 = \sum_{i=1}^{\infty} x_i^2$. If $Tx = (\sqrt{1 - \|x\|^2}, x_1, x_2, \ldots)$ for $\|x\| \leqslant 1$, and $Tx = (\sqrt{1 - \|x\|^{-2}}, x_1\|x\|^{-2}, x_2\|x\|^{-2}, \ldots)$ for $\|x\| \geqslant 1$, then T is a continuous mapping of $l_2 \to l_2$ and $\|T_x\| = 1$ for all $x \in l_2$. Setting $f(x) = x - Tx$, we observe that f is continuous and $(f(x), x) = \|x\|^2 - (Tx, x) \geqslant \|x\|^2 - \|Tx\| \, \|x\|$. Thus $(f(x), x)/\|x\| \geqslant \|x\| - 1$ as $\|x\| \to \infty$. On the other hand, f is not surjective since $f(x) = 0$ has no solution. Indeed, if $f(y) = 0$, then since $\|Ty\| = 1$, $\|y\| = 1$. Thus if $y = (y_1, y_2, \ldots)$ and $y = Ty$, then $y_1 = 0$ and $y_{i+1} = y_i$ for all i. Hence $y = 0$, which contradicts the fact that $\|y\| = 1$.

We now determine a large class of mappings that satisfy Condition (G), but are not necessarily weakly sequentially continuous. To this end, we prove

(5.2.5) The following classes of mappings satisfy Condition (G):

(i) continuous monotone mappings T of $X \to X^*$ since $(T(x) - T(y), x - y) \geqslant 0$ for all $x, y \in X$;

(ii) completely continuous perturbations of continuous monotone mappings;

(iii) mappings of the form $T(x) = Px + Rx : X \to X^*$, where the mappings T, P, and R can be written in the form $Tx = T(x, x)$ with $T(x, y) = P(x, y) + R(x, y) : X \times X \to X^*$ and satisfy:

(a) $(y - z, P(x, y) - P(x, z)) \geqslant 0$.

(b) If $x_n \to x$ weakly and $(P(x_n, x_n) - P(x_n, x), x_n - x) \to 0$, then $Rx_n \to Rx$ weakly.

(c) If $\{x_n\}$ and $\{y_n\}$ are weakly convergent in X with $\lim_{n\to\infty} y_n = 0$, then $(Rx_n, y_n) \to 0$.

(d) For fixed $x \in X$, $R(y, x)$ and $P(y, x)$ are completely continuous mappings of $X \to X^*$.

(e) For fixed $y \in X$ the mappings $P(y, x)$ and $R(y, x)$ are bounded and continuous from the strong topology of X into the weak topology of X^*, uniformly on bounded sets of the alternate variable.

Proof: (i): Suppose T is monotone and w is an arbitrary element of X. Then for each n

$$(x_n - w, Tx_n - Tw) \geqslant 0. \tag{5.2.6}$$

If $x_n \to x$, $Tx_n \to y$ weakly, and $(Tx_n, x_n) \to (y, x)$, then by letting $n \to \infty$ in (5.2.6), we find

$$(x - w, y - Tw) \geqslant 0. \tag{5.2.7}$$

Setting $w = x - \lambda z$ in (5.2.7) for $\lambda > 0$ and arbitrary $z \in X$, we find $(z, y - T(x - \lambda z)) \geqslant 0$. Consequently, letting $\lambda \to 0$, we see that $(z, y - Tx) \geqslant 0$ for arbitrary $z \in X$. Thus $Tx = y$, as required.

(ii): A completely continuous perturbation R of a monotone mapping P satisfies the hypotheses of (iii) with $P(x, y) = P(x)$ and $R(x, y) = R(y)$. Therefore (ii) will follow from the more general case.

(iii): The argument used is an extension of the proof in (i). For arbitrary $w \in X$ and each n, by hypothesis (a)

$$(x_n - w, T(x_n, x_n) - T(x_n, w)) \geqslant (x_n - w, R(x_n, x_n) - R(x_n, w)). \tag{5.2.8}$$

Set $R(x, x) = R(x)$ and $P(x, x) = P(x)$ for any $x \in X$. Then if both $x_n \to x$ and $Tx_n \to y$ weakly, while $(Tx_n, x_n) \to (y, x)$, by letting $n \to \infty$ and utilizing hypotheses (c) and (e), we find that $(x_n - x, Tx_n - Tx) \to 0$ and $(x_n - x, Rx_n - Rx) \to 0$. Subtracting, we obtain $(Px_n - Px, x_n - x) \to 0$, so that by hypothesis (d), $(x_n - x, P(x_n, x_n) - P(x_n, x)) \to 0$. Therefore by hypothesis (b), $Rx_n \to Rx$ weakly; whence by hypothesis (c), $(Rx_n, x_n) \to (R(x), x)$. As $n \to \infty$ in (5.2.8), and by virtue of hypothesis (d), we see that *for arbitrary* w

$$(x - w, y - T(x, w)) \geqslant (x - w, R(x, x) - R(x, w)). \tag{5.2.9}$$

Setting $w = x - \lambda z$ in (5.2.9) for $\lambda > 0$, we find, as in (i), after dividing by λ and letting $\lambda \to 0$, that $(z, y - T(x, x)) \geqslant 0$. Thus $y = Tx$, as required.

5.2B Application to quasilinear elliptic equations

Operators satisfying Condition (G) arise naturally in the study of general quasilinear elliptic differential operators of divergence form. Indeed, let the differential operator

$$Au = \sum_{|\alpha| \leq m} (-1)^{|\alpha|} D^{\alpha} A_{\alpha}(x, u, \ldots, D^m u)$$

be defined on a bounded domain $\Omega \subset \mathbb{R}^N$. Then, as in Section 2.2 (iii), assuming that the coeificients $A_{\alpha}(x, u, \ldots, D^m u)$ satisfy mild continuity and growth restrictions, we can associate with A an abstract operator $T\colon \mathring{W}_{m,p}(\Omega) \to W_{-m,q}(\Omega)$ defined implicitly by

$$(Tu, \varphi) = \sum_{|\alpha| \leq m} \int_{\Omega} A_{\alpha}(x, u, \ldots, D^m u) D^{\alpha}\varphi.$$

Heuristically speaking, the ellipticity of A is specified by the dependence of A on the terms involving derivatives of order $2m$, while the terms of A involving lower order derivatives can be considered as "compact" perturbations. Thus it is natural to separate the dependence of Tu on derivatives of u of order m from its dependence on the derivatives of u of smaller order. To this end, we write Tu as the sum of a principal part Pu and a remainder Ru, and in addition define

$$(P(u, v), \varphi) = \sum_{|\alpha| = m} \int_{\Omega} A_{\alpha}(x, u, \ldots, D^{m-1}u, D^m v) D^{\alpha}\varphi, \tag{5.2.10}$$

$$(R(u, v), \varphi) = \sum_{|\alpha| \leq m-1} \int_{\Omega} A_{\alpha}(x, u, \ldots, D^{m-1}u, D^m v) D^{\alpha}\varphi. \tag{5.2.11}$$

Then, formally at least, $T(u, u) = P(u, u) + R(u, u)$, where $T(u, v) = P(u, v) + R(u, v)$, $P(u, u) = Pu$, etc.

With the above paragraph as motivation, we can now interpret the hypothesis (a)–(e) of (5.2.5(iii)) in terms of the concrete differential operator A. Hypothesis (a) is an expression of the ellipticity of A, whereas (d) is an abstract statement of the fact that terms of order less than m are compact perturbations of A. Hypothesis (e), as mentioned above, imposes growth and smoothness restrictions on the coefficients A_{α} of A. Hypotheses (b) and (c) require slightly more explanation. Equation (5.2.11) and Hölder's inequality imply that for suitable conjugate exponents p_{α}, q_{α} (to be chosen)

$$|(R(u_n), y_n)| \leq \sum_{|\alpha| \leq m-1} \|A_{\alpha}(x, u_n, \ldots, D^m u_n)\|_{q_{\alpha}} \|D^{\alpha} y_n\|_{p_{\alpha}}. \tag{5.2.12}$$

Now if $y_n \to 0$ weakly in $\mathring{W}_{m,p}(\Omega)$, then for $|\alpha| \leqslant m-1$, $D^\alpha y_n \to 0$ strongly in $L_{p_\alpha}(\Omega)$, where $p_\alpha < Np/[N-(m-\alpha)p]$. Thus, since $\{u_n\}$ is weakly convergent, $\|u_n\|_{m,p}$ is uniformly bounded, and hypothesis (c) can be considered by (5.2.12) purely as a growth restriction on $A_\alpha(x, u, \ldots, D^m u)$ for $|\alpha| \leqslant m-1$. To interpret hypothesis (b) the following result is useful.

(5.2.13) Let $u_n \to u$ weakly in $\mathring{W}_{m,p}(\Omega)$, and suppose that the functions $A_\alpha(x, y, z)$ defined on $\Omega \times \mathbb{R}^n \times \mathbb{R}^q$ satisfy the Carathéodory restrictions and the ellipticity condition

$$\sum_{|\alpha|=m} \{A_\alpha(x,y,z) - A_\alpha(x,y,z')\}\{z_\alpha - z'_\alpha\} > 0 \qquad \text{(a.e. in } \Omega),$$

for $z \neq z'$. Then if

$$\sum_{|\alpha|=m} \int_\Omega \{A_\alpha(x, u_n, \ldots, D^m u_n) - A_\alpha(x, u_n, \ldots, D^{m-1}u_n, D^m u)\}\{D^\alpha u_n - D^\alpha u\} \to 0$$

as $n \to \infty$, $D^\alpha u_n \to D^\alpha u$ (for $|\alpha| = m$) in measure on Ω.

The proof of this result is easily demonstrated by first proving that if $u_n \to u$ weakly in L_p, and $f(y) > 0$ for $y > 0$, then $\int_\Omega f(u_n) \to \int_\Omega f(u)$ implies $u_n \to u$ in measure on Ω.

Now combining (5.2.11) and (5.2.13), we note that for arbitrary $\varphi \in C_0^\infty(\Omega)$ and $|\alpha| \leqslant m-1$, if $u_n \to u$ weakly in $\mathring{W}_{m,p}(\Omega)$, then

$$A_\alpha(x, u_n, \ldots, D^m u_n) D^\alpha \varphi \to A_\alpha(x, u, \ldots, D^m u) D^\alpha \varphi$$

in measure on Ω. Thus since the functions A_α satisfy the appropriate growth restrictions,

$$\lim_{n\to\infty} (R(u_n), \varphi) = \lim_{n\to\infty} \sum_{|\alpha| \leqslant m-1} \int_\Omega A_\alpha(x, u_n, \ldots, D^m u_n)\, D^\alpha \varphi$$
$$= (Ru, \varphi).$$

Consequently, $Ru_n \to Ru$ weakly, and again the restriction on A implied by hypothesis (b) is a growth restriction on A_α for $|\alpha| \leqslant m-1$.

5.2C Removal of the coerciveness restriction

Under certain circumstances, the result (5.2.3) can be substantially improved by replacing the coerciveness hypothesis $(f(x), x)\|x\|^{-1} \to \infty$ as $\|x\| \to \infty$ with less restrictive conditions. Indeed, an operator $f \in B(X, X^*)$ satisfying Condition (G) may well map f onto a proper subset of X^*, and thus fail to be surjective.

The coerciveness hypothesis just mentioned was used in (5.2.3) to prove (a) the solvability of the Galerkin approximations $(5.2.1)_n$, and (b) an a priori estimate for the solutions of these approximate equations. Thus the improvement we now state will imply both (a) and (b) in some cases.

(5.2.14) **Theorem** Suppose f is a bounded, continuous mapping of a real separable reflexive Banach space X into X^*, satisfying the following conditions:

(I) f is an odd mapping, i.e., $f(-x) = -f(x)$ for all $x \in X$.

(II) **Condition** (G′) If $x_n \to x$ weakly in X, $f(x_n) \to y$ weakly in X^*, and $(f(x_n), x_n) \to (y, x)$, then $x_n \to x$ strongly.

Then, if $\|f(x)\| \geqslant \alpha$ for $x \in \partial \Sigma_R = \{z \,|\, \|z\| = R\}$, the equation $f(x) = g$ has a solution in Σ_R for all g with $\|g\| < \alpha$.

Proof: Again the basic idea is to use the hypotheses of the theorem to ensure both the existence of the solutions of the Galerkin approximations and a priori bounds for the solutions so obtained. For then, since Condition (G′) implies Condition (G), our previous arguments imply that a subsequence of the solutions of the Galerkin approximations will converge to a solution of $f(x) = g$.

To show that the Galerkin approximations $(5.2.1)_n$ have a solution x_n on $\Sigma_R \cap X_n$ for n sufficiently large, we first note that if $g \in X$, $\|g\| < \alpha$, and $f(x) \neq tg$ for any $t \in [0, 1]$ and $x \in \partial \Sigma_R$, then for n sufficiently large, there are constants β and $N > 0$ for which $\|P_n^*(f(z) - tg)\| \geqslant \beta$ for $t \in [0, 1]$ and $z \in \partial \Sigma_R \cap X_n$ for all $n \geqslant N$. Indeed, otherwise there would be sequences $\{P_{n_k}^*\}$, $\{z_k\}$, and $\{t_k\}$ such that $z_k \in \Sigma_R \cap X_n$, $t_k \in [0, 1]$, and $\|P_{n_k}^*[f(z_k) - t_k g]\| \to 0$ as $k \to \infty$. Thus, after passing to appropriate subsequences, we may suppose that $z_k \to z_0$ weakly, $t_k \to t_0$, and $P_k f(z_k) \to t_0 g$ strongly in X^*. Hence, for any $w \in X$ $(P_{n_k}^* f(z_k), w) = (f(z_k), P_{n_k} w) \to (t_0 g, w)$ Also

$$|(f(z_k) - t_0 g, \quad P_k w - w)| \leqslant \|f(z_k) - t_0 g\| \, \|P_k w - w\| \to 0.$$

Consequently, expanding $(f(z_k) - t_0 g, P_k w - w)$, we find $f(z_k) \to t_0 g$ weakly in X^*. Thus Condition (G′) implies that $z_k \to z_0$ strongly and so $f(z_0) = t_0 g$. Consequently $\|z_0\| = R$, $f(z_0) = t_0 g$ and $\|f(z_0)\| \leqslant \|g\|$, which contradicts the hypothesis that $\|f(z)\| \geqslant \alpha > \|g\|$ for $z \in \partial \Sigma_R \cap X_n$.

This result shows that for $n \geqslant N$, first, the mappings $P_n^*(f(z) - g)$ and $P_n^* f(z)$ are homotopic on $\partial \Sigma_R \cap X_n$, and secondly, with $g = 0$ that $P_n^* f(z) \neq 0$ on $\partial \Sigma_R \cap X_n$. Thus, by (1.6.3), the Brouwer degree $d(P_n^* f(z), 0, \Sigma_R \cap X_n)$ is an odd integer and hence not zero. Consequently, by the homotopy invariance of degree, $d(P_n^*(f(z) - g), 0, \Sigma_R \cap X_n) \neq 0$. This means that for $n \geqslant N$, $P_n^*(f(z) - g) = 0$ has a solution z_n in $\Sigma_R \cap X_n$. Thus the Galerkin approximations $(5.2.1)_n$ have a solution $\{z_n\}$ for $n \geqslant N$, and these solutions automatically satisfy the a priori bound $\|z_n\| < R$. Thus the theorem is established.

As in Section 5.2A, we now state hypotheses implying Condition (G′).

(5.2.15) Let f be a bounded continuous operator mapping X into X^*. Then f satisfies Condition (G′) if there is a completely continuous operator $R : X \to X^*$ such that with $P = f - R$

(5.2.16) $\quad (Px - Pz, x - z) + f(x - z) \geqslant c(\|x - z\|) \quad$ for $\quad x, z \in X,$

where f is weakly upper semicontinuous and satisfies $f(0) = 0$, and $c(r)$ is real-valued, positive, and continuous, and $c(r) \to 0$ if and only if $r \to 0$.

Proof: If $x_n \to x$, $f(x_n) \to y$, and $(f(x_n), x_n) \to (y, x)$, then a short computation shows that $(Px_n - Px, x_n - x) \to 0$. Also, if $x_n \to x$ weakly, then $\overline{\lim}\ f(x_n - x) \leqslant 0$. Thus (5.2.16) implies that $\lim_{n\to\infty} c(\|x_n - x\|) = 0$. Since c is continuous and $c(\beta) = 0$ if and only if $\beta = 0$, $x_n \to x$ strongly in X.

For the class of quasilinear elliptic operators discussed in (5.2.3), we prove the following condition analogous to Condition (G′).

(5.2.17) **Theorem** Suppose A is a quasilinear operator satisfying the ellipticity condition (5.2.13), and in addition:

(a) the associated abstract operator $\mathcal{A} : \mathring{W}_{m,p}(\Omega) \to W_{-m,q}(\Omega)$ satisfies the conditions mentioned in Section 5.2B together with the hypothesis (*) if $u_n \to u$ weakly, then $(P(u_n, u_n) - P(u, u_n), u_n) \to 0$;

(b) for fixed y there are integrable functions $c_0(y) > 0$ and $c_1(y)$ such that

$$\sum_{|\alpha| = m} A_\alpha(x, y, z) z_\alpha \geqslant c_0(y)|z|^p - c_1(y).$$

Then the abstract operator $\mathcal{A}$ associated with A by duality satisfies Condition (G′).

Proof: First we observe that since $\mathcal{A}$ satisfies hypothesis (a), by the hypothesis of Condition (G′), $\mathcal{A}u_n \to \mathcal{A}u$ strongly in $\mathring{W}_{-m,q}(\Omega)$.

To show that $u_n \to u$ strongly in $\mathring{W}_{m,p}(\Omega)$, we shall prove that for $|\alpha| = m$, (i) the integrals $\int_\Omega |D^m u_n|^p$ are equiabsolutely continuous and that (ii) $D^\alpha u_n \to D^\alpha u$ in measure. The desired strong convergence then follows by Vitali's theorem. Now the result (ii) follows immediately from our assumptions, by virtue of (5.2.13). On the other hand, to prove (i) we use hypotheses (b) and (*) as follows:

By hypothesis (5.2.5) and the fact that $(\mathcal{A}u_n, u_n) \to (\mathcal{A}u, u)$, we deduce (after a short computation) that $(Pu_n, u_n) \to (Pu, u)$. Then hypothesis (*) implies

(5.2.18) $\quad (P(u, u_n), u_n) \to (Pu, u).$

Now by virtue of the definition and hypothesis (b),

$$(P(u, u_n), u_n) = \sum_{\substack{|\alpha| = m \\ |\gamma| < m}} \int_\Omega A_\alpha(x, D^\gamma u, D^m u_n) D^\alpha u_n$$

$$(5.2.18') \qquad c_1(D^\gamma u) + \sum_{\substack{|\alpha|=m \\ |\gamma|<m}} A_\alpha(x, D^\gamma u, D^m u_n) D^\alpha u_n \geqslant c_0(D^\gamma u)|D^m u_n|^p.$$

Using the facts concerning equiabsolutely continuous integrals, (5.2.13), the fact that $D^\alpha u_n \to D^\alpha u$ in measure, and the positivity of the expression $\sum_{|\alpha|=m} A_\alpha(x, y, z) z_\alpha$ implies that the functions $\sum_{|\alpha|=m} A_\alpha(x, D^\gamma u, D^m u_n) D^\alpha u_n$ have equiabsolutely continuous integrals over Ω. But in that case, the inequality (5.2.18′) implies the same equiabsolutely continuous property for the integrals $|D^\alpha u_n|^p$ for $|\alpha| = m$. Thus $u_n \to u$ strongly in $\mathring{W}_{m,p}(\Omega)$ and the result is proven.

5.2D Rayleigh–Ritz approximations for gradient operators

The Galerkin approximations $(5.2.1)_n$ take a particularly elegant form if the mapping f is a gradient operator; i.e., $f(x) = F'(x)$, where $F(x)$ is a C^1 real-valued functional defined on X. Indeed, in that case the solutions of $(5.2.1)_n$ are precisely the critical points of the functional $\mathcal{G}_n(x) = F(x) - (g, x)$ defined on X_n. Thus the powerful methods of finite-dimensional critical point theory discussed briefly in Section 1.6 become applicable in the study of the approximations $(5.2.1)_n$. For historical reasons this approach is known as the Rayleigh–Ritz approximation. Here we shall illustrate the Rayleigh–Ritz method by reconsidering the class of nonlinear eigenvalue problems.

Thus we study the solutions of the equation

$$(5.2.19) \qquad \lambda_1 \mathcal{A}'(x) = \lambda_2 \mathcal{B}'(x), \qquad x \in \{x \mid \mathcal{A}(x) = \text{const.}, x \in X\},$$

which can be obtained as limits of solutions of the Rayleigh–Ritz approximations

$$(5.2.20) \quad \lambda_1^{(n)} P_n^* \mathcal{A}'(x) = \lambda_2^{(n)} P_n^* \mathcal{B}'(x), \qquad x \in \{x \mid \mathcal{A}(x) = \text{const.}, x \in X_n\}.$$

First we prove a result concerning the approximations for the "first eigenvector" of (5.2.19).

(5.2.21) Suppose that $\mathcal{A}(x)$ and $\mathcal{B}(x)$ are two C^1 real-valued functionals defined on the separable reflexive Banach space X such that

(i) $\mathfrak{M} = \{x \mid \mathcal{A}(x) = \text{const.}\}$ is a bounded star-shaped set on X;

(ii) $Ax = \mathcal{A}'(x)$ is a bounded continuous mapping that satisfies Condition (G′);

(iii) $B(x) = \mathcal{B}'(x)$ is completely continuous, with $\mathcal{B}(x) = 0$ if and only if $x = 0$.

Then, $c_1 = \sup_{\mathfrak{M}} \mathcal{B}(x)$ is attained on $\mathfrak{M}$ by an element $\bar{x}$ that satisfies the equation

(5.2.22) $\lambda_1 A\bar{x} = \lambda_2 B\bar{x}$,

where λ_1 and λ_2 are two real numbers not both zero. Furthermore, $c_1 = \lim_{n\to\infty} \sup_{\mathfrak{M}\cap X_n} \mathfrak{B}(x)$, and the elements $(\bar{x}, \lambda_1, \lambda_2)$ are limits of a subsequence $(x_n, \lambda_1^{(n)}, \lambda_2^{(n)})$ satisfying (5.2.20) and the extremal property

$$\sup_{\mathfrak{M}\cap X_n} \mathfrak{B}(x) = \mathfrak{B}(x_n), \qquad x_n \in X_n \cap \mathfrak{M}.$$

Proof: First we note that by hypothesis the sets $\mathfrak{M} \cap X_n$ are compact, so that for each n, $c_{1,n} = \sup_{\mathfrak{M}\cap X_n} \mathfrak{B}(x)$ is attained by an element $x_n \in \mathfrak{M} \cap X_n$. Consequently, for each n, there are constants $\lambda_1^{(n)}$, $\lambda_2^{(n)}$ not both zero such that the triple $(x_n, \lambda_1^{(n)}, \lambda_2^{(n)})$ satisfies (5.2.20). Without loss of generality, we may suppose that $|\lambda_1^{(n)}| + |\lambda_2^{(n)}| = 1$, so that (after possibly passing to subsequences) $\lambda_1^{(n)}$ and $\lambda_2^{(n)}$ converge to λ_1 and λ_2 (say) with $|\lambda_1| + |\lambda_2| = 1$.

Since X is a reflexive Banach space and the sequences $\{\|x_n\|\}$, $\{\|Ax_n\|\}$ are uniformly bounded, we may suppose that, after again passing to suitable subsequences, $x_n \to \bar{x}$ weakly and $Ax_n \to y$ weakly. Assuming that $\lambda_1 \neq 0$, we find that $x_n \to \bar{x}$ strongly since the operator A satisfies Condition (G′), $y = (\lambda_2/\lambda_1)Bx$, and

$$\lambda_1^{(n)}(Ax_n, x_n) = \lambda_1^{(n)}(P_n^* Ax_n, x_n) = \lambda_2^{(n)}(P_n^* Bx_n, x_n) \to \lambda_2(Bx, x).$$

Thus since $\mathfrak{M}$ is closed, $\bar{x}$ belongs to $\mathfrak{M}$, and satisfies (5.2.22) by virtue of our discussion in Section 5.2.1.

Finally, we rule out the possibility that $\lambda_1 = 0$. If $\lambda_1 = 0$, then $\mathfrak{B}'(\bar{x}) = 0$, so that $\bar{x} = 0$. But since $\mathfrak{B}(x)$ is a weakly continuous functional, this implies that

$$0 = \mathfrak{B}(x) = \lim_{n\to\infty} \mathfrak{B}(x_n) = \lim_{n\to\infty} \sup_{\mathfrak{M}\cap X_n} \mathfrak{B}(x) = c_1.$$

But this fact contradicts the hypothesis (iii) of (5.2.21). Consequently the desired result is proved.

We now give brief mention of a result for more general critical points that can be obtained by finite-dimensional approximation and the Ljusternik–Schnirelmann category (see Section 6.6)

(5.2.23) **Theorem** Suppose the functionals $\mathfrak{A}(x)$ and $\mathfrak{B}(x)$ satisfy the hypothesis of (5.2.21), and in addition suppose

(a) $\mathfrak{A}(x)$ and $\mathfrak{B}(x)$ are even functions of x, and

(b) $(\mathfrak{A}'(x), x)$ and $\mathfrak{B}(x)$ are strictly positive for $x \neq 0$.

Then the real numbers

(5.2.24) $c_N = \sup_{[A]_N} \inf_A \mathfrak{B}(x)$

are critical points of $\mathcal{B}(x)$ restricted to

$$\mathfrak{M} = \{x \mid \mathcal{Q}(x) = \text{const.}, x \in X\},$$

where

(5.2.25) $$[A]_N = \{A \mid A \subset \mathfrak{M},\ A \text{ compact}, \quad \text{cat}(A/Z_2, \mathfrak{M}/Z_2) \geqslant N\}.$$

Furthermore, for each fixed N there is a sequence of pairs $(\bar{x}_{N,n}, \lambda_{N,n})$ with $c_{N,n} = \mathcal{B}(\bar{x}_{N,n}) \to c_N$ and $\bar{x}_{N,n} \in \mathfrak{M} \cap X_n$ such that $(\bar{x}_{N,n}, 1, \lambda_{N,n})$ satisfies the Rayleigh–Ritz approximations $(5.2.1)_n$ as well as the minimax characterization $c_{N,n} = \sup_{[A \cap X_n]} \inf_{A \cap X_n} \mathcal{B}(x)$. Furthermore, for each N, there is a subsequence of pairs $(\bar{x}_{N,n_j}, \lambda_{N,n_j})$ converging strongly in $X \times \mathbb{R}^1$ to $(\bar{x}_N, \lambda_N)$, where $\bar{x}_N \in \mathcal{B}^{-1}(c_n) \cap \mathfrak{M}$ is a critical point of $\mathcal{B}(x)$ restricted to $\mathfrak{M}$ and $(\bar{x}_N, \lambda_N)$ satisfies the equation $\mathcal{Q}'(\bar{x}_N) = \lambda_N \mathcal{B}'(x_N)$.

For a proof of this result we refer the reader to the paper by Rabinowitz (1973).

5.2E Steady state solutions of the Navier–Stokes equations

The Navier–Stokes equations for three-dimensional steady flow of a viscous incompressible fluid in a bounded domain Ω in $\mathbb{R}^3$ can be written

(5.2.26) $$-\nu\, \Delta u + (u \cdot \nabla)u + \nabla p = g,$$

(5.2.27) $$\text{div } u = 0,$$

(5.2.28) $$u|_{\partial\Omega} = \beta(x).$$

Here $u(x)$ denotes the velocity vector of the fluid, ν the viscosity of the fluid, p the pressure, g the external force acting on the fluid, and $\beta(x)$ the value of $u(x)$ on $\partial\Omega$. We demonstrate the existence of a solution for the system (5.2.26)–(5.2.28) by means of the Galerkin approximation scheme described in (5.2.1), for any value of ν provided $\beta(x)$ is suitably restricted. By a limiting process, one can show that the analogue of the system (5.2.26)–(5.2.28) for unbounded domains also has a solution. Actually we prove

(5.2.29) **Theorem** The system (5.2.26)–(5.2.28) possesses a generalized solution $u(x)$ in the sense of Section 1.5, provided $g \in L_2(\Omega)$ and $\beta(x)$ is the boundary value of a function $\beta_*(x)$ defined on $\bar{\Omega}$ with $\nabla\beta_*$ Hölder continuous, β_* such that either (i) $|\nabla \beta_*(x)|$ or $|\beta_*(x)|$ is sufficiently small, or (ii) $\beta_*(x) = \text{curl}\ \gamma(x)$, where $\gamma(x) \in C^1(\bar{\Omega})$ for $\partial\Omega$ of class C^2. The solution $u(x)$ is smooth in Ω and at all sufficiently smooth portions of $\partial\Omega$, provided g is Hölder continuous in $\bar{\Omega}$.

Proof: We proceed by first representing the generalized solutions of (5.2.26)–(5.2.28) by means of an operator equation of the form $f(u) = g$, where f is a mapping on the Hilbert space $\mathring{H}$ of solenoidal vectors $w(x)$ whose components $w_i(x) \in \mathring{W}_{1,2}(\Omega)$. Then we show that the operator f satisfies the hypotheses of (5.2.3) with $\beta(x)$ suitably restricted. The regularity of the generalized solution then follows from the results mentioned in Section 1.5.

First, suppose $\beta(x) \equiv 0$. Then the weak solutions in $\mathring{H}$ are in one-to-one correspondence with the solutions of the operator equation

$$\nu u - \mathfrak{B}(u, u) = \tilde{g}, \tag{5.2.30}$$

where $\mathfrak{B}(u, u)$ and $\tilde{g}$ are defined on $\mathring{H}$ by the formulas

$$(\mathfrak{B}(u, u), \phi) = \sum_{k=1}^{3} \int_\Omega u_k u \cdot D_k \phi; \qquad (\tilde{g}, \phi) = \int_\Omega g \cdot \phi. \tag{5.2.31}$$

As demonstrated in Section 4.3, the operator $\mathfrak{B}(u) = \mathfrak{B}(u, u)$ is a completely continuous mapping of $\mathring{H}$ into itself. In addition, for $u \in C_0^\infty(\Omega) \cap \mathring{H}$,

$$(\mathfrak{B}(u, u), u) = \sum_{k=1}^{3} \int_\Omega u_k u \cdot D_k u = \tfrac{1}{2} \int_\Omega u \cdot \nabla |u|^2 = 0 \tag{5.2.32}$$

since $\operatorname{div} u = 0$ and $u \mid_{\partial\Omega} = 0$. Thus for $0 < \nu < \infty$, $f_\nu(u) \equiv \nu u - \mathfrak{B}(u, u)$ satisfies Condition (G), and by (5.2.32),

$$(\nu u - \mathfrak{B} u, u)/\|u\| = \nu \|u\|.$$

so that $f_\nu(u)$ is coercive. Consequently, by Theorem (5.2.3), the operator equation (5.2.30) has a solution $\tilde{u} \in \mathring{H}$; so that (5.2.26)–(5.2.28) has the generalized solution $\tilde{u}$ in $\mathring{K}$.

More generally, if $\beta(x) \neq 0$, suppose that, by virtue of the restrictions of the theorem, there is a constant c with $\gamma > c \geqslant 0$ and

$$|(\mathfrak{B}(\beta_*, u), u)| \leqslant c\|u\|_{\mathring{H}}^2 \qquad \text{for} \quad u \in \mathring{H}. \tag{5.2.33}$$

Then if we represent a generalized solution u of (5.2.26)–(5.2.28) by $u = w + \beta_*$ with $w \in \mathring{H}$, w satisfies the equation

$$\nu w - \{\mathfrak{B}(w, w) + \mathfrak{B}(w, \beta_*) + \mathfrak{B}(\beta_*, w)\} = f_*, \tag{5.2.34}$$

where $f_* = f - \beta_* - \mathfrak{B}(\beta_*, \beta_*)$. Now we show that Theorem (5.2.3) can be applied to (5.2.34). To this end, observe that each operator $\mathfrak{B}(w, w)$, $\mathfrak{B}(w, \beta_*)$, and $\mathfrak{B}(\beta_*, w)$ is completely continuous. Thus for $0 < \nu < \infty$, the operator $f_\nu(w)$ on the left-hand side of (5.2.34) satisfies Condition (G). Furthermore, by arguing as in (5.2.32), $(\mathfrak{B}(w, \beta_*), w) = 0$, so that by (5.2.33)

$$(f_\nu(w), w)/\|w\|_{\mathring{H}} = \nu\|w\| + (\mathfrak{B}(\beta_*, w), w)/\|w\| \geqslant (\nu - c)\|w\|.$$

Consequently, $f_\nu(w)$ satisfies the coerciveness condition of (5.2.3), and therefore the equation (5.2.34) is solvable. Thus the system (5.2.26)–(5.2.28) has a generalized solution in $\mathring{H}$.

Finally, we demonstrate (5.2.33), assuming that $\beta(x)$ satisfies one of the restrictions of the theorem. First suppose that either $|\beta_*(x)| \leqslant M_0$ or $|\nabla \beta_*(x)| \leqslant M_1$ for $x \in \bar{\Omega}$. Then by (5.2.31), using the fact that $|\beta_*(x)| \leqslant M_0$, and Sobolev's inequality

$$|(\mathfrak{B}(\beta_*, w), w)| \leqslant M_0 \|w\|_{0,2} \|\nabla w\|_{0,2} \leqslant M_0 c_1 \|w\|_{\mathring{H}}^2,$$

and using the fact that $|\nabla \beta_*(x)| \leqslant M_1$,

$$|(\mathfrak{B}(\beta_*, w), w)| \leqslant M_1 \|w\|_{L_2}^2 \leqslant M_1 c_1^2 \|w\|_{\mathring{H}}^2.$$

Thus (5.2.33) is satisfied if either $\nu > M_0 c_1$ or $\nu > M_0 c_1^2$. On the other hand, if $\beta_*(x) = \operatorname{curl} \gamma(x)$ as in hypothesis (ii) of the theorem, we employ the inequality

$$\int_\Omega v^2/\rho^2 \leqslant \bar{c}_1 \int_\Omega |\nabla v|^2 \quad \text{for} \quad v \in \mathring{W}_{1,2}(D) \quad \text{with} \quad \rho = \operatorname{dist}(x, \partial\Omega) \tag{5.2.35}$$

and construct a function $h(t) \in C^{\infty}[0, \infty)$ depending on two parameters k, α such that (i) $h(t) = 1$ for $0 < t < k\alpha$ and $h(t) = 0$ for $t > (1 - k)\alpha$, and (ii) as $k \to 0$, $th'(t) \to 0$ uniformly with α and t. Then letting $\rho(x) = \text{dist}(x, \partial\Omega)$, $\beta_{**} = \text{curl}(h(\rho)\gamma)$,

$$\beta_{**} = h \text{ curl } \gamma - \gamma \times h'(\rho) \nabla \rho.$$

Thus $\beta_{**} = \beta_*$ on $\partial\Omega$, $\beta_{**} \equiv 0$ outside a small neighborhood of $\partial\Omega$; and since $\rho \in C^2$, $\beta_{**} \in C^{1,\mu}(\bar{\Omega})$. Furthermore, for any $\epsilon > 0$, $|\rho\beta_{**}(x)| < \epsilon$. By (5.2.35),

$$|(\mathfrak{B}(\beta_{**}, u), u)| \leq |\mathfrak{B}(\rho\beta_{**}, u/\rho, u)| \leq \epsilon\|u/\rho\|_{0,2}\|\nabla u\|_{0,2} \leq \epsilon c_1\|w\|^2$$

for all $u \in C_0^{\infty}(\Omega)$. Thus for ϵ sufficiently small, β_{**} satisfies (5.2.33). This completes the proof of the theorem.

5.3 Homotopy, the Degree of Mappings, and Its Generalizations

5.3A Heuristics

Many problems involving a given nonlinear operator $f \in C(X, Y)$ can be studied by focusing attention on the topological properties of the mapping defined by f. In particular, replacing f with a simpler mapping $\tilde{f}$ in the same (appropriately defined) homotopy class as f, sometimes makes it possible to solve the problem for f by solving an analogous problem for the simpler mapping $\tilde{f}$.

This procedure is well known for mappings between finite-dimensional spaces. Thus, for a given analytic function $f(z)$ defined on a closed disk $\Sigma_R = \{z | |z| \leq R\}$; the number of zeros of f inside Σ_R is a topological invariant, provided of course that $f(z) \neq 0$ on $\partial\Sigma_R = \{z \mid |z| = R\}$. In fact by Rouché's theorem, two analytic functions f and $f + g$ define on Σ_R have the same number of zeros inside Σ_R if f and $f + g$ are homotopic on $\partial\Sigma_R$, in the sense that $|f(z) + tg(z)| \neq 0$ for $z \in \partial\Sigma_R$ and $t \in [0, 1]$. More generally, by the theorem of H. Hopf mentioned in Section I.6.7, two continuous mappings f and $f + g$ defined on the ball $\Sigma_R = \{x \mid |x| \leq R\}$ in $\mathbb{R}^N$ with $f \neq 0$ on $\partial\Sigma_R = \{x \mid |x| = R\}$ have the same "algebraic" number of zeros in Σ_R if and only if they are homotopic.

Here we shall discuss this homotopy approach to nonlinear problems for operators defined on bounded domains of Banach spaces. Immediately however the following formidable obstacle arises in the infinite-dimensional case, unless we refine our notion of homotopy.

(5.3.1) Let H be a separable Hilbert space of infinite dimension. Let f and g be any two continuous mappings of the sphere $\partial\Sigma_1 = \{x \mid \|x\|_H = 1\}$ into itself. Then f and g are homotopic, i.e., there is a continuous mapping $h(x, t)$: $\partial\Sigma_1 \times [0, 1] \to \partial\Sigma_1$ such that $h(x, 0) \equiv f(x)$ and $h(x, 1) \equiv g(x)$.

Proof: The basic idea is to construct a fixed point free mapping σ of the ball $\Sigma_1 = \{x \mid \|x\|_H \leq 1\}$ into itself and thence a retraction $r(x)$ of Σ_1 onto $\partial\Sigma_1$. Then the desired homotopy can be explicitly written in the form $h(x, t) = r(tg(x) + (1-t)f(x))$.

To construct the fixed point free continuous mapping σ, let $(e_1, e_2, \ldots)$ denote a complete orthonormal basis for H so that an arbitrary element x of H can be written $x = \sum_{i=1}^{\infty} x_i e_i = (x_1, x_2, \ldots)$, with $\|x\|^2 = \sum_{i=1}^{\infty} x_i^2$. We can then define $\sigma(x) = \left(\sqrt{1 - \|x\|^2}, x_1, x_2, \ldots\right)$ for $x \in \Sigma_1$. A simple computation shows that $\|\sigma(x)\|^2 = 1$; so that if σ possesses a fixed point $y \in \Sigma_1$, $\|y\| = 1$. Consequently, if $y = (y_1, y_2, \ldots)$, then $y_i = y_{i+1}$ for $i = 1, 2, \ldots,$ while $y_1 = 0$. Thus $y = 0$, which contradicts the fact that $\|y\| \neq 0$. Hence σ is a continuous mapping of Σ_1 into Σ_1 that has no fixed point.

To construct the retraction $r(x)$ of Σ_1 onto $\partial\Sigma_1$, we proceed as follows. For $x \in \Sigma_1$, the line $L(x)$ joining x and $\sigma(x)$ does not degenerate to a point since σ is fixed point free. Thus the line $L(x)$ can be extended so as to intersect $\partial\Sigma_1$ in a point $r(x)$ on the opposite side of $\sigma(x)$ from x but distinct from $\sigma(x)$. The mapping $x \to r(x)$ is the desired retraction because it clearly maps Σ_1 onto $\partial\Sigma_1$ continuously and by construction leaves $\partial\Sigma_1$ fixed pointwise.

5.3B Compact perturbations of a continuous mapping

Because of (5.3.1), we shall in Sections 5.3B–5.3D restrict attention to the following special class of homotopic deformations.

(5.3.2) **Definition** Let S be a closed subset of a Banach space X. Suppose f is a fixed continuous mapping of $X \to Y$ (a Banach space). Then g_0 and g_1 are *compactly homotopic* on S (relative to f) if there is a *continuous* compact mapping $h(x, t)$: $S \times [0, 1] \to Y$ with $g_0(x) = f(x) + h(x, 0)$ and $g_1(x) = f(x) + h(x, 1)$, and such that $g(x, t) = f(x) + h(x, t) \neq 0$ on $S \times [0, 1]$.

Clearly compact homotopy defines an equivalence relation on the class $\mathcal{C}_f(S, Y)$ of compact perturbations of $f \in \mathcal{C}_f(S, Y) = \{g \mid g = f + K, K$ compact, $g \in C(S, Y)\}$. In the sequel we shall attempt to represent the resulting equivalence classes by computable topological invariants and to interpret these invariants in terms of the mapping properties of a given compact perturbation of the mapping f. In order to obtain a first result in this direction, suppose S is a closed subset of X and O denotes a component of $X - S$.

(I) Notation and definitions Let $\mathcal{C}_f(S, Y) = \{g \mid g = f + K\}$, $\mathcal{C}_f^0(S, Y) = \{g \mid g \in \mathcal{C}_f(S, Y), g \neq 0 \text{ on } S\}$.

Let O be a component of $X - S$. Then $g \in \mathcal{C}_f^0(S, Y)$ is called *inessential* (w.r.t. O) if g has an extension $\tilde{g} \in \mathcal{C}_f^0(O \cup S, Y)$. Otherwise, g is called *essential*. Moreover, we let $\mathcal{C}_f^p(S, Y) = \{g \mid g = f + C, g \neq p \text{ on } S\}$. Thus g is essential (with respect to O) if every extension $\tilde{g} \in \mathcal{C}_f(O \cup S, Y)$ has a zero in O. Clearly, then to prove that a given $g \in \mathcal{C}_f(O \cup S, Y)$ has a zero in O, we need only show that g is essential (with respect to O). By virtue of the next result we shall see that the same conclusion holds if some mapping $\tilde{g}$, compactly homotopic to g (on S), is essential.

(5.3.3) **Theorem** The properties essential and nonessential (relative to O) are invariant under compact homotopy.

Proof: It suffices to prove the result for an inessential mapping $g \in \mathcal{C}_f^0(S, Y)$. In that case, suppose $g, \tilde{g} \in \mathcal{C}_f^0(S, Y)$ are compactly homotopic on S and g admits an extension $G \in \mathcal{C}_f(O \cup S, Y)$. Then we shall construct an extension $\tilde{G} \in \mathcal{C}_f(O \cup S, Y)$ of $\tilde{g}$ such that G and $\tilde{G}$ are compactly homotopic on $O \cup S$.

Since g and $\tilde{g}$ are compactly homotopic, there is a continuous compact mapping $h(x, t)$: $S \times I \to Y$ satisfying the definition. Let $T_0 = S \times [0, 1] \cup (S \cup O \times \{0\})$ and define h^*: $T_0 \to Y$ by setting

$$h^*(x, t) = \begin{cases} G(x) - f(x) & \text{for } x \in S \cup O, \quad t = 0 \\ h(x, t) & \text{for } x \in S, \quad t \in [0, 1]. \end{cases}$$

Then h^* is compact and continuous on T_0, so that by the extension property (2.4.4) of compact operators, h^* can be extended to a continuous compact mapping $H^*(x, t)$ of $S \cup O \times [0, 1] \to Y$.

In order to define the desired extension $\tilde{G}$, we must ensure that $\tilde{G} \neq 0$ on $S \cup O$. Hence we let $S_1 = \{x \mid x \in S \cup O, f(x) = -H^*(x, t) \text{ for } t \in [0, 1]\}$. Now S and S_1 are disjoint closed sets and by Tietze's theorem there is a continuous function $Y(x)$: $S \cup O \to [0, 1]$ that vanishes on S_1 and takes the value 1 on S. Let $H(x, t) = H^*(x, Y(x)t)$ on $S \cup O \times [0, 1]$, and set $\tilde{G}(x) = f(x) + H(x, 1)$.

Now we note that (i) $\tilde{G}$ is an extension of g since if $t = 1$ and $x \in S$

$$H(x, 1) = H^*(x, Y(x)) = H^*(x, 1) = h(x, 1).$$

Thus $\tilde{G}(x) = f(x) + h(x, 1) = g$. (ii) $\tilde{G}(x) \neq 0$ for $x \in S \cup O$ and, in fact, more generally $\tilde{G}(x, t) = f(x) + H(x, t) \neq 0$ for $t \in [0, 1]$, $x \in S \cup O$. Indeed otherwise $x \in S_1$, and thus

$$f(x) + H(x, 0) = f(x) + H^*(x, 0) = G(x) = 0.$$

But this last equality is impossible since $G(x)$ has no zero on $S \cup O$. Thus not only does the extension $\tilde{G}$ exist, but moreover since $f(x) + H(x, 0) = G(x)$, G and $\tilde{G}$ are compactly homotopic on $S \cup O$.

To proceed further with this notion we note the following fact from finite-dimensional homotopy theory that is important in understanding our subsequent developments.

(5.3.4) A continuous mapping $f: S^n \to S^m$ is essential relative to the open ball $\Sigma_1 = \{x \mid x \in \mathbb{R}^{n+1}, |x| < 1\}$ if and only if the homotopy class $[f] \in \pi_n(S^m)$ (of Section 1.6) is nontrivial.

Proof: Suppose f is not essential, so that there is an extension F of f to $\overline{\Sigma}_1$ with $F(x) \neq 0$. Then setting $H(x, t) = F(tx)/|F(tx)|$ for $t \in [0, 1]$, we note that by this homotopy $f(x)$ is homotopic to the point $H(x, 0) = F(0)/|F(0)|$, so that $[f] \in \pi_n(S^m) = 0$. Conversely, if $[f] = 0$, there is a homotopy $h(x, t)$ of $f(x)$ for $t \in [0, 1]$ with $|h(x, t)| = 1$. Consequently $F(tx) = h(x, t)$ is the desired nonzero extension of f to $\overline{\Sigma}_1$.

To proceed further it will be necessary to establish infinite-dimensional analogues of (5.3.4) yielding criteria for a given mapping $g \in \mathcal{C}_f(S, Y)$ to be essential. This task and more will be taken up in the next subsections for special choices of f. First we shall choose f to be the identity mapping of a Banach space X onto itself; and in the second case, we suppose that f is a linear Fredholm mapping L of nonnegative index p mapping X into Y.

Indeed, example (5.3.1) shows that the infinite-dimensional result analogous to (5.3.4) is more subtle, and in fact we shall proceed in the sequel by considering only compact homotopies of a fixed linear operator. In such cases we shall, in fact, prove a generalization of (5.3.4) to the infinite-dimensional cases. But these generalizations will require, in general (for index $L > 0$), the use of the notion of *stable* homotopy class of the associated infinite-dimensional mapping. (This use of the word *stable* refers to the fact (1.6.8) that the homotopy groups $\{\pi_{n+p}(S^n), n = 1, 2, 3 \ldots\}$ are isomorphic for $p > 0$ only for sufficiently large n.) Thus, *in general*, to establish the essentialness of a given mapping in our class it *does not suffice* to examine *only* the homotopy class of a sufficiently close finite-dimensional approximation. (See Section 5.3D below.)

5.3C Compact perturbations of the identity and the Leray–Schauder degree

Let I denote the identity mapping of a Banach space into itself, D denote a bounded domain of X, and ∂D be the boundary of D. Then, as in the finite-dimensional case, it is possible to establish a one-to-one correspondence between the compact homotopy classes of $\mathcal{C}_I(\partial D, X)$ and the integers $\mathbb{Z}$ by means of a function called the Leray–Schauder degree. In addition we shall establish a necessary and sufficient condition that a mapping $g \in \mathcal{C}_I(\partial D, X)$ be essential; namely, that its Leray–Schauder degree be different from zero.

The Leray-Schauder degree of a compact perturbation of the identity $g = I + C$ relative to a point $p \in X$ and D (written $d(I + C, p, D)$) can be defined by an analogue of the Galerkin approximation procedures of Section 5.2, assuming $g(x) \neq p$ on ∂D. Assuming as known the facts (1.6.3) concerning the Brouwer degree d_B for continuous mappings between spaces of the same finite dimension, we can define the integer-valued function $d(I + C, p, D)$ in two steps:

Step 1: If the compact mapping $C: D \to X$ has finite-dimensional range (i.e., $C(D) \subset X_n$, a finite-dimensional linear subspace of X), then assuming $p \in X_n$, we define the Leray–Schauder degree of $I + C$ at p relative to D by

$$(5.3.5) \qquad d(I + C, p, D) = d_B(I + C, p, D \cap X_n) \quad \text{for} \quad I + C \in \mathcal{C}_I(\partial D, X).$$

In this case, the Brouwer degree d_B is a finite integer since $(I + C)x \neq p$ for $x \in \partial(D \cap X_n) = \partial D \cap X_n$, by hypothesis. Consequently, in this case, the Leray–Schauder degree is well defined provided the definition is independent of both the finite-dimensional subspace containing p and of $C(D)$.

Step 2: For a general compact mapping $C: D \to X$, using (2.4.2), we approximate C by a sequence of compact mappings with finite-dimensional range $C_n: D \to X_n$ (X_n a finite-dimensional subspace of X) such that $\sup_{x \in D} \|C_n x - Cx\| \leqslant 1/n$. Then again, assuming $p \in X_n$, and the Leray–Schauder degrees of the approximations (as defined in step 1) exist, we define

$$(5.3.6) \qquad d(I + C, p, D) = \lim_{n \to \infty} d(I + C_n, p, D).$$

Clearly the function $d(I + C, p, D)$ is well defined if the limit (5.3.6) exists and is independent of the approximating sequence C_n.

In order to justify the definition of $d(I + C, p, D)$ just given, we now provide the following

(5.3.7) **Justification of step 1:** We show that the integer $d(I + C, p, D)$ defined by (5.3.5) is independent of both the finite-dimensional linear subspace X_n containing p and $C(D)$. Thus, suppose the finite-dimensional subspaces X_n and X_p both contain $\{p\} \cup C(D)$. Since

$X_p \cap X_n$ is also a finite-dimensional linear subspace containing $p \cup C(D)$, it suffices to assume that (i) the domain of $I + C$ is restricted to $D \cap X_p$, (ii) $X_n \subset X_p$, and (iii) to prove that

$$d_B(I + C, p, D \cap X_n) = d_B(I + C, p, D \cap X_p)$$

(a statement involving only properties of the finite-dimensional Brouwer degree). To this end, suppose $\dim X_n = n$, $\dim X_p = n + k$, and a basis $\mathfrak{B}$ in X_p is chosen so that we may identify X_n with $\mathbb{R}^n$ and X_p with $\mathbb{R}^{n+k}$. Since the Brouwer degree will be independent of the basis $\mathfrak{B}$ chosen, (5.3.7) will be proven once we establish

(5.3.8) **Lemma** Let Δ be a bounded domain in $\mathbb{R}^{n+k}$ and f be a continuous mapping of $\Delta \to \mathbb{R}^n$. Then $d_B(I + f, p, \Delta) = d_B(I + f, p, \Delta \cap \mathbb{R}^n)$ for all $p \in \mathbb{R}^n$, provided $x + f(x) \neq p$ for $x \in \partial\Delta$.

Proof: By virtue of the definition of the Brouwer degree, it suffices to suppose that f is a C^1 mapping and that the Jacobian determinant of the mapping $x + f(x)$, $J_{n+k}(x)$, does not vanish on the set σ $\{x \mid x \in \Delta, x + f(x) = p$, for a fixed $p \in \mathbb{R}^n\}$. Then by the properties of the Brouwer degree of Section 1.6

$$\begin{aligned} d_B(I + f, p, \Delta) &= \sum_\sigma \operatorname{sgn} \det J_{n+k}(x) \\ &= \sum_\sigma \operatorname{sgn} \det \begin{pmatrix} J_n(x) & 0 \\ 0 & I_k \end{pmatrix} \\ &= \sum_\sigma \operatorname{sgn} \det (J_n(x)) \\ &= d(I + f, p, \Delta \cap \mathbb{R}^n), \end{aligned}$$

where I_k is the identity matrix in $\mathbb{R}^k$ and $J_n(x)$ is the Jacobian determinant of the mapping $x + f(x)$ on $\Delta \cap \mathbb{R}^n$.

(5.3.9) **Justification of step 2:** We begin by showing that for n sufficiently large, the integers $d(I + C_n, p, D)$ (as specified in step 1) can be defined [i.e., $(I + C_n)x \neq p$ for $x \in \partial D$. Thus, we first find a number $\alpha > 0$ such that $\inf_{x \in \partial D} \|(I + C)x - p\| \geqslant \alpha$, by virtue of the fact that $I + C \in \mathcal{C}_I(\partial D, X)$. Indeed, otherwise there would be a bounded sequence $\{x_j\} \in \partial D$ with $\|x_j + Cx_j - p\| \to 0$. By the compactness of C, after possibly passing to a subsequence, we may suppose $Cx_j \to y$ (say). Hence $\{x_j\}$ converges to z (say). Then, since ∂D is closed, $z \in \partial D$ and $z + Cz = p$, which is the desired contradiction. Now it is easy to show that $I + C_n - p \in \mathcal{C}_I(\partial D, X)$ for n sufficiently large, since for $x \in \partial D$ and $n \geqslant n_0$ (say)

$$\begin{aligned} \|x + C_n x - p\| &\geqslant \|x + Cx - p\| - \|Cx - C_n x\| \\ &\geqslant \alpha - \tfrac{1}{2}\alpha \geqslant \tfrac{1}{2}\alpha. \end{aligned}$$

Next we show that the numbers $d_n = d(I + C_n, p, D)$ stabilize for n sufficiently large, by supposing that for any integers $n, m \geqslant n_0$, $\sup_{\partial D} \|Cx - C_n x\| \leqslant \frac{1}{2}\alpha$ and proving $d_n = d_m$ for any integer $m \geqslant n_0$. This fact will ensure that the limit (5.3.6) exists and is independent of the approximations C_n. To this end, suppose the integers $n, m \geqslant n_0$, $C_n(D) \subset X_n$, $C_m(D) \subset X_m$, and X_{n+m} denotes the subspace generated by X_m and X_n. Then by Lemma (5.3.8),

(5.3.10) $$d(I + C_n, p, D \cap X_n) = d_B(I + C_n, p, D \cap X_{n+m}),$$

and similarly for the mapping $I + C_m$. Then, for $x \in D \cap X_{n+m}$, let $h(x, t) = x + tC_m x + (1 - t)C_n x - p$ so that on $\partial D \cap X_{n+m}$

$$\|h(x, t)\| \geqslant \|x + Cx - p\| - t\|C_m x - Cx\| - (1 - t)\|C_n x - Cx\|$$
$$\geqslant \alpha - \tfrac{1}{2}(t\alpha + (1 - t)\alpha) \geqslant \tfrac{1}{2}\alpha.$$

Consequently, by virtue of the homotopy invariance of the Brouwer degree and (5.3.8),

$$d_n = d_B(I + C_n, p, D \cap X_{n+m}) = d_B(I + C_m, p, D \cap X_{n+m}) = d_m.$$

We are now in a position to state and prove two major properties of the Leray–Schauder degree mentioned above.

(5.3.11) **Theorem** Let $f, g \in \mathcal{C}_I^0(\partial D, X)$, where D is a convex domain of X. Then f and g are compactly homotopic if and only if $d(f, 0, D) = d(g, 0, D)$.

(5.3.12) **Theorem** Let $f \in \mathcal{C}_I^0(\partial D, X)$. Then f is essential relative to D if and only if $d(f, 0, D) \neq 0$. Thus if $d(f, 0, D) \neq 0$ the equation $f(x) = 0$ has a solution in D.

Proof of (5.3.11): By virtue of the definition of the Leray–Schauder degree just given, if $f, g \in \mathcal{C}_I(\partial D, X)$ and are compactly homotopic, then clearly the mappings f and g have the same Leray–Schauder degree at zero.

Conversely, suppose $f, g \in \mathcal{C}_I(\partial D, X)$ and $d(f, 0, D) = d(g, 0, D)$. Then by virtue of the definition just given, we may suppose that the ranges of the compact operators $c = f - I$ and $c_1 = g - I$ are contained in the same finite-dimensional linear subspaces X_n of X. In addition, by restricting the domains of f and g to $X_n \cap D$, we may also assume that $d_B(I + c, 0, X_n \cap D) = d_B(I + c_1, 0, X_n \cap D)$. Then by the result of H. Hopf (1.6.7), f and g are homotopic on $X_n \cap \partial D$; so that there is a continuous function $h(x, t) = x + c(x, t)$ defined on the closed set $\Sigma = [X_n \cap \partial D] \times [0, 1]$ such that $h(x, 0) = f(x)|_{X_n \cap \partial D}$ and $h(x, 1) = g(x)|_{X_n \cap \partial D}$. Choosing a basis $(e_1, \ldots, e_n)$ in X_n, we can write $c(x, t)$ as an n-tuple of real-valued continuous functions $c_i(x, t)$ $(i = 1, \ldots, n)$.

By Tietze's theorem, each of these continuous functions can be continuously extended to a function C_i on $\partial D \times [0, 1]$ with $\sup_\Sigma |C_i(x, t)|$

$= \sup_{\partial D \times [0,1]} c_i(x, t)$. We then consider the function $H(x, t) = x + \sum_{i=1}^{n} C_i(x, t) e_i$. $H(x, t) - x$ is compact on $\partial D \times [0, 1]$ since it has a closed, bounded, finite-dimensional range, and in addition, for $x \in \partial D$, $H(x, 0) = f$, $H(x, 1) = g$. Furthermore, $H(x, t)$ is a compact homotopy between f, g in the class $\mathcal{C}_I(\partial D, X)$ since for $t \in [0, 1]$ and $x \in \Sigma$, $H(x, t) \neq 0$ by definition, while for $x \in \partial D \cap X_n$, $H(x, t) \neq 0$ because x and $C(x, t)$ are linearly independent. Consequently, f and g are compactly homotopic on ∂D.

Proof of (5.3.12): First suppose $f \in \mathcal{C}_I(\partial D, X)$ and $d(f, 0, D) \neq 0$. Then by (5.3.11), for any compact perturbation of the identity $g = I + C$ defined on $\bar{D}$ and coinciding with f on ∂D, $d(g, 0, D) = d(f, 0, D) \neq 0$. Indeed, the convex combination of f and g defines a compact homotopy of f into g. Thus it suffices to prove that the equation $g(x) = 0$ has a solution. By the definition of the Leray–Schauder degree, we may suppose that there is a sequence of mappings $g_n = I + C_n$, where for each n, C_n is a compact mapping with finite-dimensional range and $\sup_D \|C_n x - Cx\| \leqslant 1/n$. Furthermore, $d(I + C_n, 0, D) = d(I + C, 0, D) \neq 0$ for n sufficiently large. In addition, by restricting $I + C_n$ to the intersection of D with a finite-dimensional linear subspace X_n of X, we find

$$d_B(I + C_n, 0, D \cap X_n) = d(I + C, 0, D) \neq 0.$$

It follows then, by the properties of the Brouwer degree, that $(I + C_n)x = 0$ has a solution $x_n \in D \cap X_n$.

Next we show that $\{x_n\}$ has a convergent subsequence with limit $\bar{x}$, and $g(\bar{x}) = 0$. Indeed, $\{x_n\}$ is bounded and for an appropriate subsequence $\{x_{n_j}\}$, $\{Cx_{n_j}\}$ converges. Thus

$$\text{(5.3.13)} \qquad \|x_{n_j} + Cx_{n_j}\| \leqslant \|Cx_{n_j} - C_{n_j}x_{n_j}\| + \|x_{n_j} + C_{n_j}x_{n_j}\| \leqslant 1/n_j;$$

so that as $n \to \infty$, $\{x_{n_j}\}$ converges to $\bar{x}$ (say). Hence (5.3.13) implies that $g(\bar{x}) = 0$.

To prove the converse, we suppose that $f \in \mathcal{C}_I(\partial D, X)$, $d(f, 0, D) = 0$, and f is essential. Then by (5.3.3) and (5.3.11), all mappings $\tilde{f} \in \mathcal{C}_I(\partial D, X)$ with $d(\tilde{f}, 0, D) = 0$ must be essential. Therefore, by the definition of Leray–Schauder degree, all continuous mappings $f \in \mathcal{C}_I(\partial D, X)$ defined on $D \cap X$ with $f \neq 0$ on $\partial D \cap X$ must be essential. Hence the constant mapping is essential (a contradiction). Hence, in order for $f \in \mathcal{C}_I(\partial D, X)$ to be essential, $d(f, 0, D) \neq 0$.

Properties of the Leray–Schauder Degree We now describe the basic properties of the Leray–Schauder degree $d(f, p, D)$, when considered as a function of the three variables f, p, and D. Then we use these properties to discuss the computation of the degree for general classes of mappings in $\mathcal{C}_I(\partial D, X)$. To begin we prove:

(5.3.14) **Theorem** Suppose D is a bounded domain contained in a Banach space X, and $f - p \in \mathcal{C}_I^0(\partial D, X)$. Then the Leray–Schauder degree $d(f, p, D)$ is an integer with the following properties:

(i) (Homotopy invariance) If $(h(x, t) - p) \in \mathcal{C}_I^0(\partial D, X)$ for $t \in [0, 1]$ is a compact homotopy with $h(x, 0) = f$, then $d(f, p, D) = d(h(x, t), p, D)$ for $t \in [0, 1]$.

(ii) If p and p' are in the same component of $X - f(\partial D)$, then $d(f, p, D) = d(f, p', D)$.

(iii) $d(f, p, D)$ is uniquely determined by its values on ∂D.

(iv) (Continuity) $d(f, p, D)$ is a continuous (locally constant) function of $f \in C(\overline{D})$ (with respect to uniform convergence) and $p \in X$.

(v) (Domain decomposition) If D is the union of a finite number of open disjoint sets D_i $(i = 1, \ldots, N)$ with $\partial D_i \subset \partial D$ and $f(x) \neq p$ on $\bigcup_{i=1}^N \partial D_i$, then

$$(5.3.15) \qquad d(f, p, D) = \sum_{i=1}^{N} d(f, p, D_i).$$

(vi) (Excision) If Δ is a closed subset of $\overline{D}$ on which $f(x) \neq p$, then $d(f, p, D) = d(f, p, D - \Delta)$.

(vii) (Cartesian product formula) If $X = X_1 \oplus X_2$ with $D_i \subset X_i$, $f = (f_1, f_2)$ with $f_i\colon D_i \to X_i$ $(i = 1, 2)$, $D = D_1 \times D_2$ and $p = (p_1, p_2)$, then $d(f, p, D) = d(f_1, p_1, D_1) d(f_2, p_2, D_2)$, provided the right-hand side is defined.

(viii) (Index theorem) If the solutions of $f(x) = p$ are isolated in D, then $d(f, p, D) = \sum_i d(f, p, O_i)$, where O_i is any sufficiently small open neighborhood containing only one solution, and all solutions are contained in $\bigcup_i O_i$.

Proof: (i): This result is a restatement of (5.3.11).

(ii): First we observe that since f is proper on ∂D, $f(\partial D)$ is closed, and consequently each of the components of $X - f(\partial D)$ are open, arcwise connected sets. Let any such component be denoted D_i, then there is an arc $p(t)$ in D_i for $t \in [0, 1]$ joining p and p' and avoiding $f(\partial D)$. Thus, by (i), for $t \in [0, 1]$, $d(f, p, D) = d(f, p(t), D) = d(f, p', D)$.

(iii): Let f_0 be a compact perturbation of the identity that agrees with f on ∂D. Then by (i), $h(x, t) = tf + (1 - t)f_0$ is a compact homotopy joining f and f_0, $d(f_0, p, D) = d(f, p, D)$.

(iv): This is immediate from the definition and (i).

(v)–(vii): Since each of these facts is valid for the Brouwer degree, the definition of $d(f, p, D)$ implies the validity of each for mappings f with finite-dimensional range, and so, by approximation, for all $f \in \mathcal{C}_I^0(\partial D, X)$.

(viii): The index theorem follows immediately from the excision and domain decomposition properties of degree using the fact that if the solutions of $f(x) = p$ are isolated, they are finite in number.

We now apply the eight properties of the Leray–Schauder degree just established to investigate classes of mappings that have nonzero Leray–Schauder degree relative to a bounded domain D of a Banach space X. By virtue of (5.3.12), such mappings will be essential relative to D, and hence of particular importance.

(5.3.16) **Theorem** Suppose $f \in \mathcal{C}_I^0(\partial D, X)$. Then

(i) If D contains the origin and $f = I - C$ is a linear homeomorphism, then $d(f, 0, D) = (-1)^{\beta}$, where β is the number of eigenvalues of C (counted according to multiplicity) in the interval $(1, \infty)$. More generally, assuming f is a compact perturbation of the identity and if f is a homeomorphism of D onto itself, then $d(f, 0, D) = \pm 1$ provided D contains the origin.

(ii) If f is asymptotically linear, i.e., there is a compact linear mapping C with $\|f(x) - x + Cx\|/\|x\| \to 0$ as $\|x\| \to \infty$, and D is a sufficiently large domain containing the origin, then $d(f, 0, D) = (-1)^{\beta}$, where β is the number of eigenvalues of C (counted with multiplicity in the interval $(1, \infty)$), provided $I - C$ is a linear homeomorphism.

(iii) If f is an odd mapping and D is a symmetric domain containing the origin, then $d(f, 0, D)$ is an odd integer. More generally, if the parity assumption is weakened to $f(x) \neq tf(-x)$ for $t \in [0, 1]$ and $x \in \partial D$, then $d(f, 0, D)$ is an odd integer.

(iv) Suppose $f = I + N$, with N compact, and that all solutions of the family of equations $f_t(x) = x + tNx = 0$ for $t \in [0, 1]$ lie in some fixed bounded domain D that contains the origin, then $d(f, 0, D) = 1$.

(v) If X is a complex Banach space and f is complex analytic, then:

(a) $d(f, 0, D) \geqslant 0$;

(b) a necessary and sufficient condition for $d(f, 0, D) > 0$ is that $0 \in f(D)$; and

(c) a necessary and sufficient condition for $d(f, 0, D) \geqslant 2$ is that either the equation $f(x) = 0$ has more than one solution in D or at the unique solution x_0 of $f(x) = 0$ the linear operator $f'(x_0)$ is not invertible.

Proof: (i): Suppose $f = I - C$ is a linear homeomorphism with $f \in \mathcal{C}_I(\partial D, X)$. Let X_1 be the direct sum of the invariant subspaces of C corresponding to the eigenvalues in $(1, \infty)$. Then, by hypothesis, dim X_1

$= \beta < \infty$. Hence $X = X_1 \oplus X_2$, where X_2 is invariant under f, and by (5.3.14), if $f_i = f|_{X_i}$ $(i = 1, 2)$ and $f = (f_1, f_2)$, then

$$(5.3.17) \qquad d(f, 0, D) = d(f_1, 0, D \cap X_1) d(f_2, 0, D \cap X_2).$$

Now on the finite-dimensional space $D \cap X_1$, f_1 is (compactly) homotopic to $-I$. Indeed, setting $h(x, t) = -(1 - t)x + t(I - C)x = (2t - 1)I - tC$ for $t \in [0, 1]$, we observe that $h(x, t) \neq 0$ on $\partial D \cap X_1$ this is clear for $t = 0$ and follows by hypothesis for $t = 1$ and for other t values since otherwise C would have an eigenvalue on the interval $(-\infty, 1)$ on X_1. Thus by (1.6.3), $d(f, 0, D \cap X_1) = (-1)^\beta$. Next, we observe that f_2 is homotopic to the identity I on $\partial D \cap X_2$ by virtue of the compact homotopy $g(x, t) = x - tCx$. Clearly, $g(x, t) \neq 0$ on $(\partial D \cap X_2) \times [0, 1]$ since if it did vanish there, C would have an eigenvalue in X_2 on the interval $(1, \infty)$. Consequently, $d(f_2, 0, D \cap D_2) = 1$; and finally by (5.3.17), $d(f, 0, D) = (-1)^\beta$.

Next, suppose f is a homeomorphism with inverse f^{-1} and $f(D) = D$. Then, by definition, since D contains the origin, $d(I, 0, D) = 1$,

$$1 = d(ff^{-1}, 0, D) = d(f, 0, D) d(f^{-1}, 0, D)$$

so that $d(f, 0, D) = \pm 1$. (For a more general result, see Note C)

(ii): By the homotopy invariance of degree on ∂D, the boundary of a sufficiently large domain containing the origin, f is compactly homotopic to $I - C$. Indeed, if $f = I - C - N$ is linearly asymptotic to $I - C$, then $\|Nx\|/\|x\| \to 0$ as $\|x\| \to \infty$. Therefore, the compact homotopy $h(x, t) = x - \{Cx + tNx\}$ joining f and $I - C$ does not vanish on ∂D. Indeed, $h(x, t) = 0$ implies $\|x_0 - Cx_0\| = t_0 \|Nx_0\|$ for some $t_0 \in [0, 1]$ and $x_0 \in \partial D$; and since $I - C$ is invertible, for some constant $\beta > 0$, $\beta \|x_0\| \leqslant \|x_0 - Cx_0\| = t_0 \|Nx_0\| \leqslant \|Nx_0\|$. This last inequality contradicts the fact that $\|Nx\|/\|x\| \to 0$ as $\|x\| \to \infty$, since β is independent of x. Thus by the homotopy invariance of degree and (i) above, $d(f, 0, D) = d(I - C, 0, D) = (-1)^\beta$.

(iii): Suppose $f = I + N$ is an odd mapping with N compact. Then if N had finite-dimensional range, $d(f, 0, D)$ would be an odd integer since by virtue of (1.6.3), the Brouwer degree has this property. Thus in the general case it suffices to prove that N can be closely approximated by odd compact mappings with finite-dimensional range N_ϵ. In fact, we shall construct such approximations with $\|Nx - N_\epsilon x\| \leqslant \epsilon$ for all $x \in D$ and any $\epsilon > 0$. To this end, suppose M is a compact finite-dimensional approximation to N on D with $\|Mx - Nx\| \leqslant \epsilon$. Then the operator $N_\epsilon x = \frac{1}{2}\{M(x) - M(-x)\}$ is clearly an odd compact mapping with finite-dimensional range and

$$\|Nx - N_\epsilon x\| \leqslant \tfrac{1}{2}\{\|Mx - Nx\| + \|M(-x) - N(-x)\|\} \leqslant \epsilon.$$

To prove the more general result, we use the homotopy invariance of degree, by showing that the mapping $f = I + N'$ (say) is compactly homotopic on ∂D to $f_1(x) = \frac{1}{2}\{f(x) - f(-x)\} = x + \frac{1}{2}\{N'(x) - N'(-x)\}$. Indeed, consider the compact homotopy defined on ∂D

$$h(x, t) = (1 + t)^{-1}\{f(x) - tf(-x)\}$$
$$= x + (1 + t)^{-1}\{N'(x) - tN'(-x)\}.$$

The fact that $f(x) \neq \lambda f(-x)$ for $t \in [0, 1]$ and $x \in \partial D$ implies $h(x, t) \neq 0$ for $t \in [0, 1]$ and $x \in \partial D$. Consequently, by the result of the above paragraph,

$$d(f, 0, D) = d(h(x, t), 0, D) = d(\tfrac{1}{2}(f(x) - f(-x)), 0, D)$$
$$= \text{odd integer}.$$

(iv): The compact homotopy $f_t = I + tN$ joins f and I, and by hypothesis does not vanish on $\partial D \times [0, 1]$. Therefore, by the homotopy invariance of degree,

$$d(f, 0, D) = d(f_t, 0, D) = d(I, 0, D) = 1.$$

(v): To establish (a) and (b) for a general complex analytic operator $f \in \mathcal{C}_I(\partial D, X)$, we first apply the result (1.6.2) to conclude that the set $\sigma = \{x \mid x \in D, f(x) = 0\}$ is finite. Thus, denoting the points in σ as $x_1, \ldots, x_n$

$$(5.3.18) \qquad d(f, 0, D) = \sum_{i=1}^{n} d(f, 0, O_i),$$

where the O_i are small (pairwise disjoint) open neighborhoods of D such that $x_i \in O_i$ for $i = 1, \ldots, n$. Thus to prove (a) and (b), we need only show that each term on the right-hand side of (5.3.18) is positive. To this end, observe that by adding an arbitrarily small complex linear map $L_i(x - x_i)$ of finite rank to f, we may suppose that $f'(x_i)$ is a linear homeomorphism. Then since $f(x) = f'(x_i)(x - x_i) + O(\|x - x_i\|^2)$, the homotopy invariance of degree imply that if O_i is a sufficiently small neighborhood of x_i,

$$d(f, 0, O_i) = d(f'(x_i), 0, O_i) = (-1)^{\beta}.$$

But, since $f'(x_i)$ is a linear isomorphism defined on a complex Banach space X, β is even. Thus (a) and (b) are established.

Finally, we observe that property (c) holds for $f = I + N$, where N is a compact complex analytic operator with finite-dimensional range since it holds for the Brouwer degree (cf. (1.6.3(x))). Furthermore, by the argument used to prove (a) and (b), if the solutions of $f(x) = 0$ are not unique, then $d(f, 0, D) \geqslant 2$. Therefore, it suffices to prove (c) under the assumption that the solution $x_0 = 0$ of $f(x) = 0$ exists and is unique in D, $f'(0)$ is not invertible on X_1, $X_1 \cap \text{Ker}\, f'(0) = 0$ and $X = \text{Ker}\, f'(0) \oplus X_1$. Then $f'(0) = I + C$ with C compact, and we set $C_1 = PC$ and $C_2 = (I - P)C$, where P and $I - P$ denote the canonical projections of X onto $\text{Ker}\, f'(0)$ and X_1, respectively. Then $I + C_2$ is invertible on X. Moreover, since C_2 is compact, there is a continuous complex valued function $\alpha(t)$ defined for $t \in [0, 1]$ with $\alpha(0) = 0$ and $\alpha(1) = 1$ such that the operators $I - \alpha(t)C_2$ are invertible. (Indeed C possesses discrete eigenvalues whose only possible limit point is zero.) Then by the analytic form of the inverse function theorem (3.1.1), if $f(x) = x + Cx + Rx$, where $Rx = O(\|x\|^2)$, the operator $I + \alpha(t)[C_2 x + Rx]$ has a uniquely defined inverse $h(x, t) = x + \mu(x, t)$ joining $h(x, 0) = x$ and $h(x, 1) = [x + C_2 x + Rx]^{-1}$ defined and continuous jointly in a neighborhood U of the origin for $t \in [0, 1]$. Furthermore, since $x = h(x, t) + \alpha(t)(C_2 + R)(h(x, t))$, $x - h(x, t) = \mu(x, t)$ is compact in x and t. Also, if $f \circ h(x, t) = 0$ in $\overline{U} \times [0, 1]$, then $h(x, t) = 0$ so that $x = 0$. Thus, by virtue of the homotopy invariance of degree for $t \in [0, 1]$,

$$d(f, 0, U) = d(f \circ h(x, t), 0, U) = d(f \circ h(x, 1), 0, U)$$
$$= d(I + C_1 h(x, 1), 0, U).$$

Now the mapping $C_1 h(x, 1)$ is a compact complex analytic mapping with finite-dimensional range. In addition, the Fréchet derivative of $x + C_1 h(x, 1)$ at $x = 0$ is $I + C_1(I + C_2)^{-1}$. Thus the mapping $\tilde{f}(x) = x + C_1 h(x, 1)$ (restricted to the associated finite-dimensional subspace of X) is a compact complex analytic perturbation of the identity with finite-dimensional range and such that $\tilde{f}'(0)$ is not invertible, while $d(\tilde{f}, 0, D)$ is defined. Thus $d(f, 0, D) \geqslant 2$, by applying the analogous result (1.6.3) for the Brouwer degree.

Finally we examine the conditions under which a mapping $f \in \mathcal{C}_I(\partial D, X)$ is inessential, relative to D.

(5.3.19) **Theorem** Suppose $f = I + C$ is a compact perturbation of the identity defined on a bounded domain of a Banach space X. Then, if defined,

(i) $d(f, p, D) = 0$ if f maps D into a proper subspace X' of X and $p \in X'$;

(ii) $d(f, p, D) = 0$ if $f(x) \neq p$ in D. The converse is true provided X is a complex Banach space, and f is complex analytic.

Proof: (i): Let Δ_p be the open component of $X - f(\partial D)$ containing p. Since $f(D)$ is contained in a proper subspace X' of X, there is a point $q \in \Delta_p$ that is not in the range of f. Otherwise $\Delta_p \subset f(D)$. Now by virtue of

(5.3.14(ii)) and the argument used in proving property (5.3.12) of the Leray–Schauder degree

$$d(f, p, D) = d(f, q, D) = 0.$$

(ii): This fact is an immediate consequence of the argument used in proving (5.3.12) and (5.3.16).

5.3D Compact perturbations of a linear Fredholm mapping and stable homotopy

Let L be a fixed bounded linear Fredholm operator of nonnegative[1] index p mapping a Banach space X into itself. Let $D = \{x \mid \|x\| < 1\}$ and $\partial D = \{x \mid \|x\| = 1\}$. Then we shall represent the compact homotopy classes of $\mathcal{C}_L^0(\partial D, X)$ by means of known homotopy invariants. In addition, in certain cases, we shall determine special necessary and sufficient conditions that a mapping $g \in \mathcal{C}_L^0(\partial D, X)$ be essential, and apply this to solvability questions involving operator equations.

To accomplish these goals, notions concerning the sequence of homotopy groups of a sphere S^n, $\{\pi_{n+p}(S^n)\}$ (as n runs through the positive integers with p fixed) are of prime importance. As was mentioned in Section 1.6, for fixed $p > 0$, the groups $\pi_{n+p}(S^n)$ are isomorphic finite Abelian groups provided $n > p + 1$, and this isomorphism is given by a canonical one, the so-called Freundenthal suspension E: $\pi_{n+p}(S^n) \to \pi_{n+1+p}(S^{n+1})$. These isomorphic groups are called the pth *stable homotopy groups of* S^n. It is useful to find a simple analytic expression of the Freundenthal suspension Ef of a given representative f: $S^{n+p} \to S^n$ of $[f] \in \pi_{n+p}(S^n)$. To accomplish this let $\tilde{f}$ be any continuous extension of f to the interior of $S^n \subset \mathbb{R}^{n+1}$, and let Ef: $S^{n+p+1} \to S^{n+1}$ be the mapping

$$(*) \qquad Ef(x_1, \ldots, x_{k+1}) = (\tilde{f}(x_1, \ldots, x_{k+1}), x_{k+2})/|(\tilde{f}(x_1, \ldots, x_{k+2}))|$$

with $k = n + p$.

This is easily justified by virtue of the geometric definition of suspension given in Chapter 1 and the fact that the homotopy class of Ef depends only on $[f]$. Moreover the expression $(*)$ can be easily extended to give a simple analytical expression of the various iterates of Freundenthal suspension operator $E^k f$ $(k > 0)$.

[1] The reader should note that homotopy considerations are of little value in the case of Fredholm operators L of negative index. Indeed, in Section 3.1 we proved that C^1 operators of the form $L + C$ must have a nowhere dense range. Thus the simple idea of perturbing a point p in the range of $L + C$ to a nearby point p', without affecting the solvability of $L + C$ breaks down in total agreement with (3.1.46).

In case $p=0$, $\pi_n(S^n)\approx\mathbb{Z}$ (the additive group of integers) and the resulting homotopy classes of mapping $f\colon S^n\to S^n$ are well behaved under suspension, in the sense that the essentialness of such a mapping is preserved under suspension. This is *no longer true, in general, if* $p>0$, and indeed *interesting homotopy properties of a mapping* $f\colon S^{n+p}\to S^n$ *may be lost* upon iterated suspension. A case in point concerns the group $\pi_3(S^2)$ which is known to be isomorphic to Z, while $\pi_4(S^3)=Z_2$, thus if $[\alpha]$ is a generator of $\pi_3(S^2)$, $E[n\alpha]=0$ whenever n is even.

Given a mapping $f\colon S^{n+p}\to S^n$ we can thus associate with it its homotopy class $[f]\in\pi_{n+p}(S^n)$. Moreover, we consider its sequence of iterates by Freundenthal suspensions E^kf and associated homotopy classes $E^k[f]\in\pi_{n+p+k}(S^{n+k})$. Thus we call $E^k[f]$ the stable homotopy class of f provided the integer k is so large that $n+k>p+1$. We are now in a position to prove:

(5.3.20) **Theorem** (Svarc, 1964) Suppose L is a fixed linear Fredholm operator of nonnegative index p mapping X into Y. Then the compact homotopy classes $\mathcal{C}_L^0(\partial D, Y)$ are in one-to-one correspondence with the elements of the pth stable homotopy group $\pi_{n+p}(S^n)$ $(n>p+1)$.

Proof: The basic idea is to repeat the arguments given in the construction of the Leray–Schauder degree, substituting the properties of stable homotopy for the Brouwer degree. Thus given $f\in\mathcal{C}_L^0(\partial D, Y)$, we may suppose that $f=L+C$, where (i) C has finite-dimensional range contained in the (finite-dimensional) subspace Y_{n+1} of Y (provided $n>p+1$) by (2.4.2), and (ii) by (1.3.38) L is surjective. Since L is a linear Fredholm operator of index p, we may write $X=\operatorname{Ker} L\oplus X_1$, where $\dim\operatorname{Ker} L=p$ and $L\colon X_1\to Y$ is a bounded linear homeomorphism with inverse L^{-1}. By restricting the domain of f to $\partial D_{n+p}\cap\{\operatorname{Ker} L\oplus L^{-1}(Y_{n+1})\}$, we then obtain a natural mapping $\tilde{f}\colon\partial D_{n+p}\to Y_{n+1}$ defined by $\tilde{f}(x)=\tilde{f}(x_0+x_1)$ $=Lx_1+C(x_0+x_1)$. Furthermore, since $f\in\mathcal{C}_L^0(\partial D, Y)$, not only does $f(x)\neq 0$ for $x\in\partial D$, but also there is a positive number $\alpha>0$ with $\inf_{x\in\partial D}\|f(x)\|\geqslant\alpha>0$. Indeed, otherwise by arguing as in the proof of (5.3.9) and using the properties of linear Fredholm operators, there would be a sequence $\{x_n\}\in\partial D$ with $x_n\to\bar{x}$ and $\|f(x_n)\|\to 0$ so that $f(\bar{x})=0$ and $\bar{x}\in\partial D$. Thus $\tilde{f}(x)\neq 0$ for $x\in\partial D_{n+p}$, and we can define a natural mapping of $S^{n+p}\to S^n$; $f_0(x)=\tilde{f}(x)/\|\tilde{f}(x)\|$ for $x\in S^{n+p}$. The correspondence τ between the homotopy class of $f\in\mathcal{C}_L(\partial D, Y)$ and $\pi_{n+p}(S^n)$ is defined by setting $\tau([f])=[f_0]$. In order to show that τ is well defined, it is necessary to prove that τ is independent of the finite-dimensional subspace Y_n defining $[f_0]$. To this end, suppose Y_{n+1} and Y_{m+1} are two subspaces with $n, m>p+1$, both containing $C(\partial D)$ and let f_0 and g_0 denote the associated mappings between spheres. Then the subspace

$Y_{n+1} \cap Y_{m+1}$ contains $C(\partial D)$, and both f_0 and g_0 can be regarded as extensions by iterated suspension of the same mapping

$$\gamma_0: \partial D \cap \{\operatorname{Ker} L \cap h^{-1}(Y_{n+1} \cap Y_{m+1})\} \to Y_{n+1} \cap Y_{m+1}.$$

By a fundamental topological result, the homotopy classes $[f_0]$, $[g_0]$ of both f_0 and g_0 depend only on $[\gamma_0]$. Hence, both $[f_0]$ and $[g_0]$ are in the same stable homotopy class, by virtue of the properties of the Freudenthal suspension operator.

To prove that the correspondence τ is one-to-one, suppose that the two mappings $f, g \in \mathcal{C}_L^0(\partial D, Y)$ are compact, finite-dimensional perturbations of L with $[\tau f] = [\tau g]$. We wish to prove that f and g are compactly homotopic in $\mathcal{C}_L^0(\partial D, X)$. By arguing as in the first part of (5.3.4), we may suppose that the ranges of $f - L$ and $g - L$ are contained in the same finite-dimensional subspace $\tilde{Y}$ of Y. Then there is a compact homotopy $Lx_1 + C(x, t)$: $\tilde{X} \times [0, 1] \to \tilde{Y}$ from f to g. By repeating the argument of the proof of (5.3.11), the Tietze extension theorem ensures that the mapping $Lx_1 + C(x, t)$ can be extended to a compact homotopy in $\mathcal{C}_L^0(\partial D, Y)$.

It remains to show that if f_1 and f_2 are compactly homotopic in $\mathcal{C}_L^0(\partial D, Y)$, then $\tau[f_1] = \tau[f_2]$. To this end we first suppose that $f_1 = L + C_1$ and $f_2 = L + C_2$ are such that each compact operator C_i $(i = 1, 2)$ has finite-dimensional range. Then, as in (5.3.11), we observe that the compact homotopy $h(x, t) = L + C(x, t)$ joining f_1 and f_2 may be chosen such that $C(x, t)$ has a fixed finite-dimensional range for all t. Then by the definition of $\tau[f_1]$ and $\tau[f_2]$ given above, it follows that $\tau[f_1] = \tau[f_2]$. For the general case, we first note that the compact homotopy classes $[f_1]$ and $[f_2]$ each have representatives of the special form $\tilde{f}_i = L + \tilde{C}_i$ $(i = 1, 2)$ with $\tilde{C}_i$ compact. Then, the argument just given shows that the correspondence $\tau[f_i] = \tau[\tilde{f}_i]$ $(i = 1, 2)$ does not depend on the particular representative $\tilde{f}_i$ chosen.

(5.3.20′) **Corollary** Under the hypothesis of Svarc's theorem, $f = L + C$ is inessential if and only if the natural mapping τ constructed in the proof of (5.3.20) is such that $\tau(f) = 0$.

Proof: Let $g = L - c_0$, where c_0 is a constant map into Y, where c_0 is chosen so that the image of the set $\{x_1 \mid x = x_1 + x_0,\ X = X_1 \oplus \operatorname{Ker} L,\ x_1 \in X_1,\ \|x\| \leqslant 1\}$ under L does not contain c_0. Consequently, by our construction of τ in (5.3.20) $\tau(g) = 0$. Now suppose $\tau[f] = 0$ so that $\tau[f] = \tau[g]$. Then f and g have the same compact homotopy class. By (5.3.3) f is necessarily inessential since g is.

On the other hand, if f is inessential, then necessarily f is compactly homotopic to g but $\tau[g] = 0$ and so $\tau[f] = 0$.

A simple yet interesting construction of a mapping $g \in \mathcal{C}_L(\partial D, Y)$ with index $L = p$ can be obtained as follows. Suppose ψ is compact and $I + \psi$ is a mapping of a Banach space X onto a linear subspace Y of codimension p, while $x + \psi(x) \neq 0$ on the boundary of the unit sphere $\partial\Sigma_1$ of X. By virtue of (5.3.19), the Leray–Schauder degree $d(I + \psi, 0, \partial\Sigma_1) = 0$; and so the mapping $I + \psi$ is inessential relative to 0. However, by restricting the range of I and ψ to Y and regarding $g = I + \psi$ as a mapping from the unit sphere of X into Y, one can study the mapping properties of $I + \psi$ by means of g. Indeed, denoting by L and C the operators obtained by restricting the range of I and X to Y, $g \in \mathcal{C}_L(\partial D, Y)$; and L can be regarded as a linear operator of index p. In fact, we shall determine necessary and sufficient conditions for such a mapping g to be essential.

Application to equations involving operators with singular points In order to utilize Svarc's theorem (5.3.20) we attempt to apply it to study the *solvability* of a simple class of semilinear operator equations. The class of operators we consider consists of *uniformly bounded* compact perturbations N of a fixed linear Fredholm operator $L \in L(X, Y)$ of index p. If $p \geqslant 0$, (5.3.20) does imply a solvability criteria for the equation $Lu + Nu = 0$ provided the associated *stable* homotopy class τ of $L + N$ is nontrivial when τ is regarded as an element of $\pi_{n+p}(S^n)$ for n sufficiently large. Such a result is clearly difficult to apply in case $p > 0$. Indeed if $p = 0$, as already mentioned, the essentialness of good finite-dimensional approximations to maps of the form $L + C$ is preserved under iterated suspension. However this is no longer true, in general, if $p > 0$. Thus it is necessary to supplement the abstract theorem by using simplifying hypotheses that can often be verified in concrete problems.

Thus in the sequel we shall assume not only the uniform boundedness of the operator N but also the following restriction on the *asymptotic* behavior of N.

Hypothesis (A): Let $X = \operatorname{Ker} L \oplus X_1$ and P_0 be the canonical projection of Y onto coker L. Then $\|P_0 N(x_0 + x_1)\| \neq 0$ whenever $x_1 \in X_1$ is uniformly bounded and $x_0 \in \operatorname{Ker} L$ is sufficiently large in norm.

This assumption is generally valid for those operator equations that admit a *a priori estimates*. We are now in a position to prove the following improvement of (5.3.20).

(5.3.21) **Theorem** Let D_R be a ball in a Banach space X of sufficiently large radius R and suppose $L + N \in \mathcal{C}_L^0(\partial D_R, Y)$, where N is uniformly bounded on X and satisfies the above Hypothesis (A). Then $L + N$ is essential if and only if the stable homotopy class of the mapping $\tilde{\mu}(a) = \mu(a)/|\mu(a)|$: $S^{d-1} \to S^{d_*-1}$ is nontrivial, where $\mu(a) = P_0 N(Ra)$, $d = \dim \operatorname{Ker} L$, $d_* = \dim \operatorname{Ker} L^*$, and a is an element of norm 1 on $\operatorname{Ker} L$.

Proof: We show first that an operator $f = L + C$ satisfying the stated hypotheses can be deformed on ∂D_R by a compact homotopy to the form $\tilde{f}(x_0, x_1) = (Lx_1, P_0Nx_0)$: $X_1 \oplus \operatorname{Ker} L \to Y_1 \oplus \operatorname{coker} L$ and then apply the construction given in the proof of Švarc's result (5.3.20) to show in case $Lx_1 = x_1$, that the homotopy class of $\tilde{f}$ regarded as a singularity free map of ∂D_R corresponds to the stable homotopy class of the normalized map $\tilde{\mu}$ associated with $\mu(a) = P_0N(Ra)$. Then, the result will follow easily since the essentialness of a mapping f is invariant under linear homeomorphisms.

Step 1: Thus to demonstrate the compact homotopy joining f to $\tilde{f}$, we write f in the form $f = (P_1 f, P_0 f)$, where P_1 and P_0 are the canonical projections of Y onto Y_1 and coker L, respectively. Then we consider the compact homotopy

$$h(x, t) = (Lx_1 + tP_1Nx, P_0N(x_0 + tx_1))$$

joining f and $\tilde{f}$. Assuming the uniformly bounded mapping N satisfying Hypothesis (A), $h(x_1, t) \neq 0$ on $\partial D_R \times [0, 1]$ provided R is sufficiently large. Indeed if both $P_0N(x_0 + tx_1)$ and $Lx_1 + tP_1Nx$ are zero, both $\|x_0\|$ and $\|x_1\|$ (and consequently $\|x\|$) must be sufficiently small.

Step 2: Now we observe that since L: $X_1 \to Y_1$ is a linear homeomorphism, we may suppose, without loss of generality, that $\tilde{f}(x_0, x_1) = (x_1, P_0Nx_0)$ and that coker $L \subset \operatorname{Ker} L$, so that $\operatorname{Ker} L = \operatorname{coker} L \oplus W$. In this case we assert the homotopy class of $\tilde{f}$, $[\tilde{f}]$, corresponds *by means of the construction in the proof of* (5.3.20) to the stable homotopy class f of the normalized map $\tilde{\mu}$ associated with $P_0N(Ra)$ when $[\tilde{\mu}]$ is regarded as an element of $\pi_{d-1}(S^{d_\bullet - 1})$. To verify this following the construction of the correspondence τ in (5.3.20), we replace $Lx = x_1$ with the surjective map $Lx = x_1 + \epsilon v$, where $\epsilon > 0$ is small and $x = x_1 + v + w$ with $v \in \operatorname{coker} L = V$, $w \in W$ and $Cx = P_0N(x_0)$ with $\tilde{C}x = P_0N(x_0) - \epsilon v$. Thus Range $C \subseteq \operatorname{coker} \tilde{L}$, and we can regard C as a mapping of $X_1 \oplus \operatorname{Ker} L \to X_K \oplus \operatorname{coker} L$ (with X_K a linear subspace of dimension K so chosen that $K + n > p + 2$). Setting $S^{N+P} = \{x \mid \|x\| = 1, x \in \operatorname{Ker} L \oplus L^{-1}(V \oplus X_K)\}$, since $L^{-1}(V \oplus X_K) = V \oplus X_K$, we can identify S^{N+P} as the unit sphere in $X_K \oplus V \oplus W$. Consequently $[f]$ coincides with the homotopy class of the normalization of the mapping

$$f(x) = (x_K + \epsilon v) + P_0Nx_0 - \epsilon v = x_K + P_0Nx_0,$$

so $\tau[f]$ = the homotopy class of the Kth iterated Freudentahal suspension homomorphism of $[\tilde{\mu}(a)]$, $E^K[\tilde{\mu}]$, and by our choice of K, $E^K[\tilde{\mu}]$ is stable.

Step 3: Finally to prove our result we note by (5.3.20′) that $\tilde{f}$ (and consequently f) is inessential if and only if the stable homotopy class of $\tilde{\mu}(a)$ is zero.

The utilization of unstable homotopy groups for operator equations The result (5.3.21) just obtained can be sharpened still further in analogy with (5.3.4) by attempting to *eliminate the term "stable"* from the statement of the homotopy criterion for solvability. Indeed in case the index of L, $p = 0$, (5.3.21) implies that the operator equation $Lx + Nx = 0$ is solvable if the homotopy class of the map $\tilde{\mu}(a)$ is nontrivial. On the other hand, for $p > 0$, a short study of the tabulation of the homotopy groups $\pi_{n+p}(S^n)$ as can be found in Toda [1961] shows that much information is often lost in using the *stable* homotopy class of $\tilde{\mu}$ as a criterion for solvability. It turns out that the nontriviality of the homotopy class of $\tilde{\mu}$ does not ensure solvability, but rather the following sharp result holds.

(5.3.22) **Theorem** Suppose that $L \in L(X, Y)$ is a linear Fredholm operator of nonnegative index p and that $N \in C'(X, Y)$ satisfying Hypothesis (A) as well as the property that $\|Nx\|$ and $\|N'(x)\|$ are uniformly bounded. Suppose for some $\epsilon > 0$ the following inequalities hold off some finite-dimensional space $W = \text{Ker } L \oplus V$

(5.3.23) $\quad \|Lw\| \geqslant (c + \epsilon)\|w\|, \qquad \|PN'(u)w\| \leqslant c\|w\|,$

where P is canonical projection of Y onto $L(X/W)$. Then, if dim $V = m$, the equation $Lx + Nx = 0$ is solvable provided the mth iterate of the Freudenthal suspension homomorphism of the homotopy class of $\tilde{\mu}$, $[\tilde{\mu}]$, defined in Section 5.3D above, $E^m[\tilde{\mu}]$, is a nontrivial element if $\pi_{d+m-1}(S^{d.+m-1})$. In particular the equation is solvable provided $[\mu]$ is nontrivial if either $p = 0$, $m = 0$ or more generally E^m is an isomorphism of $\pi_{d-1}(S^{d.-1})$ into $\pi_{m+d-1}(S^{m+d.-1})$.

Proof: The fundamental idea used here is the immediate application of the reduction lemma (5.1.9) to replace the solvability question with a finite-dimensional problem. The finite-dimensional problem is then resolved by means of the properties of the Freudenthal suspension mapping.

To carry out this idea we write $X = W \oplus W_1$ and observe that the reduction lemma (5.1.9) implies that the solvability of $Lu + Nu = 0$ can be reduced to a study of the equation

(5.3.23′) $\quad Lw_0 + P_Y N(w_0 + w_1[w_0]) = 0,$

where P_Y is the canonical projection of $Y = L(W_1) \oplus Y_0 \to Y_0$. Moreover the reduction lemma and the uniform boundedness of Nu over Y implies that $\|w_1(w_0)\|$ is also uniformly bounded over x. Now we study the system (5.3.23′) by once more decomposing it into two parts, one part on Ker L and the other part on V. Thus writing $W = \text{Ker } L \oplus V$ and $w_0 \in W$ as

$w_0 = x_0 + v$, the left-hand side of (5.3.23′) can be written as the mapping $\tilde{g}(w) = (Lv + P_v N(w), P_0 N(w_0 + w_1(w_0)))$. Next we observe that on a sphere of sufficiently large radius R in W, $g(w)$ can be homotopically deformed to the mapping $g_0(w) = (Lv, P_0 N(x_0))$ by the homotopy

$$h(w, t) = \big(Lv + tP_v N(w), P_0 N(x_0 + t\{w_0 + w_1[w_0]\})\big).$$

Indeed, on the sphere ∂D_R of radius R (chosen sufficiently large), the uniform boundedness of N and Hypothesis (A) imply once again that if $h(x, t) = 0$, both $\|v\|$ and $\|x_0\|$ must be small, so that $h(x, t) \neq 0$ on ∂D_R. Finally, we observe that the essentialness of homotopy class of $g_0(w)$ is unaffected by assuming that L is the identity mapping since L is here regarded as a linear homeomorphism of V onto $L(V)$. Consequently the $g_0(w)$ can be written $g_0(w) = (v, P_0 N(Ra))$ where $a \in \{a \mid a \in \operatorname{Ker} L, \|a\| = 1\}$. Thus the homotopy class of the associated normalized map $\tilde{g}_0 = g_0/|g_0|, [g_0]$ coincides with the mth iterated Freudenthal suspension of $\tilde{\mu}$, $E^m[\tilde{\mu}]$. Consequently the desired result is established.

Remarks:

(a) The result (5.3.22) *does not require the compactness of N* for its validity. This is another significant improvement over (5.3.21).

(b) The result (5.3.22) is sharp in the sense that it is not difficult to construct two examples, (i) a nonsolvable equation $Lu + Nu = 0$, where L and N satisfy the hypothesis of Theorem (5.3.21) and the homotopy class of $[\tilde{\mu}]$ is nontrivial, and as well another (ii) a solvable equation with the *stable* homotopy class of $[\tilde{\mu}]$ trivial, but the homotopy class $[\tilde{\mu}]$ itself nontrivial. Here we sketch the fundamental ideas involved in (i) choosing the Hilbert space H to be the sequence space l_2 with the standard inner product and $L + N \in M(H, H)$. Actually we seek the desired example in the simplest possible case index $L = 1$, and attempt to exploit the interesting homotopy facts that $\pi_3(S^2) = Z$, while $\pi_4(S^3) = Z_2$ where the Freudenthal suspension of a generator α of $\pi_3(S^2)$, $E[\alpha] \neq 0$ while $E[2\alpha] = 0$. Thus if the mapping N is so chosen that the homotopy class of the associated mapping $\tilde{\mu}$, $[\tilde{\mu}] = [2\alpha] \in \pi_3(S^2)$, then $[\mu] \neq 0$ while $E^K[\tilde{\mu}] = 0$ for $K > 0$. To be more precise, we let $x = (x_1, x_2, \ldots)$ denote a typical element of l_2 and set $Lx = (0, 0, 0, x_5, x_6, x_7, \ldots)$ so that dim Ker $L = 4$, dim coker $L = 3$ and index $L = 1$. To define N we note that in terms of the complex numbers Z_1, Z_2 a representative $h(Z_1, Z_2)$ of $[2\alpha]$ is given by $h(Z_1, Z_2) = (2Z_1^2|Z_1|^{-1}Z_2, 1 - 2|Z_2|^2)$ and a representative $\phi = (\phi_1, \phi_2, \phi_3, \phi_4)$ of $E[2\alpha]$ is given for real X_5 by $\phi_E(Z_1, Z_2, X_5) = (h(Z_1, Z_2), 2X_5|Z_2|)$. Moreover, since ϕ_E is inessential, it is has nonzero extension to the interior of S^5 which we denote by $\tilde{\phi} = (\tilde{\phi}_1, \tilde{\phi}_2, \tilde{\phi}_3, \tilde{\phi}_4)$. Moreover, we extend $\tilde{\phi}$ to $\mathbb{R}^4$ by setting $\tilde{\phi}(Ra) = \phi_E(a)$ for $R \geq 1$. Maintaining the notation of (5.3.21) for generic $u \in l_2$, we set $u = ra + x$, $ra \in \operatorname{Ker} L$, $\|a\| = 1$, $x \perp \operatorname{Ker} L$, and define $N_i(Ra + x) = \tilde{\phi}_i(a)$ $(i = 1, 2, 3)$. $N_4(Ra + x)_1 = \tilde{\phi}_4(a) - a_5$ are $N_i(u) = 0$, $i \geq 5$. One easily verifies that $N = (N_1, N_2, N_3, N_4, \ldots)$ satisfies Hypothesis (A), and as mentioned above one easily shows that the mapping associated with N is such that $[\tilde{\mu}] \neq 0$, while the stable homotopy class of $E[\tilde{\mu}] = 0$. Finally, one easily shows that the equation $Lu + Nu = 0$ is not solvable in l_2. Indeed $Lu + Nu \neq 0$ by the construction of $\tilde{\phi}$. On the other hand, if $R > 1$, the fourth coordinate of $Lu + Nu$ is $a_5(2|Z_2| + R - 1) > 0$ for $a_5 \neq 0$; whereas if $a_5 = 0$, $\tilde{\phi} = (\phi_1, \phi_2, \phi_3, 0) \neq 0$ for $(a_1, a_2, a_3, a_4) \in S^3$.

(c) Operators $L + N$ as discussed in (5.3.21) arise by expanding a nonlinear Fredholm operator about a singular point.

5.3E Generalized degree for C^2 proper Fredholm operators of index zero

If the mapping $f(x) - p$ belongs to $\mathcal{C}_I^0(\partial D, X)$ and is smooth (C^2 say) on the bounded domain D of X, then its Leray–Schauder degree can be defined by differential techniques. More precisely, assuming $f(x) \neq p$ for $x \in \partial D$, we shall show that the Leray–Schauder degree $d(f, p, D)$ can be computed as follows:

Step 1: Suppose that on the set σ_p of all solutions of $f(x) = p$ in D, $f'(x)$ is invertible. Then by the inverse function theorem and the fact that f is proper on D, σ_p is finite; and we set

(5.3.24) $$d(f, p, D) = \sum_{x \in \sigma_p} d(f'(x), 0, D).$$

Step 2: If $f'(x)$ is not invertible on σ_p, by (3.1.45), we may find a sequence $p_n \to p$ in X so that $f(x) \neq p_n$ on ∂D, and on the set $\sigma_{p_n} = \{x \mid x \in D, f(x) = p_n\}$, $f'(x)$ is invertible. We then set $d(f, p, D) = \lim_{n\to\infty} d(f, p_n, D)$.

(5.3.25) The definition of $d(f, p, D)$ just given coincides with the definition given in Section 5.3C.

Proof: If $f'(x)$ is invertible on σ_p, the two definitions certainly coincide by virtue of the properties of Leray–Schauder degree. Otherwise, we observe that since the Leray–Schauder degree $d(f, p, D)$ is continuous in p and $f(x) \neq p$ on ∂D, $\lim_{n\to\infty} d(f, p_n, D)$ certainly exists as $p_n \to p$ and equals $d(f, p, D)$.

The method of defining the degree of smooth mappings described above clearly applies to a very broad class of (nonlinear) Fredholm operators. However, in any such extension of the Leray–Schauder degree, if homotopies through proper Fredholm operators of fixed index are allowed, the crucial property of homotopy invariance cannot be preserved without additional restriction. For example, Kuiper proved that the group of linear invertible operators defined on a separable infinite-dimensional Hilbert space H is contractible. Thus any two invertible operators L_1 and L_2 defined on H are homotopic on the unit sphere $\{\|x\|_H = 1\}$ through invertible linear Fredholm operators (of index zero), even though with any definition of the type (5.3.24), L_1 and L_2 may have different signs.

Hence, for proper smooth Fredholm operators of index zero defined on D, we shall define a homotopy invariant (mod 2) degree in the following manner: Let f be a proper Fredholm operator of index zero and class C^2

on a bounded domain D with values in a Banach space Y, and suppose $f(x) \neq p$ for $x \in \partial D$.

Step 1′: Suppose that at each point of the set $\sigma_p = \{x \mid x \in D, f(x) = p\}$ the mapping f is regular (i.e., $f'(x)$ is a surjective linear mapping of $X \to Y$), then the properness of f ensures that the set σ_p is compact, while the fact that $f'(x)$ is of index xero and surjective implies that $f'(x)$ is invertible and hence that $f(x)$ is a local homeomorphism on σ_p, by the inverse function theorem. Consequently, σ_p is finite, and we set the generalized degree $d_g(f, p, D)$ equal to the parity of the set σ_p.

Step 2′: If at some point of σ_p the mapping f is not regular, by (3.1.45), we may find a sequence $p_n \to p$ in Y such that $f(x) \neq p_n$ for $x \in \partial D$ and on $\sigma_{p_n} = \{x \mid x \in D, f(x) = p_n\}$ f is regular. Then we set $d_g(f, p, D) = \lim_{n\to\infty} d_g(f, p_n, D)$. Of course, the definition just given is sensible only if $d_g(f, p, D)$ is independent of the sequence p_n and if the relevant limit exists. In fact, we prove

(5.3.26) The function $d_g(f, p, D)$ discussed in steps 1′, 2′ is well defined.

Proof: It suffices to show that if p is a regular value for f and $f(x) \neq p$ for $x \in \partial D$, then the number of points in $f^{-1}(p) \cap D$ is a locally constant function in $C^2(D) \cap C(\bar{D})$. For then, if there are two sequences of regular values for y, $\{p_n\}$ and $\{q_n\}$, both tending to p in Y, for n sufficiently large, $d_g(f, p_n, D) = d_g(f, q_n, D)$. In addition, the sequence of integers $d_g(f, p_n, D)$ stabilizes. Consequently, the definition of $d_g(f, p, D)$ in step 2′ will be justified.

We prove a slightly more general result that if $f(x) \neq p$ for $x \in \partial D$, p is a regular value for f, and g is sufficiently close to f in $C^1(D) \cap C(\partial D)$, then the number of points in $f^{-1}(p) \cap D$ is equal to the number of points in $g^{-1}(p) \cap D$. As discussed in step 1′ above $f^{-1}(p) \cap D$ contains a finite number of points $x_1, \ldots, x_k$ (say). Let O_i $(i = 1, \ldots, k)$ be a family of small pairwise disjoint open neighborhoods with $x_i \in O_i$. Then $f(\bar{D} - \cup_{i=1}^k O_i)$ does not contain p; and for g sufficiently close to f, the properness of f implies that $g(\bar{D} - \cup_{i=1}^k O_i)$ also does not contain p. Since $f'(x_i)$ is a surjective linear Fredholm operator of index zero $(i = 1, \ldots, k)$, $f'(x_i)$ is a linear homeomorphism for each $i = 1, \ldots, k$. The inverse function theorem then implies that g is a diffeomorphism of O_i onto a neighborhood of p. Thus there is exactly one point $z_i \in O_i$ with $g(z_i) = p$. This means that the number of points in $f^{-1}(p) \cap D$ is the same as the number of points in $g^{-1}(p) \cap D$.

Now we show that the function $d(f, p, D)$ has the crucial properties of degree.

(5.3.27) Suppose that f is a C^2 proper Fredholm mapping of index zero defined on D (a convex open subset of X) with $f(x) \neq p$ on ∂D. Then:

(i) $d_g(f, p, D) \neq 0$ implies that the equation $f(x) = p$ has a solution in D (so if $f(x) \neq p$ in D, $d_g(f, p, D) = 0$);

(ii) $d_g(f, p, D)$ is invariant under proper C^2 homotopies $h(x, t)$ that are Fredholm operators of index zero, with $h(x, t) \neq p$ for $x \in \partial D$, $t \in [0, 1]$;

(iii) $d_g(f, p, D)$ is continuous in p, and $f \in C^2$ and depends only on the component in $Y - f(\partial D)$ that contains p.

(iv) If D is a ball with center at the origin, and f is odd, $d(f, 0, D) \neq 0$.

Proof: (i): If $d(f, p, D) \neq 0$, by definition, there is a sequence of points $p_n \to p$ and $x_n \in D$ such that $f(x_n) = p_n$. Since f is proper on $\bar{D}$, $\{x_n\}$ has a convergent subsequence with limit $\bar{x}$. Clearly the continuity of f implies that $f(\bar{x}) = p$ and $\bar{x} \in D$ since $f(x) \neq p$ for $x \in \partial D$. If $f(x) \neq p$ in $\bar{D}$, then $d_g(f, p, D) = 0$, by definition.

(ii): Suppose $h(x, t)$ is a C^2 proper Fredholm operator of index zero defined on $\bar{D} \times [0, 1]$ joining f and g, such that $h(x, t) \neq p$ for $x \in \partial D$ and $t \in [0, 1]$, and p is a regular value for $h(x, t)$. Then $h^{-1}(p)$ is a compact one-dimensional manifold[1] with boundary equal to $(f^{-1}(p), 0) \cup (g^{-1}(p), 1)$; i.e., the number of points in $f^{-1}(p)$ (denoted $\#(f^{-1}(p))$) and in $g^{-1}(p)$ (denoted $\#(g^{-1}(p))$). Since the boundary of a compact one-dimensional manifold has an even number of points,

$$\#(f^{-1}(p)) = \#(g^{-1}(p)) \pmod 2. \tag{5.3.28}$$

Suppose now that p is regular for f and g but not for h, then by the argument in the proof of (5.3.26), there is a neighborhood V of p such that

$$\#(f^{-1}(p')) = \#(f^{-1}(p)); \; \#(g^{-1}(p')) = \#(g^{-1}(p)) \qquad \text{for all} \quad p' \in V.$$

By (3.1.45), there is a regular value $\tilde{p}$ for $h(x, t)$ in V, and since (5.3.28) holds for $\tilde{p}$ by our first argument, it holds also for p.

Finally, if p is not regular for either f or g, by (3.1.45), there is a sequence $p_n \to p$ that will be regular for both and such that $h(x, t) \neq p_n$ on $\partial D \times [0, 1]$. Consequently, by the definition of $d(f, p, D)$ and the above paragraph,

$$d_g(f, p, D) = d_g(f, p_n, D) = d_g(g, p_n, D) = d_g(g, p, D).$$

(iii)–(iv): These are immediate consequences of (ii), as in Section 5.3C.

[1] For fixed t, $h(x, t)$ is a Fredholm mapping of index 0. Since $h(x, t)$ is C^2 and proper on $D \times [0, 1]$, $h(x, t)$ is Fredholm of index 1, and so at a regular value p of $h(x, t)$, $\dim h^{-1}(p) = 1$.

5.4 Homotopy and Mapping Properties of Nonlinear Operators

In this section, we shall derive general mapping properties of a nonlinear operator $f \in C(X, Y)$ from facts about (a) (generalized) degree functions relative to various bounded domains $D \subset X$, and more generally (b) the "essentialness" of f relative to D. Unless otherwise stated, we shall restrict the class of admissible mappings Δ discussed here by assuming that if D is a convex bounded domain of X with $f(x) \neq p$ on ∂D, then there is a degree function $\tilde{d}(f, p, D)$ (with values in Z or Z_2) such that:

(i) $\tilde{d}(f, p, D) \neq 0$ implies $p \in f(D)$;

(ii) $\tilde{d}(f, p, D)$ is invariant under admissable compact homotopies; and

(iii) if D is a ball with center at the origin and f is odd, then f is essential relative to D and $\tilde{d}(f, 0, D) \neq 0$.

By our discussion of the preceding section, Δ includes compact perturbations of the identity and linear Fredholm operators of nonnegative index, as well as, by Section 5.3E, compact perturbations of proper C^2 Fredholm operators of index zero.

5.4A Surjectivity properties

We first prove

(5.4.1) **Theorem** Suppose $f \in C(X, Y) \cap \Delta$ is proper and there is some point $p_* \in Y$ such that $d(f, p_*, \Sigma) \neq 0$ whenever Σ is an open ball centered at the origin containing $f^{-1}(p_*)$. Then f is surjective.

Proof: Let $p \in Y$ and let L be the straight line segment in Y joining p and p_*. Then since f is proper and L is compact, $f^{-1}(L)$ is compact and consequently bounded. Hence we can find a ball $\Sigma_R = \{x \mid \|x\| \leqslant R\}$ with sufficiently large radius R so that $f^{-1}(L) \subset \Sigma_R$. Then if we denote the points of L by $p(t) = tp + (1 - t)p_*$ for $t \in [0, 1]$, the homotopy invariance of degree implies

$$\tilde{d}(f, p, \Sigma_R) = \tilde{d}(f, p(t), \Sigma_R) = \tilde{d}(f, p_*, \Sigma_R) \neq 0. \tag{5.4.2}$$

Thus the equation $f(x) = p$ has a solution in Σ_R. Consequently f is subjective.

(5.4.3) **Corollary** Let $f \in C(X, Y) \cap \Delta$ be an odd proper mapping, then f is surjective.

Proof: Since $f \in \mathcal{K}$ and is proper, $f^{-1}(0)$ is bounded, so that $d(f, 0, \Sigma_R)$

$\neq 0$ for the ball $\Sigma_R = \{x \mid \|x\| < R\}$ with R so large that $f^{-1}(0) \subset \Sigma_R$. Thus f is surjective by the above Theorem (5.4.2).

(5.4.4) **Corollary** Let $f \in C(X, Y) \cap \Delta$ be a proper complex analytic mapping between complex Banach spaces. Suppose (as is the case for the Leray–Schauder degree) that for such mappings, $\tilde{d}(f, p, D) \neq 0$ whenever D is a ball and $p \in f(D) - f(\partial D)$. Then f is surjective.

Proof: By Theorem (5.4.2) it suffices to find a point $p_* \in Y$ such that $\tilde{d}(f, p_*, \Sigma) \neq 0$ whenever the ball Σ is so large as to contain $f^{-1}(p_*)$. Now let p_* be any point in $f(X)$. Then, by hypothesis, since $p_* \in f(\Sigma) - f(\partial \Sigma)$, $\tilde{d}(f, p_*, \Sigma) \neq 0$. Consequently, f is surjective.

(5.4.5) **Corollary** Suppose C is a compact, asymptotically linear operator defined on a Banach space X with asymptotic derivative C_1. Furthermore, suppose $L \in L(X, Y)$ is a linear Fredholm operator of index zero such that $L + C_1$ is invertible. Then $f = L + C$ is surjective.

Proof: Under the given hypotheses, we first prove that if $f = L + C$, then the inverse image of a bounded set B of Y is bounded in X, and that f is a closed mapping. Then, by repeating exactly the argument given in Theorem (5.3.16(ii)) with $\tilde{d}(f - p, 0, D)$ the Leray–Schauder degree, we prove that f is surjective.

First, we prove that if B is a bounded set in Y, then $f^{-1}(B)$ is bounded in X. Otherwise there would be a sequence $x_n \in X$ with $\|x_n\| \to \infty$ and a number M independent of n such that

$$\|(L + C_1)x_n + (C - C_1)x_n\| \leqslant M. \tag{5.4.6}$$

On the other hand, since $L + C_1$ is invertible, there is a constant $k > 0$ (independent of n) such that $\|(L + C_1)x_n\| \geqslant k\|x_n\|$. Therefore, (5.4.6) implies

$$\|x_n\| \left\{ k - \frac{\|(C - C_1)x_n\|}{\|x_n\|} \right\} \leqslant M.$$

Letting $n \to \infty$, we have the desired contradiction, since C_1 is the asymptotic derivative of C and consequently $\|(C - C_1)x_n\| / \|x_n\| < \frac{1}{2}k$ for n sufficiently large.

Next we prove that f is a closed mapping. Indeed, let C be a closed set in X, $y_n \in f(C)$ with $y_n \to y$ in Y. By the above argument there is a bounded set $\{x_n\}$ such that $f(x_n) = y_n$. Thus after possibly passing to a subsequence, we may suppose that $Cx_n \to z$ (say), so that $\{Lx_n\}$ converges strongly in Y. But, by (1.3.37), some subsequence $\{x_{n_j}\}$ converges to $\bar{x}$ (say). By continuity, $f(\bar{x}) = y$. Thus f has closed range.

5.4B Univalence and homeomorphism properties

The degree can be used to prove global univalence results from local data, as in Section 5.1. As an example, we prove the following uniqueness result.

(5.4.7) **Theorem** Let D be a bounded domain of a Banach space X and f a local homeomorphism of $D \to X$. If $f \in \mathcal{C}_I(\partial D, X)$ is a compact perturbation of the identity with $d(f, p, D) = \pm 1$, then the equation $f(x) = p$ has precisely one solution in D.

Proof: First we observe that since f is a local homeomorphism and f is proper on $\bar{D}$, the set $\sigma_p = \{x \mid f(x) = p, x \in D\}$ is discrete and hence finite. By (5.3.24),

$$\pm 1 = d(f, p, D) = \sum_{x \in f^{-1}(p)} d(f, p, O_x), \tag{5.4.8}$$

where the O_x are small disjoint open sets containing x. Thus it suffices to show that (∗) $d(f, f(p(t)), O_{p(t)}) = \text{constant}$ whenever $p(t)$ is a path in D. For then, $d(f, p, O_x) = d(f, p, O_y)$ if $x, y \in \sigma_p$; so that by (5.4.8), the number of points in σ_p is one.

To prove (∗), let $p(t)$ be any path in D. Then for fixed T, let $O_{p(T)}$ be chosen as above and let $p(t_1) \in O_{p(T)}$. Then $d(t) = d(f, f(p(t)), O_{p(T)})$ is a constant function of $t \in [t_1, T]$, by the homotopy invariance of degree. We shall prove that $d(t) = d(f, f(p(t)), O_{p(t)})$ is a constant function of $t \in [t_1, T]$. To this end, let $S_{p(t_1)}$ be an open ball about $p(t_1)$ contained in $O_{p(t_1)} \cap O_{p(T)}$. Then

$$d(f, f(p(t_1)), O_{p(t_1)}) = d(f, f(p(t_1)), S_{p(t_1)}) = d(fp(t_1), O_{p(T)}). \tag{5.4.9}$$

Thus provided $|t_1 - T|$ is sufficiently small to ensure that $p(t_1) \subset O_{p(T)}$, (5.4.9) implies that $d(t)$ is constant in $[t_1, T]$. Hence the set $\{t \mid d(t) = d(1)\}$ is open and closed in $[0, 1]$, and therefore $d(t)$ is a constant function of $t \in [0, 1]$, as required.

(5.4.10) **Corollary** Suppose the hypotheses of (5.4.7) are satisfied and that $f(D) \cap f(\partial D) = \varnothing$. Then f is a univalent mapping of D onto $f(D)$.

Proof: To prove the result, we note that $f(D) \subset (X - f(\partial D))$, so that $d(f, p', D)$ is defined for $p' \in f(D)$ and is constant since $f(D)$ is arcwise connected. Thus by hypothesis $d(f, p', D) = \pm 1$; and by Theorem (5.4.7) the equation $f(x) = p'$ has exactly one solution. Therefore f is univalent on $f(D)$.

In view of the foregoing, it is natural to ask whether a one-to-one mapping f of a bounded open set D onto $f(D)$ is indeed a homeomorphism. As in the finite-dimensional case, any such result requires a considerable proof. Actually we prove

(5.4.11) **Invariance of Domain Theorem** Let $f \in C(D, Y) \cap \Delta$ be a compact perturbation of a linear mapping L that is a Fredholm operator of zero index, where D is an open subset of the Banach space X. If f is a one-to-one mapping of D onto $f(D)$, then f is an open mapping and hence a homeomorphism of D onto $f(D)$.

Proof: We prove that an interior point x_0 of D is mapped by f into an interior point of $f(D)$ in two steps. First, we show that under the given hypotheses, if the mapping $f(x)$ is essential in a small ball Σ about x_0, then for some $\epsilon > 0$ $f(\Sigma)$ contains a sphere of radius ϵ about $f(x_0)$. Secondly, we show that under the hypotheses of the theorem, $f(x) - f(x_0)$ is essential in Σ by proving that f is homotopic to an odd mapping $\tilde{g} \in \Delta$ and by using property (iii) defining Δ.

Without loss of generality, suppose x_0 is the origin and the radius of Σ is one. Then to complete the first step of the proof just outlined, we observe that $f(\partial\Sigma)$ is a closed set since f is proper on bounded subsets of D, so that the distance $\epsilon = d(\tilde{f}(\partial\Sigma), 0) > 0$ since $\tilde{f}(x) = f(x) - f(0) \neq 0$ on $\partial\Sigma$. Now let $\|y - f(0)\| < \epsilon$, then we show that $g(x) = f(x) - y$ and $\tilde{f}(x)$ are compactly homotopic on $\partial\Sigma$, by setting $h(x, t) = f(x) - ty - (1 - t)f(0)$ for $t \in [0, 1]$. Then on $\partial\Sigma$,

$$\|h(x, t)\| = \|f(x) - f(0)\| + t\|f(0) - ty\| > \epsilon - t\epsilon \geqslant 0.$$

By (5.3.3), since $\tilde{f}(x)$ is essential on $\partial\Sigma$ by hypothesis, g is also essential on $\partial\Sigma$, and so $f(x) = y$ has a solution in Σ for $\|y - f(0)\| < \epsilon$. Consequently $f(\Sigma)$ covers an open ϵ-sphere about $f(0)$, and the first step is completed.

Next we show that under the hypotheses, the mapping $\tilde{f}(x) = f(x) - f(0)$ is essential relative to Σ. We do this by showing that $\tilde{f}$ is compactly homotopic to an odd mapping $\tilde{g} \in \mathcal{K}$.

Indeed, if $\tilde{f} = L + C$ with C compact, let $h(x, t) = Lx + \{C(x/(1 + t)) - C(-tx/(1 + t))\}$. Clearly $h(x, t)$ is the desired compact homotopy since on $\partial\Sigma$, if $h(x_0, t_0) = 0$ for $\|x_0\| = 1$ and some $t_0 \in [0, 1]$, then $f(x/(1 + t)) = f(-tx/(1 + t))$ which is not possible since f is one-to-one on Σ. Now $h(x, t)$ joins $\tilde{f}$ and the odd mapping $h(x, 1) = Lx - \{C(x/2) - C(-x/2)\}$, which is essential on Σ since $h(x, 1) \in \mathcal{K}$. Thus f is essential on Σ, as required.

5.4C Fixed point theorems

As was mentioned in Chapter 3, it is often important to give precise conditions under which a mapping f of a Banach space X into itself possesses a fixed point. By virtue of example (5.3.1), the direct extension of the Brouwer fixed point theorem (1.6.4), based only on the continuity of f, is false. Consequently, it is natural to attempt to solve the equation $x = f(x)$ by supposing that the mapping $I - f \in \Delta$. As a first result we prove the following version of Schauder's fixed point theorem (2.4.3).

(5.4.12) (Rothe) Let f be a compact mapping defined on the closed unit ball $\bar{\Sigma} = \{x \mid \|x\| \leqslant 1\}$ of a Banach space X. Suppose f maps $\partial\Sigma = \{x \mid \|x\| = 1\}$ into $\bar{\Sigma}$. Then f has a fixed point in $\bar{\Sigma}$.

Proof: Suppose not. Then, by virtue of the homotopy invariance of the Leray–Schauder degree, since the compact homotopy $h(x, t) = I - tf \neq 0$ for $t \in [0, 1]$ and $x \in \partial\Sigma$,

$$d(I - f, 0, \Sigma) = d(h(x, t), 0, \Sigma) = d(I, 0, \Sigma) = 1.$$

Consequently, the equation $x = f(x)$ has a solution in Σ, contradicting the fact that f has no fixed point in $\bar{\Sigma}$.

For complex analytic mappings f, (5.4.12) can be considerably sharpened, as follows.

(5.4.13) **Corollary**[1] Suppose that f is a compact complex analytic mapping defined on the closed unit ball $\bar{\Sigma}_1$ of a complex Banach space X. In addition, suppose that f maps $\partial\Sigma_1 = \{x \mid \|x\| = 1\}$ into the interior of Σ_1. Then f has one and only one fixed point in Σ_1.

Proof: The proof of (5.4.12) above shows that $d(I - f, 0, \Sigma_1) = 1$. Thus the result (5.4.7) implies that the fixed point x_0 of f must be unique.

The argument just given can be easily extended to prove the following *a priori bound principle*.

(5.4.14) **Theorem** Let $f(x, t)$ be a one-parameter family of compact operators defined on a Banach space X for $t \in [0, 1]$, with $f(x, t)$ uniformly continuous in t for fixed $x \in X$. Furthermore, suppose that *every* solution of $x = f(x, t)$ for some $t \in [0, 1]$, is contained in the fixed open ball $\Sigma = \{x \mid \|x\| < M\}$. Then, assuming $f(x, 0) \equiv 0$, the compact operator $f(x, 1)$ has a fixed point $x \in \Sigma$.

Proof: Since $f(x, t)$ is compact for fixed t on $\bar{\Sigma}$ and uniformly continuous in t for fixed x, $f(x, t)$ is compact on $X \times [0, 1]$. Furthermore, $h(x, t) = x - f(x, t)$ is a compact homotopy on $\partial\Sigma = \{x \mid \|x\| = M\}$, since by assumption, $x \neq f(x, t)$ for $x \in \partial\Sigma$ and $t \in [0, 1]$. Thus by the homotopy invariance of degree, since by hypothesis the equation $x = f(t, x)$ has no solutions on $\partial\Sigma$ for any $t \in [0, 1]$,

$$d(x - f(x, 1), 0, \Sigma) = d(h(x, t), 0, \Sigma) = d(I, 0, \Sigma) = 1.$$

[1] Earle and Hamilton have shown that the compactness hypothesis in this result may be removed.

Consequently $f(x, 1)$ has a fixed point in Σ.

In the same circle of ideas, we prove

(5.4.15) Let f be a compact mapping of the closed unit sphere $\overline{\Sigma}_1 = \{x \mid \|x\| \leqslant 1\}$ into the Banach space X such that the mapping $g = I - f$ satisfies

(5.4.16) $\qquad g(x) \neq \beta g(-x) \qquad$ for any $\beta > 0$ and $x \in \partial \Sigma_1$.

Then f has a fixed point in $\overline{\Sigma}_1$.

Proof: Again, suppose f has no fixed point in $\overline{\Sigma}_1$. Then setting

$$g_t(x) = x - \frac{1}{1+t}\{f(x) - tf(-x)\}, \qquad t \in [0, 1],$$

we note that the hypothesis (5.4.16) above implies that $g_t(x) \neq 0$ for $x \in \Sigma_1$, and that $g_0(x) = g$. Thus $d(g_t, 0, \Sigma_1)$ for $t \in [0, 1]$ is defined, and by virtue of the odd parity property of g_1 of the Leray–Schauder degree, $d(g, 0, \Sigma_1) =$ odd integer. Consequently, by the homotopy invariance of degree $d(g, 0, \Sigma_1) = d(g_1, 0, \Sigma_1) \neq 0$, $g(x) = 0$ is solvable in Σ_1, so that f has a fixed point in Σ_1. This fact, however, contradicts the assumption that f has no fixed points in $\overline{\Sigma}_1$, and so the proof is completed.

Another interesting result in this connection is the following analogue of (5.4.12).

(5.4.17) Let f be a compact mapping defined on a bounded domain D, such that ∂D does not contain the origin of a Hilbert space X. Moreover, suppose that

(5.4.18) $\qquad (f(x), x) \leqslant \|x\|^2 \qquad$ for each $x \in \partial D$.

Then f has a fixed point in $\overline{D}$.

Proof: Suppose f has no fixed point in $\overline{D}$, then the mapping $g = I - f \in \mathcal{C}_I(\partial D, X)$ and the Leray–Schauder degree $d(I - f, 0, D) = 0$. Consequently g cannot be compactly homotopic to I on ∂D. Thus for some $\lambda_0 \in (0, 1]$ and $x_0 \in \partial D$, $x_0 = \lambda_0 f(x_0)$. But (5.4.18) implies that $\lambda_0 \geqslant 1$. Hence $\lambda_0 = 1$, and so f has a fixed point on ∂D. This is the desired contradiction, and the theorem is established.

An interesting Banach space analogue of (5.4.17) is

(5.4.19) Let T be a compact mapping in $C(D, X)$, where $D = \{x \mid \|x\| < 1\}$. Furthermore, for each $x \in \partial D$, $\|x - Tx\|^2 \geqslant \|Tx\|^2 - \|x\|^2$. Then T has a fixed point in $\overline{D}$.

Proof: We consider the compact homotopy $h(x, t) = x - tTx$ defined on

$\partial D \times [0, 1]$, and repeat the argument of (5.4.14), assuming T has no fixed point in $\bar{D}$ and obtaining a contradiction by showing that $h(x, t) \neq 0$ on $\partial D \times [0, 1]$. Indeed, if $x_0 \in \partial D$, $t_0 \in (0, 1)$, and $h(x_0, t_0) = 0$, then

$$\|Tx_0\| = \frac{1}{t_0} \quad \text{and} \quad \|x_0 - Tx_0\|^2 = (1 - t_0)^2 \|Tx_0\|^2 = \frac{(1 - t_0)^2}{t_0^2}$$

implying that

$$\|Tx_0\|^2 - \|x_0\|^2 = 1/t_0^2 - 1 = (1 - t_0^2)/t_0^2.$$

Thus the hypothesis of (5.4.19) implies that

$$(1 - t_0)^2 \|Tx_0\|^2 \geqslant (1 - t_0^2) t_0^{-2}$$

$$(1 - t_0) \|Tx_0\|^2 \geqslant (1 + t_0) t_0^{-2} \qquad \text{(since } t_0 \neq 1)$$

$$(1 - t_0)[t_0^2 \|Tx_0\|^2] \geqslant 1 + t_0 \qquad \text{(an impossibility)}.$$

Thus we have obtained the desired contradiction and the result is established.

5.4D Spectral properties and nonlinear eigenvalue problems

Suppose $f(x, \lambda)$ is a one-parameter family of compact perturbations of the identity defined on $\bar{D} \times \mathbb{R}^1$ and depending (uniform) continuously on the real parameter λ, where D is a domain of a Banach space X. Furthermore, suppose that $f(0, \lambda) \equiv 0$. Then the Leray–Schauder degree can be used to great advantage in the study of solutions (x, λ) of the equation $f(x, \lambda) = 0$ other than the obvious "trivial" solutions $(0, \lambda)$. As a simple example, we prove

(5.4.20) **Theorem** Suppose that the one-parameter family $f(x, \lambda)$ satisfies the above restrictions, and that for two distinct values of λ, λ_0, and λ_1, the Leray–Schauder degree is defined and $d(f(x, \lambda_0), 0, D) \neq d(f(x, \lambda_1), 0, D)$. Then the equation $f(x, \lambda) = 0$ has a solution $(\bar{x}, \bar{\lambda})$ with $\bar{x} \in \partial D$ and $\bar{\lambda} \in [\lambda_0, \lambda_1]$.

Proof: Assume that the equation $f(x, \lambda) = 0$ has no solution $(\bar{x}, \bar{\lambda})$ with $\bar{x} \in \partial D$ and $\bar{\lambda} \in [\lambda_0, \lambda_1]$. Then $h(x, t) = f(x, t\lambda_1 + (1 - t)\lambda_0)$ for $t \in [0, 1]$ defines a compact homotopy joining $f(x, \lambda_0)$ and $f(x, \lambda_1)$. By the homotopy invariance of degree, $d(f(x, \lambda_0), 0, D) = d(f(x, \lambda_1), 0, D)$, which contradicts the hypotheses of the theorem. Consequently, $h(x_0, t_0) = 0$ for some $x_0 \in \partial D$ and $t_0 \in [0, 1]$.

As a simple but interesting consequence of (5.4.20) we mention

(5.4.21) **Corollary** Suppose N is a compact, asymptotically linear operator with asymptotic derivative C, defined on a Banach space X. If λ_0^{-1} is an eigenvalue of odd multiplicity of C, then for any $\epsilon > 0$ there is a ball Σ of X such that for every open set D containing Σ, the equation $x = \lambda N x$ has a solution $(\bar{x}, \lambda)$ with $\bar{x} \in \partial D$ and $\lambda \in [\lambda_0 - \epsilon, \lambda_0 + \epsilon]$.

Proof: Let $\epsilon > 0$ be given. Then we shall calculate the Leray–Schauder degree of the operators $I - (\lambda_0 + \epsilon)N$ and $I - (\lambda_0 - \epsilon)N$, at zero, relative to any bounded set D of X that contains a ball $\Sigma_\epsilon = \{x \mid \|x\| \leqslant R_\epsilon\}$ with R_ϵ sufficiently large. We shall show that these degrees are different; so that by Theorem (5.4.20), the equation $x = \lambda N x$ has the type of solution $(\bar{x}, \lambda)$ described above.

To calculate the Leray–Schauder degree $d(I - (\lambda_0 + \epsilon)N, 0, D)$, we shall show that $I - (\lambda_0 + \epsilon)N$ is compactly homotopic on ∂D to the linear operator $L_\epsilon = I - (\lambda_0 + \epsilon)C$, provided $d(\partial D, 0)$ is sufficiently large. Indeed, suppose $\epsilon > 0$ is sufficiently small, then since C is compact and L_ϵ is invertible, there is a constant β (independent of x) such that $\|L_\epsilon x\| \geqslant \beta \|x\|$. Thus, since C is the asymptotic derivative of N, for $t \in [0, 1]$ and $\|x\| \geqslant R_\epsilon$ so large that $\|Nx - Cx\| < \beta \|x\| / 2(|\lambda_0| + 1)$,

$$\|L_\epsilon x - (\lambda_0 + \epsilon)t(Nx - Cx)\| \geqslant \|L_\epsilon x\| - t(\lambda_0 + \epsilon)\|Nx - Cx\|$$
$$\geqslant \{\beta - \tfrac{1}{2}\beta\}\|x\| = \tfrac{1}{2}\beta\|x\| > 0.$$

Hence, by the homotopy invariance of degree and (5.3.16), if D contains Σ_{R_ϵ},

(5.4.22) $$d(I - (\lambda_0 + \epsilon)N, 0, D) = d(L_\epsilon, 0, D) = (-1)^\mu$$

where μ is the number of eigenvalues of C greater than $(\lambda_0 + \epsilon)^{-1}$. Similarly, if $L_{-\epsilon} = I - (\lambda_0 - \epsilon)C$, then

(5.4.23) $$d(I - (\lambda_0 - \epsilon)N, 0, D) = d(L_{-\epsilon}, 0, D) = (-1)^{\mu_1},$$

where μ_1 is the number of eigenvalues of C greater than $(\lambda_0 - \epsilon)^{-1}$. Since the multiplicity of λ_0^{-1} is an odd number, $\mu \neq \mu_1 \pmod 2$, and so the Leray–Schauder degrees of $I - (\lambda_0 \pm \epsilon)N$ on D relative to zero are different, as desired. Thus the corollary is established.

In the same way, we prove

(5.4.24) **Corollary** Suppose D is a bounded open set containing the origin in the Banach space X. Suppose N is a compact operator mapping ∂D into X with X infinite dimensional and

(5.4.25) $$\|Nx\| > 0 \qquad \text{for each} \quad x \in \partial D.$$

Then the equation $x = \lambda Nx$ has a solution $(\bar{x}, \lambda)$ with $\bar{x} \in \partial D$.

Proof: Suppose the corollary is false. Then by (5.4.20), the function $d(\lambda) = d(I - \lambda N, 0, D)$ is defined for all $\lambda \in \mathbb{R}^1$, and is, in fact, a constant function. Clearly, for $\lambda = 0$, $d(\lambda) = 1$ since $d(\lambda) = d(I, 0, D)$ in that case. We shall obtain a contradiction, by using (5.4.25) to show that $d(\lambda) \neq 1$ for some λ of sufficiently large absolute value. (By virtue of (5.3.14), this fact is independent of any compact extension of N into D.)

To this end, observe that the compactness of N implies the existence of a number $\alpha > 0$ such that $\|Nx\| \geq \alpha > 0$ for $x \in \partial D$; while the fact that D is bounded implies that $\|x\| \leq R$ (say). Then for λ sufficiently large, $\lambda > 2R/\alpha$ (say), $\|\lambda N - I\| \geq \lambda\alpha - \|x\| \geq R$. Thus by the definition of the Leray–Schauder degree, there is a finite-dimensional subspace X_n of X and a compact mapping N_n: $D \to X_n$ approximating N such that $d(I - \lambda N, 0, D) = d(I - \lambda N_n, 0, D)$. Moreover, if the restriction of N_n to $D \cap X_n$ is denoted $\tilde{N}_n$, then $d(I - \lambda N_n, 0, D) = d_B(I - \lambda \tilde{N}_n, 0, D \cap X_n)$. Without loss of generality we may suppose that dim X_n is odd, and that on $\partial D \cap X_n$, $\|N_n x\| \geq \frac{1}{2}\alpha$.

Next, we demonstrate (∗) $I \pm \lambda \tilde{N}_n$ is homotopic to $\pm \lambda \tilde{N}_n$ on $\partial D \cap X_n$ (avoiding zero), for $|\lambda|$ sufficiently large. Once (∗) is established, the theorem will be proved. The fact that X_n is odd dimensional implies that (for $\lambda \neq 0$) the Brouwer degrees $d_B(\lambda \tilde{N}_n, 0, D \cap X_n)$ and $d_B(-\lambda \tilde{N}_n, 0, D \cap X_n)$ are either both zero or of opposite sign, so that one of them must be different from unity. Then by the conclusion of the last paragraph, for some $\lambda \in (-\infty, \infty)$, $d(I - \lambda N, 0, D) \neq 1$, as required.

To prove (∗), let $|\beta|$ be sufficiently large and $x \in \partial D \cap X_n$. Then for $t \in [0, 1]$,

$$\|t(x + \beta \tilde{N}_n x) + (1 - t)\beta \tilde{N}_n x\| = \|\beta N_n x + tx\|$$
$$\geq |\beta| \, \|N_n x\| - \|x\| > \tfrac{1}{2}|\beta|\alpha - R > 0.$$

Thus (∗) is established and the theorem is proven.

For complex analytic mappings, we have the following important consequence of (5.3.16(v)).

(5.4.25′) **Corollary** Let D be a bounded domain of a complex Banach space X (with $0 \notin \partial D$), and let $f(x, \lambda)$ be a one-parameter family of complex analytic mappings defined on $\bar{D} \times \mathbb{R}^1$, which are compact on the product of $\bar{D}$ with any bounded interval of $\mathbb{R}^1$, and such that $f(x, 0) \equiv 0$. Suppose, in addition, that $(x_0, \lambda_0) \in D \times \mathbb{R}^1$ is a point of bifurcation for the equation $g(x, \lambda) = x - f(x, \lambda) = 0$. Then the equation $g(x, \lambda) = 0$ has a solution $(\bar{x}, \bar{\lambda})$ with $\bar{x}$ on ∂D and $\bar{\lambda} \in (0, \lambda_0]$.

Proof: Since (x_0, λ_0) is a point of bifurcation of $g(x, \lambda) = 0$, $g_x(x_0, \lambda_0)$ is not invertible; and so by virtue of (5.3.16(v)), if $g(x, \lambda_0) = 0$ has no solutions on ∂D, then $d(g(x, \lambda_0), 0, D) \geq 2$. However, $d(g(x, 0), 0, D) = 0$ or 1 depending on whether or not $0 \in D$. In either case, the conclusion of the corollary follows from Theorem (5.4.20).

We now use the results just established to study two related questions:

(i) spectral problems for the equation $x = f(x, \lambda)$, where $f(0, \lambda) \equiv 0$, in which one studies the "spectrum" σ_p of the solutions $\mathbb{S}$ of $x = f(x, \lambda)$ as λ varies over the real numbers, where $\sigma_p = \{ \mu \mid \mu \in \mathbb{R}^1, (x, \mu) \in \mathbb{S}, x \neq 0\}$;

(ii) continuation problems for $x = f(x, \lambda)$; in which one supposes (x_0, λ_0) is a point of bifurcation (in the sense of Chapter 4) for the equation $x = f(x, \lambda)$, and one studies the component of the closure of the nontrivial solutions $(\bar{x}, \bar{\lambda}) \in \mathbb{S}$ that contains (x_0, λ_0).

As a first result concerning the set σ_p we let $d(Z, Y)$ denote the distance between the sets Z, Y and prove

(5.4.26) **Theorem** Let $f(x, \lambda)$ be a compact operator defined on $X \times (-\infty, \infty)$ with $f(0, \lambda) \equiv 0$, and such that $\|f(x, \lambda)\| \to \infty$ as $\lambda \to \infty$ uniformly on every bounded set Σ of X with $d(\Sigma, 0) > 0$. Suppose that for every open set U containing the origin, the equation $x = f(x, \lambda)$ has a solution $(x(u), \lambda_u)$ with $x(u) \in \partial U$ and $\lambda_u \in \mathbb{R}^1$ such that as $\|x(u)\| \to \infty$, $\lambda_u \to \lambda_\infty$, and as $\|x(u)\| \to 0$, $\lambda_u \to \lambda_0$. Then the equation $x = f(x, \lambda)$ has solutions $(\bar{x}, \bar{\lambda})$ with $\bar{x} \neq 0$ for any $\bar{\lambda} \in (\lambda_0, \lambda_\infty) - \{0\}$, i.e., $\bar{\lambda} \in \sigma_p$.

Proof: Suppose $\mu \in (\lambda_0, \lambda_\infty) - \{0\}$ is not in the set σ_p. Then we shall obtain a contradiction by constructing a bounded open set V containing the origin but such that $g(x, \lambda) = x - f(x, \lambda)$ has no nontrivial solution on ∂V. To this end, let the two components of $\mathbb{R}^1 - \{\mu\}$ be E_∞ and E_0, where $\lambda_\infty \in E_\infty$ and $\lambda_0 \in E_0$. Moreover, suppose $F_\infty = \{x_u \mid x_u = f(x_u, \lambda_u), \lambda_u \in E_\infty\}$ and $F_0 = \{x_u \mid x_u = f(x_u, \lambda_u), \lambda_u \in E_0\} \cup \{0\}$. Clearly the compactness of $f(x, \lambda)$ implies that the disjoint sets F_∞ and F_0 are closed, while $F_0 \cup F_\infty$ contains all the nontrivial solutions of $x = f(x, \lambda)$ mentioned in the theorem. Thus $d(F_\infty, 0) > 0$, while the elements in F_0 are uniformly bounded. Next we can show that $d(F_0, F_\infty) > 0$. Indeed, otherwise there would be sequences $\{x_n\} \in F_\infty$ and $\{y_n\} \in F_0$ that are uniformly bounded away from zero and infinity, but such that $\|x_n - y_n\| \to 0$, while $x_n = f(x_n, \lambda_n)$ and $y_n = f(y_n, \lambda_n')$ for $\lambda_n \in E_\infty$, $\lambda_n' \in E_0$. By hypothesis, we may then suppose that $|\lambda_n|$ and $|\lambda_n'|$ are uniformly bounded, and after possibly passing to subsequences, we may suppose $\lambda_n \to \bar{\lambda}$ and $\lambda_n' \to \bar{\lambda}'$. Hence, again possibly after passing to subsequences, we may assume that $\{x_n\}$, and consequently $\{y_n\}$, converges strongly to $z \neq 0$, and that $z \in F_0 \cap F_\infty$.

This last fact is the desired contradiction, and consequently there is a number $\beta > 0$ with $d(F_0, F_\infty) = \beta$. Now let $O(F_0)$ be the bounded open set obtained as the union of open balls of radius $\frac{1}{2}\beta$ centered at F_0. Then $\partial O(F_0)$ is disjoint from F_0 and F_∞, contradicting the hypothesis of the theorem, so that $\mu \in (\lambda_0, \lambda_\infty) - \{0\}$.

Next we turn to the continuation problem mentioned in Section 4.1 and prove the following global analogue of (4.2.3) for the equation

(5.4.27) $\quad (I - \lambda L)x + g(x, \lambda) = 0.$

(5.4.28) **Theorem** (Rabinowitz) Suppose L is a linear compact operator mapping of a Banach space X into itself with λ_0^{-1} an eigenvalue of odd multiplicity, while $g(x, \lambda)$ defined on a domain $D \times V$ of $X \times \mathbb{R}^1$ is continuous and compact in x, continuous in λ and of higher order in x at the origin, i.e.,

$$\|g(x, \lambda)\| = o(\|x\|) \qquad \text{as} \quad \|x\| \to 0$$

uniformly for bounded λ. Then, if $\bar{C}$ denotes the closure of the component C of nontrivial solutions of (5.4.27) containing $(0, \lambda_0)$, one of the following alternatives hold: either (i) $\bar{C}$ is noncompact in $D \times U$ (so that if $D \times \mathbb{R}^1$ coincides with $X \times \mathbb{R}^1$, $\bar{C}$ is unbounded) or (ii) $\bar{C}$ contains at least one but at most a finite number of points $(0, \lambda_i)$ with λ_i^{-1} an eigenvalue of L distinct from λ_0 and the number of such λ_i of odd multiplicity (*excluding* λ_0) must be odd.

Proof: Suppose $\bar{C}$ is compact in $D \times U$. Then the number of distinct points of the form $(0, \lambda_i)$ (as described in the theorem) must be finite since otherwise the compactness of L would imply the noncompactness of $\bar{C}$. Thus to prove the theorem we need only prove the evenness of the number of points $(0, \lambda_k)$ with λ_k^{-1} an eigenvalue of odd multiplicity for L contained $\bar{C}$.

To this end we choose a bounded open subset Ω of $D \times U$ containing $\bar{C}$ and such that there are no solutions (x, λ) of (5.4.27) on $\partial\Omega$ and so that Ω contains no points of the form $(0, \lambda_k)$ that differ from those of $\bar{C}$. Then to measure the nontrivial solutions of $f(x, \lambda) = (I - \lambda L)x + g(x, \lambda)$ on $\|x\| = \rho$, we consider the Leray–Schauder degree at $(0, 0)$ of the mapping $f_\rho(x, \lambda) = (f(x, \lambda), \|x\|^2 - \rho^2)$ relative to Ω. By our construction of Ω this degree $d_\rho = d(f_\rho, (0, 0), \Omega)$ is defined.

We shall obtain the parity result desired in three simple steps: (i) d_ρ *is independent of* ρ *by the homotopy invariance of degree and is fact equal to* 0 as one can choose ρ so large that $f_\rho(x, \lambda) = 0$ has no solutions; (ii) then by choosing ρ small and showing the fact that *the only contributions to* d_ρ *for* ρ *small came from local contributions* near points of the form $(0, \lambda_k)$. This follows from (5.3.14); and finally (iii) computing $d(f_\rho, (0, 0), \Omega)$ *for* ρ *small,* *which we assert equals the right-hand side of* (*) *below, so together with* (i) *and* (ii) *we find*

$$(*)\quad 0 = \sum_{\lambda_k} \{d(I - (\lambda_k - \epsilon)L, 0, \|x\| < \rho) - d(I - (\lambda_k + \epsilon)L, 0, \|x\| < \rho)\}$$
$$= \sum_{\lambda_k \text{ (odd)}} \pm 2$$

(where summation occurs only over λ_k of odd multiplicity). From which we conclude the parity of the points λ_k of odd multiplicity is even.

To prove (iii) we compute the contribution to d_ρ near each $(0, \lambda_k)$ $d_\rho(k)$, for small $\epsilon \neq 0$ and small ρ we consider the homotopy

$$h(x, t) = tf_\rho(x, \lambda + \epsilon) + (1 - t)\{I - (\lambda_k + \epsilon)L, \epsilon_0^2 - \epsilon^2\}$$

relative to $(0, 0)$ on the set $\Sigma = \{(x, \epsilon) \mid \|x\|^2 + \epsilon^2 < \rho^2 + \epsilon_0^2\}$. Clearly $h(x, t) \neq 0$ on $\partial\Sigma$, provided (r, ϵ) is chosen sufficiently small for in such a case $\epsilon = \pm\epsilon_0$ and so $x = 0$. Thus by the homotopy invariance of degree $d_\rho(k) = d(I - (\lambda_k + \epsilon)L, \epsilon_0^2 - \epsilon^2\}$. To compute this latter degree we use (5.3.16) since the only solutions of $h(x, 0) = 0$ are $\rho = 0$, $\epsilon = \pm\epsilon_0$, and the Fréchet derivative of $h(x, 1)$ at $(0, \epsilon)$ is given by

$$h'(0, \epsilon)[\tilde{x}, \tilde{\epsilon}] = ((I - \lambda L)x, -2\epsilon\tilde{\epsilon}).$$

Thus at $\epsilon = \pm\epsilon_0$ the local index is $-d(I - (\lambda_k + \epsilon)L, 0, \|x\| < \rho)$ for $\epsilon > 0$, while at $\lambda = (\lambda_k - \epsilon)$ is $d(I - (\lambda_k - \epsilon)L, 0, \|x\| < \rho)$. Consequently the fact (iii) follows from (5.3.25) and the additive property of Leray–Schauder degree.

5.4E Necessary and sufficient conditions for solvability and its consequences

In the case of operator equations of the form $Lu + Nu = f$, where $L \in L(X, Y)$ is a linear Fredholm operator of nonnegative index p mapping a Banach space X into a Banach space Y, and N is a compact mapping that satisfies the hypotheses of (5.3.21), we can sharpen our previous results. Indeed we now prove (a) necessary and sufficient conditions for solvability and (b) the openness of the range of $L + N$. Indeed we begin with the index zero case with dim Ker $L > 0$ and demonstrate:

(5.4.29) **Theorem** Suppose L is a linear Fredholm self-adjoint operator mapping a Hilbert space H into itself, and that Nu is a uniformly bounded compact continuous mapping of H into itself, such that the following limit exists uniformly for $\|x\|$ uniformly bounded, $\phi(a) = \lim_{r\to\infty} P_0N(ra + x)$, $a \in \text{Ker } L \cap \{\|x\| = 1\}$. Moreover suppose for all positive r,

(5.4.30) $\quad (N(ra + x), a) < (\phi(a), a) \qquad x \perp \text{Ker } L.$

Then (i) a necessary and sufficient condition for the solvability of the equation $Lu + Nu = f$ is that $(f, a) < (\phi(a), a)$, and (ii) the mapping $L + N$ has open range.

Proof: The necessity of the condition $(f, a) < (\phi(a), a)$ follows immediately from (5.4.30) by taking the inner product of $Lu + Nu = f$ with $a \in \operatorname{Ker} L$ and using the self-adjointness of L. To derive the sufficiency of the condition we first observe if it is satisfied, Hypothesis (A) of (5.3.21) is satisfied for the operator $L + N$ via (5.4.30), and moreover on a sufficiently large sphere $\partial \Sigma_r$ of radius r in Ker L, the Brouwer degree $d_B(\phi(a), f, \Sigma_r) = 1$. Consequently the criterion of (5.3.21) implies $f \in \operatorname{Range}(L + N)$.

To derive the openness of the range of $L + N$ we show that if $f_0 \in \operatorname{Range}(L + N)$, the range of $L + N$ also contains a ball of positive radius about f. This end we argue as follows. First if $f - f_0 \in (\operatorname{Ker} L)^{\perp}$ the solvability of the equation $Lu + Nu = f$ is an immediate consequence of part (i) just established. On the other hand, if the projection of $f - f_0$ onto Ker L is sufficiently small in norm, f is also in the range of $L + N$, by virtue of the strict inequality of the necessary and sufficient condition (as stated) and the finiteness of dim Ker L. Thus (5.4.29) is established.

As an application of this result we give an alternative proof of the solvability of the partial differential equation already discussed in Theorem (5.1.8), concerning negative constant Gaussian curvature metrics on compact two-manifolds M. The equation involved can be written:

$$(5.4.31) \qquad \Delta u - e^{2u} = K(x), \qquad \text{with} \quad \operatorname{vol}(\mathfrak{M}, g) = 1$$

where Δ denotes the Laplace–Beltrami operator defined on the manifold $(\mathfrak{M}, g)$. Clearly the hypotheses of (5.4.29) do not seem to be satisfied by (5.4.31) since the nonlinear term $\exp 2u$ is not uniformly bounded in any Banach space so far mentioned. To overcome this difficulty we apply the maximum principle for Δ on $\mathfrak{M}$, which states that if $u(x)$ is a smooth solution of (5.4.31), then at a positive maximum x_0 of $u(x)$, $\Delta u(x_0) = \exp 2u(x_0) + K(x_0) < 0$ so that $u(x_0) \leqslant c_0$ (c_0 an absolute constant). This justifies our replacing the equation (5.3.31) by

$$(5.4.31') \qquad \Delta u - f_0(u) = K(x),$$

where $f_0(u) = e^{2u}$ for $u \leqslant c_0$ and $f_0(u)$ strictly increasing with limit $f(\infty)$ for $u \geqslant c_0$. Now $f_0(u)$ is uniformly bounded and we can give an alternate proof for the result (5.1.8).

A necessary and sufficient condition for the solvability of (5.4.31) *is that* $\bar{K} < 0$, *i.e., the mean value of* $K(x)$ *over* (M, g) *be negative.*

Proof: Clearly it suffices to consider only the truncated equation

(5.4.31′). This equation can clearly be written in the Sobolev space $W_{1,2}(\mathfrak{M}, g)$ in the form

$$Lu + Nu = -g,$$

where L is the operator naturally associated with Δ, Nu with $f_0(u)$ and g with $K(x)$. Here L is self-adjoint and (5.4.30) is satisfied with

$$(Lu, u) = \int_{\mathfrak{M}} |\nabla u|^2, \quad (Nu, v) = \int_{\mathfrak{M}} f_0(u)v, \quad \int_{\mathfrak{M}} K(x)v = (g, v),$$

Ker L consists of the constants. Thus (5.4.31′) is transformed into operator form and the result (5.4.29) is applicable. We observe that the solvability criteria of (5.4.29) becomes for $a = -1$

$$(-g, -1) = \int_{\mathfrak{M}} K(x) < \lim_{r \to -\infty} \int_{\mathfrak{M}} f_0(-r)(-1) = 0.$$

Thus the necessary and sufficient conditions for solvability are satisfied provided the constant c_0 in the definition of $f_0(u)$ is chosen sufficiently large so that the inequality for $a = 1$ is automatically satisfied.

Remark: We shall take up a higher dimensional analogue of the equation (5.4.31) again in Chapter 6, where it is easily solved via minimization techniques. Moreover, other examples of elliptic boundary value problems amenable to (5.4.29) are given in Section 5.5E.

We now proceed to the more general case of a mapping $f = L + N$ of a Banach space X into a Banach space Y, where L is a linear Fredholm operator of index $p > 0$.

(5.4.32) **Theorem** Let P denote the canonical projection of Y onto coker L, where $L \in \Phi_p(X, Y)$ as mentioned above. Then if $N \in M(X, Y)$ is compact and uniformly bounded and the following hypotheses are satisfied:

(i) $\lim_{R\to\infty} PN(Ra + x) = \eta(a) \neq 0$ holds uniformly for $x \in (\text{Ker } L)^{\perp}$ and $\|x\|$ uniformly bounded where $a \in \text{Ker } L$ with $\|a\| = 1$;

(ii) $\|PN(Ra + x)\| < \|\eta(a)\|$;

(iii) the *stable homotopy class* of $\eta(a)$, $\lim_{k\to\infty} E^k[\eta(a)] \neq 0$;

then a necessary and sufficient condition for the solvability of the equation $Lu + Nu = f$ is $\|Pf\| < \|\eta(a)\|$ for all a. Moreover, the mapping $L + N$ has open range in Y.

Proof: Suppose $f \in \text{Range}(L + N)$, then for some $u \in X$

$$PN(u) = Pf, \qquad \text{so by (ii)} \qquad \|Pf\| < \|\eta(a)\|.$$

Thus the condition stated in the above theorem is necessary. On the other hand, if this latter condition holds, we note that by virtue of (5.3.21) it suffices to prove that the stable homotopy class of $P[N(Ra) - f] \neq 0$ (for R sufficiently large). But the conditions (i), (ii) imply that the two maps $f_1 = P[N(Ra) - f]$ and $f_2 = P[N(Ra)]$ are homotopic on a large enough sphere (i.e., for R sufficiently large) via the simple homotopy $tf_1 + (1 - t)$ f_2. Thus since the stable homotopy class of $P[N(Ra)] \neq 0$, by virtue of the hypothesis (iii) above, the stable homotopy class of $P(N(Ra) - f) \neq 0$ and the first part of theorem is proved by applying (5.3.21).

The proof that $L + N$ has open range now follows as in (5.4.29) above.

An interesting consequence of (5.4.29) is the finiteness of the number of solutions of the equation $Lu + Nu = f$ for almost all f in the range of $L + N$ in the case index $L = 0$ and (provided N is sufficiently smooth), a corresponding result holds in case index $L \geqslant 0$. In fact we prove

(5.4.32′) **Theorem** Suppose the hypotheses of (5.4.32) hold and in addition X is reflexive while N is compact, then for any regular value $f \in \text{Range}(L + N)$ (i.e., by (3.1.45) apart from a possible set of first Baire category) the number of solutions of $Lu + Nu = f$ is either finite if index $L = 0$ or a compact submanifold of Y of dimension p if index $L = p > 0$, and N is sufficiently smooth.

Proof: We first suppose $f \in \text{Range}(L + N)$ and index $L = 0$, and we shall prove that (5.4.32) implies the fact: (∗) any sequence $\{u_n\}$ such that $\|Lu_n + Nu_n - f\| \to 0$ is uniformly bounded in norm.

Assuming the validity (∗) for the moment, we note that if f is a regular value of $L + N$ the solutions $\tilde{u}$ of $Lu + Nu = f$ are isolated (since $L + N'$ is invertible), while (∗) implies that if the solutions $\{\tilde{u}\}$ of $Lu + Nu = f$ are infinite, they are certainly uniformly bounded. In that case the weak compactness of bounded sets in a reflexive Banach space and the compactness of N imply for some weakly subsequence $\{u_{n_j}\}$ that Lu_{n_j} is strongly convergent. Consequently by the property (1.3.27) of Fredholm operators $\{u_{n_j}\}$ is strongly convergent. This contradicts the fact that the solutions of $Lu + Nu = f$ are isolated. Consequently, these solutions must be finite in number. The case for index $L > 0$ follows from the remark after Smale's theorem (3.1.29).

Finally, we demonstrate the fact (∗), and to this end we suppose the sequence u_n satisfies the condition that $F(u_n) = Lu_n + Nu_n - f$ tend to zero in norm. Then, decomposing u_n in the form $u_n = v_n + z_n$ with $z_n \in \text{Ker } L$ and $v_n \in X_1$, we find

$$\|Lv_n\| \leqslant \|F(u_n)\| + \|Nu_n - f\|.$$

Thus, the uniform boundedness of N implies the uniform boundedness of $\|Lv_n\|$. On the other hand, since L is invertible off Ker L, $\{\|v_n\|\}$ is also uniformly bounded. Thus to verify (*) it suffices to show that $\{\|z_n\|\}$ is uniformly bounded. But this follows by supposing otherwise and obtaining a contradiction. For then, from the fact that $f \in \text{Range}(L + N)$, (5.4.29) implies that if $\|z_n\| \to \infty$, $\|P_0 f\| < \|P_0 N(z_n + v_n)\|$; but since $F(u_n) \to 0$, $P_0(N(z_n + v_n) - f) \to 0$, so that we have the desired contradiction.

5.4F Properties of cone preserving operators

A real Banach space X admits a cone K if K is a closed convex subset of X such that (i) $x \in K$ implies $\alpha x \in K$ for any nonnegative real number α, and (ii) $x \in K$ implies $-x \notin K$, unless $x \equiv 0$. Many of the results proved in earlier subsections can be considerably sharpened for mappings f that map a cone K of X into itself. Such mappings are called cone preserving. As an example of such a sharpening, we prove the following extension of (5.4.24).

(5.4.33) **Theorem** Let D be a bounded open domain (containing the origin) in a Banach space X that admits a cone K. Suppose N is a compact cone preserving mapping such that

(5.4.34) $$\|Nx\| > 0 \quad \text{for} \quad x \in (K \cap \partial D).$$

Then the equation $x = \lambda Nx$ has a solution (x_0, λ_0) such that $\lambda_0 > 0$ and $x_0 \in (\partial D \cap K)$.

Proof: We argue as in (5.4.24), by reducing the problem to a finite-dimensional one. Indeed, if $x = \lambda Nx$ has no solutions as described in the theorem, then by (5.4.34) and the compactness of N, there is a number $\alpha > 0$ such that

(5.4.35) $$\|Nx - tx\| \geq \alpha > 0 \quad \text{for} \quad t > 0 \quad \text{and} \quad x \in \partial D \cap K.$$

Then by (2.4.2), there is a compact operator N_α with odd finite-dimensional range X_k such that $\|N_\alpha x - Nx\| \leq \frac{1}{2}\alpha$ for $x \in \partial D \cap K$. Consequently, by (5.4.34) for $t > 0$ and any $x \in \partial D \cap K$, $\|N_\alpha x - tx\| \geq \frac{1}{2}\alpha$; so that the equation $x = \lambda N_\alpha x$ has no solutions (x, λ) with $x \in \partial D \cap K$ and $\lambda > 0$.

We shall contradict this last statement by providing the theorem for the case when X is an odd finite-dimensional Banach space; for then the equation $x = \lambda N_\alpha x$ will have a solution (x, λ) with $x \in X_k \cap (\partial D \cap K)$ and $\lambda > 0$. In case X has an odd finite dimension, suppose for the moment that N has an extension $\tilde{N}$ onto ∂D such that (i) $\tilde{N}$ maps ∂D onto K and (ii) $\|\tilde{N}x\| > 0$ for $x \in \partial D$. Then by the argument of (5.4.24), the equation $x = \lambda \tilde{N}x$ has a solution (x_0, λ_0) with $x_0 \in \partial D$. We shall prove that $d_B(I - \lambda\tilde{N}, 0, D) = 0$ for λ sufficiently large, so that $\lambda_0 > 0$. Once this fact is established, $x_0 = \lambda_0 \tilde{N} x_0 \in \partial D \cap K$ so that $\tilde{N} x_0 = N x_0$; whence the equation $x = \lambda N x$ has the solution (x_0, λ_0) with $x_0 \in \partial D \cap K$ and $\lambda_0 > 0$.

To prove that $d_{\mathrm{B}}(I - \lambda \tilde{N}, 0, D) = 0$ for λ sufficiently large, we show that the vector field $x - \lambda \tilde{N} x$ omits a direction for λ sufficiently large. Indeed, let $u \in K$ ($u \not\equiv 0$) and suppose that as $\lambda_n \to \infty$ there is a sequence (t_n, x_n) such that

$$(5.4.36) \qquad x_n - \lambda_n \tilde{N} x_n = t_n u \qquad \text{with} \quad x_n \in \partial D \quad \text{and} \quad t_n > 0.$$

Then (after possibly passing to subsequences), we may suppose that $\tilde{N} x_n \to v$; and since $\inf \|\tilde{N} x\| > 0$ for $x \in \partial D$, $v \neq 0$ and $v \in K$. Consequently $x_n/\lambda_n - \tilde{N} x_n \to -v$, but by (5.4.36), $x_n/\lambda_n - \tilde{N} x_n = t_n/\lambda_n u$. This is a contradiction since $-v \not\in K$, while $t_n u/\lambda_n \in K$ for all n.

Finally, we show that the extension $\tilde{N}$ exists, where $N(\partial D \cap K)$ lies in X_k, a k-dimensional linear subspace of X and $\|Nx\| \geqslant \beta > 0$ for $x \in (\partial D \cap K)$. Then $\overline{\text{co}}\ N(\partial D \cap K)$ does not contain 0. Furthermore, by choosing an orthonormal basis in X_k, we may write

$$N(x) = (n_1(x), \ldots, n_k(x)) \qquad \text{for} \quad x \in \partial D \cap K,$$

and by Urysohn's lemma, we may suppose that the functions $n_i(x)$ are extended with preservation of continuity to functions $\tilde{n}_i(x)$ on $\bar{D}$. Now to construct $\tilde{N}$, let $y \in \text{int co}\ N(\partial D \cap K)$ and let $r(x)$ be the continuous retraction of X_k onto $\text{co}\ N(\partial D \cap K)$ defined for $x \not\in \text{co}\ N(\partial D \cap K)$ by constructing the line segment $L(x, y)$ joining x and y in X_k and letting $r(x)$ be the point of intersection $L(x, y) \cap \partial\, \overline{\text{co}}\ N(\partial D \cap K)$. Now we define $\tilde{N} x$ on $\bar{D}$, by setting $\tilde{N} x = P\tilde{n}(x)$, where $\tilde{n}(x) = (\tilde{n}_1(i), \ldots, \tilde{n}_k(x))$. Clearly for $x \in \partial D \cap K$, $\tilde{N} x = Nx$, so that $\tilde{N}$ is a continuous extension of N to $\bar{D}$. In addition, $\tilde{N}(\bar{D}) \subset \overline{\text{co}}\ N(\partial D \cap K) \subset K$ and for $x \in \bar{D}$,

$$\|\tilde{N} x\| \geqslant d(\overline{\text{co}}\ N(\partial D \cap K), 0) > 0.$$

Thus $\tilde{N}$ is the desired extension.

An important consequence of (5.4.33) is the following important theorem of Krasnoselski (1964).

(5.4.37) **Theorem on Monotone Minorants** Suppose N is a compact cone preserving operator defined on the cone K, such that there is a linear cone preserving operator L (i.e., a monotone operator in the sense that it preserves the order relation in K) and a nonzero $x_0 \in K$ such that

$$(5.4.38) \qquad Nx \geqslant Lx \quad \text{and} \quad Lx_0 \geqslant \alpha x_0, \qquad \text{where} \quad \alpha > 0.$$

Then for any bounded domain D (containing 0), the equation $x = \lambda N x$ has solutions (x, λ) with $x \in \partial D \cap K$ and $\lambda > 0$.

Proof: Let $N_\epsilon x = Nx + \epsilon x_0$ for any $\epsilon > 0$. Then N_ϵ is compact, and for

$x \in \partial D \cap K$, $\|N_\epsilon x\| \geqslant \inf_{y \geqslant \epsilon x_0} \|y\| > 0$. Hence by (5.4.33), there is a pair $(\lambda_\epsilon, x_\epsilon)$ satisfying

(5.4.39) $$Nx_\epsilon + \epsilon x_0 = \lambda_\epsilon x_\epsilon, \qquad \text{with } \lambda_\epsilon > 0 \text{ and } x_\epsilon \in \partial D \cap K.$$

Now as $\epsilon \to 0$ (after possibly passing to subsequences), we may suppose that $Nx_\epsilon \to y$ and $\lambda_\epsilon \to \lambda_0$. Clearly $y \in \partial D \cap K$; so that it remains only to prove that $\lambda_0 \neq 0$.

To this end, we first note that (5.4.38) implies

(5.4.40) $$Lx_\epsilon + \epsilon x_0 \leqslant \lambda_\epsilon x_\epsilon \qquad \text{and} \qquad x_\epsilon \geqslant \lambda_\epsilon^{-1} \epsilon x_0.$$

Thus, there is a largest number $t_\epsilon > 0$ such that $x_\epsilon \geqslant t_\epsilon x_0$, implying that $Lx_\epsilon \geqslant t_\epsilon \alpha x_0$. But (5.4.39) implies that $Lx_\epsilon \leqslant \lambda_\epsilon x_\epsilon$. Therefore, $x_\epsilon \geqslant t_\epsilon \alpha \lambda_\epsilon^{-1} x_0$, and by the maximality of t_ϵ, $t_\epsilon \geqslant t_\epsilon \alpha / \lambda_\epsilon$ so that $\lambda_\epsilon \geqslant \alpha > 0$. Consequently, $\lambda_0 > 0$, as required.

5.5 Applications to Nonlinear Boundary Value Problems

The results of Sections 5.3 and 5.4 are of immense value in proving qualitative results on the structure of the solutions of boundary value problems for nonlinear partial and ordinary differential equations. In particular, questions that we shall consider here include (a) existence (or nonexistence), (b) uniqueness (or nonuniqueness), (c) continuous dependence on a parameter, as well as (d) continuation of solutions of problems depending on a parameter.

In general, the following steps are necessary to apply the abstract results proven in the previous sections to concrete problems. First, any parameters implicit in the nonlinear systems should be introduced explicitly by appropriate coordinate transformations. Secondly, suitable Banach spaces X and Y must be chosen so that the differential system under consideration can be represented as a well-defined mapping f defined on a domain in X with range in Y. Next one must prove the basic boundedness, continuity, and differentiability properties of f that are necessary to apply the appropriate degree theory to the problem at hand. Finally, one must prove the analytical estimates necessary to calculate the degree of f.

We begin by considering an analogue of the original problem discussed by Leray and Schauder in their fundamental paper (1934).

5.5A The Dirichlet problem for quasilinear elliptic equations

Let Ω be a bounded domain in $\mathbb{R}^N$ with boundary $\partial\Omega$, and consider the following system of equations defined on $\bar{\Omega}$,

(5.5.1) $$\sum_{|\alpha|+|\beta|=2} A_{\alpha\beta}(x, u, Du)\, D^\alpha D^\beta u + A_0(x, u, Du) = 0 \quad \text{in} \quad \Omega,$$

(5.5.2) $$u|_{\partial\Omega} = g.$$

The classic Dirichlet problem for (5.5.2)–(5.5.3) consists in determining a function $u \in [C^2(\Omega) \cap C(\bar{\Omega})]$ that satisfies (5.5.1)–(5.5.2) in the pointwise sense. The differential operator on the left-hand side of (5.5.2) is elliptic if there is a constant $\mu > 0$ such that $A_{\alpha\beta}(x, y, z)\xi_\alpha\xi_\beta \geqslant \mu|\xi|^2$ for $|y|, |z| \leqslant M$, $x \in \bar{\Omega}$. The examples given in Section 1.2 show that such quasilinear elliptic Dirchlet problems may not be solvable for a variety of reasons, including the shape and size of Ω or the rapidity of growth and sign of $A_0(x, y, z)$.

The question of the solvability of Dirichlet problems of this class was posed by Hilbert in 1900 in his famous address (Hilbert, 1900), and was studied extensively by S. Bernstein thereafter. In $N = 2$, Bernstein attempted to solve (5.5.1)–(5.5.2) by (a) introducing a parameter t explicitly into the system (5.5.1)–(5.5.2) to obtain a one-parameter family of systems P_t so that for $t = 0$, the system P_0 is solvable, while for $t = 1$, the system P_1 coincides with (5.5.1)–(5.5.2); and (b) showing that each P_t is solvable for $t \in (0, 1]$, by continuation. The continuation method was greatly extended in 1934 by Leray and Schauder, who transformed it into a homotopy argument by means of the degree. This approach, however, requires difficult analytic a priori estimates in order to ensure that the degree of the mapping f can be defined. Once these estimates have been established, the basic idea is to apply the a priori bound principle discussed in (5.4.14).

As a simple example, we mention

(5.5.3) **Theorem** Suppose $\partial\Omega$ and g are of class C^3, while the functions $A_{\alpha\beta}(x, y, z)$ and $A_0(x, y, z)$ are C^1 in x, y, z. Then the Dirichlet problem for (5.5.1)–(5.5.2) is solvable provided any solution v_t of the system obtained from (5.5.1) by replacing A_0 by tA_0 and g by tg for $t \in [0, 1]$ satisfies the a priori estimates

(5.5.4) $$\sup_\Omega |v_t| \leqslant M_1; \qquad \sup_\Omega |\nabla v_t| \leqslant M_2,$$

where M_1 and M_2 are constants independent of t and v_t.

Proof Sketch: As mentioned above, we apply (5.4.14) to prove the result, but first it is necessary to determine an appropriate Banach space X for the operator. To this end, we follow the Schauder inversion method discussed in Section 2.2D. Indeed, the a priori estimates (5.5.4) show that any solution v_t of a member of the adjusted system has a Hölder continuous gradient with exponent $\alpha \in (0, 1)$ and independent of t and v_t. We let $X = C^{1,\alpha}(\bar{\Omega})$, and define a mapping T: $C^{1,\alpha}(\bar{\Omega})$ into itself by fixing $u \in X$

and considering the solution U of the linear elliptic Dirichlet problem

$$\sum_{|\alpha|+|\beta|=2} A_{\alpha\beta}(x, u, Du)\, D^{\alpha}D^{\beta}U + A_0(x, u, Du) = 0 \qquad \text{on} \quad \Omega, \qquad u|_{\partial\Omega} = g.$$

By the results of Section 2.2D, $Tu = U$ maps X into itself, is bounded, and in fact, maps bounded sets of X into bounded sets of $C^{2,\alpha}(\overline{\Omega})$. Since $C^{2,\alpha}(\overline{\Omega})$ is a compact subset of $C^{1,\alpha}(\overline{\Omega})$, the mapping T is compact.

Now we apply the a priori bound principle (5.4.14) with $f(u, t) = tTu$ and $X = C^{1,\alpha}(\overline{\Omega})$. By hypothesis, if v satisfies $u = tTu$, then $v \in C^{2,\alpha}(\overline{\Omega})$, and also satisfies the adjusted system. Consequently, $|v\|_{C^1(\overline{\Omega})} \leqslant M_1 + M_2$. Furthermore, by the a priori estimate (5.5.4) mentioned at the beginning of the proof, and a Hölder continuity of Ladyhenskaya and Uralsteva (1968), there is a number $\alpha \in (0, 1)$,

$$|\nabla v(x) - \nabla v(y)| \leqslant M_3|x - y|^{\alpha},$$

where M_3 is independent of $t \in [0, 1]$ and v. Thus

$$\|v\|_{C^{1,\alpha}(\overline{\Omega})} \leqslant M_1 + M_2 + M_3$$

implying that, by (5.4.14), (5.5.1)–(5.5.2) has a solution $u \in C^2(\overline{\Omega})$. See Ladyhenskaya and Uralsteva (1968) for more details.

5.5B Positive solutions for the Dirichlet problem for $\Delta u + f(x, u) = 0$

An interesting application of the Schauder fixed point theorem is concerned with the positive solutions of the following Dirichlet problem defined on a bounded domain $\Omega \subset \mathbb{R}^N$,

(5.5.5)
$$\Delta u + \lambda^2 f(x, u) = 0, \qquad f(x, u) \geqslant \beta > 0 \quad \text{for} \quad u \geqslant 0,$$
$$u\,|_{\partial\Omega} = 0.$$

We prove the following extension of the result established for system (1.2.3)–(1.2.4):

(5.5.6) Suppose in addition to the fact that $f(x, u) \geqslant \beta > 0$ for $u \geqslant 0$, that $f(x, u)$ is nondecreasing in u for fixed x and that $f(x, u) \geqslant g(x)u$ for $u \geqslant 0$. Then there is a finite (critical) number $\lambda_c > 0$ such that for $\lambda < \lambda_c$ (5.5.5) has at least one positive solution, while for $\lambda > \lambda_c$ (5.5.5) has no positive solution.

Proof: The argument can be divided in a natural way into three parts. First it is shown that under the given hypotheses (5.5.5) has a solution for some $\lambda > 0$. Next it is shown that if (5.5.5) has a positive solution for $\lambda_0 > 0$, then it has a positive solution for all λ in the interval $(0, \lambda_0]$. Finally, we show that for all λ sufficiently large, (5.5.5) has no positive solution.

(i) (5.5.5) *has a positive solution for some* λ: Observe that the positive solutions of (5.5.5) are in one-to-one correspondence with the positive solutions of the integral equation

(5.5.7)
$$u = \lambda^2 \int_{\Omega} G(x, y) f(x, u).$$

Now observing that Green's function $G(x,y)>0$ in and is integrable over Ω, we conclude that for $u \geqslant 0$ there is a constant $\gamma>0$ with

$$Tu = \int_\Omega G(x,y)f(x,u) \geqslant \beta \int_\Omega G(x,y) = \gamma \qquad \text{(say)}. \tag{5.5.8}$$

Let $C(\Omega)$ denote the Banach space of continuous functions defined on Ω with the sup norm, then the operator Tu just defined is clearly a continuous and compact mapping of the positive cone of $C(\Omega)$ into itself. The inequality (5.5.8) ensures that the same statement can be made for the mapping $Su = Tu/\|Tu\|_{C(\Omega)}$. In fact, S is a continuous compact mapping from the bounded closed convex set $\Sigma^+ = \{u \mid u \geqslant 0, \|u\|_C \leqslant 1\}$ into $\partial\Sigma^+ = \{u \mid u \geqslant 0, \|u\| = 1$ and a fixed point $\bar{u}$ of S is a solution of (5.5.5) with $\lambda^2 = 1/\|T\bar{u}\|$. Now the Schauder fixed point theorem (2.4.3) implies that S has a fixed point $\bar{u} \in \partial\Sigma^+$, and therefore (5.5.5) has a positive solution with $\lambda > 0$.

(ii) *If* (5.5.5) *has a positive solution* u_0 *for* $\lambda_0 > 0$, *then (5.5.5) has a positive solution for all* λ *in the interval* $(0, \lambda_0]$: Let $T_\lambda u = \lambda^2 \int_\Omega G(x,y)f(x,u)$. Then for $\lambda \in (0, \lambda_0]$, we shall show that T_λ maps the closed convex bounded set $\Sigma_0 = \{u \mid 0 \leqslant u \leqslant u_0, u \in C(\Omega)\}$ into itself. Since T_λ is continuous and compact, Schauder's fixed point theorem will again imply that T_λ has a fixed point u_λ in Σ_0 and u_λ will satisfy (5.5.5). To show that T_λ maps Σ_0 into itself, we note that since $f(x,u)$ is nondecreasing in u and $G(x,y) > 0$ in Ω, for $u \in \Sigma_0$

$$f(x,0) \leqslant f(x,u) \leqslant f(x,u_0)$$

and

$$\int_\Omega Gf(x,0) \leqslant \int_\Omega Gf(x,u) \leqslant \int_\Omega Gf(x,u_0).$$

Thus for $\lambda \in (0, \lambda_0]$ and $u \in \Sigma_0$, $0 < T_\lambda(0) \leqslant T_\lambda(u) \leqslant T_{\lambda_0}(u_0) = u_0$, as $T_\lambda(u) \in \Sigma_0$, as required.

(iii) *For* λ *sufficiently large*, (5.5.5) *has no positive solution*: If $(u_1, \lambda_1{}^2)$ denotes the first eigenfunction and eigenvalue of $\Delta u + \lambda^2 g(x)u = 0$ subject to the Dirichlet boundary condition $u_1 \mid_{\partial\Omega} = 0$, then $u_1 > 0$ in Ω. Thus multiplying (5.5.5) by u_1 and integrating by parts twice, we find that if u satisfies (5.5.5), then

$$0 = \int_\Omega \{\Delta u + \lambda^2 f(x,u)\}u_1 = \int_\Omega \{-\lambda_1^2 g(x)u_1 u + \lambda^2 f(x,u)\}$$
$$\geqslant \int_\Omega (\lambda^2 - \lambda_1^2) g(x) u_1 u.$$

Thus we arrive at a contradiction of $\lambda > \lambda_1$.

5.5C Periodic water waves

Here we consider the classic problem of proving the existence of steady periodic waves at the free surface $\partial\Gamma$ of an ideal incompressible fluid, under gravity. Because of their precision and relative simplicity, the results described here represent one of the most successful attempts to apply our analysis to a given difficult nonlinear eigenvalue problem. We suppose the flow is steady, irrotational, and two dimensional, the fluid occupying a domain Γ in $\mathbb{R}^2$. The points in $\mathbb{R}^2$ are denoted by Cartesian coordinates (x, y). Euler's equation of motion and the equation of continuity for this problem then become

$$\Delta\zeta = 0 \qquad \text{in } \Gamma, \tag{5.5.9}$$

$$\tfrac{1}{2}|\nabla\zeta|^2 + gy = \text{const.} \qquad \text{on } \partial\Gamma, \tag{5.5.10}$$

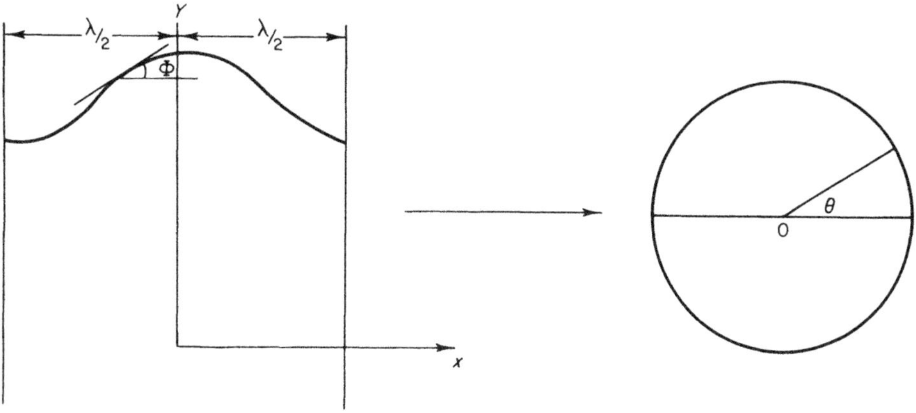

FIG. 5.3 Notation for periodic water wave problem.

where ζ denotes the velocity potential for the flow. Hence we are forced to solve a nonlinear free boundary value problem. Following an argument due to Levi-Civita, one introduces the complex variable $z = x + iy$ and two analytic functions of z,

$$u(z) = \zeta + i\psi \quad \text{and} \quad \omega = \log\left\{ \frac{\partial\zeta}{\partial x} - i\frac{\partial\zeta}{\partial y} \right\} = C(\Phi) + i\Phi.$$

Here ψ is the stream function for ζ, Φ is the angle formed by the velocity vector V at the point (x, y), and $C(\Phi)$ is the harmonic conjugate of Φ. In order to work in a known domain, one chooses $u = \zeta + i\psi$ as an independent variable and regards ω as a function of u. Assuming, for simplicity, that that the fluid is at infinite depth, and after performing the recommended period transformation, the desired periodic solutions are in one-to-one correspondence with the nontrivial solution of the nonlinear integral equation

$$(5.5.11) \qquad \Phi(\theta) = \lambda \int_0^{\pi} K(\theta', \theta) e^{3C(\Phi)} \sin \Phi \, d\theta',$$

where $\lambda = (gv)/2\pi c^2$, v is the wavelength, and c denotes the constant horizontal velocity of the moving wave. $K(\theta', \theta)$ is the Green function associated with the Neumann problem for Δ in a circle, and the additive constant in the definition of $C(\Phi)$ is so chosen that $\int_0^{2\pi} C(\Phi(\theta))\, d\theta = 0$. Note that (5.5.11) is in the form of a nonlinear eigenvalue problem. See Fig. 5.3 for notation.

There are basically two types of problems associated with (5.5.11): (i) a local bifurcation problem for Φ very small, and (ii) a general global problem for $|\Phi|$ unrestricted. The local problem was "solved" in 1925 by Levi-Civita, but the global problem (which we discuss here) remained only

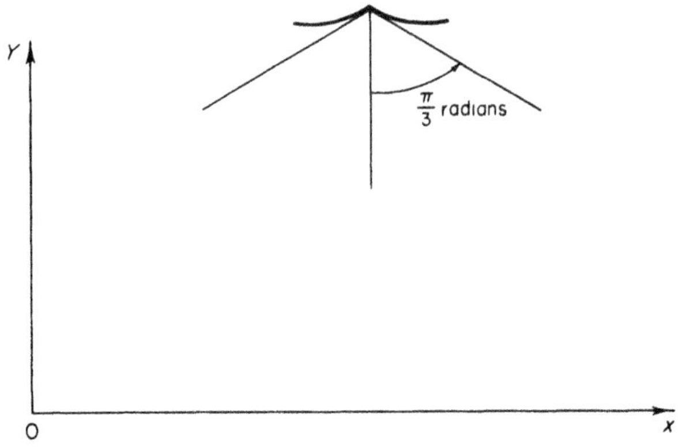

FIG. 5.4 Limiting form at the crest of a periodic water wave.

partially solved until 1961 when the Russian mathematician Y. P. Krasovskii proved the following results.

(5.5.12) **Theorem** There exist steady periodic wave satisfying (5.5.9) and (5.5.10) for which the maximum angle of inclination of the tangent to the wave profile takes any value in the open interval $(0, \pi/6)$. The wave is symmetric relative to a vertical axis passing through the peak of the wave. Furthermore, waves of this type with arbitrarily large Froude number λ cannot exist.

Before sketching the proof of this interesting result, we note that the number $\pi/6$ appearing in the theorem is sharp in the sense that (i) Stokes' periodic "limit" waves have $\max |\Phi| = \pi/6$ (see Fig. 5.4) and possess cusps, (ii) the solutions of (5.5.9), (5.5.10) show that steady periodic waves with $\max |\Phi| > \pi/6$ do not exist (see Wehausen (1969) for further information). Actually, Krasovskii proved a sharp analogue of Theorem 5.5.12 for waves of finite depth and periodic bottom by slightly modifying the proof given below.

Sketch of the proof of Theorem (5.5.12): The proof breaks down into the following steps:

(1) representation of the equation (5.5.11) as an operator equation of the form $x = \lambda Ax$ in a suitable Banach space X;

(2) proof of complete continuity of the map A in X;

(3) application of the Leray–Schauder degree to the operator equation;

(4) proof of the estimates necessary to calculate the Leray–Schauder degree.

In order to carry out the steps (1)–(4), we need to know the following analytic facts concerning the conjugation operator C of a harmonic function and the kernel $K(\theta', \theta)$. (Interestingly, the limiting number $\pi/6$ arises naturally from these facts and the requirement that the operator A be completely continuous.)

L_p estimates for the boundary values of conjugate harmonic functions Let $u(z)$ be a harmonic function defined in the unit circle $|z| < 1$ of the complex plane, with boundary values $u(e^{i\theta}) \in L_p[0, 2\pi]$ $(1 \leqslant p \leqslant \infty)$. Then, if $v(z)$ denotes the harmonic function conjugate to $u(z)$ in $|z| < 1$ and normalized by setting $\int_0^{2\pi} v(t)\, d\theta = 0$, $f(z) = u(z) + iv(z)$ is analytic in $|z| < 1$, and $f(0)$ is real. Now we define the linear mapping $C(u(e^{i\theta})) = v(e^{i\theta})$, and inquire about L_p boundedness of C. In this connection, we have the following results:

Fact 1 Theorem of M. Riesz For $1 < p < \infty$, C is a bounded map of $L_p[0, 2\pi] \to L_p[0, 2\pi]$, and thus there is a constant c_p independent of u such that

$$\|Cu\|_{L_p} \leqslant c_p \|u\|_{L_p}.$$

Fact 2 Zygmund's Theorem If $|u| \leqslant 1$, then

$$\int_0^{2\pi} \exp(\lambda |Cu|)\, d\theta \leqslant \frac{4\pi}{\cos \lambda} \qquad \text{for} \quad 0 \leqslant \lambda < \frac{\pi}{2}.$$

These results can again be proven by the method of singular integral operators, by extending (1.3.18) "to the periodic case," see Zygmund, (1934).

Fact 3 L_p estimates for K $\max_\theta \int_0^\pi |K(\theta', \theta)|^p\, d\theta' \leqslant C_p$, and for $1 < p < \infty$ and fixed θ, $\partial K/\partial\theta$ maps $L_p[0, 2\pi] \to L_p[0, 2\pi]$ boundedly. This third fact is a well-known property of Green's function for Δ.

Steps 1 and 2: Now let $X = C_0[0, \pi]$, i.e., the continuous functions on $[0, \pi]$ that vanish at 0 and π. Let $\|\Phi\|_X = \sup_{[0, \pi]} |\Phi(\theta)|$ and define the operator

$$A\Phi(\theta) = \int_0^\pi K_1(\theta', \theta) e^{3C(\Phi)} \sin \Phi\, d\theta'. \tag{5.5.13}$$

One shows that A is a completely continuous map defined on the sphere $S(0, \rho)$ of radius $\rho < \pi/6$ in X. Note that $\pi/6$ comes up naturally by combining (5.5.13) with Fact 2. Clearly, by Facts 1 and 3 above, A is a well-defined and continuous map from $S(0, \rho) \to X$ for $\rho < \pi/6$. In fact, under Hölder's inequality, one easily shows that for $\Phi_1, \Phi_2 \in S(0, \rho)$, $\rho = \pi/6 - d$ $(d > 0)$,

$$\|A\Phi_1 - A\Phi_2\| \leqslant K_d \|\Phi_1 - \Phi_2\|.$$

To verify the compactness of A, we again use Facts 1 and 3 to show that if $\tilde{\Phi}(\theta) = A\Phi$, then for some $s > 1$, $\|d\tilde{\Phi}/d\theta\|_{L_s} \leqslant K_{d,s}$ for $\Phi \in S(0, \rho)$ with $\rho = \pi/6 - d$ (as above). Consequently, $\|\tilde{\Phi}\|_{C_{0,\mu}} \leqslant M_\rho$ for some $\mu > 0$. The desired compactness of A thus follows. (Here $C_{0,\mu}$ is the Banach space of Hölder continuous functions of exponent μ.)

Step 3: In order to apply the Leray–Schauder degree to prove the existence of a solution of (5.5.11), we let

$$A_{\epsilon,\lambda}\Phi = \lambda\left[A\Phi + \epsilon\int_0^\pi K(\theta', \theta) \sin\theta' \, d\theta'\right].$$

Note that A_ϵ is compact (and positive). We prove (†): the Leray–Schauder degree of $I - A_{\epsilon,\lambda}$ on the positive cone $K_\beta = \{\Phi(\theta) \mid \Phi \in C_0[0, \pi], \Phi \geqslant 0, \|\Phi\|_{C_0} \leqslant \beta\}$, $0 < \beta < \pi/6$, is different for large and for small λ. The last part of Theorem (5.5.12) and (†) suffice to prove the existence part of Theorem (5.5.12). To see this, we first note that (†) implies that there are sequences $\{\lambda_n\}$, $\{\epsilon_n\}$, $\{\Phi_n\}$ with $\lambda_n > 0$, $\epsilon_n \to 0$, and $\Phi_n \in C_0[0, \pi]$ such that

$$\Phi_n = \lambda_n A_{\epsilon_n}\Phi_n, \qquad \|\Phi_n\|_C = \beta.$$

By the compactness of A and the boundedness of $|\lambda_n|$ (due to the nonexistence part of Theorem (5.5.12) there is a (strongly) convergent subsequence $\{\lambda_{n_j}\}$ and $\{\Phi_{n_j}\}$ with limits $(\lambda_\beta, \Phi_\beta)$ such that

(5.5.14) $$\Phi_\beta = \lambda_\beta A\Phi_\beta, \qquad \|\Phi_\beta\|_C = \beta, \qquad \Phi_\beta(\theta) \geqslant 0 \quad \text{on} \quad [0, \pi].$$

Thus one can extend $\Phi(\theta)$ to an odd 2π-periodic function of θ.

Step 4: First we prove (†) mentioned in Step 3 above. For λ very small, $d(I - A_{\epsilon,\lambda}, 0, K_\beta) = 1$ since for $\lambda = 0$, $A_{\epsilon,\lambda} \equiv 0$. On the other hand, for λ very large, $\Phi(\theta) - A_{\epsilon,\lambda}\Phi(\theta)$ cannot be positive for $\max|\Phi(\theta)| \leqslant \beta$, so $d(I - A_{\epsilon,\lambda}, 0, K_\beta) = 0$ in that case.

The nonexistence result of Theorem (5.5.12) is somewhat more difficult. It is based on the following two a priori estimates for solution $\Phi(\theta)$ of (5.5.11):

(5.5.15) There are absolute positive constants γ and δ such that $\Phi^\gamma(\theta) \geqslant (\lambda/\delta)^\gamma L(\Phi^\gamma)$, where $L\Phi = \int_0^\pi K(\theta', \theta)\Phi(\theta')\, d\theta'$ provided $\|\Phi\|_{C_0} \leqslant \pi/2$.

(5.5.16) There is an absolute constant $\beta > 0$ such that $\Phi^\gamma(\theta) \geqslant \beta \sin\theta$.

Assuming (5.5.15) and (5.5.16) with β maximal, the proof of nonexistence is as follows. Applying the operator L to (5.5.16) and using (5.5.15), we have

$$\left(\frac{\delta}{\lambda}\right)^\gamma \Phi^\gamma \geqslant L\Phi^\gamma(\theta) \geqslant \beta L(\sin\theta) = \beta \sin\theta,$$

i.e., $\Phi^\gamma \geqslant (\lambda/\delta)^\gamma\beta \sin\theta$, so that $(\lambda/\delta)^\gamma \leqslant 1$. Hence for $\lambda > \delta$, (5.5.11) can have no solution. To end our sketch of the proof of Theorem (5.5.12), we prove (5.5.15) and (5.5.16). To demonstrate (5.5.15), it suffices to show that for $\Phi \in K_\beta$

$$(5.5.17) \qquad L(e^{3C(\Phi)} \sin \Phi) \geqslant \frac{1}{\delta} L(\Phi^\gamma)^{1/\gamma}.$$

Now (5.5.17) follows from Hölder's (inverse) inequality since

$$L(e^{3C(\Phi)} \sin \Phi) \geqslant L(e^{3C(\Phi)}\Phi) \geqslant \{L(e^{3qC(\Phi)})\}^{1/q}\{L(\Phi^\gamma)\}^{1/\gamma}$$

with $1/p+1/\gamma=1$ $(q<0)$. Then the basic Facts 2 and 3 imply that for $q=-1/10$ and $|\Phi| \leqslant \pi/2$, $|L(e^{3|q|C(\Phi)})| < \delta^{-|q|}$. Finally, we prove (5.5.16). Applying the inequality (5.5.15) k times and letting $\Phi^\gamma(\theta)=\sum_{n=1}^\infty a_n \sin n\theta$, we find

$$(5.5.18) \qquad L^k\Phi^\gamma = \sum_{n=1}^{\infty} \frac{a_n}{n^k} \sin n\theta \leqslant \left(\frac{\delta}{\lambda}\right)^{\gamma k} \Phi^\gamma(\theta.)$$

Furthermore $\sum(a_n/n^k) \sin n\theta \geqslant a_1 \sin\theta - |\sum_{n=2}^\infty \cdots|$ and

$$\left|\sum_{n=2}^{\infty} \frac{a_n}{n^k} \sin n\theta\right| \leqslant \max_n |a_n| \sum_{n=2}^{\infty} \frac{|\sin\theta|}{n^{k-1}} .$$

Thus (5.5.18) implies that

$$\Phi^\gamma(\theta) \geqslant \left(\frac{\lambda}{\delta}\right)^{\gamma k} \left\{a_1 - \max_n |a_n| \sum_{n=2}^{\infty} \frac{1}{n^{k-1}}\right\} \sin\theta.$$

Since $\Phi(\theta) \geqslant 0$ on $[0, \pi]$, $a_1 > 0$ and choosing k sufficiently large we can choose $\{a_1 - \max |a_n| \sum_{n=2}^\infty (1/n^{k-1})\} \geqslant a_1/2$.

5.5D The continuation of periodic motions of autonomous systems

We consider the periodic solutions of the second-order system

$$(5.5.19) \qquad \ddot{x} + Ax + f(x) = 0; \qquad |f(x)| = o(|x|).$$

Here $x(t)$ is an N-vector function of t, A is an $N \times N$ positive definite matrix, and $f(x)$ is an odd, C^2, N-vector function of higher order in x. In Section 4.1, we investigated the periodic solution of (5.5.19) near the singular point $x = 0$ by means of bifurcation theory. Here we shall focus attention on the global structure of the periodic solutions of (5.5.19).

As a first result, we consider the global analogue of Liapunov's theorem (4.1.4).

(5.5.20) **Theorem** Suppose the positive eigenvalues of the nonsingular matrix A, $\lambda_1^2, \lambda_2^2, \ldots, \lambda_N^2$, are such that for some integer j $(1 \leqslant j \leqslant N)$,

$$\text{(5.5.21)} \qquad \lambda_i/\lambda_j \neq \text{integer} \qquad \text{for} \quad i = 1, \ldots, N \quad (i \neq j).$$

Then (5.5.19) has a family of periodic solutions $x(\epsilon)$ of period $\tau(\epsilon)$ depending continuously on a real parameter ϵ, and such that

(i) as $\epsilon \to 0$, $x(\epsilon) \to 0$ and $\tau(\epsilon) \to 2\pi/\lambda_j$; while
(ii) as $\epsilon \to \infty$, either $\sup |x(\epsilon)| + \tau(\epsilon) \to \infty$ or $x(\epsilon) \to 0$ and $\tau(\epsilon) \to 2\pi n/\lambda_k$ for $k = 1, \ldots, N$ and $n = 1, 2, \ldots$; but if $n = 1$, then $k \neq j$ (i.e., as either the amplitude of $x(\epsilon) \to \infty$, the period $\tau(\epsilon) \to \infty$, or $x(\epsilon)$ tends to a covering (possibly multiple) of the periodic solution of the linearized equation).

Proof: By repeating the argument of Section 4.1 and setting $t = \lambda s$, the odd periodic solutions of (5.5.19) are in one-to-one correspondence with the solutions of the operator equation

$$\text{(5.5.22)} \qquad x = \lambda^2\{\mathcal{G}x + \mathcal{N}(x)\}$$

in the Sobolev space $H = \mathring{W}_{1,2}\{[0, \pi]; \mathbb{R}^N\}$. Here the operators $\mathcal{G}$ and $\mathcal{N}$ are completely continuous and defined implicitly for $x, y \in H$ by the formulas

$$\text{(5.5.23)} \qquad (\mathcal{G}x, y) = \int_0^\pi Ax(s)\cdot y(s)\, ds; \quad (\mathcal{N}x, y) = \int_0^\pi f(x(s))y(s)\, ds.$$

The condition (5.5.21) implies that the eigenvalue λ_j^2 of $\mathcal{G}$ is simple on an appropriately chosen closed subspace of H (see Section 4.1.C). (Cf. Section 4.1.) Consequently, by Theorem (5.4.28), there is a continuum of solutions of (5.5.22), $(x(\epsilon), \tau(\epsilon))$, joining $(0, 1/\lambda_j^2)$ either to ∞ or to $(0, N^2/\lambda_k^2)$, where $N = 1, 2, \ldots$ and $k = 1, \ldots, N$, with $k \neq j$ in case $N = 1$. Thus the theorem is proved.

We sharpen the result just obtained by imposing other restrictions on the vector function $f(x)$ in (5.5.19).

One important class of results can be obtained by using the result (5.4.37) on monotone minorants. Suppose, for example, that we set $g(x) = Ax + f(x)$ and write $g(x) = (g_1(x), \ldots, g_N(x))$, where $x = (x_1, \ldots, x_N)$. Then we prove

(5.5.24) **Theorem** Suppose $g(x)$ is an odd function of x with the properties:

(i) $g_i(x) \geqslant 0$, whenever $x_i \geqslant 0$ $(i = 1, \ldots, N)$; and
(ii) there is a constant $k > 0$ and an integer j $(1 \leqslant j \leqslant N)$ such that $g_j(x) \geqslant kx_j$ for all nonnegative vectors x.

Then the system (5.5.19) has a one-parameter family of periodic solutions $x(\epsilon)$ with period $\tau(\epsilon)$ for $\epsilon \in (0, \infty)$ with $\epsilon = \sup_{[0, \tau(\epsilon)]} |x(\epsilon)|$.

Proof: To apply (5.4.37), let the Banach space $X = \Pi_{i=1}^{N} C[0, 2\pi]$, and consider the two-point boundary value problem

$$(5.5.25) \qquad x_{ss} + \lambda^2\{Ax + f(x)\} = 0,$$

$$(5.5.26) \qquad x(0) = x(1) = 0.$$

Arguing as in (5.5.20), we may suppose that the λ-periodic solutions of (5.5.25)–(5.5.26) correspond to the solutions of (5.5.22). The solutions of the latter system, however, can be written as the solutions of the integral equation

$$(5.5.27) \qquad x(s) = \lambda^2 \int_0^1 G(t, s) g(x(t))\, dt,$$

where G is the Green function for the operator x_{ss} and the boundary condition (5.5.26). Clearly, if K is the cone of the nonnegative vector functions on X, then G is a compact linear mapping of K into itself. By hypothesis, $g(x(s))$ also maps $K \to K$, and so

$$Tx(s) = \int_0^1 G(t, s) g(x(t))\, dt$$

is also cone preserving and a compact map. The hypotheses of (5.5.24) then imply that $Tx \geqslant \mathfrak{B}x$, where

$$(5.5.28) \qquad \mathfrak{B}x(s) = \left\{0, 0, \ldots, k\int_0^1 G(t, s) x_i(t)\, dt, 0, \ldots\right\}.$$

Clearly $\mathfrak{B}$ is monotone on K and cone preserving. Thus by (5.4.37), (5.5.25) and (5.5.26) have a family of solutions $(x(\epsilon), \lambda(\epsilon))$ with $\|x(\epsilon)\| = \epsilon$ and $\lambda(\epsilon) > 0$. Consequently, by extending these solutions to $(-\infty, \infty)$, as odd periodic functions of s, this family corresponds to the desired family of periodic solutions of (5.5.19).

5.5E Necessary and sufficient conditions for the solvability of coercive semilinear elliptic boundary value problems

We begin by considering the following semilinear Dirichlet problem defined on a bounded domain $\Omega \subset \mathbb{R}^N$:

$$(5.5.29) \qquad \begin{aligned} &\mathfrak{L}u + f(u) = g \\ &D^\alpha u \mid_{\partial\Omega} = 0, \qquad |\alpha| \leqslant m - 1, \end{aligned}$$

where $\mathfrak{L}$ is the formally self-adjoint operator

$$\mathfrak{L}u = \sum_{|\alpha|,\,|\beta| \leqslant m} (-1)^{|\alpha|} D^{\alpha}\{a_{\alpha\beta}(x)D^{\beta}u\},$$

and the function f satisfies the following hypothesis:

$(*)$ $$\lim_{s\to\pm\infty} f(s) = f(\pm\infty) < \infty$$

exist and moreover $f(-\infty) < f(s) < f(+\infty)$.

We prove the following result:

(5.5.30) **Theorem** A necessary and sufficient condition for the solvability of (5.5.29) is for each z_0 of $L_2(\Omega)$ of norm 1 in Ker $\mathfrak{L}$ the following inequalities hold

$$\int_{\Omega} gz_0 < f(+\infty)\int_{\{z_0>0\}}|z_0| - f(-\infty)\int_{\{z_0<0\}}|z_0|$$

Proof: Using the duality method Section 2.2D of representing (5.5.29), we obtain a Hilbert space reformulation of (5.5.29) as the operation equation

(5.5.31) $$Lu + Nu = \tilde{g}$$

with L and N mapping the Sobolev space $\mathring{W}_{m,2}(\Omega)$ into itself defined implicitly by the formulas

$$(Lu, v) = \sum_{|\alpha|,\,|\beta| \leqslant m} \int_{\Omega} a_{\alpha\beta}(x)\, D^{\alpha}u\, D^{\beta}v,$$

$$(Nu, v) = \int_{\Omega} f(u)v, \quad (\tilde{g}, v) = \int gv$$

Now our result (5.4.29) applies directly to (5.5.31) since, as we shall see, the hypothesis $(*)$ implies that the condition (5.4.30) is automatically satisfied. In fact we show that

(5.5.32) $$\lim_{R\to\infty} (N(Rz_0 + v), z_0) = f(+\infty)\int_{z_0>0}|z_0| - f(-\infty)\int_{z_0<0}|z_0|$$

uniformly for $v \perp$ Ker L and $\|v\|$ uniformly bounded. Once we demonstrate (5.5.32), our result will follow directly from the result (5.4.29). To this end we denote the right hand side of (5.5.31) by $n(z_0)$ and note that by definition

$$(N(Rz_0 + v), z_0) = \int_{\Omega} f(Rz_0 + v)z_0.$$

Thus given $\epsilon > 0$, we shall show that

(5.5.33) $$\left|n(z_0) - \int_{\Omega} f(Rz_0 + v)z_0\, dx\right| < \epsilon$$

for R sufficiently large. First, we note that there is a $\delta > 0$, such that

$$\int_A f(\infty)|z_0|\,dx < \epsilon/4$$

for any measurable set A, with $m(A) < \delta$. $m(A)$ here denotes the Lebesgue measure of the set A. We note that δ can be chosen independent of $z_0 \in \operatorname{Ker} L$ of L_2 norm 1 since $\operatorname{Ker} L$ is a finite-dimensional space.

Now, for any v with $\|v\| \leqslant k$, letting $\Omega_N = \{x \in \Omega,\ |v(x)| \leqslant N\}$, we can choose N sufficiently large so that $m(\Omega - \Omega_N) < \delta$ for all such v. Thus the left-hand side of (5.5.33) is less than or equal to

$$\left|\int_{\Omega_+ \cap \Omega_N} [f(\infty) - f(Rz_0 + v)]|z_0|\,dx\right|$$
$$+ \left|\int_{\Omega_- \cap \Omega_N} [f(-\infty) - f(Rz_0 + v)]|z_0|\,dx\right|$$
$$+ \left|\int_{\Omega - \Omega_N} f(Rz_0 + v)z_0\,dx\right| + \left|\int_{\Omega - \Omega_N} f(\infty)z_0\,dx\right|.$$

The last two terms of the above are less than $\epsilon/4$ for the reasons just stated. Next we show that for R sufficiently large, the first two terms are each $< \epsilon/4$. This follows by the Lebesgue convergence theorem since for $|v| \leqslant N$, on $\Omega_+ \cap \Omega_N$, the integrand in each tends to zero pointwise and is bounded by an integrable function. Thus we have shown that (5.5.33) holds, and we have completed the proof.

The result (5.5.30) concerning the Dirichlet problem (5.5.29) can be substantially extended by utilizing (5.3.22) (5.4.32). Indeed let Pu denote an mth order elliptic system of equation k unknowns with "coercive" boundary conditions Bu expressed in terms of differential operators of order less than m. Then let $f(x, D^\alpha u)$ denote a continuous bounded vector-valued function of the vector u such that $\lim_{R\to\infty} f(x, Ru_0)$ exists uniformly. Then on the basis of (5.4.32) we can find necessary and sufficient conditions for the solvability of the elliptic system

$$Pu + f(x, D^\alpha u) = 0 \qquad \text{on} \quad \Omega \subset \mathbb{R}^N,$$
$$Bu\,|_{\partial\Omega} = 0.$$

Such a system can be represented by an operator equation with domain a Banach space X of vector-valued functions each component of which is an element of the Sobolev space $W_{m,p}(\Omega)$ and satisfies the boundary conditions B. The range of this mapping is an L_2 space of vector-valued functions. Moreover the abstract associated linear operator L defined by (P, B) defined on such a Banach space X is well known to be a Fredholm operator with discrete spectrum, so that the hypothesis (5.3.23) of (5.3.22) is generally satisfied.

NOTES

A Further linearization results for proper nonlinear Fredholm operators

It is natural to attempt to extend the results of Section 5.1 on linearization to a more general context. In this connection, the following result related to the Banach-Mazur theorem holds:

(1) Let f be a nonlinear Fredholm operator of index $p > 0$ acting between Banach spaces X and Y. Then if f is a proper mapping, f must possess singular points (cf. Berger and Plastock, 1977).

The proof of this result is based on the observation that if the mapping f possessed no singular points, f would necessarily determine a "fibration" between the spaces X and Y (see Spanier, 1966). Then by a corollary of the covering homotopy theorem the contractibility of Y would imply the contractibility of $f^{-1}(y)$ for each $y \in Y$, which would contradict the fact that $f^{-1}(y)$ is a compact orientable p-dimensional manifold.

B Additional results on mappings with singular points

The arguments used in the proof of (5.1.14) to determine the structure of the range of the operator A defined over a bounded domain Ω of $\mathbb{R}^N$ by

$$Au = \Delta u + f(u), \qquad u \mid \partial\Omega = 0,$$

can be generalized in various directions. Among these are the results of Podolak (1976):

(1) Suppose the asymptotic condition (5.1.13′) is replaced by

$$\lambda_{k-1} < \lim_{t \to -\infty} f'(t) < \lambda_k < \lim_{t \to +\infty} f'(t) < \lambda_{k+1}, \tag{5.1.13^k}$$

where $\lambda_{k-1}, \lambda_k, \lambda_{k+1}$ denote three successive eigenvalues of Δ. Then a result analogous to (5.1.14c) holds, provided we suppose

$$\dim \operatorname{Ker}(\Delta + \lambda_k) = 1 \qquad \text{and} \qquad \int_\Omega u_k |u_k| \neq 0,$$

where u_k is an eigenfunction associated with λ_k. In particular, $g \in O_2$ in (5.1.14c) means that the boundary value problem

$$\Delta u + f(u) = g, \qquad u|_{\partial\Omega} = 0,$$

has at least two solutions.

(2) More generally, suppose $L \in \Phi_p(X, Y)$ is a linear Fredholm operator of index $p \geqslant 0$ with dim coker $L = 1$ and N is a compact nonlinear mapping of X into Y, satisfying a global Lipschitz condition (with a sufficiently small Lipschitz constant) and the asymptotic condition

$$\lim_{t \to \infty} \frac{N(tu)}{t} = n(u).$$

Then if P_0 denotes a projection of Y onto coker L and $P_0 n(x_0) \neq 0$ for all elements $x_0 \in \operatorname{Ker} L$ of norm 1, a result analogous to (5.1.14c) also holds provided we interpret $g \in O_2$ as meaning that the equation $Lu + Nu = g$ has multiple solutions (in the present case a compact submanifold of dimensions p).

C Further properties and applications of the Leray–Schauder degree

(1) Let D be a bounded domain of a Banach space X and suppose f and g denote

compact perturbations of the identity. Then the following composition theorem holds for the Leray–Schauder degree of fg

$$d(fg, p, D) = \sum_i d(f, p, \Delta_i) d(g, \delta_i, \Delta_i)$$

for any $\delta_i \in \Delta_i$, where Δ_i denotes the bounded components of $X - g(\partial D)$.

As an application of this result the following generalization of the Jordon separation theorem can be established.

(2) Let D and D' be bounded open sets of a Banach space X such that there is a homeomorphism (i.e., a compact perturbation of the identity) between $\overline{D}$ and $\overline{D}'$, then the number of components of $X - D$ and $X - D'$ are equal.

D Further results on the Dirichlet problem for quasilinear elliptic partial differential equations

Let Ω be a bounded domain in $\mathbb{R}^N$ with smooth bonndary $\partial\Omega$. Then we consider the solvability of the following quasilinear elliptic boundary value problem

$$\text{(i)} \qquad \sum_{i,j=1}^{n} a_{ij}(x, u, \nabla u) \frac{\partial^2 u}{\partial x_i \, \partial x_j} = B(x, u, \nabla u),$$

$$\text{(ii)} \qquad u \mid_{\partial\Omega} = 0.$$

Here f is a given continuous function, and we seek a smooth function u satisfying (i) at points of Ω, and assuming the boundary condition (ii). Examples of geometric problems giving rise to (i)–(ii) include finding a nonparametric minimal surface or more generally a surface of prescribed mean curvature.

Then the Leray–Schauder degree together with some rather subtle a priori bounds yields existence theorems for (i)–(ii). When applied to the equation

$$\{(1 + |\nabla u|^2)I - \nabla u \, \nabla u\} \frac{\partial^2 u}{\partial x_i \, \partial x_j} = nk(1 + |\nabla u|^2)^{3/2}$$

defined a hypersurface of constant mean curvature k over a smoothly bounded domain Ω the following result is obtained:

Theorem The Dirichlet problem in Ω for hypersurfaces of constant mean curvature is solvable for arbitrary C^2 boundary data if and only if the mean curvature H of the boundary surfaces satisfies the inequality $H \geqslant [n/(n-1)]k$ for each point of the boundary. Moreover the solution is unique if it exists.

For a full discussion of such results we refer the reader to the paper by Serrin (1969).

E Bibliographic notes

Section 5.1: The material discussed in this chapter section has an interesting history; see, for example, Hadamard (1904) and discussions of the monodromy theorem in complex variables texts. Our discussion of (5.1.1) is due to Plastock (1974). We also refer to papers of Browder (1954) and John (1968). The theory of operators satisfying conditions as in (5.1.6) are called strongly monotone operators and have been the subject of numerous recent papers and monographs. We refer the reader to Brezis (1968), Lions (1969), and Browder (1976). The reader will find the result (5.1.4) in the paper by Banach and Mazur (1934). Our discussion of Section 5.1B is based on papers of Berger and Podolak (1976, 1975). The nonlinear Dirichlet problem (5.1.13) together with the restrictions (5.1.13′) was originally studied in Ambrosetti and Prodi (1972).

Section 5.2: Our discussion of this section is based on the work of Lions (1969) and Pohozaev (1967). The Rayleigh–Ritz approximation for nonlinear eigenvalue problems are well discussed in Browder (1968) and Rabinowitz (1973). The result (5.2.29) on the steady-state solutions of the Navier–Stokes equations follows Fujita (1961), which in turn is based on papers of Hopf (1951) and Leray (1933).

Section 5.3: The homotopy arguments in nonlinear problems of analysis began with Schauder. Our discussion of essential and inessential mappings follows Granas (1961). The Brouwer degree was extended to compact perturbations of the identity in the paper by Leray and Schauder (1934), where applications to the solvability of the Dirichlet problem for nonlinear elliptic equations can be found. Excellent treatments of the Leray–Schauder degree and its applications can be found in many books, including Krasnoselski (1964), Cronin (1964), Schwartz (1969), Nirenberg (1974), and Bers (1957), to mention only a few. The application of homotopy theory to compact perturbations of a linear Fredholm operator of positive index can be found in the papers of Svarc (1964) and Geba (1964). Applications of this theory to elliptic boundary value problems was given first by Nirenberg (1972); see also Berger and Podolak (1977), where the result (5.3.22) can be found. Our discussion of the generalized degree for proper Fredholm operators of index zero is adapted from Elworthy and Tromba (1970), and analogous results for operators of higher index can be found in Smale (1965); see also Palais (1967).

Section 5.4: The result (5.4.1) can be found in Plastock (1974), while the result (5.4.5) is due to Krasnoselski (1964). The invariance of domain theorem is originally due to Schauder. The proof given here is adapted from Granas (1961). Rothe's theorem (5.4.12) can be found in Rothe (1953). The result (5.4.24) can be found in Cronin (1973). The result (5.4.28) was originally proved in Rabinowitz (1973), the proof given here is due Ize (1975). The result (5.4.29) can be found in Berger and Podolak (1975), and its generalization (5.4.32) in Berger and Podolak (1977). The study of cone preserving mappings as discussed in Section 5.4F is based on arguments of Krasnoselski (1964) and has been the subject of a great deal of contemporary research; see Amann (1976) for a survey of recent work.

Section 5.5: As mentioned in the text, the solvability of the Dirichlet problem for quasilinear elliptic equations was one of Hilbert's problems in his address (Hilbert, 1900). An excellent recent survey with many new results can be found in Serrin (1969) and in the book by Ladyhenzskaya and Uralsteva (1968). The study of positive solutions of semilinear elliptic boundary value problems has been the focus of much recent research going far beyond the result (5.5.6); see Amann (1976), and Krasnoselski (1964) for good surveys. Levi-Civita's early result on periodic water waves can be found in Levi-Civita (1925). A good survey from a physical point of view can be found in Wehausen (1968). Krasovskii's paper containing the result (5.5.12) can be found in Krasovskii (1961). The study of continuation problems for periodic solutions of Hamiltonian systems dates back to Poincaré. Interesting experimental results in this connection were obtained by Stromgren and his colleagues; see Stromgren (1932). The result (5.5.24) can be found in Krasnoselski (1964). The result (5.5.30) is due to Landesmann and Lazar (1970) and to Williams (1972). The proof given here is due to Podolak (1974).

F Complete integrability for nonlinear elliptic boundary value problems.

The result (5.1.14) can be completed in a certain sense. Indeed, the mapping A defined by (5.1.13) can be shown to be canonically topologically equivalent (in an obvious notation) to the mapping A_0: $(t,w)\to(t^2,w)$, independent of the dimension of the domain Ω. Assuming more smoothness for the function f of (5.1.13), it is possible to prove the topological equivalence is actually differentiable so that one can say that all the critical points of A are "folds" in an appropriate infinite-dimensional generalization of Whitney's notion. The details of this work are contained in a paper of Berger and Church (*Indiana J. Math*).

Important in this direction is the fact that, contrary to other completely integrable problems, the methods of study used are "stable." Indeed if the mapping A is perturbed, interesting results for the perturbed problem can still be obtained even though the perturbed map is not topologically equivalent to A_0.

CHAPTER 6

CRITICAL POINT THEORY FOR GRADIENT MAPPINGS

In this chapter we shall discuss some basic properties of operator equations involving gradient mappings. Since the zeros of a gradient mapping F' are precisely the critical points of the real-valued functional F, we shall focus attention on those properties of F' that can be discussed in terms of the geometry of the graph of F. The abstract results obtained have fundamental importance for many classic problems that can be formulated in terms of the calculus of variations. In fact, we illustrate the application of our results by solving a number of problems in differential geometry and mathematical physics, as discussed in Chapter 1.

The special properties of gradient operators and their associated critical point theory lend great insight to the general problems described in Chapter 5. Here we shall see how these same problems can be studied by variational methods. These methods generally provide sharper information than can be obtained by more general techniques of Chapter 5.

First, we study those critical points that can be obtained as the absolute minima of a functional F over a linear space X. Then we turn to isoperimetric problems, i.e., absolute minima of F over curved subspaces of X. Isoperimetric problems can be regarded as a simple analytic method of studying saddle points of F when regarded as a functional on X. Finally, we discuss a more far reaching approach to saddle points, namely their classification and association with topological invariants related to the graph of F. The nonvanishing of these topological invariants ensures the existence of critical points of various types. Moreover, in each case we discuss applications of the results obtained to some pertinent problems of geometry and physics.

6.1 Minimization Problems

A fundamental heuristic principle of scientific understanding can be formulated as follows: "Many phenomena can be understood in terms of the minimization of an energy functional $\mathcal{G}(u)$ over an appropriate class of objects C." Thus in Chapter 1, we described geodesics and minimal surfaces from this vantage. For the problems of mathematical physics, phase transitions, elastic instability, and diffraction of light are among the phenomena that can be studied from this point of view. Indeed, the characterization of phenomena by variational principles has been a cornerstone in the transition from classical to contemporary physics.

Thus it is natural mathematically to study the simple and important class of critical points of a real-valued C^1 functional $\mathcal{G}(u)$ defined on an open set U of a Banach space X, namely the relative minima, i.e., points $u_0 \in U$ at which $\mathcal{G}(u) \geqslant \mathcal{G}(u_0)$ for all u near u_0.

Apart from the heuristic principle of the above paragraph, the importance of this class lies in the fact that such points not only are solutions of the gradient operator equation $\mathcal{G}'(u_0)=0$, but also possess remarkable stability properties. Roughly speaking, these stability properties are of two kinds: the first asserting that smooth perturbations of a functional $\mathcal{G}(u)$ possessing a relative minimum at u_0 must also possess a relative minimum nearby; the second asserting that oscillations starting near a strict relative minimum u_0 of $\mathcal{G}(u)$ always remain near u_0. We remark that such stability properties are crucial both for the actual computation of relative minima and in the interpretation of their meaning in nature. In this and the next section, we take up the study of the minima of functionals and the applicability of our results to specific problems of general interest.

6.1A Attainment of infima

In a finite-dimensional Banach space X, any continuous functional defined on a closed bounded set M attains its infimum. As is well known, this property *need not hold*[1] for infinite-dimensional spaces since closed bounded sets there are not necessarily compact. Thus in a Hilbert space H of infinite dimension, a bounded self-adjoint linear operator L with no point spectrum has the property that $\alpha = \inf(Lu, u)$ over the unit sphere $\partial\Sigma = \{u \mid \|u\| = 1\}$ is not attained on $\partial\Sigma$. Indeed, if α were attained on $\partial\Omega$, α would be an eigenvalue of L and hence in the point spectrum of L.

The actual problem of determining restrictions on the closed subset M of a Banach space X and the functional $\mathcal{G}(u)$ to assure the attainment of the desired infima has been studied since Weierstrass first pointed out that such a functional may not achieve its infima. The basic restrictions involved center around the various notions of the compactness of the set $M^\alpha = \{u \mid u \in M,\ \mathcal{G}(u) \leqslant \alpha\}$ and the lower semicontinuity of $\mathcal{G}$. The concept of lower semicontinuity was introduced in our discussion of minimal surfaces in Chapter 1, and is a familiar property of Lebesgue integration (cf. Fatou's theorem). Weak lower semicontinuity of the functional $\mathcal{G}(u)$ is understood to mean that whenever $u_n \to u$ weakly in X, $\mathcal{G}(u) \leqslant \underline{\lim}\ \mathcal{G}(u_n)$. Thus, in accord with (1.3.11) the norm of a Banach space X is weakly lower semicontinuous on X.

A simple result in this direction is

[1] Historically the point proved to be crucial for the justification of Riemann's approach to potential theory (see the introductory chapter of Courant (1950)).

(6.1.1) **Theorem** Suppose $\mathcal{G}(u)$ is a bounded functional defined on a (sequentially) weakly closed and nonempty subset M of a reflexive Banach X. Then if $\mathcal{G}(u)$ is coercive on M (in the sense that $\mathcal{G}(u) \to \infty$ whenever $\|u\| \to \infty$ with $u \in M$), and in addition $\mathcal{G}(u)$ is weak lower semicontinuous on M, then $c = \inf \mathcal{G}(u)$ over M is finite and attained at a point $u_0 \in M$.

In particular, if $M = X$ and $\mathcal{G}(u)$ is C^1, then $\mathcal{G}'(u_0) = 0$ so that $c = \mathcal{G}(u_0)$ is a critical value of $\mathcal{G}(u)$, and any element in $\mathcal{G}^{-1}(c)$ is a critical point of $\mathcal{G}(u)$.

Proof: By the coerciveness of $\mathcal{G}$ on M, the set $M^\alpha = \{u \mid u \in M, \mathcal{G}(u) \leqslant \alpha\}$ is bounded for any finite number α. Consequently, $c = \inf_M \mathcal{G}(u)$ is bounded above $-\infty$ since the functional $\mathcal{G}(u)$ itself is bounded. Moreover, any minimizing sequence $\{u_n\} \in M^{\alpha+1}$ is bounded and so has a weakly convergent subsequence (which we relabel $\{u_n\}$) with weak limit $\bar{u}$. The weak lower semicontinuity of $\mathcal{G}(u)$ then implies that $c = \mathcal{G}(\bar{u})$ since

$$\mathcal{G}(\bar{u}) \leqslant \lim \mathcal{G}(u_n) = c = \inf_M \mathcal{G}(u).$$

Moreover, $\bar{u} \in M$ since M is weakly closed, and so $\bar{u} = u_0$ is the desired minimum. If $M = X$, and $\mathcal{G}(u)$ is C^1, then for any point $u \in \mathcal{G}^{-1}(c)$, $\mathcal{G}(u + th) \geqslant \mathcal{G}(u)$. Thus for any $t \in \mathbb{R}^1$ and $h \in X$,

$$(6.1.2) \qquad (d/dt)\mathcal{G}(u + th)|_{t=0} = (\mathcal{G}'(u), h) = 0,$$

so that $\mathcal{G}'(u) = 0$.

In order to investigate the applicability of this result, it is essential to derive criteria for a functional $\mathcal{G}(u)$ to be (i) weak lower semicontinuous, and (ii) coercive. The next two lemmas provide fairly general criteria, which will prove useful in the sequel.

(6.1.3) **Criterion for Weak Lower Semicontinuity** A functional $\mathcal{G}(u)$ is weakly lower semicontinuous on a reflexive Banach space X if it can be represented as the sum $\mathcal{G}(u) = \mathcal{G}_1(u) + \mathcal{G}_2(u)$, where $\mathcal{G}_1(u)$ is convex and $\mathcal{G}_2(u)$ is sequentially weakly continuous (i.e., continuous with respect to weak convergence). More generally, $\mathcal{G}(u)$ is weak lower semicontinuous if $\mathcal{G}(u) = \mathcal{G}(u, u)$, where $\mathcal{G}(x, y)$ is a function defined on $X \times X$ with the properties that $\mathcal{G}(x, y)$ is convex in x for fixed y, and sequentially weakly continuous in y uniformly over bounded sets in $x \in X$.

Proof: First we verify that a convex functional $\mathcal{J}(x)$ is weakly lower semicontinuous. By (2.5.2), if $x_n \to x$ weakly in X,

$$\mathcal{J}(x_n) - \mathcal{J}(x) = \int_0^1 (x_n - x, \mathcal{J}'(x_n(s)))\, ds, \quad \text{where} \quad x_n(s) = sx_n + (1 - s)x$$

$$= \int_0^1 (x_n - x, \mathcal{J}'(x))\, ds + \int_0^1 (x_n - x, \mathcal{J}'(x_n(s)) - \mathcal{J}'(x))\, ds.$$

By the convexity of $\mathcal{J}$, the last integral on the right is nonnegative (since the integrand itself is nonnegative). On the other hand, since $x_n \to x$ weakly in X, the first term tends to zero. Thus $\underline{\lim}[\mathcal{J}(x_n) - \mathcal{J}(x)] \geqslant 0$.

Next suppose that $\mathcal{G}(x)$ satisfies the more general properties mentioned in the theorem. Then if $x_n \to x$ weakly in X, writing

$$\mathcal{G}(x_n, x_n) = \mathcal{G}(x_n, x) + \{\mathcal{G}(x_n, x_n) - \mathcal{G}(x_n, x)\} \tag{6.1.4}$$

we find that $\mathcal{G}(x_n, x) = \mathcal{J}(x_n)$ (say) being convex, is weakly lower semicontinuous. Consequently as $n \to \infty$, $\mathcal{G}(x, x) \leqslant \underline{\lim}\ \mathcal{G}(x_n, x)$. Furthermore, since $x_n \to x$ weakly, $\{\|x_n\|\}$ is uniformly bounded. Thus by the hypotheses, $|\mathcal{G}(x_n, x_n) - \mathcal{G}(x_n, x)| \to 0$ uniformly. Hence $\mathcal{G}(x) = \mathcal{G}(x, x) \leqslant \underline{\lim}\ \mathcal{G}(x_n, x_n) = \underline{\lim}\ \mathcal{G}(x)$, and the result is proved.

(6.1.5) **Criteria for Coerciveness** Suppose the C^1 functional $\mathcal{G}(u)$ is defined on a reflexive Banach space X and satisfies either of the two conditions:

(i) $(\mathcal{G}'(u), u) \geqslant g(\|u\|)$ for some continuous function $g(r)$ such that $\int^\infty g(r)/r\, dr = \infty$;

(ii) $\mathcal{G}'(u) = Lu + R'(u)$ is a semilinear operator such that $m = \inf_{\|u\|=1} (Lu, u) \not\in \sigma_e(L)$, and for $\|u\|_X$ sufficiently large,

$$R(u) + \frac{m}{2}\|u\|^2 \geqslant \eta(\|u\|_{\tilde{X}}),$$

where X is continuously imbedded in a Banach space $\tilde{X}$ with $\eta(r)$ a continuous function satisfying $\eta(r) \to \infty$ as $r \to \infty$.

Then $\mathcal{G}(u)$ is coercive on X.

Proof: (i): We will show that $\underline{\lim}\ \mathcal{G}(u) = \infty$ as $\|u\| \to \infty$. For any w of norm 1 in X and $s \geqslant 0$,

$$\mathcal{G}(sw) - \mathcal{G}(0) = \int_0^s (w, \mathcal{G}'(tw))\, dt = \int_0^s (tw, \mathcal{G}'(tw))\, \frac{dt}{t}.$$

Thus by hypothesis, there is a positive number $R > 0$ (independent of w) such that

$$\mathcal{G}(sw) \geqslant \mathcal{G}(0) + \mathcal{G}(Rw) + \int_R^s \frac{g(t)}{t}\, dt.$$

As $\|u\| \to \infty$, $\underline{\lim}\ \mathcal{G}(u) \geqslant \int_R^\infty g(t)/t\, dt + \text{const}$. Consequently, $\mathcal{G}(u)$ is coercive.

(ii): Let N be the *finite-dimensional* null space of $L - m$, and let $\{u_k\}$ be a sequence such that $\|u_k\|_X \to \infty$. We have $u_k = u_k' + u_k''$, where $u_k' \perp N$ and $u_k'' \in N$. Now since $m \not\in \sigma_e(L)$, there is an absolute constant c_0 with

$$\begin{aligned} \mathcal{G}(u_k) &= \tfrac{1}{2}(Lu_k, u_k) + R(u_k) \\ &> c_0\|u_k'\|_X^2 + \eta(\|u_k\|_{\tilde{X}}), \qquad c_0 > 0. \end{aligned}$$

Thus if $\|u_k'\|_X \to \infty$, we have $\mathcal{G}(u_k) \to \infty$. Otherwise, we must have $\|u_k'\|_X \leqslant C$ and $\|u_k''\|_X \to \infty$. This implies that $\|u_k'\|_{\tilde{X}} \leqslant C'$, and so $\|u_k''\|_{\tilde{X}} \to \infty$ since $\dim \operatorname{Ker}(L - m) < \infty$ and all Banach spaces defined on finite-dimensional spaces have equivalent norms. This gives $\|u_k\|_{\tilde{X}} \to \infty$. Hence $\mathcal{G}(u_k) \to \infty$ in this case as well.

(6.1.6) **Examples**

(i) *If $\mathcal{G}'(u) = Lu + \mathfrak{N}'(u)$ is a semilinear operator with the essential spectrum of L, $\sigma_e(L)$, nonnegative, then $\mathcal{G}(u)$ is weak lower semicontinuous on X.* Indeed, in this case, $\mathcal{G}(u) = \frac{1}{2}(L, u) + \mathfrak{N}(u)$, and by (1.3.37), $L = L_1 + L_2$ with $L_1 \geqslant 0$ and L_2 compact. Thus $\mathcal{G}(u)$ can be written as the sum of the convex functional $\frac{1}{2}(L_1 u, u)$ and the functional $\mathcal{G}_2(u) = \frac{1}{2}(L_2 u, u) + \mathfrak{N}(u)$, which is clearly continuous with respect to weak convergence. Thus (5.1.3) applies.

(ii) Let $\{ g_{ij}(x) \mid i, j = 1, 2, \ldots, N\}$ be smooth functions, and consider the functional

$$\mathcal{G}(x) = \sum_{i,j=1}^{N} \int_a^b g_{ij}(x)\dot{x}_i\dot{x}_j$$

over the space of functions $W_N = \mathring{W}_{1,2}[(a, b), \mathbb{R}^N]$, where $\sum_{i=1}^N g_{ij}(\xi)\xi_i\xi_j$ is a positive definite quadratic form in ξ. To apply (6.1.3), let

$$\mathcal{G}(x, y) = \sum_{i,j} \int_a^b g_{ij}(y)\dot{x}_i\dot{x}_j .$$

For fixed y, $\mathcal{G}(x, y)$ is defined on W_N and convex in X. On the other hand, if $y_n \to y$ weakly in W_N, $g_{ij}(y_n) \to g_{ij}(y)$ uniformly on $[a, b]$. Thus for an absolute constant $K > 0$

$$|\mathcal{G}(x, y_n) - \mathcal{G}(x, y)| = |\sum \int_a^b \{ g_{ij}(y_n) - g_{ij}(y)\} \dot{x}_i\dot{x}_j|$$

$$\leqslant K \sup_{i,j} \sup_{[a,b]} | g_{ij}(y_n) - g_{ij}(y)| \|x\|^2_{W_N} \to 0$$

where the convergence is uniform over bounded sets in X.

Another useful criterion for the attainment of the infimum of a C^1 functional $\mathcal{G}(u)$ defined on a Hilbert space X, *that can be stated independently of semicontinuity assumptions* is obtained by requiring $\mathcal{G}(u)$ to satisfy the following "compactness condition."

Condition (C) If a sequence $\{x_n\} \in X$ is such that $\mathcal{G}(x_n)$ is uniformly bounded and $\mathcal{G}'(x_n) \to 0$, then x_n has a convergent subsequence.

In fact the following result holds:

(6.1.1′) **Theorem** Suppose that the C^1 functional $\mathcal{G}(x)$ defined on a

Hilbert space X is such that $\mathcal{G}'(x)$ satisfies a uniform Lipshitz condition and is bounded below. Then, if $\mathcal{G}(x)$ satisfies Condition (C), $\inf_X \mathcal{G}(x)$ is attained at $\bar{x}$ and $\mathcal{G}'(\bar{x}) = 0$.

Proof: Suppose $c = \inf_X \mathcal{G}(x)$ is not attained so that c is not a critical value for $\mathcal{G}(x)$. Then Condition (C) implies for some $\epsilon > 0$, $I^{c+\epsilon} = \{x \mid \mathcal{G}(x) \leqslant c + \epsilon\}$ also contains no critical points. (Indeed otherwise there would be a convergent sequence of critical points $\{x_n\}$ with $\mathcal{G}(x_n)$ tending to c and so Condition (C) would imply $\lim_{n\to\infty} x_n = \bar{x}$ is a critical point with $\mathcal{G}(\bar{x}) = c$.) We now apply the method of steepest descent of Section 3.2 to this situation by considering the initial value problem

$$\frac{dx}{dt} = -\mathcal{G}'(x), \qquad x(0) = x_0,$$

where x_0 is an arbitrary point of $\mathcal{G}^{c+\epsilon}$. By the results of (3.1.27), the solution $x(t)$ of this initial value problem exists for all t, provided $x(t)$ stays uniformly bounded, and moreover along $x(t)$

$$(*) \qquad \frac{d}{dt}\mathcal{G}(x(t)) = -\|\mathcal{G}'(x(t))\|^2;$$

so that, since $\mathcal{G}(x(t))$ is bounded from below, $\|\mathcal{G}'(x(t))\| \to 0$ as $t \to \infty$. Next we make use of the fact that $\mathcal{G}(x)$ satisfies Condition (C) to conclude that for any sequence $t_n \to \infty$, since $\mathcal{G}(x(t_n))$ must be uniformly bounded $x(t_n)$ has a convergent subsequence, $x(t_{n_j})$ with limit $\bar{x}$ and by the continuity of $\mathcal{G}'(x)$, $\mathcal{G}'(\bar{x}) = 0$. Consequently $\bar{x}$ is a critical point of $\mathcal{G}$. This is the desired contradiction since $\bar{x} \in \mathcal{G}^{c+\epsilon}$, by virtue of $(*)$.

6.1B An illustration

A simple (yet nontrivial) example of Theorem (6.1.1) is obtained by considering the T-periodic solutions of the nonautonomous Hamiltonian system

$$(6.1.7) \qquad \ddot{x} = \nabla U(x, t),$$

where $x(t)$ is an N-vector and $U(x, t)$ is a C^1 real-valued function of x and t. Supposing that $U(x, t)$ is T-periodic in t, we seek T-periodic solutions of (6.1.7). In fact we can prove

(6.1.8) **Theorem** If the T-periodic function $U(x, t)$ has the coercive property that $U(x, t) \to \infty$ as $|x| \to \infty$ uniformly in t, then (6.1.7) has a T-periodic solution that can be obtained as the minimum of the functional

$$(6.1.9) \qquad \mathcal{G}(x) = \int_0^T \{\tfrac{1}{2}\dot{x}^2 + U(x, t)\}\, dt$$

over all T-periodic C^1 N-vector functions $x(t)$.

Proof: Let W_N denote the space of all absolutely continuous T-periodic N-vector functions $x(t)$ such that $|\dot{x}(t)|^2 \in L_2[0, T]$. $W_N = W_{1,2}[(0, T), \mathbb{R}^N]$ is a Hilbert space with respect to the inner product

(6.1.10) $$(x, y)_{W_N} = \int_0^T \{\dot{x}(t)\cdot\dot{y}(t) + x(t)\cdot y(t)\}\, dt.$$

We shall show that

(6.1.11) $$\inf_{W_N} \mathcal{G}(x) = \inf_{W_N} \int_0^T \{\tfrac{1}{2}\dot{x}^2(t) + U(x, t)\}\, dt$$

is attained at $\tilde{x}(t) \in W_N$. Then by the results mentioned in Section 1.5, $\tilde{x}(t)$ will be a C^2 function. Consequently, $\tilde{x}(t)$ will be the desired T-periodic solution of (6.1.7). To verify that $\inf \mathcal{G}(x)$ over W_N is finite and is attained at $\tilde{x}(t)$, we apply Theorem (6.1.1), and verify the weak lower semicontinuity and coerciveness of $\mathcal{G}(x)$. First we observe that if $x_n \to x$ weakly in W_N, $x_n \to x$ uniformly on $[0, T]$, by Sobolev's imbedding theorem. Thus $\mathcal{G}(x)$ is the sum of the convex quadratic functional $\int_0^T \dot{x}^2(t)$ and the (sequentially) weakly continuous functional $\int_0^T U(x, t)$. (6.1.3) ensures that $\mathcal{G}(x)$ is weak lower semicontinuous. To prove the coerciveness of $\mathcal{G}(x)$, we let

(6.1.12) $$(Lx, x) = \int_0^T \dot{x}^2 \qquad \text{and} \qquad R(x) = \int_0^T U(x, t)\, dt.$$

Clearly the operator L so defined is a self-adjoint Fredholm operator and $m = \inf(Lx, x)$ over $\|x\| = 1$ is zero. Furthermore, Ker L consists of the constant N-vectors $\{c\}$, so that a general element $x(t)$ of W_N can be written uniquely as $x(t) = y(t) + c$, where $y(t)$ has mean value zero on $(0, T)$. Now

(6.1.13) $$\|y(t)\|_{W_N} < \sqrt{T}\, \|\dot{y}^2(t)\|_{L_2[0,T]}$$

and $U(x, t)$ is uniformly bounded from below (by $-K_0$, say) on $[0, T]$ since $U(x, t)$ tends uniformly to ∞ as $|x| \to \infty$. Thus

(6.1.14) $$\mathcal{G}(x(t)) = \mathcal{G}(c + y(t)) \geq \tfrac{1}{2}\|y\|^2_{W_N} - K_0 T,$$

so that $\mathcal{G}(x(t)) \to \infty$ as $\|y(t)\|_{W_N} \to \infty$. Hence it suffices to consider the possibility that the sequence $x_n(t) = y_n(t) + c_n$ is such that $\{\|y_n\|_{W_N}\}$ is uniformly bounded while the constants $|c_n| \to \infty$. To this end, let

(6.1.15) $$\Omega_n = \{t \mid t \in [0, T], |y_n(t)| > \tfrac{1}{2}|c_n|\}.$$

Then $\int_{\Omega_n} \frac{1}{4} c_n^2 \leq \|y_n(t)\|^2_{L_2} \leq$ const. C_0 (say), so that $\mu(\Omega_n) \leq 4C_0/c_n^2$. The complement Ω_n' of Ω_n has measure $\mu(\Omega_n') \geq T - 4C_0/c_n^2$. Let n be so large that $\mu(\Omega_n') > T/2$. Then on Ω',

$$|x_n| = |y_n + c_n| \geq |c_n| - |y_n| \geq |c_n|/2.$$

By hypothesis there is a function $\eta(r)$ such that $\eta(r) \to \infty$ as $r \to \infty$, with $U(x, t) \geq \eta(|x|)$, and

(6.1.16) $$R(x_n) = \int_0^T U(y_n(t) + c_n, t)\, dt = \int_\Omega + \int_{\Omega'}$$
$$\geq -TK_0 + \eta\left(\frac{|c_n|}{2}\right)\frac{T}{2}.$$

Thus $R(x_n) \to \infty$ as $c_n \to \infty$. We conclude that $\mathcal{G}(x)$ is coercive on W_N and the theorem is proved.

6.1C Minimization problems associated with quasilinear elliptic equations

The abstract considerations of the previous subsection do not distinguish between the well-known significant differences in concrete variational problems involving single integrals on the one hand and multiple integrals on the other. Indeed, for a large class of "regular" variational problems of the form $\mathcal{I}(u) = \int_a^b F(x, u, u_x)\,dx$ (with $u(x)$ an N-vector function of x), one can establish both the existence and minimizing property of any critical point $\tilde{u}(x)$ of $\mathcal{I}(u)$ in the small. Consequently the search for critical points of $\mathcal{I}(u)$ in the large can be decomposed into a succession of local problems. For variational problems involving multiple integrals, this cannot be done. Thus the length of a curve may be defined by polygonal approximation, while the area of surfaces may not be approximated by the analogous simplices, as was mentioned in Section 1.1.

Furthermore, the regularity properties of an absolute minimum of functionals involving single integrals are relatively easy to establish, while the analogous regularity properties of minima of multiple integrals remain only partially proven, to date.

Critical points of functionals associated with quasilinear elliptic operators, however, do possess a certain interesting "local" minimizing property. Indeed, suppose that Ω is a bounded domain and $\tilde{u}(x)$ is a smooth critical point of

(6.1.17) $$\mathcal{I}(u) = \int_\Omega F(x, D^\alpha u, D^\beta u)\,dx,$$

where $|\alpha| \leqslant m - 1$ and $|\beta| = m$ over the class of functions in $C^m(\Omega)$ that vanish on $\partial\Omega$ together with all their derivatives of order $\leqslant m - 1$; while the function $F(x, y, z)$ is of class C^m and is strictly convex in z for fixed x, y (so that the Euler–Lagrange equations associated with $\mathcal{I}$ are elliptic). Then $\tilde{u}$ has the following minimizing property:

(6.1.18) Suppose $\eta(x)$ $(\neq 0)$ is a C^∞ function vanishing outside of a sufficiently small neighborhood Ω_{x_0} of an arbitrary point $x_0 \in \Omega$. Then $\mathcal{I}(\tilde{u} + \eta) > \mathcal{I}(\tilde{u})$. Consequently, $\tilde{u}$ can never be a relative maximum.

Proof: Since $\mathcal{I}'(u) = 0$, Taylor's theorem shows that for some $t \in [0, 1)$

(6.1.19) $$\mathcal{I}(\tilde{u} + \eta) = \mathcal{I}(\tilde{u}) + \tfrac{1}{2}(\mathcal{I}''(u + t\eta)\eta, \eta),$$

where

$$(\mathcal{I}''(v)\eta, \eta) = \sum_{|\alpha'|, |\beta'| \leqslant m} \int_\Omega F_{X_{\alpha'} X_{\beta'}}(x, D^\alpha v, D^\beta v)\, D^{\alpha'}\eta\, D^{\beta'}\eta.$$

We show that $\mathcal{I}(\tilde{u} + \eta) > \mathcal{I}(\tilde{u})$ by proving that the second term in (6.1.19) is strictly positive. To this end, we note for $|\alpha| < m$, $\|\eta\|_\alpha \leqslant \epsilon(\Omega_{x_0})\|\eta\|_m$, where $\epsilon(\Omega_{x_0}) \to 0$ as $\mu(\Omega_{x_0}) \to 0$. Thus a simple computation shows that for $\mu(\Omega_{x_0})$ sufficiently small, there are constants $c_1, c_2 > 0$, independent of n such that

$$(\mathcal{G}''(u + t\eta)\eta, \eta) \geqslant c_1\|\eta\|_m^2 - c_2\|\eta\|_{m-1}^2$$
$$\geqslant \{c_1 - c_2\epsilon(\Omega_{x_0})\}\|\eta\|_m^2 > 0.$$

Next we take up the problem of finding a function attaining the infimum of the functional $\mathcal{G}(u) = \int_\Omega F(x, D^\alpha u, D^m u)\,dx$, $|\alpha| < m$, defined over a bounded domain $\Omega \subset \mathbb{R}^N$ with smooth boundary $\partial\Omega$ that satisfies the Dirichlet boundary condition $D^\alpha u|_{\partial\Omega} = f_\alpha(x)$ for $|\alpha| \leqslant m - 1$. A well-known result in this direction is

(6.1.20) **Theorem** Suppose that there is a function $f(x)$ in $W_{m,p}(\Omega)$ such that the trace of $D^\alpha f$ on $\partial\Omega$ coincides with f_α, $|\alpha| \leqslant m - 1$. Furthermore, suppose that the function $F(x, y, z)$ and its partial derivatives $\partial F/\partial y$, $\partial F/\partial z$ are continuous, and F satisfies the two conditions:

(6.1.21a) $F(x, y, z) \geqslant c_0|z|^p - c_1$, where c_0, c_1 are constants > 0;

(6.1.21b) $F(x, y, z)$ is convex in z for fixed x, y.

Then $\inf \mathcal{G}(u)$ over the class $\mathcal{C} = \{u \mid u \in W_{m,p}(\Omega)$ such that $D^\alpha u \mid_{\partial\Omega} = f_\alpha, |\alpha| \leqslant m - 1\}$ is finite and attained by a function $\tilde{u}(x) \in \mathcal{C}$.

Proof: First we note that the class $\mathcal{C}$ is nonempty since by hypothesis $f(x) \in \mathcal{C}$. Secondly, since $\mathcal{G}(u)$ and $\mathcal{G}(u) + c_1$ have the same critical points, we may suppose that (6.1.21a) holds with $c_1 = 0$. Also, hypothesis (6.1.21a) implies that $\mathcal{G}(u)$ is coercive on $\mathcal{C}$ since $\mathcal{G}(u) \geqslant c_0\int_\Omega |D^m u|^p$, while by Sobolev's theorem, there is a positive constant c_α independent of u for $|\alpha| \leqslant m - 1$ such that

$$\|D^\alpha u\|_{L_p} \leqslant \|D^\alpha(u - f)\|_{L_p} + \|D^\alpha f\|_{L_p} \leqslant c_\alpha\|D^m(u - f)\|_{L_p} + \|D^m f\|_{L_p}$$
$$\leqslant c_\alpha\|D^m u\|_{L_p} + \text{const.}$$

Hence $\mathcal{G}(u) \to \infty$ whenever $\|u\|_{m,p} = \{\sum_{|\beta| \leqslant m}\|D^\beta u\|_{L_p}^p\}^{1/p} \to \infty$, and in addition, $\mathcal{G}(u)$ is bounded from below. Thus the argument of (6.1.1) is applicable once we show that by hypothesis (6.1.21b), $\mathcal{G}(u)$ is lower semicontinuous with respect to weak convergence in $W_{m,p}(\Omega)$. Let $u_n \to u$ weakly in $W_{m,p}(\Omega)$, with $u_n \in \mathcal{C}$. Then $u \in \mathcal{C}$, and for $|\alpha| < m$, $D^\alpha u_n \to D^\alpha u$ strongly in $L_p(\Omega)$ so that $\{u_n\}$ has a weakly convergent subsequence (which we relabel $\{u_n\}$)converging almost everywhere to u in Ω. By Egorov's theorem, given $\epsilon > 0$, there is a set $\Omega_\epsilon \subset \Omega$ such that $D^\alpha u_n \to D^\alpha u$ for $|\alpha| \leqslant m - 1$ uniformly on Ω_ϵ, while $\mu(\Omega_\epsilon) \geqslant \mu(\Omega) - \epsilon$. Let

$$\Omega_{\epsilon,N} = \left\{x \mid x \in \Omega_\epsilon, \sum_{|\alpha| \leqslant m} |D^\alpha u| \leqslant N\right\}.$$

Since $u \in W_{m,p}(\Omega)$, $\mu(\Omega - \Omega_{\epsilon, N}) \to 0$ as $\epsilon \to 0$ and $N \to \infty$. With these preliminaries, we now apply the convexity hypothesis (6.1.21b) to prove the desired lower semicontinuity. Defining

$$\mathcal{I}_{\epsilon, N}(u, v) = \int_{\Omega_{\epsilon, N}} F(x, D^{\alpha}u, D^{m}v) \, dx \qquad \text{and} \qquad \mathcal{I}_{\epsilon, N}(u, u) = \mathcal{I}_{\epsilon, N}(u),$$

we have

$$\mathcal{I}_{\epsilon, N}(u_n) - \mathcal{I}_{\epsilon, N}(u) = \{ \mathcal{I}_{\epsilon, N}(u_n, u_n) - \mathcal{I}_{\epsilon, N}(u_n, u) \}$$
$$+ \{ \mathcal{I}_{\epsilon, N}(u_n, u) - \mathcal{I}_{\epsilon, N}(u, u) \}.$$

By hypothesis (6.1.21b),

$$\mathcal{I}_{\epsilon, N}(u_n, u_n) - \mathcal{I}_{\epsilon, N}(u_n, u) \geqslant \int_{\Omega_{\epsilon, N}} F_z(x, D^{\alpha}u_n, D^{m}u)\{ D^{m}u_n - D^{m}u \},$$

while on $\Omega_{\epsilon, N}$,

$$F(x, D^{\alpha}u_n, D^{m}u) \to F(x, D^{\alpha}u, D^{m}u) \quad \text{uniformly}$$

and

$$F_z(x, D^{\alpha}u_n, D^{m}u) \to F_z(x, D^{\alpha}u, D^{m}u) \quad \text{uniformly}.$$

Consequently, since $D^{m}u_n \to D^{m}u$ weakly in $L_p(\Omega)$, $\lim \mathcal{I}_{\epsilon, N}(u_n) = \mathcal{I}_{\epsilon, N}(u)$ as $n \to \infty$. Now since $F(x, y, z)$ is nonnegative, $\mathcal{I}_{\epsilon, N}(u_n) \leqslant \mathcal{I}(u_n)$; and because ϵ and N are arbitrary, $\underline{\lim} \mathcal{I}(u_n) \geqslant \mathcal{I}(u)$ as $n \to \infty$. Thus (6.1.20) is established.

As already mentioned, general results ensuring the regularity of the minimum $\tilde{u}(x)$ obtained in (6.1.20) have not been found if both m, $N > 1$. Thus, in general, one cannot say that $\tilde{u}(x)$ satisfies the resulting Euler–Lagrange system

$$\sum_{|\alpha| \leqslant m} (-1)^{|\alpha|} D^{\alpha}F_{\alpha}(x, Du, \ldots, D^{m}u) = 0, \qquad D^{\alpha}u|_{\partial\Omega} = f_{\alpha}, \quad |\alpha| \leqslant m - 1,$$

where

$$F_{\alpha} = \partial F(x, X^{\beta})/\partial X^{\alpha}.$$

Recently, the case $m = 1$ has been successfully resolved for all N, and has been the subject of several books (for example, Morrey (1966) and Ladyhenskaya and Uraltseva (1968)). For the case $N = 1$, the problem is considerably simplified. Indeed, $\tilde{u}(x)$ is absolutely continuous and in $W_{m,p}(a, b)$ for $p > 1$, and $\tilde{u}(x)$ is Hölder continuous with exponent $m - 1/p$. This result was obtained in Section 1.5.

To proceed further two simplifying assumptions are useful: first that the Euler–Lagrange equation $\mathcal{I}'(u) = 0$ associated with the functional $\mathcal{I}(u)$ is semilinear, and secondly that the equation $\mathcal{I}'(u) = 0$ is of second order. The significance of the semilinearity is twofold in that (i) the smoothness of any generalized solution can often be reduced to the regularity theory for *linear* elliptic equation (as described in Section 1.5) and moreover (ii) the lower semicontinuity and coerciveness criteria of (6.1.3) and (6.1.5) are readily applicable. Thus, for example, if

$$\mathcal{G}(u) = \tfrac{1}{2}(Lu, u) + \mathcal{N}(u); \qquad u \in \mathring{W}_{m,2}(\Omega),$$

where

$$(Lu, u) = \int_\Omega \sum_{|\alpha|, |\beta| \leqslant m} a_{\alpha\beta}(x) D^\alpha u \, D^\beta u$$

is a quadratic form associated with the linear elliptic operator L and $\mathcal{N}(u) = \int_\Omega F(x, u)$ with $F(x, u)$ a C^1 nonnegative function bounded below by a fixed parabola $P(u) = c_1 u^2 + c_2$ with $c_1 > 0$. Then assuming $\Omega \subset \mathbb{R}^N$ a bounded domain, the coefficient $a_{\alpha\beta}(x)$ smooth in Ω, the weak lower semicontinuity of $\mathcal{G}(u)$ on $\mathring{W}_{m,2}(\Omega)$ will follow from Gårding's inequality (1.4.22) applied to (Lu, u) and Fatou's theorem applied to $\mathcal{N}(u)$. To see this latter fact, observe that if $u_n \to u$ weakly in $\mathring{W}_{m,2}(\Omega)$, $u_n \to u$ in measure on Ω, and the nonnegativity of $F(x, u)$ implies $\underline{\lim}_{n\to\infty} \int_\Omega F(x, u_n) \geqslant \int_\Omega F(x, u)$ by Fatou's theorem. The coerciveness of $\mathcal{G}(u)$ is more delicate, but will follow from (6.1.5(ii)), $X = L_2(\Omega)$, provided the positive constant c_1 associated with $P(u)$ dominates the negative spectrum of L. A key point here is that no growth restrictions need be imposed on the function $F(x, u)$. The reason for this fact is the nonnegative assumption on $F(x, u)$ which ensures the integrability of $F(x, u)$ for $u \in \mathring{W}_{m,2}(\Omega)$ and the fact that if inf $\mathcal{G}(u)$ over $\mathring{W}_{m,2}(\Omega)$ is some finite number c, say. We need only consider $\mathcal{G}(u)$ defined on the set $\mathring{W}_{m,2}(\Omega) \cap \mathcal{G}^{-1}(c, c+\epsilon)$ for any fixed $\epsilon > 0$.

Variational problems associated with second-order quasilinear elliptic equations possess special simplifying features. These are primarily due to special properties of the functions in $W_{1,p}(\Omega)$ and to well-known methods (such as the maximum principle) for obtaining a priori bounds for the solutions of such systems. We end this subsection by pointing out a simple (yet useful) instance of this simplification.

(6.1.22) **On A Priori Bounds** Suppose that $\tilde{u}(x)$ minimizes the functional $\mathcal{G}(u) = \int_\Omega F(x, u, Du)\, dx$ over the class $\mathcal{C}$ of functions u in $W_{1,p}(\Omega)$ $(p \geqslant 1)$ that have the prescribed boundary condition $u|_{\partial\Omega} = f$. If there is a number $k > 0$ such that for all $x \in \Omega$ and $|z| > 0$:

(i) $F(x, y, z) > F(x, k, 0)$ whenever $y > k$;
(ii) $F(x, y, z) > F(x, -k, 0)$ whenever $y < -k$;
(iii) ess sup $|f| \leqslant k$ over $\partial\Omega$;

then ess sup $|\tilde{u}(x)| \leqslant k$ over Ω. Furthermore, if

(iv) $F_z(x, y, z) = O(|z|^{p-1})$ and $F_y(x, y, z) = O(|z|^p)$ as $|z| \to \infty$ for $|x| + |y|$ bounded,

then $\tilde{u}$ is a generalized solution of the resulting Euler–Lagrange system.

Proof: First we note that the special properties of $W_{1,p}(\Omega)$ imply that truncation of functions in $\mathcal{C}$ remain in $\mathcal{C}$. Indeed, suppose that $v \in \mathcal{C}$ with ess sup $|v| > k$, and let $\tilde{v}$ denote the truncated function

$$(6.1.23) \quad \tilde{v}(x) = \begin{cases} \inf(v, k) & \text{for} \quad v > 0, \\ \sup(v, -k) & \text{for} \quad v < 0. \end{cases}$$

Then $v \in W_{1,p}(\Omega)$ and $v|_{\partial\Omega} = f$ by (iii), so that $v \in \mathcal{C}$.

Next we prove the first part of (6.1.22) by using (i) and (ii) to prove that $\mathcal{I}(\tilde{v}) < \mathcal{I}(v)$. Indeed, (i) and (ii) imply

$$(6.1.24) \quad \mathcal{I}(v) - \mathcal{I}(\tilde{v}) = \int_{|v| \geq k} \{F(x, v, v_x) - F(x, \tilde{v}, \tilde{v}_x)\} > 0.$$

Consequently, ess sup $|\tilde{u}(x)| \leq k$ over Ω.

To finish the proof, we note that $\tilde{u}(x)$ is then also the minimum of $\mathcal{I}(u)$ over the restricted class $\mathcal{C}_{k+1} = \{u \mid u \in \mathcal{C}, |u|_{L_\infty} \leq k + 1\}$. Thus (iv) implies that $\mathcal{I}(\tilde{u} + t\zeta)$ is differentiable in $\mathcal{C}_{k+1}$ for arbitrary $\zeta \in C_0^\infty(\Omega)$. Whence, $d/dt\, \mathcal{I}(\tilde{u} + t\zeta)|_{t=0} = 0$ and for all $\zeta \in C_0^\infty(\Omega)$,

$$\int_\Omega \{F_z(x, \tilde{u}, \tilde{u}_x)\zeta_x + F_y(x, \tilde{u}, \tilde{u}_x)\zeta\}\, dx = 0.$$

Consequently $\tilde{u}$ is a generalized solution of the Euler–Lagrange equations for $\mathcal{I}(u)$, as required.

We now consider the following second-order semilinear boundary value problem defined on the bounded domain $\Omega \subset \mathbb{R}^N$

$$(6.1.25) \quad Lu + f(x, u) = 0, \qquad u|_{\partial\Omega} = g(x).$$

Here L is a linear second-order elliptic differential operator with smooth coefficients.

The argument used in the second part of (6.1.22) applies whenever we can find an a priori bound for $|\tilde{u}(x)|$. In fact, we prove

(6.1.26) The system (6.1.25) always has a solution $\tilde{u}(x)$ minimizing the associated functional $\mathcal{I}$, independent of any growth restriction on the function $f(x, u)$ provided $g(x)$ is continuous,

$$Lu = \sum_{|\alpha|, |\beta| = 1} D^\alpha \{a_{\alpha\beta}(x) D^\beta u\} - cu$$

with $c > 0$, while for some number $M > 0$

(6.1.27) $$(\operatorname{sgn} y)f(x, y) < 0 \qquad \text{for} \quad |y| > M > \max_{\partial\Omega} |g(x)|.$$

Furthermore, under this proviso, all solutions $w(x)$ in Ω of the system (6.1.25) satisfy $\sup |w(x)| \leq M$.

Proof: We first establish the second part of (6.1.26) by using the maximum principle for second-order elliptic equations. Let $f_M(x, y)$ be the Lipschitz continuous function defined by

(6.1.28) $$f_M(x, y) = \begin{cases} f(x, M) & \text{for} \quad y > M, \\ f(x, y) & \text{for} \quad -M \leq y \leq M, \\ f(x, -M) & \text{for} \quad y < -M. \end{cases}$$

Then any solutions $w_M(x)$ in Ω of

(6.1.29) $$Lu + f_M(x, u) = 0, \qquad u|_{\partial\Omega} = g$$

coincide with those of (6.1.25). Indeed, at a positive maximum $\overline{w}$ of $w_M(x)$ in Ω, $Lu = 0$. Thus $f_M(x, u) \geq 0$ so that by the maximum principle for L (see Protter and Weinberger, 1967), $\overline{w} \leq M$. On the other hand, at a negative minimum $\tilde{w}$, the same argument yields $\tilde{w} \geq -M$. Thus $|w_M(x)| \leq M$ and $f_M(x, w_M(x)) = f(x, w_M(x))$.

Now since $f_M(x, u)$ automatically satisfies the growth restrictions needed to satisfy (6.1.3), the resulting minimization problem associated with (6.1.29) is easily solved, by $\bar{u}(x)$ (say). On the other hand, $|\bar{u}(x)| \leq M$ and $\bar{u}(x)$ satisfies (6.1.25). The hypothesis (6.1.27) also implies that with $F_M(x, u) = \int_0^u f_M(x, s)\, ds$, $\int_\Omega F_M(x, u) \geq \int_\Omega F(x, u)$. Hence $\bar{u}(x)$ also minimizes $\mathcal{I}(u) = \{\frac{1}{2}(Lu, u) - \int_\Omega F(x, u)\}$ since

$$\mathcal{I}_M(u) = \left\{ \tfrac{1}{2}(Lu, u) - \int_\Omega F_M(x, u) \right\} \leq \mathcal{I}(u),$$

over the class of functions in $W_{1,2}(\Omega)$ with prescribed boundary values $g(u)$.

The importance of (6.1.26) resides in its independence of nonuniqueness considerations. This point is well illustrated by considering the Dirichlet problem over Ω

(6.1.30) $$\epsilon^2 \Delta u + u - g^2(x)u^3 = 0, \qquad u|_{\partial\Omega} = 0$$

discussed briefly in Section 4.4. We now prove a definitive result on the continuation of the solution $u_1(x, \epsilon)$ described in Section 4.4C, even though for sufficiently small ϵ this system will have an arbitrarily large number of distinct solutions (as we shall prove in Section 6.6).

(6.1.31) The solution $u_1(x, \epsilon)$ of (6.1.30) described in Section 4.4 can be uniquely continued as a continuous positive function of ϵ to the open interval $(0, \lambda_1^{-1/2})$ and no further, where λ_1 is the lowest eigenvalue of the Laplacian on Ω subject to null boundary conditions.

Proof: First we observe that for each $\epsilon > 0$ the equation satisfies the hypotheses of (6.1.22) with $K = \sup_\Omega(1/|g(x)|)$ so that if $\mathcal{I}_\epsilon(u) = \int_\Omega(\epsilon^2|\nabla u|^2 - u^2 + \frac{1}{2}g^2u^4)$, then inf $\mathcal{I}(u)$ over $\mathring{W}_{1,2}(\Omega)$ is attained at $v_1(x, \epsilon) \in C^2(\Omega)$ (say) and $v_1(x, \epsilon)$ satisfies (6.1.30). Furthermore, for $\epsilon^2 > 1/\lambda_1$, by the variational characterization of λ_1, $\mathcal{I}_\epsilon(u) > 0$ for smooth $u(x) \not\equiv 0$ in $\mathring{W}_{1,2}(\Omega)$. Thus for $\epsilon^2 > 1/\lambda_1$, $v_1(x, \epsilon) \equiv 0$. On the other hand, for $\epsilon^2 < 1/\lambda_1$, $\mathcal{I}_\epsilon(u_1) < 0$ for u_1 the positive eigenfunction associated with λ_1, while for $u \in \mathring{W}_{1,2}(\Omega)$, $|u| \in \mathring{W}_{1,2}(\Omega)$ and $\mathcal{I}_\epsilon(u) = \mathcal{I}_\epsilon(|u|)$. Consequently, (we may suppose) $v_1(x, \epsilon) > 0$ in Ω for $\epsilon < 1/\lambda_1$. Thus it remains to prove that $v_1(x, \epsilon)$ is the unique positive solution of (6.1.30) and that $v_1(x, \epsilon)$ depends continuously on ϵ.

(6.1.32) **Lemma** The solution $u_1(x, \epsilon)$ of (6.1.30) is the only positive solution and depends continuously on ϵ for $\epsilon \in (0, \lambda_1^{-1/2})$.

Proof: Assume that u_1 and u_2 are distinct positive solutions of (6.1.30) for fixed $\epsilon^2 \in (0, \lambda_1^{-1})$. Then, with $\lambda = \epsilon^{-2}$, the difference $v = u_1 - u_2$ satisfies the system

$$(6.1.33)\qquad \Delta v - \lambda g^2\{u_1^2 + u_1u_2 + u_2^2\}v + \mu v = 0 \quad \text{in } \Omega, \qquad v|_{\partial\Omega} = 0,$$

for $\mu = \lambda$. Regarding (6.1.33) as an eigenvalue problem for μ, with λ fixed, denote the smallest eigenvalue of (6.1.33) by μ_1. Similarly, regarding the positive solution u_1 as an eigenfunction of the system

$$(6.1.34)\qquad \Delta w - \lambda g^2u_1^2w + \nu w = 0 \quad \text{in } \Omega, \qquad w|_{\partial\Omega} = 0,$$

for $\nu = \lambda$, denote by ν_1 the smallest eigenvalue of (6.1.34) in which λ is fixed. The variational characterization of the lowest eigenvalues μ_1 and ν_1 implies that $\mu_1 > \nu_1$. On the other hand, positive eigenfunctions of (6.1.34) belong to the lowest eigenvalue, we have $\lambda = \nu_1$; and, by the definition of μ_1 as a lowest eigenvalue, $\mu_1 \leqslant \lambda$. Therefore, $\mu_1 \leqslant \nu_1$, and we have a contradiction and the conclusion is that $v \equiv 0$, i.e., that $u_1 \equiv u_2$.

Finally, we prove in two steps that $u_1(x, \epsilon)$ depends continuously on ϵ. First, suppose $\epsilon_n \to \bar{\epsilon}, \in (0, \lambda_1^{-1/2})$. Then $u_1(x, \epsilon_n)$ satisfies (6.1.30), and consequently the two sequences $\{|\Delta u_1(x, \epsilon_n)|\}$ and $\{|u_1(x, \epsilon_n)|\}$ are uniformly bounded over Ω by an absolute constant M, say. Again, there is a positive constant c_Ω such that

$$\sup_\Omega |\nabla u_1(x, \epsilon_n)| \leqslant c_\Omega, \qquad \sup_\Omega |\Delta u_1(x, \epsilon_n)| \leqslant c_\Omega M.$$

Thus the sequence $\{u_1(x, \epsilon_n)\}$ is uniformly bounded and equicontinuous. By an application of the Arzela-Ascoli theorem, $\{u_1(x, \epsilon_n)\}$ has a uniformly convergent subsequence $\{u_1(x, \epsilon_{n_j})\}$ with limit $\bar{u}(x)$. Clearly $\bar{u}$ satisfies (6.1.30) for $\epsilon = \bar{\epsilon}$, is nonnegative since it is the uniform limit of nonnegative functions, and is not identically zero since $\mathcal{I}_{\epsilon_{n_j}}(u_1(x, \epsilon_{n_j})) \to \mathcal{I}_{\bar{\epsilon}}(\bar{u})$ and min $\mathcal{I}_\epsilon(u)$ is a monotone decreasing function of ϵ that is strictly positive for $\epsilon < \lambda_1^{-1/2}$. Thus $\bar{u} = u_1(x, \bar{\epsilon})$ by uniqueness. Secondly, if we suppose that $\sup_\Omega|u_1(x, \epsilon_n) - u_1(x, \bar{\epsilon})| \not\to 0$, then for some subsequence, which we relabel $\{\epsilon_n\}$, there is an absolute constant such that

$$\sup_\Omega |u_1(x, \epsilon_n) - u_1(x, \bar{\epsilon})| \geqslant \alpha > 0,$$

which contradicts the above result.

6.2 Specific Minimization Problems from Geometry and Physics

In order to sharpen the abstract results of the preceding section, we now take up some specific minimization problems of importance in differential geometry and mathematical physics.

6.2A Hermitian metrics of constant negative Hermitian scalar curvature

The classic uniformization theorem for Riemann surfaces discussed in Section 1.2 implies that any C^∞ compact Riemannian 2-manifold $(\mathfrak{M}, g)$ admits a conformally equivalent metric $\tilde{g}$ of constant negative Gaussian curvature if and only if the Euler characteristic of M, $\chi(\mathfrak{M})$ is negative. We proved this result in Chapter 5 as an application of Hadamord's theorem (5.1.5). Here we shall prove an analogous result (for which (5.1.5) is not applicable) for compact complex Kähler manifolds *of higher dimension*, by replacing the Gaussian curvature with an appropriate scalar curvature function.

For such a complex manifold $\mathfrak{M}$, the necessary and sufficient condition we find depends only on the sign of the integral of the appropriate scalar curvature function for $(\mathfrak{M}, g)$. Consequently, the result is a direct generalization of the case for complex dimension 1, by the classical Gauss–Bonnet theorem since our result is generally independent of g.

The proof we give rests heavily on the minimization methods and the global theory of semilinear elliptic partial differential equations on manifolds. The main difficulty in the higher dimensional cases consists in finding substitutes for the appropriate Sobolev imbedding theorems which fail in this context, for $\dim_{\mathbf{C}} \mathfrak{M} > 1$ since then $\exp u$ is not integrable for arbitrary $u \in W_{1,2}(\mathfrak{M}, g)$.

(I) Formulation of the problem as partial differential equations Let $\mathfrak{M}$ be a C^∞ complex compact manifold of complex dimension N with a Kähler metric g defined (in local coordinates) by setting $ds^2 = \sum_{\alpha,\beta} g_{\alpha,\beta}\, dz^\alpha\, d\bar{z}^\beta$. Then if σ is a real C^∞ function defined on M, we consider the Hermitian metric $\tilde{g}$ defined by setting $\overline{ds}^2 = e^{2\sigma}\, ds^2$. Then we shall find that the Hermitian scalar curvatures R and $\tilde{R}$ of $(\mathfrak{M}, g)$ and $(\mathfrak{M}, \tilde{g})$, respectively, are related by the formula

$$(6.2.1) \qquad \tilde{R} = e^{-2\sigma}\{R - N\,\Delta\sigma\},$$

where Δ denotes the associated real Laplace–Beltrami operator defined on

(M, g). This formula is derived as follows. Relative to the *Hermitian connection*,[1] the components of the Ricci tensor $\tilde{R}_{\alpha\bar{\beta}}$ relative to $(M, \tilde{g})$ are given by the expression

$$\text{(6.2.2)} \qquad \tilde{R}_{\alpha\bar{\beta}} = -\frac{\partial^2 \log \tilde{G}}{\partial z^\alpha \partial \bar{z}^\beta} \qquad \text{with} \quad \tilde{G} = \det |\tilde{g}_{\alpha\bar{\beta}}|.$$

Since $\tilde{G} = e^{2N\sigma}G$, we find that in terms of the components of the Ricci tensor $R_{\alpha\beta}$ for $(\mathfrak{M}, g)$

$$\tilde{R}_{\alpha\beta} = R_{\alpha\beta} - 2N \frac{\partial^2 \sigma}{\partial z^\alpha \, \partial \bar{z}^\beta}.$$

Since the desired scalar curvatures are the traces of their respective Ricci tensors, we find (as in Appendix B)

$$\text{(6.2.3)} \qquad \tilde{R}e^{2\sigma} = R - 2N\Box\sigma,$$

where $\Box = \frac{1}{2}\Delta$ is the "complex Laplacian" relative to $(\mathfrak{M}, g)$. Clearly (6.2.3) yields (6.2.1). Thus if c^2 is some positive (nonzero) constant, the determination of the conformal metric with (Hermitian) scalar curvature $-c^2$ can be effected by proving that the partial differential equation

$$\text{(6.2.4)} \qquad N\,\Delta\sigma - R - c^2 e^{2\sigma} = 0$$

has a globally defined C^∞ solution σ over $(\mathfrak{M}, g)$. Clearly, if $\sigma(x)$ satisfies (6.2.4), we find

$$\int_{\mathfrak{M}} R(x)\, dV = -c^2 \int_{\mathfrak{M}} e^{2\sigma}\, dV < 0$$

by integrating (6.2.4) over $\mathfrak{M}$. Thus an immediate necessary condition for the solvability of (6.2.4) is that $\int_{\mathfrak{M}} R(x)\, dV < 0$. In fact, this condition is also sufficient and we state

(6.2.5) **Theorem** A necessary and sufficient condition for a compact Kähler manifold $(\mathfrak{M}, g)$ to admit a conformally equivalent (Hermitian) metric with scalar curvature a negative constant is that $\int_{\mathfrak{M}} R(x)\, dV < 0$.

Proof: Since the necessity of the condition was proved above, it suffices to prove that (6.2.4) is solvable for some $c \neq 0$ provided the mean value of $R(x)$ over $\mathfrak{M}$ is negative. To this end, we proceed to prove the following three lemmas.

Lemma (α) The infimum of the functional

$$\text{(6.2.6)} \qquad \mathcal{G}(u) = \int_{\mathfrak{M}} \left\{ \frac{N}{2} |\nabla u|^2 + R(x)u + \tfrac{1}{2} c^2 e^{2u} \right\} dV$$

[1] This is the unique connection that is compatible with the complex structure of $\mathfrak{M}$ and differs from the Levi-Civita connection associated with the Riemannian structure of $(\mathfrak{M}, g)$ if $\dim_{\mathbb{C}} \mathfrak{M} > 1$.

over the class of functions $W_{1,2}(\mathfrak{M}, g) = \{u \mid \int_{\mathfrak{M}}(|\nabla u|^2 + u^2)\, dV_g < \infty\}$ is attained by an element $\bar{u} \in W_{1,2}(\mathfrak{M}, g)$.

Lemma (β) The function $\bar{u}$ described in Lemma (α) can be chosen to be essentially bounded provided $\sup_M R(x) < 0$.

Lemma (γ) The function $\bar{u}$ described in Lemma (α) can be chosen to be a C^∞ function and $\bar{u}$ satisfies the equation (6.2.4).

We now proceed to prove these three results. Clearly on combining these lemmas we shall obtain a proof of the theorem whether or not $\sup_M R(x) < 0$.

Proof of Lemma (α) To prove inf $\mathcal{I}(u)$ over $W_{1,2}(\mathfrak{M}, g)$ is attained, we use (6.1.1). To achieve this we use the Hilbert space structure of $W_{1,2}(\mathfrak{M}, g)$ in the following steps;

(i): We first show $\mathcal{I}(u)$ defined by (6.2.6) is bounded below on $H = W_{1,2}(\mathfrak{M}, g)$ by η (say), $\eta > -\infty$. Then, if $\delta = \inf_H \mathcal{I}(u)$, set $\mathcal{I}_{\delta+1} = \{u \mid u \in H, \mathcal{I}(u) \leqslant \delta + 1\}$ and apply (6.1.1) to $\mathcal{I}_{\delta+1}$ by verifying that (ii) $\mathcal{I}(u)$ is coercive on $\mathcal{I}_{\delta+1}$, and (iii) weakly lower semicontinuous on $\mathcal{I}_{\delta+1}$, to show that $\mathcal{I}_{\delta+1}$ is sequentially weakly closed.

To show that $\mathcal{I}(u)$ is bounded below on $W_{1,2}(\mathfrak{M}, g)$ we note that on setting $u = u_0 + \bar{u}$, where u_0 has mean value zero over $(\mathfrak{M}, g)$ and $\bar{u} = (\mathrm{vol}(\mathfrak{M}, g))^{-1}\int_{\mathfrak{M}} u\, dV$,

$$\mathcal{I}(u) = \int \frac{N}{2} |\nabla u_0|^2 + \int R(x) u_0 + \bar{u} \int R(x) + \tfrac{1}{2} c^2 e^{2\bar{u}} \int e^{2u_0}\, dV.$$

Thus by combining Poincaré's inequality $\|\nabla u_0\|_{0,2} \geqslant c\|u_0\|_{0,2}$ and the fact that $e^{2u_0} \geqslant 1 + 2u_0$, we find from the Cauchy–Schwarz inequality that for any $\epsilon > 0$,

$$(6.2.7) \qquad \mathcal{I}(u) \geqslant \frac{N}{2}\|\nabla u\|_{0,2}^2 - \frac{c}{\epsilon}\|R(x)\|_{0,2}^2 - c\epsilon\|\nabla u\|_{0,2}^2 + \bar{u}\int_{\mathfrak{M}} R(x)\, dV + \tfrac{1}{2} c^2 e^{2\bar{u}}\, \mathrm{vol}(\mathfrak{M}, g).$$

Thus setting $\epsilon = N/2c$, we find

$$(*) \qquad \mathcal{I}(u) \geqslant \frac{c^2}{N}\|R(x)\|_{0,2}^2 + \eta(\bar{u}) \quad \text{with } \eta(\bar{u}) \to \infty \text{ as } |\bar{u}| \to \infty,$$

since $\int R(x) < 0$. Consequently, $\inf_H \mathcal{I}(u) > -\infty$.

(ii): In the same way we can verify that $\mathcal{I}(u)$ is coercive on $\mathcal{I}_{\delta+1}$, where $\delta = \inf_H \mathcal{I}(u)$. To this end, let $u_n \in W_{1,2}(\mathfrak{M}, g)$ be such that $\|u_n\|_{1,2} \to \infty$, then we show that $\mathcal{I}(u_n) \to \infty$. Thus we note that an equivalent norm on $W_{1,2}(\mathfrak{M}, g)$ can be chosen to be $\|u\|_{1,2}^2 = \|\nabla u_0\|^2 + |\bar{u}|^2$. Hence we need only investigate the behavior of $\mathcal{I}(u_n)$ as either $\|\nabla u\|_{0,2}$

$\to \infty$ with $|\bar{u}_n|^2$ bounded or $|\bar{u}_n|^2 \to \infty$. In the former case the result follows from (6.2.7) by setting $c\epsilon = \frac{1}{4}$. In the latter case we use (*).

(iii): The weak sequential lower semicontinuity of $\mathcal{I}(u)$ on $\mathcal{I}_{\delta+1}$ follows easily. Indeed, if $u_n \to u$ weakly in $W_{1,2}(\mathfrak{M}, g)$, $u_n \to u$ strongly in $L_2(\mathfrak{M}, g)$ by Rellich's lemma. Thus $\int R(x)u_n \to \int R(x)u$. Consequently, by Fatou's theorem and (6.1.3), $\underline{\lim}\ \mathcal{I}(u_n) \geqslant \mathcal{I}(u)$. From this fact one also observes that in fact $\mathcal{I}_{\delta+1}$ is weakly sequentially closed.

Proof of Lemma (β) We shall prove the existence of a finite real number $k > 0$ such that for any element $u \in \mathcal{I}_{\delta+1}$ the truncated function

(6.2.8) $$u_k = \begin{cases} \inf(u, k) & \text{for } u \geqslant 0, \\ \sup(u, -k) & \text{for } u \leqslant 0, \end{cases}$$

is such that $\mathcal{I}(u_k) \leqslant \mathcal{I}(u)$. For then $u_k \in \mathcal{I}_{\delta+1} \cap W_{1,2}(\mathfrak{M}, g)$, and so the minimizing sequence $\{u_n\}$ must also. Hence $\bar{u}$ of Lemma (α) can be chosen to consist of functions with $\text{ess sup}_M |u| \leqslant k$. To find such a real number k we first observe that if $u \in W_{1,2}(\mathfrak{M}, g)$, $u_k \in W_{1,2}(\mathfrak{M},g)$, and in fact $\|\nabla u_k\|_{0,2} \leqslant \|\nabla u\|_{0,2}$. Therefore, it suffices to consider the effect of altering the function u to u_k for the functional

$$J(u) = \int_{\mathfrak{M}} \{R(x)u + \tfrac{1}{2}c^2e^{2u}\}\, dV.$$

Since $\mathfrak{M}$ is compact, and $\sup_{\mathfrak{M}} R(x) < 0$, there are two positive numbers a_1 and a_2 such that $-a_1 \leqslant R(x) \leqslant -a_2$. Hence as $u \to +\infty$, the integrand $f(u) = R(x)u + \frac{1}{2}c^2e^{2u} \to \infty$. Therefore there is a positive number k_1 such that $f(k_1) \leqslant f(u)$ for $u \geqslant k_1$. On the other hand, if $u \to -\infty$, $f(u)$ also tends to ∞ since $\sup_{\mathfrak{M}} R(x) < 0$. Thus there is a positive number k_2 such that $f(k_2) \leqslant f(u)$ for $u \leqslant -k_2$. Consequently, the desired positive number k in (6.2.8) can be chosen to be $\sup(k_1, k_2)$.

Proof of Lemma (γ) We divide the proof into two cases:

Case I: $\sup_{\mathfrak{M}} R(x) < 0$. We use Lemma ($\beta$) to assert that $\bar{u}$ is essentially bounded. Hence if v is any C^∞ function defined on $(\mathfrak{M}, g)$, the minimality of $\bar{u}$ implies that

(6.2.9) $$\lim_{\epsilon \to 0} \frac{1}{\epsilon}\{\mathcal{I}(\bar{u} + \epsilon v) - I(\bar{u})\}$$
$$\equiv \int_{\mathfrak{M}} \{N\, \nabla \bar{u}_1\, \nabla v + R(x)v + c^2e^{2\bar{u}}v\} = 0.$$

Since the C^∞ functions are dense in $W_{1,2}(\mathfrak{M}, g)$, we find that the integral identity on the right-hand side of (6.2.9) holds for all $v \in W_{1,2}(\mathfrak{M},g)$. Thus $\bar{u}$ can be considered as a weak solution of the *linear* nonhomogeneous equation in ω

(6.2.10) $$N\,\Delta\omega = R + c^2e^{2\bar{u}}.$$

Since the right-hand side of (6.2.10) is in $L_p(\mathfrak{M}, g)$ for finite $p > 1$, the L_p regularity theory for linear elliptic equations implies that $\bar{u} \in W_{2,p}(\mathfrak{M}, g)$ for all $1 < p < \infty$. Consequently by the Sobolev imbedding theorem (after a possible redefinition on a set of measure zero) $\bar{u} \in C_{1,\alpha}(\mathfrak{M}, g)$ (the space of functions with Hölder continuous first derivatives of exponent α). Thus, the Schauder regularity theory applies to (6.2.10) since now $\omega = \bar{u}$ satisfies (6.2.10) (in the weak sense) with the right-hand side in $C_{1,\alpha}(\mathfrak{M}, g)$. Consequently, $\bar{u} \in C^{2,\alpha}(\mathfrak{M}, g)$. Thus $\bar{u}$ satisfies (6.2.4) in the classical sense and iteration of the Schauder regularity theory yields the fact that $\bar{u} \in C^{\infty}(\mathfrak{M}, g)$.

Case II: $\sup_M R(x) \geqslant 0$. We reduce this case to Case I by the following device. Write a tentative solution u of (6.2.4) in the form $u = v + w$, where

(6.2.11) $$N\,\Delta v - \bar{R} - c^2 e^{2w} e^{2v} = 0 \quad \text{with} \quad \bar{R} = \{\mathrm{vol}(\mathfrak{M}, g)\}^{-1} \int_{\mathfrak{M}} R\, dV_g,$$

(6.2.12) $$N\,\Delta w - R(x) + \bar{R} = 0.$$

Clearly, (6.2.12) is uniquely solvable up to an additive constant, and since $\bar{R} < 0$ by hypothesis, the arguments used in Lemmas (α) and (β) apply to (6.2.11) with w fixed. Furthermore, once a minimizing essentially bounded $\bar{v}$ is found for the functional $I(v)$ related (6.2.11), the regularity argument of Case I holds. Since the solution w of (6.2.12) is also a C^{∞} function, $u = \bar{v} + w$ is a C^{∞} function, and clearly satisfies (6.2.4), as can easily be seen by adding (6.2.11) and (6.2.12).

Now we interpret the condition $\int_{\mathfrak{M}} R(x)\, dV < 0$ in terms of analytic invariants of $(\mathfrak{M}, g)$. In the case of one complex dimension, the Gauss–Bonnet theorem implies the Euler–Poincaré characteristic of $\mathfrak{M}$, $\chi(\mathfrak{M})$, is negative. More generally, an interesting formula from the theory of Kähler manifolds implies that

$$\int_{\mathfrak{M}} R(x)\, dV = k_N \int_{\mathfrak{M}} c_1 \wedge \omega^{N-1}, \qquad k_N = \text{positive const.},$$

where c_1 is the first Chern class of $\mathfrak{M}$ and ω is the fundamental form of $(\mathfrak{M}, g)$. Thus we have

(6.2.13) **Corollary** A necessary and sufficient condition for a compact Kähler manifold $(\mathfrak{M}, g)$ of complex dimension N to admit a metric $\tilde{g}$, conformal to g, with constant negative Hermitian scalar curvature is $\int_{\mathfrak{M}} c_1 \wedge \omega^{N-1} < 0$.

For algebraic varieties of $\mathfrak{M}$ this condition is expressible in terms of the sign of the degree of the canonical divisor of $\mathfrak{M}$.

6.2B. Stable equilibrium states in nonlinear elasticity

Generally speaking the equilibrium states of an elastic body B acted on by given conservative forces can be determined as the critical points of an appropriate smooth potential energy functional $\mathcal{I}(u)$. Because of the great complexity of the possible equilibrium states, it is important to determine those states that correspond to absolute minima of $\mathcal{I}(u)$. Indeed, by the remarks of Section 4.3B, such states will be *stable*. Here we take up the problem of demonstrating the actual attainment of the infimum of $\mathcal{I}(u)$ for various problems in elasticity.

Case I: Deformable plates: We suppose a thin flexible elastic plate is clamped along its edge and is acted on by the combined action of forces acting along its edge and a force $\bar{f}$ acting normal to the plane of the plate. The resulting equilibrium states are governed by the von Kármán equations (1.1.12). Our discussion of Chapter 2 shows that these equilibrium states are solutions of the operator equation $\mathcal{A}_\lambda(u) = f$, where $f \in \mathring{W}_{2,2}(\Omega)$ is a representation of $\bar{f}$ and is proportional to the magnitude of $\bar{f}$. Moreover in (2.5.7) and (2.7.18) we showed that the operator

$$\mathcal{A}_\lambda(u) = u + Cu - \lambda Lu$$

is (i) a gradient mapping of $\mathring{W}_{2,2}(\Omega)$ into itself with

$$\mathcal{I}_\lambda(u) = \tfrac{1}{2}\|u\|_{2,2}^2 - (\lambda/2)(Lu, u) + \tfrac{1}{4}(Cu, u)$$

so that $\mathcal{I}'_\lambda(u) = \mathcal{A}_\lambda(u)$, and (ii) $\mathcal{A}_\lambda(u)$ is a *proper* mapping (in the sense of Section 2.7). Analogous results clearly hold for $\tilde{\mathcal{I}}_\lambda(u) = 2\mathcal{I}_\lambda(u) - (f, u)$.

Thus to demonstrate the coerciveness of $\tilde{\mathcal{I}}_\lambda(u)$ on $\mathring{W}_{2,2}(\Omega)$ let $\|u_n\| \to \infty$. Then

$$\begin{aligned}\tilde{\mathcal{I}}_\lambda(u_n) &= \|u_n\|^2 + \tfrac{1}{2}(Cu_n, u_n) - \lambda(Lu_n, u_n), - (f, u_n)\\ &= \|u_n\|^2 + \tfrac{1}{2}\|C(u_n, u_n)\|^2 - \lambda(C(u_n, u_n), F_0) - (f, u_n).\end{aligned}$$

Consequently for any $\epsilon > 0$, by the Cauchy–Schwarz inequality

$$\tilde{\mathcal{I}}_\lambda(u_n) \geqslant \|u_n\|^2 + (\tfrac{1}{2} - \lambda\epsilon)\|C(u_n, u_n)\|^2 - (\lambda/\epsilon)\|F_0\|^2 - 2\|f\|\,\|u_n\|.$$

Choosing $\tfrac{1}{2} = \lambda\epsilon$ we find

$$\tilde{\mathcal{I}}_\lambda(u_n) \geqslant \|u_n\|^2 - 2\|f\|\,\|u_n\| - 2\lambda^2\|F_0\|^2.$$

Thus $\tilde{\mathcal{I}}_\lambda(u_n) \to \infty$ for fixed λ as $n \to \infty$. The weak lower semicontinuity of $\tilde{\mathcal{I}}_\lambda(u)$ is an immediate consequence of the criterion (6.1.3) and the complete continuity of the operators L and C. Thus from (6.1.1) we conclude

(6.2.14) For *any fixed* λ *and* f, *the functional* $\tilde{\mathcal{I}}_\lambda(u)$ *is bounded below on* $\mathring{W}_{2,2}(\Omega)$ *and attains its infimum on that set.* Moreover this infimum yields a smooth solution of the associated von Kármán equations.

Case II: Deformable shallow shells: We now generalize the result for plates just obtained by supposing that instead of being flat, the thin elastic structure S under consideration has some initial curvature described by the functions $k_1(x, y)$, $k_2(x, y)$ that measure the Gaussian curvature of S. We suppose that the shell is acted upon by forces on its boundary as well as by forces Z acting normal to the shell. The resulting deformations can be determined by the nonlinear von Kármán equations (4.3.1)–(4.3.2) together with the boundary conditions

(4.3.23)–(4.3.24). Again, for simplicity, we suppose that $Z = \lambda\psi_0$, and that λF_0 is a solution of the linear problem $\Delta^2 F = 0$ together with the boundary conditions (4.3.23)–(4.3.24). Then solutions (w, f) can be sought in the form $(w, F + \lambda F_0)$, where w and F satisfy

(6.2.15) $$\Delta^2 F = -\tfrac{1}{2}[w, w] - (k_1 w_x)_x - (k_2 w_y)_y,$$

(6.2.16) $$\Delta^2 w = [F, w] + \lambda[F_0, w] + (k_1 F)_x + (k_1 F)_y + \lambda Z',$$

where $Z' = Z + (k_1 F_0)_x + (k_1 F_0)_y$ together with the homogeneous boundary conditions

(6.2.17) $$D^\alpha F|_{\partial\Omega} = D^\alpha w|_{\partial\Omega} = 0 \qquad \text{for } |\alpha| \leqslant 1.$$

As in (2.5.7), these equations can be written in the form of operator equations in $\mathring{W}_{2,2}(\Omega)$:

(i) $$F = -\tfrac{1}{2}C(w, w) - Lw,$$

(ii) $$w = C(F, w) + \lambda C(F_0, w) + Lw + \lambda Z'.$$

Substituting (i) into (ii), and using the results of (2.5.7), we find that the associated potential energy functional can be chosen to be

(6.2.18) $$\mathcal{G}(w, \lambda) = \|w\|^2 + \|\tfrac{1}{2}C(w, w) + L_1 w\|^2 - \lambda(Lw, w) - \lambda(Z', w).$$

We now will show

(6.2.19) **Theorem** For all Z, ψ and functions ψ_0, k_1, k_2, inf $\mathcal{G}(w, \lambda)$ over $\mathring{W}_{2,2}(\Omega)$ is finite and attained by an element of $\mathring{W}_{2,2}(\Omega)$ that can be associated with a solution (w, F) of (6.2.15)–(6.2.17).

Proof: We apply (6.1.1) and show that $\mathcal{G}(w, \lambda)$ defined by (6.2.18) is weakly lower semicontinuous and coercive on $\mathring{W}_{2,2}(\Omega)$. The weak lower semicontinuity on $\mathring{W}_{2,2}(\Omega)$ is obvious by (6.1.3) since by (2.5.7) the operators $C(w, w)$, $L_1(w)$, and L are completely continuous on $\mathring{W}_{2,2}(\Omega)$, which means that $\mathcal{G}(w, \lambda)$ is the sum of a convex and a weakly sequentially continuous functional. Again it remains to prove the coerciveness of $\mathcal{G}(w, \lambda)$. To this end, consider the set

(6.2.20) $$E_1 = \{ w \mid \|w\| = 1, \|w\|^2 - \lambda(Lw, w) < \tfrac{1}{2} \}$$

Clearly on the weak closure of E_1, E_1', $\inf_{E_1'}\|w\| > 0$, for otherwise there would be a sequence $w_n \in E_1$ such that $w_n \to 0$ weakly, so that by the complete continuity of L, $(Lw_n, w_n) \to 0$ and hence $\|w_n\| < 1$ for sufficiently large n. Thus on E_1', inf $\|C(\bar{w}, \bar{w})\|^2 \geqslant \alpha^2 > 0$ (since otherwise there would be a $\bar{w} \in E_{1'}$ such that $C(\bar{w}, \bar{w}) = 0$, so that the surface $\bar{w} = \bar{w}(x, y)$ has zero Gaussian curvature a.e. and on $\partial\Omega$, $w = 0$, which implies $\bar{w} \equiv 0$). Now setting $w = \|w\|v$ with $\|v\| = 1$, we have for $v \in E_1$, as $\|w\| \to \infty$.

(6.2.21) $$G_2(w, \lambda) \geqslant \|w\|^2 \left\{ 1 - \lambda(Lv, v) + \frac{\alpha^2}{4}\|w\|^2 - \|L_1 v\|^2 \right\}$$
$$- \lambda(Z', w) \geqslant \|w\|^2 \left\{ \frac{\alpha^2}{4}\|w\|^2 - K \right\},$$

where K is a constant independent of w and v; and for $v \notin E_1$,

(6.2.22) $$G_2(w, \lambda) \geqslant \|w\|^2\{\|v\|^2 - \lambda(Lv, v)\} - \lambda(Z', w) \geqslant \tfrac{1}{4}\|w\|^2 - K.$$

Hence in both cases, $G_2(w, \lambda) \to \infty$ as $\|w\| \to \infty$.

6.2C Plateau's problem

Here we solve the (simply connected) Plateau's problem for recitifiable Jordan curves $\Gamma \subset \mathbb{R}^3$, (cf. Section 1.1A) as a modification of the ideas discussed in Section 6.1. More precisely, we seek a smooth, simply connected parametric surface S in $\mathbb{R}^3$ spanning Γ such that the area of S is minimal. Thus if we let Ω be the open unit disk in $\mathbb{R}^2$, we seek a vector $r(x, y) = (u_1(x, y), u_2(x, y), u_3(x, y))$ that represents a surface S spanning Γ in such a way that (1) $\partial\Omega$ is continuously mapped onto Γ in a one-to-one manner, and (2) the area of S,

$$A(S) = \iint_\Omega (|J(u_1, u_2)|^2 + |J(u_2, u_3)|^2 + |J(u_1, u_3)|^2)^{1/2} dx\, dy \tag{6.2.23}$$

is minimized, where $|J(u, v)|$ is the Jacobian determinant of u and v with respect to x, y.

An important simplification of this problem results from differential geometric considerations. For any surface $S = \{r \mid r = [u_1(x, y), u_2(x, y), u_3(x, y)]\}$, we write the first fundamental form as

$$ds^2 = dr \cdot dr = g_{11}\, dx^2 + 2g_{12}\, dx\, dy + g_{22}\, dy^2,$$

where

$$g_{11} = r_x \cdot r_x, \qquad g_{12} = r_x \cdot r_y, \quad \text{and} \qquad g_{22} = r_y \cdot r_y,$$

and the area of S as

$$A(S) = \iint_\Omega (g_{11}g_{22} - g_{12}^2)^{1/2} dx\, dy.$$

Since for any three numbers α, β, γ (all positive)

$$\sqrt{\alpha\gamma - \beta^2} \leqslant \sqrt{\alpha\gamma} \leqslant \tfrac{1}{2}\{\alpha + \gamma\},$$

with equality holding if and only if $\alpha = \gamma$, $\beta = 0$, we find from this fact that

$$A(S) \leqslant \tfrac{1}{2} \iint_\Omega (g_{11} + g_{22})\, dx\, dy$$

with equality holding if and only if $g_{12} = 0$ and $g_{11} = g_{22}$. Consequently, if isothermal parameters are introduced on S,

$$A(S) = \iint_\Omega \tfrac{1}{2}(g_{11} + g_{22})\, dx\, dy = \tfrac{1}{2} \iint_\Omega (|r_x|^2 + |r_y|^2)\, dx\, dy.$$

Since the surface area of S is independent of parametrization, by choosing isothermal parameters on S, $A(S)$ is minimized by minimizing the Dirichlet integral

$$D[r(x, y)] = \iint_\Omega \{|\nabla u_1|^2 + |\nabla u_2|^2 + |\nabla u_3|^2\}\, dx\, dy \tag{6.2.24}$$

over all vectors $r = (u_1, u_2, u_3)$ that satisfy the boundary condition (1) on $\partial\Omega$. Since the resulting Euler–Lagrange equation is simply

$$\Delta u_1 = \Delta u_2 = \Delta u_3 = 0,$$

we may restrict attention to vectors $r = (u_1, u_2, u_3)$ whose components are harmonic functions in Ω (i.e., harmonic vectors) and satisfy the conditions

$$(6.2.24') \qquad |r_x|^2 = |r_y|^2 \qquad \text{and} \qquad r_x \cdot r_y = 0.$$

We outline the main steps in the proof of

(6.2.25) **Theorem** Suppose γ is a rectifiable Jordan curve in $\mathbb{R}^3$. Then there is a harmonic vector r_∞ that minimizes the Dirichlet integral (6.2.24) over the class $\mathcal{C}$ of all vectors $r(x, y)$ whose components lie in $W_{1,2}(\Omega)$ and satisfy:

(i) $D[r(x, y)] \leqslant M$, where M is sufficiently large;
(ii) $r(x, y)$ satisfies the boundary condition (6.2.24′).

The vector r_∞ is harmonic, and consequently is a solution of Plateau's problem.

Sketch of Proof: First, we observe that the natural admissible class $\mathcal{C}$ for the variational problem posed in (6.2.24) is nonvacuous. Indeed, if $r = (u_1, u_2, u_3)$ and the rectifiable Jordan curve is parametrized by arc length s, then in terms of polar coordinates (r, θ), we may write u_i in the Fourier series

$$u_i = a_0 + \sum_{k=1}^{\infty} r^k \{\alpha_k \cos ks + \beta_k \sin ks\} \qquad \text{for} \quad r < 1,$$

so that

$$\int_\Omega |\nabla u_i|^2 = \pi \sum_{k=1}^{\infty} k(\alpha_k^2 + \beta_k^2).$$

Consequently, $D[r] < \infty$ since the rectifiability of Γ implies that $\sum k(\alpha_k^2 + \beta_k^2)$ converges for some vector r in $\mathcal{C}$.

Next we show why $\inf D[r]$ is attained in $\mathcal{C}$. To this end, we note that the abstract arguments of (6.1.1) are not directly applicable since $D[r]$ is not defined on a natural reflexive Banach space X. However, we can argue as follows: For any minimizing sequence $r_n \in \mathcal{C}$ with $D[r_n] \to \inf_{\mathcal{C}} D[r]$, we can replace r_n by a sequence of harmonic vectors r_n' with the same boundary values as r_n, so that $r_n' \in \mathcal{C}$, and we thereby obtain an improved minimizing sequence since $D[r_n'] \leqslant D[r_n]$. In the same way, since the Dirichlet integral $D[r]$ is invariant under conformal transformations of Ω into itself, we suppose that harmonic vectors $\{r_n'\}$ are normalized by the requirement that they map three distinct points p_1, p_2, p_3, on $\partial\Omega$ onto three distinct points q_1, q_2, q_3 of γ. Then we use the compactness lemma for the resulting admissable class $\mathcal{C}$ proven by Garabedian (1964), viz., the existence of a uniformly convergent subsequence of $\{r_n'\}$ and the equicontinuity of the boundary values of $\{r_n'\}$. Hence (after possibly passing to a subsequence) $\{r_n'\}$ converges uniformly to a harmonic function r_∞ that satisfies the boundary condition (1); so that $r_\infty \in \mathcal{C}$. Therefore the lower semicontinuity of $D[r]$ over $\mathcal{C}$ implies that $\inf_{\mathcal{C}} D[r] = D[r_\infty]$.

Finally, we show that the harmonic vector r_∞ satisfies (6.2.24′). For this purpose it suffices to vary only the parametric representation of r_∞, while leaving the geometric surface fixed since the Dirichlet integral is not invariant under such changes of coordinates. Thus for small $\epsilon > 0$, let f: $(x, y) \to (x, y')$ as follows

$$x' = x + \epsilon\lambda(x, y), \qquad y' = y + \epsilon\mu(x, y)$$

be a diffeomorphism of Ω onto a domain Ω'. Then for any admissible harmonic vector $r(x, y)$, we set $Z_\epsilon(x', y') = r(x, y)$. Denoting by D' the Dirichlet integral taken over Ω' with respect to x', y', we find, after a short calculation that

$$\text{(6.2.26)} \qquad \frac{d}{d\epsilon} D'(Z_\epsilon)|_{\epsilon=0} = \int\int_\Omega \{r_x^2 - r_y^2\}\{\lambda_u - \mu_v\} + 2r_x \cdot r_y\{\lambda_v + \mu_u\}.$$

Now the Riemann mapping theorem for simply connected domains in the plane allows us to assert that the functions λ and μ may be chosen to be arbitrary continuous functions with piecewise continuous first derivatives (provided ϵ is small enough). Consequently, the functions $\lambda_u - \mu_v$ and $\lambda_v + \mu_u$ may be arbitrarily chosen in the class of piecewise continuous functions. Thus for $r = r_\infty$, we find that $(d/d\epsilon)D'(Z_\epsilon)|_{\epsilon=0} = 0$. and so (by the fundamental lemma of the calculus of variations) (6.2.26) implies that $r_x^2 = r_y^2$ and $r_x \cdot r_y = 0$ (i.e., the equations (6.2.24′) are satisfied). Hence $r_\infty(x, y)$ defines a parametric representation of the desired minimal surface.

6.2D Dynamic instability in Euclidean quantum field theory (in the mean field approximation)

A. Wightman (1974) has suggested the following device to describe certain nonuniqueness questions concerning simple quantum field theory models. We consider the following semilinear elliptic partial differential equation defined on $\mathbb{R}^2$ (see Section 1.1B for more motivation)

$$\text{(6.2.27)} \qquad \Delta u - Q'(u) = f, \qquad f \in C_0^\infty(\mathbb{R}^2),$$

where $Q(u) = \frac{1}{2} m^2u^2 + P(u)$ is a polynomial of even degree (> 2) that tends to ∞ as $|u| \to \infty$ and such that $Q(u) \geqslant \alpha u^2$ (for some fixed $\alpha > 0$). Then we prove the following result concerning the solutions of (6.2.27) in $W_{1,2}(\mathbb{R}^2)$ regarded as critical points of the functional

$$\mathcal{G}_f(u) = \int_{\mathbb{R}^2} \{ \tfrac{1}{2} |\nabla u|^2 + Q(u) + fu\} \, dV$$

(and later interpret the result for the quantum field theory in question).

(6.2.28) **Theorem** Under the given hypotheses on $Q(u)$, $c(f) = \inf \mathcal{I}_f(u)$ over $W_{1,2}(\mathbb{R}^2)$ is finite and attained by an element $u(f) \in C_0^\infty(\mathbb{R}^2)$. Moreover $u(f) \to 0$ as $|x| \to \infty$ and, if $\|f\|_{L_2}$ is sufficiently small, the absolute minimizer $u(f)$ is unique and tends to 0 as $\|f\|_{L_2} \to 0$.

Proof: The fact that $c(f)$ is finite and attained follows immediately from (6.1.1) and the fact that $f \in C_0^\infty(\mathbb{R}^2)$. Indeed for $|u|$ sufficiently large, $Q(u) + fu \geqslant (\alpha/2)u^2$ so the coerciveness and weakly lower semicontinuity follow easily from the criteria of Section 6.1. To prove the equation (6.2.27) has a unique solution near $w = 0$ for $\|f\|_{L_2}$ small enough, we use the contraction mapping theorem. Indeed, writing (6.2.27) in the form

$$(6.2.29) \qquad \Delta u - m^2 u - P'(u) = f; \qquad \deg P'(u) \geqslant 2,$$

we note that in $W_{1,2}(\mathbb{R}^2)$ for $m > 0$, $(\Delta - m)$ is invertible while $P'(u)$ is of higher order, thus for $\|f\|_{L_2(\mathbb{R}^2)}$ sufficiently small, (6.2.29) has a unique solution w_f in a small neighborhood of 0 that depends continuously on $\|f\|_{L_2}$. Moreover a simple computation of the second variation of $\mathcal{I}_f(u)$ at w_f shows for some $\beta > 0$

$$\delta^2 \mathcal{I}(w_f, v) = \int_{\mathbb{R}^2} \left[\tfrac{1}{2} |\nabla v|^2 + \left\{ \tfrac{1}{2} m^2 + P''(w_f) \right\} v^2 \right] \geqslant \beta \int_{\mathbb{R}^2} (|\nabla v|^2 + v^2),$$

so w_f is an isolated relative minimum of $\mathcal{I}(f)$. The fact that any minimizer $u_f \in C^\infty(\mathbb{R}^2)$ is also immediate from the regularity theory of Section 1.5.

To prove the minimizer $u(f)$ is unique for $\|f\|_{L_2(\mathbb{R}^2)}$ sufficiently small, we argue as follows: (a) the equation (6.2.29) has a unique solution w_f near $w = 0$, (b) we then identify $u(f)$ with w_f, provided $\|f\|_{L_2}$ is sufficiently small, (c) we show that 0 is the isolated critical value of $\mathcal{I}_0(u)$ corresponding to $c(0)$, and finally (d) that this absolute minima is stable in the sense that by perturbing $\mathcal{I}_0(u)$ to $\mathcal{I}_f(u)$, $c(f)$ is near $c(0)$. Thus to conclude (b), (c), and (d), it suffices to show that if $f = 0$, $c(0) = 0$ is an isolated critical value of $\mathcal{I}_0$ and that for sufficiently small $\epsilon > 0$ the critical points of $\mathcal{I}_f(u)$ in $\mathcal{I}_f^{-1}[c(f), c(f) + \epsilon]$ are contained in a sphere of radius $\delta(\epsilon)$ which tends to zero as $\|f\|_{L_2} \to 0$.

To show $c(0) = 0$ is an isolated critical value of $\mathcal{I}_0(u)$, suppose for some sequence $\{u_n\}$, $\mathcal{I}_0'(u_n) = 0$ and $\mathcal{I}_0(u_n) \to 0$; then by the coerciveness of $\mathcal{I}_0(u)$ we may suppose that $\{u_n\}$ is weakly convergent and since $\mathcal{I}_0(u_n) \to 0$, $\|u_n\|_{1,2} \to 0$. Consequently by (1.3.11), $\{u_n\}$ is strongly convergent to zero. This however contradicts (a).

Finally, we observe that if we consider the set $\mathcal{I}_f^{-1}[c(f), c(f) + \epsilon]$, we find for any critical point u on this set

$$\alpha \|u\|_{1,2}^2 \leqslant \epsilon + \|f\|_{L_2} \|u\|_{1,2}$$

which implies $\|u\|_{1,2} \to 0$ as ϵ and $\|f\|_{L_2} \to 0$.

In order to interpret (6.2.28) for quantum field theory we assume m^2 represents the mass of a single particle state and we wish to interpret the qualitative features of the quantum field theory predicted on the model (6.2.27) in terms of the polynomial $Q(v) = \frac{1}{2} m^2 v^2 + P(v)$, where we merely suppose $Q(v)$ is bounded below, so that $\inf Q(v) = \beta$ is attained by a finite number of values $c_1, c_2, \ldots, c_k$. If $k = 1$, (6.2.28) implies that (6.2.27) has one and only one absolute minima u_f such that $u_f - c_1 \in W_{1,2}(\mathbb{R}^2)$ provided $\|f\|_{L_2}$ is sufficiently small, and as this L_2 norm tends to zero, $u_f \to c_1$. If $k \geqslant 2$, (6.2.28) implies that after setting $u = v - c_i$ $(i = 1, 2, \ldots, k)$ a dynamical instability exists in the sense that by different perturbations f, we can find solutions $u_1(f)$, $u_2(f)$ whose properties and asymptotic behavior are quite distinct. See Fig. 1.2.

6.3 Isoperimetric Problems

In many problems involving critical points, both implicit parameters and constraints are present. In such cases, the methods described in Section 6.1 can succeed only if these parameters and constraints are introduced explicitly into the equations defining the problem.

The effect of the presence of these entities on a given functional $\mathcal{G}(u)$ is to produce solutions of "saddle point" type for the associated operator equation $\mathcal{G}'(u) = 0$. Thus, for example, a normal mode $x(t)$ with period β, of the linear Hamiltonian system $\ddot{x} + Ax = 0$ (described in Section 1.1) is a saddle point of the energy functional $\int_0^\beta (\dot{x}^2(t) - Ax(t) \cdot x(t))\, dt$ since normal modes must be orthogonal to the constant N-vectors.

In these cases, arguments based on minimization of $\mathcal{G}(u)$ need to be implemented by introducing the appropriate parameters and constraints explicitly into the equations defining the problem. By so doing, the problem of finding the critical points of the functional $\mathcal{G}(u)$ is often converted into finding the critical points of a functional $G_0(u)$ subject to the constraints $G_j(u) = \alpha_j$ $(j = 1, \ldots, N)$. *The class of isoperimetric variational problems so described constitutes a natural way of studying critical points of "saddle point type" by purely analytic means.*

By an isoperimetric problem, we mean the determination of extremals of a functional $G_0(x)$ (defined on a Banach space X) restricted to a proper subset $\mathcal{C}$ of X. Recall from (3.1.31) that if x_0 is an extremum of the C^1 functional $G_0(x)$ subject to the constraint

$$\mathcal{C} = \{x \mid G_i(x) = 0, \quad G_i(x) \in C^1(X, \mathbb{R})\ (i = 1, \ldots, N)\},$$

then the vectors $G_i'(x)$ $(i = 0, \ldots, N)$ are linearly dependent. Hence by

(3.1.31) there are real numbers λ_i (not all zero) such that

(6.3.1) $$\sum_{i=0}^{N} \lambda_i G_i'(x_0) = 0.$$

The equation (6.3.1) can be considered from different points of view. First, if the set $\mathcal{C}$ is weakly closed, the result (6.1.1) can be used to prove the existence of an $x_0 \in \mathcal{C}$ that will satisfy (6.3.1), where the parameters λ_i are to be determined. This simple observation is important in case a given problem contains implicit parameters as mentioned earlier. On the other hand, if it can be shown that $\lambda_i = 0$ ($i = 1, \ldots, N$), then $\lambda_0 \neq 0$ and the extremum x_0 satisfies the operator equation $G_0'(x) = 0$. Thus, in this case, we say that the point x_0 is a critical point of $G_0(x)$ relative to the "natural" constraint C (see Section 6.3B). The term "natural" is used here to emphasize the fact that the critical point x_0 of $G_0(x)$ restricted to $C \subset X$ is also a critical point of $G_0(x)$ on X itself. We now proceed to a closer investigation of these two possibilities.

6.3A Nonlinear eigenvalue problems for gradient mappings

An important situation involving implicit parameters is the problem of finding the nontrivial solutions (u, λ) of the operator equation $Au = \lambda Bu$, where $A(0) = B(0) = 0$. As already mentioned, this problem can be considered as a nonlinear analogue of determining the point spectrum of a linear operator. Clearly, if A and B are gradient operators with $Ax = \mathcal{A}'(x)$ and $Bx = \mathcal{B}'(x)$, then the solutions of (6.3.1) will be contained in the set of critical points of $\mathcal{A}(x)$ restricted to the level sets $\sigma_c = \{x \mid \mathcal{B}(x) = c$, a constant, $x \in X\}$ as c varies over the real numbers.

As a first result on the existence of nontrivial solutions of (6.3.1), we find

(6.3.2) **Theorem** Suppose that the C^1 functionals $\mathcal{A}$ and $\mathcal{B}$ defined on the reflexive Banach space X have the following properties:

(i) $\mathcal{A}(x)$ is weakly lower semicontinuous and coercive on $X \cap \{\mathcal{B}(x) \leqslant \text{const.}\}$;

(ii) $\mathcal{B}(x)$ is continuous with respect to weak sequential convergence and $\mathcal{B}'(x) = 0$ only at $x = 0$.

Then the equation $\mathcal{A}'(x) = \lambda \mathcal{B}'(x)$ has a one-parameter family of nontrivial solutions (x_R, λ_R) for all R in the range of $\mathcal{B}(x)$ such that $\mathcal{B}(x_R) = R$; and x_R is characterized as the minimum of $\mathcal{A}(x)$ over the set $\mathcal{B}(x) = R$.

Proof: The result is an immediate consequence of (6.1.1). Indeed, since

$\mathcal{B}(x)$ is weakly continuous on X, the set $\mathcal{B}_R = \{x \mid \mathcal{B}(x) = R\}$ is weakly closed. If this set $\mathcal{B}_R$ is not empty, (6.1.1) then ensures the existence of an $x_R \in \mathcal{B}_R$ such that $\mathcal{A}(x_R) = \inf \mathcal{A}(x)$ over $x \in \mathcal{B}_R$. Since $\mathcal{A}$ and $\mathcal{B}$ are differentiable, (6.3.1) implies the existence of numbers λ_1, λ_2 (not both zero) satisfying $\lambda_1 \mathcal{A}'(x_R) + \lambda_2 \mathcal{B}'(x_R) = 0$. Now $\lambda_1 \neq 0$ since otherwise $\mathcal{B}'(x_R) \equiv 0$ implies $x_R \equiv 0$ (by hypothesis). Thus x_R satisfies the equation $\mathcal{A}'(x_R) = \lambda \mathcal{B}'(x_R)$, where $\lambda = -\lambda_2/\lambda_1$.

(6.3.3) **Remark** Suppose that the C^2 functional $\mathcal{B}(x)$ is such that $\mathcal{B}'(x)$ is a Fredholm operator defined on a Banach space X' that is dense in X. Then (6.3.2) can be strengthened by noting that (3.1.47) implies that the set of real numbers $Z = \{R \mid \mathcal{B}'(x) = 0, \mathcal{B}(x) = R\}$ is of Lebesgue measure zero. Consequently (6.3.2) is valid for those real numbers in the range of $\mathcal{B}$ apart from a possible set of measure zero.

Now if the operator $\mathcal{A}'(x) = L$ is linear (as in many of our applications) we can sharpen (6.3.2) as follows.

(6.3.4) **Theorem** Suppose L is a bounded linear self-adjoint Fredholm operator mapping a Hilbert space H into itself with nonnegative essential spectrum, while $\mathcal{N}(x)$ is a C^2 strictly convex functional that is continuous with respect to weak convergence in H such that $\mathcal{N}(x) \geqslant \mathcal{N}(0) = 0$. Then if the quadratic form is "coercive" on C_R, i.e.,

$$(Lx, x) \to \infty \qquad \text{as} \quad \|x\| \to \infty$$

for

$$x \in C_R = [x \mid \mathcal{N}(x) = R, \ \mathcal{N}'(x) \perp \operatorname{Ker} L],$$

then the equation $Lx = \lambda \mathcal{N}'(x)$ has a nontrivial solution (x_R, λ_R) with $\lambda_R \neq 0$ and $\sup(Lx, x)$ over C_R attained at $x = x_R$.

Proof: The result is obtained in three steps. First, we determine a "natural" constraint set C_R so that the solution x_R of the minimization problem, minimize $G_0(x)$ over C_R, is a solution of $Lx = \lambda \mathcal{N}'(x)$. Secondly, we prove that C_R is nonvacuous for each $R > 0$. Finally we show that $\inf G_0(x)$ over C_R is, in fact, attained by $x_R \in C_R$.

Step 1: We show that any element u_0 that achieves the minimum of $G_0(x) = \frac{1}{2}(Lx, x)$ subject to the constraint

$$C_R = \{x \mid x \in H, \ \mathcal{N}(x) = R, \ (\mathcal{N}'(x), w) = 0 \qquad \text{for all} \qquad w \in \operatorname{Ker} L\}$$

is a solution of the equation $Lx = \lambda \mathcal{N}'(x)$, which obviously has the property that $\mathcal{N}(u_0) = R$. Indeed, by (3.1.31), u_0 must satisfy an equation of the form

(6.3.5) $$\beta_0 Lu = \beta_1 \mathcal{N}'(u) + \mathcal{N}''(u)\overline{w},$$

where $\bar{w} \in \operatorname{Ker} L$. We show that $\bar{w} = 0$, while β_0 and β_1 are not zero. Taking the inner product of (6.3.5) with $\bar{w}$ and using the facts that $Lw = 0$ and $(\mathcal{N}'(u_0), w) = 0$ for any $w \in \operatorname{Ker} L$, we find that $(\mathcal{N}''(u_0)\bar{w}, \bar{w}) = 0$. Since $\mathcal{N}''(u_0)$ is positive self-adjoint, $\mathcal{N}''(u_0)\bar{w} = 0$. Since $\mathcal{N}$ is strictly convex, Ker $\mathcal{N}''(u_0) = 0$, so that $\bar{w} = 0$. Now $\beta_0 \neq 0$ since if it were zero, $\beta_1 \neq 0$ and $\mathcal{N}'(u_0) = 0$. In this case $u = 0$, since $\mathcal{N}$ is strictly convex, which contradicts the fact that $\mathcal{N}(u_0) = R > 0$. To demonstrate that $\beta_1 \neq 0$, we suppose the contrary, then $Lu_0 = 0$ so that $u_0 \in \operatorname{Ker} L$. Thus, since $u_0 \in C_R$, $(\mathcal{N}'(u_0), u_0) = 0$; and again by the convexity of $\mathcal{N}(u)$, $u_0 = 0$, contradicting again the fact that $\mathcal{N}(u_0) = R > 0$.

Step 2: Now we prove

(6.3.6) **Lemma** Suppose that $\mathcal{N}(x)$ satisfies the hypotheses of (6.3.4), then the constraint set

$$C_R = \{x \mid \mathcal{N}(x) = R, (\mathcal{N}'(x), w) = 0, w \in \operatorname{Ker} L\}$$

is nonvacuous for each $R > 0$.

Proof: Let $\mathcal{S}_i = \{x \mid (\mathcal{N}'(x), w_i) = 0\}$ $(i = 1, \ldots, N)$. Then the constraints $\mathcal{S}_i$ can be written

$$\left(\mathcal{N}'\left(y + \sum_{i=1}^{N} \beta_i w_i\right), w_j\right) = 0 \qquad (j = 1, \ldots, N),$$

where $(w_1, \ldots, w_N)$ is an orthonormal basis for Ker L and $y \in [\operatorname{Ker} L]^{\perp}$. Regarding the left-hand side of the above equation as a function of $\beta = (\beta_1, \ldots, \beta_N)$ and fixing $y \in [\operatorname{Ker} L]^{\perp}$, $(\mathcal{N}'(x), w_i) = 0$ $(i = 1, \ldots, N)$ is satisfied by those N-vectors β that are critical points of the functional $F(\beta) = \mathcal{N}(y + \sum_{i=1}^{N}(\beta_i w_i)$. Thus by the strict convexity of the functional $\mathcal{N}$, there is one and only one critical point $\beta(y) = (\beta_1(y), \ldots, \beta_N(y))$ of $F(\beta) = \mathcal{N}(y + \sum_{i=1}^{N} \beta_i w_i)$ for fixed $y \in [\operatorname{Ker} L]^{\perp}$. Thus for each positive s, there is an element $y(s) = sy + \sum_{i=1}^{N} \beta_i(sy)w_i$ such that $(\mathcal{N}'(y(s)), w) = 0$ for all $w \in \operatorname{Ker} L$. Now for fixed nonzero $y \in [\operatorname{Ker} L]^{\perp}$ the function $G(s, \beta_1, \ldots, \beta_N) = \mathcal{N}(sy + \sum_{i=1}^{N} \beta_i w_i)$ is a strictly convex function defined on $\mathbb{R}^{N+1}$, such that if $|s| + \sum_{i=1}^{N} |\beta_i| \to \infty$, $G(x, \beta_1, \ldots, \beta_N) \to \infty$. Hence, as a function of s, $g(s) = \mathcal{N}(sy + \sum_{i=1}^{N} \beta_i(sy)w_i) \to \infty$ as $|s| \to \infty$. Therefore the lemma is proved provided we show that the function $g(s)$ is a continuous function of s. This is immediate from convexity theory since $g(s) = \inf G(s, \beta_1, \ldots, \beta_N)$ over $\beta_1, \ldots, \beta_N$ and G is strictly convex.

Step 3: We show that $\inf \frac{1}{2}(Lx, x)$ over C_R is attained by an element $u_0 \in C_R$. Since C_R is a weakly closed set in the Hilbert space H, by (6.1.1) it suffices to prove that the functional (Lx, x) is weakly lower semicontinuous on H. The weak lower semicontinuity of (Lx, x) is plainly an immediate consequence of (6.1.3) since L can be written as the sum of a self-adjoint positive operator L_1 and a compact self-adjoint operator L_2.

The variational problem conjugate to the one mentioned in (6.3.2) is also useful in the study of nonlinear eigenvalue problems. Indeed for each fixed number R, we consider the level set $\partial A_R = \{A(x) = R\}$, and the number $C_R = \sup_{\partial A_R} \mathcal{B}(x)$, and prove

(6.3.7) **Theorem** Suppose the C' functionals $\mathcal{A}(x)$ and $\mathcal{B}(x)$ defined on the reflexive Banach space X satisfy the properties: (i) $\mathcal{A}(x)$ is coercive, weakly lower semicontinuous on X, and for each fixed nonzero $x \in X$, the real function $f(t) = \mathcal{A}(tx)$ is a nonzero, increasing function of t; (ii) $\mathcal{B}(x)$ is continuous with respect to weak convergence and such that $\mathcal{B}'(x) = 0$ implies $x = 0$, and for each nonzero $x \in X$, $g(t) = \mathcal{B}(tx)$ is a strictly increasing function of t; then the number C_R defined above is a critical point of $\mathcal{B}(x)$ restricted to ∂A_R for each $R \neq 0$ in the range of $\mathcal{B}(x)$, and moreover if $\mathcal{B}(x_R) = C_R$ for $x_R \in \partial A_R$ (x_R, λ_R) is a nontrivial solution of the equation $A'(x) = \lambda B'(x)$.

Proof: Since $\mathcal{A}(x)$ is coercive, the level set ∂A_R is bounded and so the functional $\mathcal{B}(x)$, being continuous with respect to weak convergence, is bounded on ∂A_R. Consequently $C_R = \sup \mathcal{B}(x)$ over ∂A_R is finite. Thus any maximizing sequence $\{x_n\} \in \partial A_R$ is bounded and so (after possibly passing to a subsequence) x_n is weakly convergent with weak limit $\bar{x}$. The continuity of $\mathcal{B}(x)$ with respect to weak convergence implies $C_R = \mathcal{B}(\bar{x})$, and we set $\bar{x} = x_R$. Thus to establish the desired result we show $x_R \in \partial A_R$ and that x_R satisfies the equation $\mathcal{A}'(x) = \lambda_R \mathcal{B}'(x_R)$ for some finite number λ_R. To this end suppose $x_R \not\in \partial A_R$, then by the weak lower semicontinuity of $\mathcal{A}(x)$ we may suppose $\mathcal{A}(x_R) < R$, and moreover hypothesis (i) of the theorem implies for some $t > 1$, $\mathcal{A}(tx_R) = R$. Now hypothesis (ii) implies that $\mathcal{B}(tx_R) > c_R = \sup_{A_R} \mathcal{B}(x)$ (a contradiction). Consequently $x_R \in \partial A_R$ and by (3.1.31), there are numbers λ_1, λ_2 such that $\lambda_1 \mathcal{A}'(x_R) + \lambda_2 \mathcal{B}'(x_R) = 0$. Now $\lambda_1 \neq 0$ since otherwise $\mathcal{B}'(x_R) = 0$ which implies by hypothesis (ii) that $x_R = 0$, which in turn contradicts the fact that $R \neq 0$.

The isoperimetric methods for studying nonlinear eigenvalue problems presented above has a distinct limitation, namely they yield only *the analogue of the first eigenvector for an operator A*. In the case of linear operators, one need not proceed to discuss any further critical point theory since the notions of orthogonality and orthogonal complement enable an *iterative* isoperimetric approach to yield a complete set of eigenvectors for A. However for nonlinear operators we cannot proceed to the full analogue of this completeness result until we develop deeper critical point theories. Indeed it will be necessary to find topological "constraints" that enable one to discuss "higher" eigenvectors for nonlinear operators.

Example: In order to understand the content and precision of (6.3.2), consider the following semilinear Dirichlet problem defined on a bounded domain[1] $\Omega \subset \mathbb{R}^N$:

(6.3.8) $$\Delta u + k(x)u + \lambda g(x)u^\sigma = 0, \qquad u|_{\partial\Omega} = 0,$$

where $k(x)$ and $g(x)$ are smooth functions (say Hölder continuous with $g(x) > 0$ on $\bar{\Omega}$). As an application of (6.3.2) we shall prove

(6.3.9) **Theorem** If σ lies in the *open* interval $(1, (N+2)/(N-2))$, then the system (6.3.8) possesses a one-parameter family (u_R, λ_R) of nontrivial smooth solutions for each positive number $R \neq 0$ with $\int_\Omega u_R^{\sigma+1} = R$ and $u_R > 0$ in Ω.

Proof: Clearly the appropriate functionals $\mathcal{A}$ and $\mathcal{B}$ used in applying (6.3.2) can be defined by setting

$$\mathcal{A}(u) = \int_\Omega \{|\nabla u|^2 - k(x)u^2\}\, dx, \qquad \mathcal{B}(u) = \int_\Omega g(x)u^{\sigma+1}\, dx.$$

Then assuming for the moment that σ is an odd integer, one easily verifies that (i) the functional $\mathcal{A}(u)$ is weakly lower semicontinuous on $\mathring{W}_{1,2}(\Omega)$, and (ii) for $\sigma \in (0, (N+2)/(N-2))$, the functional $\mathcal{B}(u)$ is weakly continuous and strictly convex on $\mathring{W}_{1,2}(\Omega)$. Thus to verify that $\mathcal{A}(u)$ is coercive relative to $\mathcal{B}$ on $\mathring{W}_{1,2}(\Omega)$, we first note that by Jensen's inequality for $\alpha = \inf g(x)$ over $\bar{\Omega}$,

$$\left\{\int_\Omega |u|\, dx\right\}^{\sigma+1} \leqslant \int_\Omega |u|^{\sigma+1}\, dx = \int_\Omega u^{\sigma+1}\, dx \leqslant \frac{1}{\alpha}\int_\Omega g(x)u^{\sigma+1}\, dx.$$

Hence $\|u\|_{L_1(\Omega)}$ is uniformly bounded, if $\mathcal{B}(u) \leqslant$ const. Furthermore, by utilizing (1.3.28), for example, there are absolute constants c_1, c_2 with $c_1 > 0$ such that

$$\mathcal{A}(u) = \int_\Omega \{|\nabla u|^2 - k(x)u^2\}\, dx \geqslant c_1\|u\|_{1,2}^2 - c_2\|u\|_{0,1}^2.$$

Consequently, $\mathcal{A}(u) \to \infty$ as $\|u\| \to \infty$ with $\mathcal{B}(u) \leqslant$ const. Since the range of the functional $\mathcal{B}(u)$ is $[0, \infty)$ and $\mathcal{B}'(u) = 0$ if and only if $u = 0$, $R = 0$ is the only excluded point in the range of $\mathcal{B}(u)$. Thus for every R in the open interval $(0, \infty)$, the system (6.3.8) has a family of nontrivial weak solutions (u_R, λ_R) with $\int_\Omega g(x)u^{\sigma+1}\, dx = R$. Furthermore, both functionals $\mathcal{A}(u)$ and $\mathcal{B}(u)$ are unchanged by replacing u by $|u|$ and $|u| \in \mathring{W}_{1,2}(\Omega)$ if $u \in \mathring{W}_{1,2}(\Omega)$. Hence we may conclude that any minimizing sequence $\{u_n\} \in \mathcal{B}_R$ for $\inf \mathcal{A}(u)$ over $\mathcal{B}_R$ will be minimizing if $\{u_n\}$ is replaced by $\{|u_n|\}$. Since a subsequence of $\{u_n\}$ converges almost everywhere to u_R, we may suppose that $u_R \geqslant 0$ (a.e. in Ω). Furthermore, the linear regularity

[1] Later in a geometric context this problem will be considered again with Ω replaced by a compact Riemannian manifold $\mathfrak{M}^N$ of dimension N.

theory, as mentioned in Section 1.5, permits us to suppose that $u_R \geqslant 0$ (everywhere in $\bar{\Omega}$) and smooth enough to satisfy (6.3.8) pointwise in Ω and at all sufficiently smooth portions of $\partial\Omega$. To show that $u_R > 0$ in Ω, we use the maximum principle. (See Protter and Weinberger, 1967.) Indeed, if $u_R = 0$ for some $x \in \Omega$, (6.3.8) implies that $u_R \equiv 0$ in Ω.

Finally, to remove the restriction of oddness on σ, we note that if we replace the term $g(x)u^\sigma$ with $g(x)|u|^{\sigma-1}u$ in (6.3.8), and repeat the argument just given, we again find a solution $u_R \geqslant 0$ for the amended system. However, in this case, $g(x)u_R^\sigma = g(x)|u_R|^{\sigma-1}u_R$; and the result is proven.

There are two possible important extensions of (6.3.9) which immediately come to mind:

(i) possible validity of (6.3.9) for $\sigma \geqslant (N+2)/(N-2)$;
(ii) removal of the restriction that Ω be a *bounded* domain of $\mathbb{R}^N$.

In both cases, the main difficulty is that the functional $\mathcal{B}(u)$ is no longer weakly continuous (or equivalently, $\mathcal{B}'(u)$ is no longer completely continuous).

Concerning (i), we note that (6.3.9) is sharp in the following sense, as follows from our discussion in (1.2.7).

(6.3.10) Suppose that in the equation (6.3.9) $k(x) = 0$, $g(x) = \beta^2 > 0$, and $\sigma \geqslant (N+2)/(N-2)$. Then (6.3.8) possesses no nontrivial smooth positive solutions.

Concerning the extension (ii), we shall prove later in (6.7.25)

(6.3.11) **Theorem** Suppose $\Omega = \mathbb{R}^N$, $g(x) = g\ (\neq 0)$ a constant, and $k(x) = k$ a negative constant. Then the system (6.3.8) has nontrivial smooth positive solutions in $W_{1,2}(\mathbb{R}^N)$ if and only if $\sigma < (N+2)/(N-2)$.

In the general case we consider functionals (with $|\alpha| \leqslant m$)

$$\mathcal{A}(u) = \int_\Omega F(x, D^\alpha u) \qquad \text{and} \qquad \mathcal{B}(u) = \int_\Omega G(x, D^\beta u)$$

with $|\beta| \leqslant m-1$ defined over $\mathring{W}_{m,p}(\Omega)$, with Ω a bounded domain in $\mathbb{R}^N$. Then provided the functions $F(x, y, z)$ and $G(x, y)$ satisfy the appropriate (Sobolev) growth conditions to ensure that $\mathcal{A}(u)$ and $\mathcal{B}(u)$ are C^1 functions defined on $\mathring{W}_{m,p}(\Omega)$ Theorem (6.3.7) yields conditions for the existence of nontrivial solutions (u_R, λ_R) for $|\gamma| \leqslant m$

$$\sum_{|\alpha| \leqslant m} (-1)^{|\alpha|} D^\alpha F_\alpha(x, D^\gamma u) = \lambda \sum_{|\beta| \leqslant m-1} (-1)^{|\beta|} D^\beta G_\beta(x, D^\gamma u),$$

$$D^\alpha u|_{\partial\Omega} = 0, \qquad |\alpha| \leqslant m-1.$$

The major hypotheses to be verified include (a) the weak lower semicontinuity in $\mathring{W}_{m,p}(\Omega)$ of $\mathcal{A}(u)$, (b) the continuity of $\mathcal{B}(u)$ with respect to weak convergence on $\mathring{W}_{m,p}(\Omega)$, and (c) the coerciveness of $\mathcal{Q}(u)$. The properties of the functions $F(x, D^\gamma u)$, $G(x, D^\gamma u)$ necessary to check these hypotheses have been discussed earlier.

6.3B Solvability of semilinear gradient operator equations

We now turn to the second area of applicability of isoperimetric variational problems mentioned earlier: exact solvability criteria (analogous to those of 5.4E) for inhomogeneous gradient operator equations of the form

$$(6.3.12) \qquad \mathcal{G}'(u) = f$$

that do not require the linear growth of $\mathcal{G}'(u)$ as in Section 5.4E. Here $\mathcal{G}(x)$ is assumed to be a C^1 real-valued functional defined on a Hilbert space H. Moreover, $\mathcal{G}'$ is assumed to be a semilinear gradient operator mapping H into itself so that $\mathcal{G}' = L + \mathcal{N}'$, where L is a self-adjoint Fredholm operator and $\mathcal{N}'$ is a (nonlinear) completely continuous mapping of H into itself. Since in the most interesting cases the operator L will have either negative spectrum or a nontrivial kernel, a solution u_0 of (6.3.12) *will not correspond to an absolute minimum* of the functional $I(u) = \mathcal{G}(u) - (f, u)$. Indeed, a solution u_0 will, generally speaking, correspond to a *saddle point* of $I(u)$. For quadratic functionals $\mathcal{G}(u)$, such critical points can be reduced to absolute minima by a judicious use of orthogonality. Here we show that in certain circumstances we can find a "nonlinear extension" of orthogonality that is suitable for studying the solvability of (6.3.12) by analytic means. In later sections, we shall show how saddle points of functionals $\mathcal{G}(u)$ can be studied by topological methods.

In order to find suitable necessary and sufficient conditions for the solution of (6.3.12), we introduce the notion of "natural constraint." Suppose a critical value of a C^1 functional $\mathcal{G}(u)$ is not an absolute minimum over a Hilbert space H, but there is a submanifold $\mathfrak{M}$ of H such that:

(i) $c = \inf_{\mathfrak{M}} \mathcal{G}(u)$ is attained by an element $\bar{u} \in \mathfrak{M}$;

(ii) for any $\bar{u} \in \mathfrak{M} \cap \mathcal{G}^{-1}(c)$, $\mathcal{G}'(\bar{u}) = 0$, so that $\bar{u}$ is not only a critical point of $\mathcal{G}$ restricted to $\mathfrak{M}$, but also of $\mathcal{G}$ considered as defined on H itself;

(iii) every critical point of $\mathcal{G}(u)$ lies on $\mathfrak{M}$.

Such situations are common in geometric problems whose solutions must satisfy fixed geometric side conditions (see for example Section 6.4B below). The following result gives a fairly general construction for natural constraints associated with a C^2 functional $\mathcal{G}(u)$.

(6.3.13) **Theorem** Let N be a closed linear subspace of a Hilbert space H, and suppose $\mathcal{G}(u)$ is a C^2 functional defined on H. Set

$$S = \{u \mid u \in H,\ \mathcal{G}'(u) \perp N\},$$

and suppose

(a) S is closed with respect to weak sequential convergence and nonvacuous,
(b) $\mathcal{G}(u)$ is coercive on S and weakly lower semicontinuous there,
(c) $\mathcal{G}''(u)$ is definite on N for each $u \in S$.

Then, (i) $c = \inf_S \mathcal{G}(u)$ is finite and attained by an element $\bar{u} \in S$, and (ii) S is a natural constraint for the functional $\mathcal{G}(u)$ (on H).

Proof: (i) follows immediately from hypotheses (a), (b), and the result (6.1.1). Consequently, we need only establish (ii). To this end, we observe that the elements of S are the zeros of the operator $P\mathcal{G}'(u)$ for $u \in H$, where P is the canonical projection of H onto N. Now the operator $P\mathcal{G}'(u)$ is a mapping of H into N whose derivative $P\mathcal{G}''(u)$ is surjective. Indeed, by hypothesis (c), $P\mathcal{G}''(u)$ maps N onto itself for each $u \in H$ (by virtue of the Lax–Milgram theorem (1.3.21)). Therefore, the result (3.1.37) implies that for an extremal $\bar{u}$ of $\mathcal{G}(u)$ restricted to S, there is a fixed element $w \in N$, such that $\bar{u}$ is a critical point of

$$\mathcal{G}(u) - (P\mathcal{G}'(\bar{u}), w) \qquad \text{defined on } H.$$

Hence $\bar{u}$ satisfies

$$\mathcal{G}'(\bar{u}) = \mathcal{G}''(\bar{u})w.$$

Taking the inner product of this equation with w, we find that, by virtue of the definition on S, $(\mathcal{G}''(\bar{u})w, w) = 0$. Consequently, hypothesis (c) implies that $w = 0$, so that $\mathcal{G}'(\bar{u}) = 0$. Thus to verify the fact that S is a natural constraint for $\mathcal{G}(u)$ on H, we need only show that every critical point v of $\mathcal{G}(u)$ (considered as defined on H) lies on S. However, this last fact is immediate since $\mathcal{G}'(v) = 0$ and so $\mathcal{G}'(v)$ is necessarily orthogonal to N.

To utilize the above general prescription for constructing natural constraints relative to the operator equation (6.3.12), we make use of the fact that $\mathcal{G}'$ is semilinear (i.e., $\mathcal{G}'(u) = Lu + \mathcal{N}'(u)$). Accordingly we choose the subspace N carefully to take account of the spectral properties of L. More precisely, we shall determine a characterization of a tentative critical value c for the functional $\mathcal{G}(u)$ based on the spectral properties of L, and then utilize (6.3.13) and hypotheses concerning the mapping $\mathcal{N}'(u)$ to ensure that the number c is actually a critical value for $\mathcal{G}(u)$.

Thus if $\mathcal{N}'(u) \equiv 0$, it is easily verified that: (a) the only possible critical value $\bar{c}$ of the functional $J(u) = \frac{1}{2}(Lu, u) - (f, u)$ can be characterized by the formula:

(6.3.14) $$\bar{c} = \inf_{x \in H_+} \sup_{y \in H_-} \sup_{z \in \operatorname{Ker} L} J(x + y + z),$$

where H_+ is the linear subspace of H on which L is positive definite, and H_- is the subspace of H on which L is negative definite; (b) if L is a Fredholm self-adjoint operator, $\bar{c}$ is attained and finite if and only if f is orthogonal to Ker L. Now (a) and (b) together imply the exact solvability criterion for the operator equation $Lu = f$ defined on H, and the formula (6.3.14) exhibits the fact that the associated critical value of $J(u)$ is generally not an absolute minimum.

We now establish (a) and (b). We first observe that if f is not orthogonal to Ker L, $J(z) = -(f, z)$ can take any real value (positive or negative). Now the orthogonality of H_+, H_-, and Ker L implies

$$J(x + y + z) = J(x) + J(y) + J(z).$$

Consequently, if f is not orthogonal to Ker L, $\bar{c}$ cannot be finite; and conversely, if $f \perp \text{Ker } L$, $J(x + y + z) = J(x) + J(y)$, so that

(6.3.15) $$\bar{c} = \inf_{H_+} J(x) + \sup_{H_-} J(y).$$

Since the operator L is self-adjoint and Fredholm, there is an absolute constant $\alpha > 0$ such that for $x \in H_+$ any $y \in H_-$

(6.3.16) $$(Lx, x) \geqslant \alpha\|x\|^2, \qquad (Ly, y) \leqslant -\alpha\|y\|^2.$$

Thus by (6.3.16), $\inf_{H_+} J(x)$ is attained at some $\bar{x} \in H_+$ and the strict convexity of $J(x)$ implies that this $\bar{x}$ is unique. A similar statement holds for $\sup_{H_-} J(y)$ and a unique point $\bar{y} \in H_-$ since $\sup_{H_-} J(y) = \inf_{H_-}\{-J(y)\}$. Consequently (6.3.15) implies that $\bar{c}$ is attained at any point of the form $\bar{u} = \bar{x} + \bar{y} + z$ for any $z \in \text{Ker } L$. Let P_+ and P_- be the canonical orthogonal projections of H onto H_+ and H_- respectively. Then

$$J'(\bar{u}) = L\bar{u} - f = (L\bar{x} - P_+ f) + (L\bar{y} - P_- f)$$
$$= P_+ J'(\bar{x}) + P_- J'(\bar{y}) = 0.$$

In the general case, we consider first a functional $\mathcal{G}(u) = \frac{1}{2}(Lu, u) + R(u)$ and define a number c analogous to $\bar{c}$ by setting

(6.3.17) $$c = \inf_{x \in H_+} \sup_{y \in H_-} \sup_{z \in \text{Ker } L} \mathcal{G}(x + y + z),$$

and determine precise conditions on the functional $R(u)$ for c to be finite and a critical value for $\mathcal{G}(u)$ on H. Then we set $R(u) = \mathcal{N}(u) - (f, u)$ and show that these conditions yield necessary and sufficient conditions for the solvability of the equation

(6.3.18) $$Lu + \mathcal{N}'(u) = f,$$

which reduce to the usual orthogonality condition of Fredholm in case $\mathcal{N}'(u) \equiv 0$. To this end, we prove the following two results.

(6.3.19) **Lemma** Let $R'(u)$ be completely continuous. Suppose the C^2 functional $\mathcal{G}(u) = \frac{1}{2}(Lu, u) + R(u)$ satisfies the following conditions:

(i) the set $\mathcal{S}_0 = \{u \mid \mathcal{G}'(u) \perp \text{Ker } L\} \neq \varnothing$;

(ii) for fixed $u \in \mathcal{S}_0$, the functional $\mathcal{G}(u+w)$ is strictly concave in $w \in \operatorname{Ker} L \cup H_-$ and $\mathcal{G}(u+w) \to -\infty$ as $\|w\| \to \infty$;
(iii) for $u \in \mathcal{S} = \{u \mid \mathcal{G}'(u) \perp \{H_- \cup \operatorname{Ker} L\}\}$, $\mathcal{G}(u) \to \infty$ as $\|u\| \to \infty$.

Then the number c defined by (6.3.17) is finite and is a critical value of $\mathcal{G}(u)$ on H.

Using the same notation and setting $\mathcal{N}'(u) - f = R'(u)$,

(6.3.20) **Theorem** Suppose the operator $\mathcal{N}'(u)$ satisfies the following conditions:

(i) for $u \in \mathfrak{M} = \{u \mid \mathcal{N}'(u) - f \perp \operatorname{Ker} L\}$, $L + \mathcal{N}''(u)$ is negative definite on $H_- \cup \operatorname{Ker} L$ and $\mathcal{G}(u+w) \to -\infty$ as $\|w\| \to +\infty$, $w \in H_- \cup \operatorname{Ker} L$;
(ii) for $u \in \mathcal{S} = \{u \mid (Lu + \mathcal{N}'(u) - f) \perp H_- \cup \operatorname{Ker} L\}$, $\mathcal{G}(u) \to \infty$ as $\|u\| \to \infty$.

Then the equation (6.3.18) is solvable if and only if the set $\mathfrak{M}$ is nonempty. Moreover, if solvable, a critical value of $\mathcal{G}(u)$ is given by (6.3.17).

Proof of (6.3.20) **using** (6.3.19): Indeed, if (6.3.18) is solvable, the set $\mathfrak{M}$ is necessarily nonvacuous since any solution of (6.3.18) is an element of $\mathfrak{M}$. On the other hand, if $\mathfrak{M} \neq \varnothing$, Lemma (6.3.19) implies that the number c defined by (6.3.17) is a critical value for $I(u)$ defined on H, and so (6.3.12) is necessarily solvable.

Proof of (6.3.19): We shall show that the result follows from (6.3.13) by verifying that the set $\mathcal{S} = \{u \mid \mathcal{G}'(u) \perp [H_- \cup \operatorname{Ker} L]\}$ and the functional $\mathcal{G}(u)$ satisfy hypotheses (a)–(c) mentioned in the statement of (6.3.13).

First we show that $\mathcal{S}$ is closed with respect to weak convergence and is nonvacuous. Since $\mathcal{G}'(u) = Lu + R'(u)$, $u_n \to u$ weakly in H implies $Lu_n \to Lu$ weakly in H and $R'(u_n) \to R'(u)$ strongly in H. Thus $u_n \in \mathcal{S}$ implies $u \in S$. We show $\mathcal{S}$ is nonvacuous by using the fact that $\mathcal{S}_0$ is. Since $\mathcal{S}_0$ is nonvacuous, for some $\bar{x}, \bar{y}$

$$\delta = \sup_{\operatorname{Ker} L} \mathcal{G}(\bar{x} + \bar{y} + z) < \infty,$$

while the finite dimensionality of Ker L implies that for some $\bar{z} \in \operatorname{Ker} L$, $\delta = \mathcal{G}(\bar{x} + \bar{y} + \bar{z})$. Setting $\bar{u} = \bar{x} + \bar{y} + \bar{z}$, we consider the functional $\mathcal{G}_1(\bar{u} + y + z) = -\mathcal{G}(\bar{u} + y + z)$ for $y \in H_-$ and $z \in \operatorname{Ker} L$. By hypothesis (ii), $\mathcal{G}_1$ is coercive on $H_- \cup \operatorname{Ker} L$ and moreover $\mathcal{G}_1$ is certainly weakly lower semicontinuous there. Now (6.1.1) implies that for $\tilde{y} + \tilde{z}$,

$$P\,\mathcal{G}_1'(\bar{u} + \tilde{y} + \tilde{z}) = 0,$$

where P is the projection of H onto $H_- \cup \operatorname{Ker} L$. Consequently, $\bar{u} + \tilde{y} + \tilde{z} \in \mathcal{S}$.

To establish the weak lower semicontinuity of $\mathcal{I}(u)$ on $\mathcal{S}$, we again uniquely decompose an arbitrary element $v \in S$ into $v = x + y + z$ (as above) and observe that $PI'(v) = 0$ implies that $Ly = PR'(v)$. Thus if $u_n \to u$ weakly, Ly_n is strongly convergent since $PR'(u_n)$ is. Thus for $u \in \mathcal{S}$, the quadratic form is continuous with respect to weak convergence. Accordingly, for $u \in \mathcal{S}$, the functional $\mathcal{I}(u)$ can be written

$$(6.3.21) \qquad \mathcal{I}(u) = \tfrac{1}{2}(Lx, x) + (Ly, y) + R(u),$$

and is thus weak lower semicontinuous, by (6.1.3), since the latter two terms in (6.3.21) are both continuous with respect to weak convergence.

The coerciveness of $\mathcal{I}(u)$ on $\mathcal{S}$ is an immediate consequence of hypothesis (ii) of (6.3.19). Finally the definiteness of $\mathcal{I}''(u)$ on N for $u \in \mathcal{S}$ is an immediate consequence of the strict concavity of $\mathcal{I}(u + w)$ also given in hypothesis (ii).

Therefore, by (6.3.13), $\mathcal{S}$ is a natural constraint for $\mathcal{I}(u)$ on H, and moreover, $\inf_{\mathcal{S}} \mathcal{I}(u)$ is a critical value of $\mathcal{I}(u)$ on H. It remains to show that the number c defined by (6.3.14) actually equals $\inf_{\mathcal{S}} \mathcal{I}(u)$. This follows from the fact that the points of S are precisely the solutions of the equation

$$P\mathcal{I}'(x + y + z) = 0, \qquad u = x + y + z,$$

determined by letting x vary over H_+ and characterized for each fixed $x \in H_+$ by the critical value $\sup_{w \in \operatorname{Ker} L \cup H_-} \mathcal{I}(x + w)$. Thus Lemma (6.3.19) is proven.

Before deriving some further consequences of (6.3.12), we first show that in certain cases the hypotheses of that result can be considerably weakened by restricting the spectrum of L.

(6.3.22) **Corollary** Suppose $\sigma_e(L)$ is nonnegative, then the coerciveness part of hypothesis (i) of (6.3.19) can be removed provided we suppose that for $u \in \mathfrak{M}$ and some $\epsilon > 0$, $\mathcal{I}''(u) - \epsilon L$ is negative definite on $\operatorname{Ker} L \cup H_-$.

Proof: Let $u \in \mathfrak{M}$, and decompose $w = y + z$ with $y \in H_-$, $z \in \operatorname{Ker} L$. Then by Taylor's theorem,

$$(6.3.23) \qquad \mathcal{I}(u + w) = \mathcal{I}(u) + (\mathcal{I}'(u), w) + \int_0^1 (1 - s)(\mathcal{I}''(u + sw)w, w)\, ds.$$

Since $u \in \mathfrak{M}$, $(\mathcal{I}'(u), z) = 0$; and since L is Fredholm, there is a finite absolute constant γ such that $(Ly, y) \leqslant -\tilde{r}\|y\|^2$ for all $y \in H_-$. By hypothesis, for some $\epsilon > 0$, $\tilde{\mathcal{I}}''(u) = \mathcal{I}''(u) - \epsilon L$ is such that $(\tilde{I}''(u)w, w) < 0$. Moreover, $\sigma_e(L) \subset [0, \infty)$ implies $\dim(H_- \cup \operatorname{Ker} L) < \infty$, so that there is a positive constant α such that for fixed $u \in \mathfrak{M}$, $\max_{\|w\|=1} (\tilde{\mathcal{I}}''(u)w, w) = -2\alpha$. Consequently, by continuity, there is a $\rho > 0$ such that

$$\|\mathcal{I}''(h) - \mathcal{I}''(u)\| \leqslant \alpha \qquad \text{for} \quad \|h - u\| \leqslant \rho.$$

Thus for s sufficiently small, $\|sw\| \leq \rho$ say, and $h = u + sw$,

$$(\tilde{\mathcal{G}}''(h)w, w) = (\tilde{\mathcal{G}}''(u)w, w) + ([\tilde{\mathcal{G}}''(u) - \tilde{\mathcal{G}}''(h)]w, w)$$
$$\leq -\alpha\|w\|^2.$$

Combining these facts with (6.3.23), we find

$$\mathcal{G}(u+w) \leq \mathcal{G}(u) + (\mathcal{G}'(u), y) + \frac{\epsilon}{2}(Lw, w) + \int_0^{\rho/\|w\|}(1-s)\{-\alpha\|w\|^2\}\, ds.$$

Thus the Cauchy–Schwarz inequality implies that for any $\delta > 0$,

$$\mathcal{G}(u+w) \leq \text{const.} + \delta\|y\|^2 - \frac{\gamma\epsilon}{2}\|y\|^2 - \alpha\rho\|w\|.$$

Choosing $\delta = \gamma\epsilon/2$, we find $\mathcal{G}(u+w) \to -\infty$ as $\|w\| \to \infty$, as required.

As a useful application of the above results, we consider the semilinear gradient operator equation

(6.3.24) $$Lu = \mathfrak{N}'(u) + g$$

with $\|\mathfrak{N}'(u)\| \leq$ const. for all $u \in H$, where $\mathfrak{N}(u)$ is a C^2 strictly convex function of u, that is, continuous with respect to weak sequential convergence and the essential spectrum of L is nonnegative. Then the following result, analogous to (5.4.29), holds.

(6.3.25) **Theorem** A necessary and sufficient condition for the solvability of (6.3.24) is that the set $\mathfrak{M}_g = \{u \mid \mathfrak{N}'(u) - g \perp \operatorname{Ker} L\}$ be nonempty. Moreover, the operator $L + \mathfrak{N}'$ regarded as a mapping from H into itself has an open range.

Proof: We leave the proof of the first part of this result to the interested reader since it follows routinely from the facts just established.

We show that the mapping $L + \mathfrak{N}'$ has open range, by showing that if $\mathfrak{M}_{g_0}$ is nonvacuous, so is $\mathfrak{M}_{g'}$ for $\|g_0 - g'\| \leq \epsilon$ for some fixed $\epsilon > 0$. Let $g_0 \in \operatorname{Range}(L + \mathfrak{N}')$, then by the first part of the theorem, the set

$$\mathfrak{M}_{g_0} = \{u \mid u \in H, (\mathfrak{N}'(u) - g_0) \perp \operatorname{Ker} L\}$$

contains a point $u_0 \in H$. Moreover, for any $u \in [\operatorname{Ker} L]^\perp$, the set

$$\mathfrak{M}_\omega = \{u \mid u \in H, (\mathfrak{N}'(u) - g_0 - \omega) \perp \operatorname{Ker} L\}$$

is nonvacuous since it contains u_0. Finally, suppose $z \in \operatorname{Ker} L$ has sufficiently small norm, then we shall show that the set

(6.3.26) $$\mathfrak{M}_z = \{u \mid u \in H, (\mathfrak{N}'(u) - g_0 - z) \perp \operatorname{Ker} L\}$$

is nonvacuous since it also contains an element $u(z)$ near u_0. Indeed, $u(z) \in \mathfrak{M}_z$ if we can solve the operator equation

(6.3.27) $$P_0(\mathfrak{N}'(u(z))) = P_0(z + g_0),$$

where P_0 is the canonical projection of H onto Ker L. If $z = 0$, u_0 satisfies (6.3.27). Thus for $\|z\|$ sufficiently small, an application of the inverse function theorem shows that if the left-hand side of (6.3.27) is regarded as a mapping of a small neighborhood of the origin of Ker L into Ker L, then (6.3.7) is solvable. Indeed, the inverse function theorem is applicable here since L is a Fredholm operator, dim Ker $L < \infty$, so that the strict concavity of $\mathfrak{N}(u)$ implies that $P_0\mathfrak{N}''(u_0)$ is an injective linear mapping of Ker L into itself, and so is invertible.

Finally, applying the first part of the our result we find, on the basis of the above paragraph, that for all $g \in H$ with $\|g - g_0\|$ sufficiently small, the equation $Lu + \mathfrak{N}'(u) = g$ is solvable. Thus the map $L + \mathfrak{N}'$ has open range, as required.

6.4 Isoperimetric Problems in Geometry and Physics

Here we shall indicate how the abstract results of the previous section can be used to solve various concrete problems in mathematical physics and differential geometry.

6.4A Families of large amplitude periodic solutions of nonlinear Hamiltonian systems

We wish to prove the existence of a family of periodic solutions of the N-dimensional dynamical autonomous system $\mathfrak{S}_N$

(6.4.1) $$x_{tt} + \nabla U(x) = 0$$

parametrized by the mean value of $U(x(t))$ over a period by suitably restricting the potential function $U(x)$. In this problem as in Liapunov's criterion of Section 4.1, the period of a possible periodic solution is the implicit parameter in question. In contrast to the discussion of Section 4.1 in this case there is no obvious first approximation to the desired family of periodic solutions. Moreover, the obvious isoperimetric variational problem (π_R) of maximizing the potential energy of $\mathfrak{S}_N$, $\int_0^1 U(x(s))\,ds$ over the 1-periodic N-vector functions $x(t)$ with fixed nonzero kinetic energy $\int_0^1 |\dot{x}|^2\,ds = R$ and its dual, does not yield the desired family. Indeed, the maximum in (π_R) is easily seen to be infinite for "coercive" potential functions $U(x)$. While the minimum in the isoperimetric problem dual to (π_R) is zero. However, on the basis of (6.3.4), we shall prove

(6.4.2) **Theorem** Suppose that the function $U(x)$ is a C^1 convex function defined on $\mathbb{R}^N$ such that

(i) $0 = U(0) \leq U(x)$, and (ii) $U(x) \to \infty$ as $|x| \to \infty$.

Then (6.4.1) possesses a one-parameter family of distinct periodic solutions $x_R(t)$ for each $R > 0$ such that the mean value of $U(x_R(t))$ over a period is R. Moreover if $U(x) = \frac{1}{2} Ax \cdot x +$ higher order terms near $x = 0$ with A a positive definite matrix, then as $R \to 0$ the period of $x_R(t)$ tends to the smallest nonzero period associated with the linearized system.

Proof: The result is proven by first establishing its validity for strictly convex C^2 potential functions $U(x)$ using (6.3.4), and then for a general C^1 convex potential function $U(x)$ satisfying (i) and (ii), by approximating $U(x)$ with suitably chosen strictly convex functions $U_N(x)$.

Step 1: We first prove (6.4.2) in case $U(x)$ is a C^2 strictly convex function $U(x)$ satisfying (i) and (ii). To this end, we exhibit the period parameter explicitly by the change of variables $t = \lambda s$ in (6.4.1), and seek a 1-periodic solution $x(s)$ ($\not\equiv 0$) of the system

(6.4.3) $$x_{ss} + \lambda^2 \nabla U(x) = 0$$

since such solutions $x(s)$ correspond to λ-periodic solutions of (6.4.1).

Clearly, by (6.3.4), under our hypotheses such a solution can be found solving the minimization problem

(π_N) Minimize $\int_0^1 |x_s|^2\, ds$ over the class of N-vector, absolutely continuous, and square integrable functions $W_N(0, 1)$ described in (6.1.8) subject to the constraints

(6.4.4)

$$\mathcal{C}_R = \left\{ x(s) \mid x(s) \in W_N,\ \int_0^1 U(x(s))\, ds = R,\ \int_0^1 \nabla U(x(s))\, ds = 0 \right\}.$$

Indeed, when regarded as an operator on $W_N(0, 1)$, the kernel of x_{ss} consists of the constant N-vectors, so that the constraint $\mathcal{C}_R$ in (6.3.4) coincides precisely with the constraint (6.4.4). Thus to complete Step 1, on the basis of (6.3.4), we need only note its hypothesis that clearly $\mathfrak{N}(x) = \int_0^1 U(x(s))\, ds$ is a C^2 strictly convex functional on $W_N(0, 1)$. The result (2.5.6) implies that $\mathfrak{N}(x)$ is weakly sequentially continuous, while the coerciveness of (Lx, x) follows as in (6.3.9) from Jensen's inequality. Indeed, if $x(s) = (x_1(s), \ldots, x_N(s))$,

$$U\left(\int_0^1 x_1(s)\, ds, \ldots, \int_0^1 x_N(s)\, ds \right) \leqslant \int_0^1 U(x(s))\, ds = R.$$

Consequently on $\mathcal{C}_R$, $|\int x(s)\, ds|$ is uniformly bounded by hypothesis (ii) above. Therefore, if $\|x(s)\|_{W_N} \to \infty$ and $x(s) \in \mathcal{C}_R$, $\int_0^1 |x_s|^2 \to \infty$ as in the proof of (6.1.8). The behavior as $R \to 0$ follows directly from the results of Section 4.2.

Step 2: We now remove the strict convexity and C^2 restrictions on $U(x)$. There exists a sequence of functions $U_1(x), U_2(x), \ldots$ such that:

(α) $0 = U_k(0) \leqslant U_k(x)$ for $x \in \mathbb{R}^N$;
(β) $U_k \in C^2(\mathbb{R}^N)$ and the Hessian of U_k is positive definite for all x;
(γ) $U_k \to U$ and grad $U_k \to$ grad U, as $k \to \infty$, uniformly on compact subsets of $\mathbb{R}^N$;
(δ) $U_k(x) \to \infty$ as $|x| \to \infty$ uniformly in k.

Such functions U_k can be obtained, for example, by smoothing U by convolutions, translating x slightly (to keep the minimum at $x = 0$), adding a small constant plus $|x|^2/k$ to assure conditions (α), (β). Applying Step 1, we find a sequence $\{x_k(s)\}$ of even 1-periodic elements of H satisfying

$$(6.4.5) \qquad \int_0^1 \dot{x}_k \cdot \dot{\phi}\, ds = \lambda_k^2 \int_0^1 \text{grad } U_k(x_k) \cdot \phi\, ds \qquad \text{for all} \quad \phi \in W_N$$

and such that $\int_0^1 |\dot{x}_k(s)|^2\, ds = \inf \int_0^1 |\dot{x}(s)|^2\, ds$ over the set

$$S_{k,R} = \left\{ x(s) \mid x(s) \in H,\ \int_0^1 \text{grad } U_k(x(s))\, ds = 0,\ \int_0^1 U_k(x(s))\, ds = R \right\}.$$

Next, we note that it suffices to prove that $\{\int_0^1 |\dot{x}_k(s)|^2\, ds\}$ is uniformly bounded. For in that case, since $\int_0^1 U_k(x_k(s))\, ds = R$, property ($\delta$) of $\{U_k(x)\}$ implies that the sequences $\{\sup |x_k(s)|\}$ and $\{\|x_k\|_H\}$ are uniformly bounded. Hence $\{x_k\}$ has a weakly convergent subsequence (which we relabel $\{x_k\}$) in W_N with weak limit $\bar{x}$; in addition, $x_k \to \bar{x}$ uniformly so that

$$\int_0^1 \text{grad } U(\bar{x}(s))\, ds = 0, \qquad \int_0^1 U(\bar{x}(s))\, ds = R > 0.$$

If $\bar{x}(s)$ is identically constant, say $\bar{x}(s) \equiv c$, then grad $U(c) = 0$. Hence $x \cdot \text{grad } U(x) = 0$ if x is on the line segment joining $x = 0$ and $x = c$, so that $U(x) = 0$ on this segment. In particular, $U(c) = 0$, which contradicts the last formula. Thus $\bar{x}(s) \not\equiv$ const. Furthermore, setting $\phi = x_k$ in (6.4.5), and taking limits, we find $\{\lambda_k^2\}$ is uniformly bounded (since $x \cdot \text{grad } U(x) > 0$ whenever $U(x) > 0$). Hence $\{\lambda_k^2\}$ has a convergent subsequence with limit $\lambda^2 \neq 0$. Thus $\bar{x}$ satisfies

$$\int_0^1 \dot{\phi} \cdot \dot{\bar{x}}\, ds = \lambda^2 \int_0^1 \text{grad } U(\bar{x}) \cdot \phi\, ds \qquad \text{for all} \quad \phi \in W_N.$$

Thus $\bar{x}$ is smooth and also satisfies $\ddot{\bar{x}} + \lambda^2 \text{ grad } U(\bar{x}) = 0$ and is clearly the desired critical point of the isoperimetric problem.

Now we show that $\{\int_0^1 |\dot{x}_k(s)|^2\, ds\}$ is uniformly bounded, to complete the proof of the theorem. To this end, let $x(s) = (\sin 2\pi s, 0, \ldots, 0)$. Then there exists a number $t_k > 0$ and a vector $c_k \in \mathbb{R}^N$ such that (by virtue of the properties of U_k)

$$y_k(s) = t_k x(s) + c_k \in S_{k,R},$$

that is,

$$\int_0^1 \text{grad } U_k(y_k(s))\, ds = 0, \qquad \int_0^1 U_k(y_k(s))\, ds = R;$$

Let $C > 0$ be so large that $|x| > C$ implies that $U_k(x) \geqslant 2R$ for $k = 1, 2, \ldots$. Then

$$2R \text{ meas}\{s \mid 0 \leqslant s \leqslant 1, |y_k(s)| > C\} \leqslant R,$$

so that

$$\text{meas}\{s \mid 0 \leqslant s \leqslant 1, |y_k(s)| \leqslant C\} \geqslant \tfrac{1}{2}.$$

Thus there is an interval $[\zeta, \eta] \subset [0, 1]$ on which $|y_k(s)| \leqslant C$, $\eta - \zeta \geqslant \frac{1}{4}$, and $|\sin 2\eta - \sin 2\zeta| \geqslant \theta \equiv 1 - \sin \frac{1}{4} > 0$. Since $|y_k(\eta) - y_k(\zeta)| \leqslant 2C$, it follows that $t_k \leqslant 2C/\theta$, and hence

$$\int_0^1 |\dot{y}_k(s)|^2\, ds \leqslant 4\left(\frac{2C}{\theta}\right)^2.$$

Consequently, the sequence $\{\int_0^1 |\dot{x}_k(s)|^2\, ds\}$ is also bounded by $4(2C/\theta)^2$.

Remark: Theorem (6.4.2) can be considerably weakened if the potential function $U(x)$ is even. Indeed, the convexity hypothesis can be replaced by the assumption that $\nabla U(x) \cdot x > 0$ for $x \neq 0$. This result is easily obtained by considering the closed subspace of $W_N(0, 1)$ consisting of those functions $x(s)$ that vanish at 0 and 1 and are odd in s.

The importance of such global results is well illustrated by considering the preservation of periodic orbits of the well-known Kepler two-body problem under small autonomous Hamiltonian perturbations. More precisely, for $N = 2$ or 3, in appropriate Cartesian coordinates, this problem can be written in the form

$$\ddot{x} = \frac{-x}{|x|^3} + \epsilon \nabla V(x), \tag{6.4.6}$$

where $V(x)$ is a C^1 real-valued function with $|\nabla V(x)| = o(1)$ for $|x|$ small, and ϵ is a small parameter. We suppose that for $\epsilon = 0$, the system described by (6.4.6) has negative total energy; i.e., $|\dot{x}|^2 - (1/|x|) < 0$ for any solution $x(t)$. Then all the solutions of (6.4.6) with $\epsilon = 0$ will be periodic, and we seek those periodic orbits near $x = 0$ that are preserved for $\epsilon \neq 0$. Clearly an important problem to be surmounted is the behavior of the term $(x/|x|^3)$ near the singularity $x = 0$. To overcome this difficulty for $N = 2$, we apply the useful regularization theory of Levi-Civita, mentioned in Section 1.1, and show the importance of our global results on the periodic solutions of (6.4.6).

To this end, suppose $x = (x_1, x_2) \in \mathbb{R}^2$ is written in complex notation $x = x_1 + x_2 = u^2$ with $u = u_1 + iu_2$ and $r = |x|$. Also, make a change of independent variable $s = \int_0^t (dt/r)$, so that $ds/dt = 1/r$. Then on the fixed energy surface $H = \frac{1}{2}|\dot{x}|^2 - (1/r) - \epsilon V = c$, we find that $(1/r)x' = x$, so that restricted to this surface the Lagrangian becomes

$$L^* = rL = (r/2)\{|x'|^2/r^2 + 1 + r(c + \epsilon V)\}.$$

Since $|x'|^2 = 4|u|^2|u'|^2$ and $r = |u|^2$, in terms of u, L^* becomes $L = 2|u'|^2 + 1 + u\bar{u}(c + \epsilon V)$. Consequently, the transformed equations can be written as

(6.4.7) $\qquad \ddot{u} + \nabla U(u) = 0, \qquad$ where $\quad U(u) = -\frac{1}{4}|u|^2\{c + \epsilon V(u^2)\}.$

Thus the periodic solutions of (6.4.7) with $\frac{1}{2}|\dot{x}|^2 - (1/|x|) - \epsilon V = c$ are in one-to-one correspondence with the periodic solutions of (6.4.7) such that

(6.4.8) $\qquad |\dot{u}|^2 + U(u) = 1.$

Clearly to find periodic solutions of (6.4.6) that satisfy (6.4.7) and (6.4.8), results of a global nature are necessary. Indeed, it does not suffice to know the existence of solutions of (6.4.7) near $u = 0$ since such solutions do not satisfy (6.4.8). Our result (6.4.2) yields a family $u_R(t)$ of periodic solutions of (6.4.7) for suitable functions V, one of which will satisfy (6.4.8).

6.4B Riemannian structures of prescribed Gaussian curvature for compact 2-manifolds with vanishing Euler–Poincaré characteristic

Here we consider the problem of determining necessary and sufficient conditions on a given Hölder continuous function $K(x)$ that will ensure the solvability of the following semilinear elliptic partial differential equation defined on a compact two-dimensional Riemannian manifold $(\mathfrak{M}, g)$ with $\chi(\mathfrak{M}) = 0$,

(6.4.9) $\qquad \Delta u - k(x) + K(x)e^{2u} = 0,$

where Δ denotes the Laplace–Beltrami operator on $(\mathfrak{M}, g)$ and $k(x)$ is a smooth function with $\int_{\mathfrak{M}} k(x)\, dV_g = 0$. As mentioned earlier, (6.2.9), an affirmative solution of this problem has the following geometric meaning: $K(x)$ is the Gaussian curvature of a Riemannian metric $(\mathfrak{M}, \bar{g})$ with $\bar{g} = e^{2u}g$ (pointwise) conformally equivalent[1] to g. A more general notion of conformal mapping involves the composition of a pointwise conformed mapping with a diffeomorphism of $\mathfrak{M}$ (see Fig. 6.1). From this geometric point of view, the immediate necessary condition for the solvability of

[1] In Kazdan and Warner (1974) the result (6.4.10) is used to prove the converse of the Gauss–Bonnet theorem for $\chi(\mathfrak{M}) = 0$.

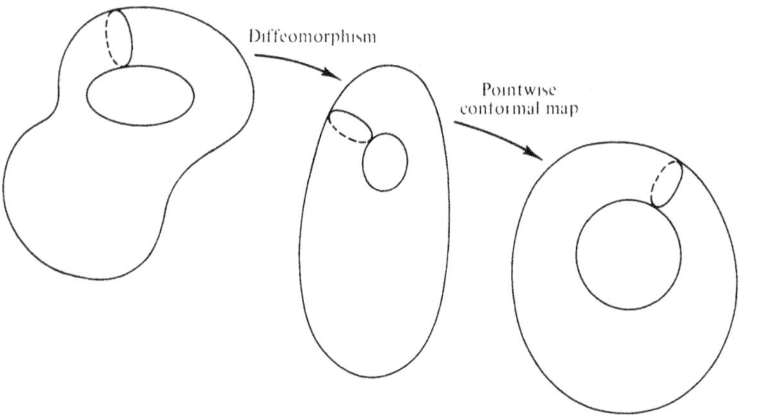

FIG. 6.1 The conformal mapping necessary to solve the general converse Gauss–Bonnet theorem for $\chi(\mathfrak{M}) = 0$.

(6.4.9), viz. $0 = \int_{\mathfrak{M}} K(x)e^{2u}\, dV_g$, obtained by integrating (6.4.9) over $\mathfrak{M}$, means that the "*integra curvatura*" of $\mathfrak{M}$ relative to the metric $\bar{g}$ must satisfy the Gauss–Bonnet theorem. This is precisely the kind of "natural" constraint we mentioned in Section 6.3B.

On the other hand, from the point of view adopted in Section 6.3, the equation (6.4.1) occupies a singular position since the functional $\mathcal{G}(u) = \int_{\mathfrak{M}} \{ \frac{1}{2} |\nabla u|^2 + k(x)u^2 \}\, dV$ is clearly *not* coercive on the set $S = \{u \mid u \in W_{1,2}(\mathfrak{M}, g), \int_{\mathfrak{M}} K(x)e^{2u}\, dV = 0\}$. This implies that in order to represent a solution of (6.4.9) as a minimum, the set S must be supplemented. We shall show that we can regain coerciveness by the addition of a simple explicit constraint in the natural isoperimetric problem for (6.4.9). In fact, we shall prove the following sharp result.

(6.4.10) Theorem Suppose the Euler–Poincaré characteristic of $\mathfrak{M}$, $\chi(\mathfrak{M}) = 0$. Then the equation (6.4.9) is solvable if and only if either $K(x) \equiv 0$ or $K(x)$ changes sign on $\mathfrak{M}$ and $\int_{\mathfrak{M}} K(x)\, e^{2u_0}\, dV < 0$, where u_0 is any solution of $\Delta u = k(x)$ on $\mathfrak{M}$.

Proof of necessity: If u satisfies (6.4.9) and $\chi(\mathfrak{M}) = 0$, then $\int_{\mathfrak{M}} K(x) \cdot e^{2u}\, dV = 0$. Thus if $K(x)$ is not identically zero, $K(x)$ must change sign on $\mathfrak{M}$. On the other hand, if we set $u = u_0 + w$, the function w satisfies the equation

$$\Delta w + K(x)e^{2u_0 + 2w} = 0.$$

Multiplying this equation by e^{-2w}, integrating over $\mathfrak{M}$, and integrating by parts, we find

$$2\int_{\mathfrak{M}} e^{-2w} |\nabla w|^2\, dV = -\int_{\mathfrak{M}} K(x)e^{2u_0}\, dV > 0.$$

Before proceeding further, we state an isoperimetric problem for the solutions of (6.4.9).

Lemma Suppose $\chi(\mathfrak{M}) = 0$ and $K(x)$ is a given function defined on $\mathfrak{M}$ such that relative to some Riemannian metric g defined on $\mathfrak{M}$, $\int_{\mathfrak{M}} K(x) e^{2u_0}\, dV < 0$. Then the (smooth) critical points of the functional $\mathcal{G}(u) = \int_{\mathfrak{M}} \{ \frac{1}{2} |\nabla u|^2 + k(x)u \}\, dV$ subject to the constraint

$$S' = \left\{ u \mid u \in W_{1,2}(\mathfrak{M}, g),\ \int_{\mathfrak{M}} u\, dV = 0,\ \int_{\mathfrak{M}} K(x)e^{2u}\, dV = 0 \right\}$$

are (apart from a constant) solutions of the equation (6.3.9).

Proof: The proof of (3.1.31) shows that a smooth critical point u of the isoperimetric variational problem (defined above) satisfies

$$(\dagger) \qquad \beta_0\{\Delta u - k(x)\} + \beta_1 K(x)e^{2u} = \beta_2,$$

where β_i $(i = 0, 1, 2)$ are constants (not all zero). Clearly, $\beta_0 \neq 0$ since otherwise $u \in S'$ implies $\beta_1 = \beta_2 = 0$, and so we set $\beta_0 = 1$. Since $\int_{\mathfrak{M}} K(x) \cdot e^{2u_0}\, dV \neq 0$, no solution of $\Delta u - k(x) = 0$ lies on S'. Thus both β_1 and β_2 cannot be zero. To show that $\beta_2 = 0$, we integrate ($\dagger$) over and find

$$\int_{\mathfrak{M}} k(x)\, dV + \beta_1 \int_{\mathfrak{M}} K(x)e^{2u}\, dV = \beta_2 \mu(\mathfrak{M}).$$

Since $\int_{\mathfrak{M}} k(x)\, dV = 0$, and $u \in S'$, $\beta_2 = 0$. Since $\beta_1 \neq 0$, there is a constant c such that $\pm e^{2c} = \beta_1$. So $\bar{u} = u + c$ satisfies $\Delta \bar{u} - k(x) \pm K(x)e^{2\bar{u}} = 0$. Now we show the $\beta_1 > 0$, so that $\beta_1 = e^{2c}$, and consequently $\bar{u} = u + c$ satisfies equation (6.4.9). Set $u = u_0 + w$ in ($\dagger$). Then by hypothesis, since $\beta_2 = 0$,

$$\Delta w + \beta_1 K(x)e^{2u_0}e^{2w} = 0.$$

Again multiplying by e^{-2w}, integrating over $\mathfrak{M}$, and integrating by parts, we find

$$\int_{\mathfrak{M}} e^{-2w} |\nabla w|^2\, dV = -\beta_1 \int_{\mathfrak{M}} K(x)e^{2u_0}\, dV.$$

Thus $\beta_1 > 0$ since $w \neq 0$.

Proof of sufficiency: To prove the existence of a critical point for this variational problem, we set $\sigma = \sigma_0 + c$, where $\int_{\mathfrak{M}} \sigma_0\, dV = 0$ so that

$$\mathcal{G}(\sigma_0) = \frac{1}{2} \int_{\mathfrak{M}} (|\nabla \sigma_0|^2 + k(x)\sigma_0)\, dV \qquad \left(\text{since} \int_{\mathfrak{M}} k(x)\, dV = 0\right)$$

$$\geqslant \frac{1}{2} \|\sigma_0\|^2 - c_1 \|k(x)\|\, \|\sigma_0\|.$$

Consequently, $\mathcal{G}(\sigma_0)$ is coercive on S', and $\mathcal{G}(\sigma)$ is weakly lower semicontinuous. Furthermore, S' is closed with respect to weak convergence on $W_{1,2}(\mathfrak{M}, g)$, and so by (6.1.1), inf $\mathcal{G}(\sigma)$ over S' is attained by an element $u \in S'$. Thus u is a weak solution of the equation (6.4.9) in the space $W_{1,2}(\mathfrak{M}, g)$. Since u is a weak solution of a linear equation of the form $\Delta u = f$ with $f \in L_p$ for all finite $p > 1$, u is smooth enough to satisfy equation (6.4.9) in the classical sense, and the theorem is thereby proved.

6.4C Riemannian manifolds with prescribed scalar curvature

Let $(\mathfrak{M}^N, g)$ be a given compact Riemannian manifold of dimension $N > 2$. On such a manifold we seek a new metric defined on $\mathfrak{M}^N$ of the form $\tilde{g} = e^{2v}g$ such that the new Riemannian manifold $(\mathfrak{M}^N, g)$ has prescribed curvature $g(x) < 0$ on $\mathfrak{M}^N$. As mentioned in Section 1.1A, if $k(x)$ denotes the scalar curvature of $(\mathfrak{M}^N, g)$, then the partial differential equation defining v can be written in the form

$$\text{(6.4.11)} \qquad \frac{4(N-1)}{N-2}\Delta u - k(x)u + g(x)u^\sigma = 0,$$

where $\sigma = (N+2)/(N-2)$ and $u = \exp \frac{1}{2}(N-2)v$ must be strictly positive on $\mathfrak{M}^N$. Here Δ is the Laplace–Beltrami operator relative to $(\mathfrak{M}^N, g)$. From our discussion of (6.3.8), it follows that special properties of the differential geometric problem must be used to solve (6.4.11) since the exponent σ is the *critical* value relative to (6.3.8). Actually we shall prove

(6.4.12) Suppose $\int_{\mathfrak{M}^N} k(x)\, dV < 0$, then (6.4.11) has a strictly positive smooth solution $u(x)$ defined on $(\mathfrak{M}^N, g)$. Consequently $\mathfrak{M}^N$ admits a Riemannian metric $\tilde{g}$ (pointwise conformally equivalent to g) with prescribed scalar curvature $g(x) < 0$.

Proof: By repeating the argument given in (6.3.8), we can find a positive smooth solution $(u_\epsilon, \lambda_\epsilon)$ of the system

$$\text{(6.4.13)} \qquad \frac{4(N-1)}{N-2}\Delta u - k(x)u + \lambda_\epsilon |g(x)| u^{\sigma-\epsilon} = 0$$

for any small $\epsilon > 0$. Furthermore $\lambda_\epsilon < 0$ since λ_ϵ is the minimum of the functional $\mathcal{Q}(u) = \int_{\mathfrak{M}^N}\{2(N-1)/(N-2)|\nabla u|^2 + k(x)u^2\}\, dV$ over the functions in $W_{1,2}(\mathfrak{M}^N, g)$ with $\mathcal{B}_\epsilon(u) = \int_{\mathfrak{M}^N}|g(x)|u^{\sigma+1-\epsilon}\, dV = 1$. For if c is the positive constant such that $\mathcal{B}(c) = 1$, then $\lambda_\epsilon < \mathcal{Q}(c) = c^2\int_{\mathfrak{M}^N} k(x)\, dV < 0$.

Now we shall show that as $\epsilon \to 0$, we can find a strongly convergent subsequence $\{u_{\epsilon_n}\}$ in $L_{(2N/(N-2))}(\mathfrak{M}^N, g)$. To this end, we first show that u_ϵ is uniformly bounded. Suppose that $M_\epsilon = \max u_\epsilon$ over $\mathfrak{M}^N$ is attained at x_0. Then $k(x_0)M_\epsilon - \lambda_\epsilon|g(x_0)|M_\epsilon^{\sigma-\epsilon} \leqslant 0$. Thus if $\lambda_0 = \inf \mathcal{A}(u)$ over $\mathcal{B}_0(u) = 1$, then

$$M_\epsilon^{\sigma-\epsilon-1} \leqslant \frac{\sup|k(x)|}{(-\lambda_\epsilon)\inf|g(x)|} \leqslant \frac{\sup|k(x)|}{\lambda_0 \inf|g(x)|},$$

so that M_ϵ is uniformly bounded. Thus $|\Delta u_\epsilon|$ is uniformly bounded, so that u_ϵ has a uniformly convergent subsequence with limit u_0. Clearly $u_0 \not\equiv 0$ satisfies (6.4.11) with $\epsilon = 0$, and $u_0 \geqslant 0$. We show that (i) $u_0 > 0$ on $\mathfrak{M}^N$, and (ii) λ_0 can be chosen to be -1. Clearly (i) follows from the maximum principle for Δ, for if $u_0 = 0$ on $\mathfrak{M}^N$ since $u_0 \geqslant 0$ on $\mathfrak{M}^N$, $u_0 \equiv 0$ on $\mathfrak{M}^N$. On the other hand, (ii) follows immediately from the fact that $\sigma \neq 1$, so that we can set $u = kw$ (for k a positive constant) in (6.4.11) and choose $k^{\sigma-1}\lambda_\epsilon = -1$. Thus $u_0 \not\equiv 0$ will satisfy the equation (6.4.11) everywhere on $\mathfrak{M}^N$.

Remark: It is interesting to note that the analogue of Theorem (6.4.12) is false for an arbitrary smooth positive function $g(x)$ as scalar curvature.

Remarks on symmetrization and isoperimetric problems In the next two isoperimetric problems we shall utilize a process called symmetrization of functions to sharpen our knowledge of the solution of isoperimetric problems. As an example, suppose D is a ball in $\mathbb{R}^N$ with the origin 0 as center. Then the symmetrization of a nonnegative function $g(x)$ (with respect to 0) is a function $g_s(x)$ that depends only on $|x|$ and is uniquely defined by the Lebesgue measure theoretic property that for each $\alpha \geqslant 0$

$$\mu(x \mid g_s(x) \geqslant \alpha) = \mu(x \mid g(x) \geqslant \alpha).$$

Thus $g_s(x)$ is a decreasing function of x and is continuous if $g(x)$ is. Moreover it can be shown that for an arbitrary C' function $F(t)$ we find symmetrization leaves integrals of $F(g)$ over D invariant while it decreases integrals of $F(|\nabla g|)$ over D. Consequently, if we wish to minimize $\int_D |\nabla u|^2$ over the class of functions $u \in W_{1,2}(D)$ with $\|u\|_{L_\sigma(D)} = 1$, we can assume a priori that the minimizer $\bar{u}(x)$ (if it exists) is of the form $g_s(x)$ for some $g \in W_{1,2}(D)$, i.e., $u(x)$ depends only on $|x|$ and is a decreasing function of $|x|$. The effect of this is to reduce the variational problem to a one-dimensional one.

We shall use this idea in the next problem in order to *sharpen* the estimate (1.4.6) in the case of the Sobolev space $W_{1,2}(S^2, g_1)$. More precisely, let S^2 be given the canonical metric of constant Gauss curvature

1, then we wish to determine the maximal constant k such that

$$(*) \qquad \sup_C \int_{S^2} e^{ku^2} < \infty \quad \text{where} \quad C = \left\{ u \mid \int_{S^2} |\nabla u|^2 = 1, \int_{S^2} u = 0 \right\}.$$

Assuming the sphere S^2 is parametrized by the coordinates (θ, ζ), where θ denotes latitude on the sphere with $\theta = \pm\pi/2$ corresponding to the poles and ζ measures longitude on the sphere, we observe that the integrals in question are independent of ζ. Thus symmetrization justifies the assumption that the maximizer in $(*)$ $\bar{u}$ is independent of ζ. By this device it turns out that one can show the constant $k = 4\pi$. However, if one adds to the constraint C the extra condition that $u(x) = u(-x)$ on S^2, the constant in question can be increased to 8π (see Moser, 1973a).

In the second problem (discussed in Section 6.4D) we suppose $\Pi(a, b)$ is a domain in the (x, y) plane symmetric about the line $y = 0$. Then it is useful to utilize the notion of Steiner symmetrization of a nonnegative function g about the line $y = 0$. $g_s(x, y)$ is defined by two properties that for fixed x, $g_s(x, y)$ depends only on y^2 and (relative to Lebesgue measure)

$$\{ y \mid f(x, y) \geqslant \alpha \} = \{ y \mid f_s(x, Y) \geqslant \alpha \}.$$

Steiner symmetrization leaves integrals of $F(f)$ over $\Pi(a, b)$ invariant while decreasing integrals of $F(|\nabla f|)$ over $\Pi(a, b)$. We utilize these properties in approximating the solutions to certain isoperimetric problems on unbounded domains by solutions of analogous isoperimetric problems on bounded domains.

6.4D Conformal metrics of prescribed Gaussian curvature on S^2

We now consider the following differential geometric problem as an application of (6.3.20):

(II) Let (S^2, g_1) denote the 2-sphere in $\mathbb{R}^3$ with its standard metric of constant Gaussian curvature 1. Then, given a C^∞ function $K(x)$ on S^2, find a metric g on S^2 (pointwise) conformally equivalent to g_1 with prescribed Gaussian curvature $K(x)$ (so that $g = e^{2u}g_1$ for some C^∞ function u).

In order to solve the problem (II) for a function $K(x)$, we first observe that if $g = e^{2u}g_1$ is the required metric, then the Gauss–Bonnet theorem implies that

$$(6.4.14) \qquad \int_{S^2} K(x) e^{2u} \, dV = 4\pi.$$

Consequently, if $\sup_{S^2} K(x) \leqslant 0$, no metric g exists.

In order to solve the problem (Π), we shall write down a semilinear elliptic partial differential equation for the mapping function u, and assuming $K(x) = K(-x)$, we shall show using (6.3.13) that (6.4.14) is a necessary and sufficient condition for the solvability of this equation, in the sense that (6.4.9) be satisfied for *some* C^∞ function u. The partial differential equation defining the mapping function u was discussed in Section 1.1, and can be written in the form

(6.4.15) $$\Delta u - 1 + Ke^{2u} = 0,$$

where Δ is the Laplace–Beltrami operator relative to (S^2, g_1). Note that the general notion of conformal mapping depicted in Fig. 6.1 is not needed in this case. We shall prove

(6.4.16) **Theorem** Suppose $K(x)$ is a Hölder continuous function on (S^2, g_1) such that $K(x) = K(-x)$. Then a necessary and sufficient condition for the solvability of (6.4.15) on S^2 is the existence of a $u \in W_{1,2}(S^2, g_1)$ such that (6.4.14) holds. Consequently, any such function $K(x)$ with $\sup_{S^2} K(x) > 0$ is the Gaussian curvature for a metric g (pointwise) conformal to (S^2, g_1). Moreover, there is a smooth function $K(x)$ on S^2 with $\max_{S^2} K(x) > 0$ such that (6.4.15) is insolvable.

Proof: We shall show that the difficulty in solving (6.4.15) is associated with the fact that any solution u is a saddle point of the associated functional

(6.4.17) $$\mathcal{I}(u) = \int_{S^2} \{ \tfrac{1}{2} |\nabla u|^2 + u - \tfrac{1}{2} Ke^{2u} \} \, dV.$$

Consequently, it is natural to attempt to apply the result to the semilinear gradient operator equation (on the Hilbert space $W_{1,2}(S^2, g_1)$) associated with (6.4.15). As usual, we set $(Lu, v) = \int_{S^2} \nabla u \cdot \nabla v$ and $(\mathcal{N}'(u), v) = \int_{S^2} K(x) e^{2u} v$. We find using the results of Section 2.5 that $L + \mathcal{N}'$ is a semilinear gradient operator equation mapping $W_{1,2}(S^2, g_1)$ into itself. In addition, since $K(x) = K(-x)$, $L + \mathcal{N}'$ maps the subspace $H = \{u \mid u \in W_{1,2}(S^2, g_1), u(x) = u(-x)\}$ of $W_{1,2}(S^2, g_1)$ into itself. On H, the operator L is nonnegative and has a one-dimensional kernel consisting of the constant functions. If we set $(f, v) = \int_{S^2} 1 \cdot v \, dV$, the solutions of the partial differential equation (6.4.15) are in one-to-one correspondence with the solutions of the semilinear operator equation $Lu + \mathcal{N}'(u) = f$. The virtue of this approach consists in the possibility of invoking Theorem (6.3.20). Indeed, the solvability criterion stated there, viz. $\mathfrak{M} = \{u \mid \mathcal{N}'(u) - f \perp \operatorname{Ker} L\} \neq \emptyset$ is equivalent to the existence of a function $u \in H$ satisfying the above equation (6.4.14). Therefore, by virtue of the result (6.3.20), we need only verify (a) the negative definiteness of $L +$

$\mathcal{N}''(u)$ (for $u \in \mathcal{M}$) on Ker L and (b) the coerciveness of $\mathcal{G}(u)$ on $\mathcal{M}$. The verification of (a) is easy. For $u \in \mathcal{M}$ and c a constant

$$\mathcal{N}(u_0 + c) = -e^{2c}\int K(x)e^{2u_0} < 0.$$

Thus $(d^2/dc^2)\mathcal{N}(u_0 + c) < 0$.

However, the verification of (b) is subtle, and proceeds as follows: For $u \in \mathcal{M}$, we write $u = u_0 + \bar{u}$, where $\bar{u}$ is the mean value of u over (S^2, g_1) and $\int u_0 = 0$. Since $\int K(x)e^{2u} = 4\pi$, we find

$$2\bar{u} = \log 4\pi - \log \int K(x)e^{2u_0}. \tag{6.4.18}$$

Thus for $u \in \mathcal{M}$, (6.4.18) implies that

$$\begin{aligned}\mathcal{G}(u) &= \int_{S^2} \tfrac{1}{2}|\nabla u_0|^2 + 4\pi\bar{u} - 2\pi\\ &\geqslant \text{const.} + \tfrac{1}{2}\int|\nabla u_0|^2 - 2\pi\log\int K(x)e^{2u_0}.\end{aligned}$$

Now setting $u_0 = v\|u_0\|$, where $\|v\| = 1$ with $\|u\| = \int_{S^2}|\nabla u|^2$, and noting that for every $\epsilon > 0$, $2u_0 \leqslant (1/\epsilon)\|u_0\|^2 + \epsilon v^2$, we find

$$\mathcal{G}(u) \geqslant \text{const.} + \left(\tfrac{1}{2} - \frac{2\pi}{\epsilon}\right)\|u_0\|^2 - 2\pi\log\int_{S^2} K(x)e^{\epsilon v^2}. \tag{6.4.19}$$

Next we use an inequality of Moser referred to in the remarks on symmetrization that implies

$$\sup_{\|v\|=1}\int_{S^2} e^{8\pi v^2} < \infty \qquad \text{for} \quad v \in H \quad \text{with} \quad \int_{S^2} v = 0.$$

Hence, choosing $\epsilon = 8\pi$ in (6.4.19), we find that $\mathcal{G}(u) \to \infty$ as $\|u_0\| \to \infty$. Thus $\mathcal{G}(u) \to \infty$ whenever $\|u\| \to \infty$ for $u \in S$, by virtue of (6.4.19). Consequently, the first part of (6.4.16) is established.

To establish the second part of the theorem, we give a brief argument that may be justified by the reader. Suppose u satisfies (6.4.15), then multiplying by ∇u and integrating over S^2, we find

$$\int_{S^2}\{\tfrac{1}{2}\nabla(|\nabla u|^2) - \nabla u + K(x)e^{2u}\,\nabla u\}\,dV = 0.$$

Using the fact that the first two terms in the brackets vanish when integrated over S^2, we find after an integration by parts that

$$0 = \int_{S^2} K(x)e^{2u}\,\nabla u = \int_{S^2} K(x)\,\nabla(e^{2u}) = \int_{S^2}(\nabla K(x))e^{2u}. \tag{6.4.20}$$

Thus for example, $K(x) = 2 + \sin\theta$ is a positive function on S^2 that does not satisfy (6.4.20) and for which (6.4.15) is insolvable.

6.4E A global free boundary problem—steady vortex rings of permanent form in an ideal fluid

The notion of a steady vortex ring was discussed briefly in Chapter 1. Here we establish the existence of a family of such axisymmetric vortex rings moving with *constant velocity and permanent form* in an ideal fluid. Two examples are of interest in this connection: (a) a "singular" vortex ring of "infinitely small" cross section, discussed in Helmholtz's pioneering paper of 1857; and (b) the vortex ring of Hill in which the vorticity is supported by a solid sphere. These examples represent extreme cases and we now describe a global existence theory that interpolates a family of vortex rings between these two extremes. (See Fig. 1.1.) The proof given is based on characterizing a vortex ring as the solution of an isoperimetric variational problem so that no a priori restriction is placed on the size of the cross section of the ring. Moreover the methods used here can be used in many other free boundary problems such as the classic problem of equilibrium shapes of a rotating fluid mentioned in Section 4.1 and equilibrium confinement figures of magneto hydrodynamics.

(I) Governing equations We begin by deriving the semilinear elliptic partial differential equation (mentioned in (1.1.17)) for the "Stokes stream function" ψ of the velocity field $\mathbf{v}$ associated with the vortex ring. Taking axes fixed in the ring and assuming the ideal fluid occupies the space $\mathbb{R}^3$, we suppose that in terms of cylindrical polar coordinates $\mathbf{v} = \mathbf{v}(r, z)$. Then since $\operatorname{div} \mathbf{v} = 0$, there is a vector $\mathbf{w} = (0, \psi/r, 0)$ such that $\mathbf{v} = \operatorname{curl} \mathbf{w}$. Thus the vorticity $\omega = \operatorname{curl} \mathbf{v}$ satisfies the relation

$$(6.4.21) \qquad \omega = \operatorname{curl}\operatorname{curl} \mathbf{w} = \Delta(0, \psi/r, 0),$$

where Δ denotes the Laplacean relative to the cylindrical polar coordinates. On the other hand, using an interesting observation of Stokes, the Euler equation of motion is satisfied by means of the vorticity equation

$$(6.4.22) \qquad \omega = \lambda r^2 f(\psi),$$

which expresses the fact that ω/r is constant on each stream surface. Here f is a prescribed function measuring the distribution of the vorticity inside the ring and λ is a positive constant measuring the actual magnitude of the vorticity inside the vortex ring. Thus if we suppose that A is a cross section of the vortex ring in a meridian plane (see Fig. 6.2) ($\theta = \text{const.}$) and Π denotes the half-plane $\{(r, z) \mid r > 0\}$, we find that ψ must satisfy

$$(6.4.23) \qquad \psi_{rr} - \frac{1}{r}\psi_r + \psi_{zz} = \begin{cases} -\lambda r^2 f(\psi) & \text{in} \quad A, \\ 0 & \text{in} \quad \Pi - \bar{A}, \end{cases}$$

On the unknown boundary ∂A of the vortex ring, we suppose (i) $\operatorname{grad} \psi$ is continuous, (ii) $\psi = 0$; while on the axis of symmetry $r = 0$ we set $\psi = -k$

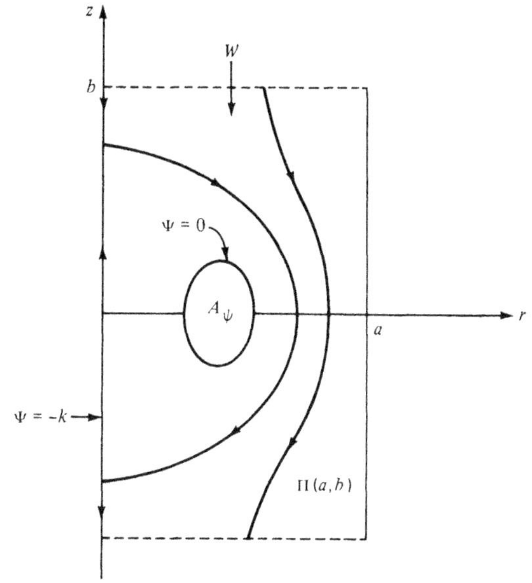

FIG. 6.2 Notation and expected streamline pattern for a steady vortex ring.

$\leqslant 0$, where k is a prescribed flux constant. Finally, we suppose that the velocity field of the ring tends to a constant vector $(0, W, 0)$ at infinity. This fact can be obtained by requiring

$$(6.4.24) \qquad \psi + \tfrac{1}{2} Wr^2 + k \to 0 \qquad \text{as} \quad r^2 + z^2 \to \infty.$$

(ii) Reformulations The free boundary value problem formulated in (i) is difficult to treat both because of the unknown domain A and because of the nonlinearity of (6.4.23). To separate these difficulties, we reformulate the problem as a semilinear Dirichlet problem on Π, which is untroubled by the unknown domain A. To this end, we extend the function $f(t)$ to the entire real axis by setting $f(t) = 0$ for $t \leqslant 0$, and $\Psi = \psi - \frac{1}{2} Wr^2 - k$. Then the desired vortex ring can be found by solving

$$(6.4.25) \qquad \psi_{rr} - \frac{1}{r} \psi_r + \psi_{zz} - \lambda r^2 f(\psi - \tfrac{1}{2} Wr^2 - k) = 0 \qquad \text{on } \Pi,$$

$$(6.4.26) \qquad \psi \mid_{\partial\Pi} = 0.$$

The maximum principle for second order linear elliptic equations will imply that the cross section A of the vortex ring can be found from a solution of (6.4.25)–(6.4.26) by setting

$$(6.4.27) \qquad A_\psi = \{(r, z) \mid \psi(r, z) > \tfrac{1}{2} Wr^2 - k\}.$$

This reformulation then requires us to find a nontrivial solution for the system (6.4.26)–(6.4.27) on the *unbounded domain* Π, and moreover, unless

$f(s) \to 0$ as $s \to 0$, the system has the added complication that the extended function $f(s)$ may *be discontinuous at* $s = 0$. Fortunately, these two difficulties can both be overcome by limiting arguments[1]: The domain Π can be approximated by large rectangles $\Pi(a, b)$ with vertices $(a, \pm b)$, $(0, \pm b)$, with a, b large, and a function $f(s)$ discontinuous at $s = 0$ can easily be approximated with a Lipschitz continuous one. Consequently it suffices to solve the system (6.4.25)–(6.4.26) with Π replaced by $\Pi(a, b)$, assuming f is Lipschitz continuous, and then to investigate the set A_ψ.

(iii) Resolution of the problem (6.4.25)–(6.4.26) on $\Pi(a, b)$ We begin by regarding the system (6.4.25)–(6.4.26) as a gradient operator equation on an appropriate Hilbert space H. Here the space H can be conveniently chosen as the closure of $C_0^\infty(\Pi(a, b))$ relative to the inner product

$$(u, v) = \int\int_{\Pi(a, b)} (1/r^2)(u_r v_r + u_z v_z)\, d\tau,$$

where $d\tau = r\, dr\, dz$. Relative to this inner product the appropriate generalized solution of the adjusted system (6.4.25) can be written conveniently as

$$(u, \phi) = \lambda \int\int_{\Pi(a, b)} f(\Psi)\phi\, d\tau \qquad \text{for all} \quad \phi \in H.$$

Now we are in a position to prove the following with $D = \Pi(a, b)$:

(6.4.28) **Theorem** Suppose $f(t)$ is a Lipschitz continuous nondecreasing function defined on $[0, \infty)$ with $f(0) = 0$ and with polynomial growth. Set $F(s) = \int_0^s f(t)\, dt$. Then the system (6.4.25) has a smooth solution $\psi(a, b)$ on $\Pi(a, b)$, for each $k \geq 0$. Moreover, $\psi(a, b)$ has the following properties: $\psi(a, b)$ is even in z, strictly positive in $\Pi(a, b)$, and is an extremal of the functional $J(u) = \int_{\Pi(a, b)} F(u - \frac{1}{2} Wr^2 - k) r\, dr\, dz$ over $\|u\|_H = 1$. Also, if $f \in C^1$ and is convex, then the set A is simply connected.

Proof: We first note that Hilbert space H can be continuously imbedded in the standard Sobolev space $\mathring{W}_{1,2}(\Pi(a, b))$, so that the imbedding theorem of Section 1.4 applies equally well to H or to $\mathring{W}_{1,2}(\Pi(a, b))$. To this end, we simply note that for $u \in H$,

$$\|u\|_{1,2}^2 = \int_{\Pi(a, b)} (u_r^2 + u_z^2)\, dr\, dz \leq \int \frac{a}{r^2} \{u_r^2 + u_z^2\} r\, dr\, dz$$

$$\leq a\|u\|_H^2 \qquad (\text{since } r \leq a).$$

We are now in a position to prove the existence of the desired solution

[1]The interested reader will find these limiting arguments discussed fully in Fraenkel and Berger (1974). D. Kinderlehrer recently demonstrated the smoothness of the free boundary.

$\psi(a, b)$, and its isoperimetric characterization on the basis of (6.3.7). First it is necessary to show that

$$\beta = \sup_{\|u\|_H = 1} \int F(u - \tfrac{1}{2} Wr^2 - k) r\, dr\, dz > 0. \tag{6.4.29}$$

(This would not be true for the analogous one-dimensional problem.) Since $F(t)$ is strictly increasing, it suffices to observe that functions with small norm in H can have arbitrarily large values on a small set. For example, given a point $x_0 \in \Pi(a, b)$, the function

$$v_\beta(x) = \begin{cases} 0 & \text{for} \quad |x - x_0| \geqslant \delta, \\ \beta \log\left(1 - \log \dfrac{\delta}{|x - x_0|}\right) & \text{for} \quad \delta \geqslant |x - x_0|, \end{cases}$$

has $\|v_\beta(x)\| = 1$ for a certain value β sufficiently large, but tends to ∞ as $|x - x_0| \to 0$.

Clearly the functional $J(u)$ is continuous with respect to weak convergence in H, by the Sobolev imbedding theorem, and the polynomial growth condition on $f(u)$. Indeed,

$$H \subset \mathring{W}_{1,2}(\Pi(a, b)) \subset L_p(\Pi(a, b)) \subset L_p(\Pi(a, b), \tau),$$

where $L_p(D, \tau)$ denotes the L_p functions over $\Pi(a, b)$ with the volume element $d\tau = r\, dr\, dz$. Now the argument of (6.3.1) shows that β is attained by an element $\psi \in H$. Moreover, $\psi \geqslant 0$ in $\Pi(a, b)$ since the nonnegative part of ψ, ψ_+, has the property that $J(\psi_+) = J(\psi)$, while $\|\psi_+\| < \|\psi\|$ if $\psi < 0$ on a set of positive measure. Thus a scaling of ψ_+ would be the desired extremal element. By virtue of (3.1.31), we therefore find that there are constants μ_1 and μ_2 (not both zero) such that

$$\mu_1(\psi, w) = \mu_2 \int_{\Pi(a, b)} f(\Psi) \dot{w} \qquad \text{for all } w \in H. \tag{6.4.30}$$

To show that $\mu_1 = 0$ and $\lambda = \mu_2/\mu_1 > 0$, it suffices to set $w = \psi$ in (6.4.30) and to observe that then the integral on the left-hand side of (6.4.30) is positive. Since $F(t) \leqslant tf(t)$, $0 < \Psi(x) < \psi(x)$ for $\psi(x) > 0$, and we have

$$\int_{\Pi(a, b)} \psi f(\psi) > \int_{\Pi(a, b)} \Psi f(\Psi) \geqslant \int_{\Pi(a, b)} F(\Psi) = \beta > 0.$$

The fact that $\psi(a, b)$ is even in z follows from Steiner symmetrization with respect to the line $z = 0$ in $\Pi(a, b)$. The regularity of $\Pi(a, b)$ follows by standard regularity theory despite the apparent singularity of the coefficients of the equation at $r = 0$.

To complete the proof of the theorem, we show that the set A_ψ defined by (6.4.27) is *simply connected* under the convexity hypothesis of f. To this end, we consider the second variation formula of (3.1.40) for arbitrary v satisfying $(v, \psi) = 0$

$$\delta^2 J(\psi, v) = \int_{\Pi(a,b)} v^2 f'(\Psi)\, d\tau - \frac{1}{\lambda}\|v\|^2.$$

We suppose that A_ψ has at least two components E_1, E_2. Let the functions w_1 and w_2 be defined by setting $w_i = \Psi_i$ on E_i ($i = 1, 2$) and zero elsewhere. Then if $v = c_1 w_1 + c_2 w_2$, where c_1, c_2 are chosen so that $c_1\|w_1\|^2 = -c_2\|w_2\|^2$, we find that, from the above, contrary to (6.4.29),

$$\delta^2 J(\psi, v) = \sum_{j=1}^{2} c_j^2 \int_{E_j} \Psi_j\{\Psi_j f'(\Psi_j) - f(\Psi_j)\}$$
$$> 0 \qquad \text{(by the convexity of } f\text{)}.$$

(iv) The limiting procedure as $\Pi(a, b) \to \Pi$ We now show that a limiting procedure applied to the results of (iii) for $\Pi\ (a, b)$ solves the problem (6.4.25)–(6.4.26) as formulated in Π. To this end we let the rectangles $\Pi\ (a, b)$ tend to the half-plane Π by letting $(a, b) \to \infty$. The results of (iii) then yield solutions $(\psi(a, b), \lambda(a, b))$ with $\lambda(a, b)$ contained in a bounded set of real numbers (uniformly bounded above zero) of the modified problem as stated in Theorem (6.4.28). We seek a convergent subsequence $\psi_n = \psi(a_n, b_n)$ in $H(\pi) = \mathring{W}_{1,2}(\Pi)$ obtained by setting $\psi(a_n, b_n) \equiv 0$ outside $\Pi(a_n, b_n)$ with associated convergent eigenvalues $\lambda(a_n, b_n) = \lambda_n$. To this end we shall prove the existence on a *fixed* domain Ω such that the associated vortex cores $A(\psi(a, b)) \subset \Omega$ no matter how large a and b become. This domain Ω will have the property that $\text{vol}[\Omega \cap \{y \mid |x - y| < 1\}] \to 0$ as $|x| \to \infty$ so that the imbedding $H(\Omega) \to L_2(\Omega)$ is compact, and consequently one can pass to the limit in the integral equation

(6.4.31) $$\lambda(a, b) \int_{A(\Pi(a,b))} G_{a,b} f(\psi(a, b)) = \psi(a, b).$$

$G_{a,b}(x, y)$ is Green's function on $\Pi(a, b)$ associated with the linear operator of (6.4.25).

To find the domain Ω we first prove the following a priori bound for the solution of (6.4.25)–(6.4.26) on $\Pi(a, b)$.

Lemma Let $l(u)$ be the length of the projection of the set

$$A(u) = \{(r, z) \mid (r, z) \in \Pi(a, b),\ u(r, z) > \tfrac{1}{2}r^2 + k\}$$

on the z axis, where $u \subset C'(\Pi(a, b))$ and

$$Y(u) = \{(r, z) \mid (\rho, z), A(u), \rho \geqslant r > 0\}$$

(i.e., $Y(u)$ is a set in $\Pi(a, b)$ containing $A(u)$ and all points (r, z) in $\Pi(a, b)$ that lie between a point (ρ, z) of $A(u)$ and the z-axis). Then

(6.4.32) $$\iint_{Y(u)} r\, dr\, dz + 2kl(u) \leqslant \|u\|^2_{H(a,b)}.$$

Proof: Let $\frac{1}{2}r^2 = y$ and assume the boundary of $Y(u)$ is smooth so that the divergence theorem can be applied there. Then a simple computation, using the divergence theorem, shows

$$\|u\|^2_{H(\Pi(a,b))} \geqslant \iint_{Y(u)} \{u_y^2\} \geqslant \iint_{Y(u)} \{u_y^2 - (u_y - 1)^2\}$$
$$\geqslant \iint_{Y(u)} (2u_y - 1) = \oint_{Y(u)} 2u - \iint_{Y(u)} r\, dr\, dz.$$

Since $u = 0$ on the z axis, and on $\partial A(u)$, $u = y + k$, we find from the last set of inequalities the desired bound (6.4.32).

We find, applying the lemma just obtained to $u = \psi(a, b)$ that the set Ω can be chosen as $\Omega = \{(r, z) \mid |z| < (r^2 + 4k)^{-1}\}$. Indeed since $\|\psi(a, b)\|_H = 1$ and by Steiner symmetrization the projection of $A(\psi)$ on the z axis must contain the interval of the form $I_R = (-h(R), h(R))$, so that the open rectangle $(0, r) \times I_r$ must be contained in $Y(u)$. Then applying (6.4.32) and noting that $2h(r) \leqslant l(\psi)$ we find $\frac{1}{2} r^2(2h(r)) + (2k)(2h(r)) < 1$, whence $h(r) < (r^2 + 4k)^{-1}$. Consequently we can find a weakly convergent subsequence (ψ_n, λ_n) in $H(\Pi)$, and pass to the limit in (6.4.25), as mentioned earlier. The weak limit (ψ, λ) then yields the desired nontrivial solution.

(v) Historical background In 1858, Helmholtz considered vortex rings of very small cross section as one of two examples of his theory. Basing himself on Helmholtz's results on the indestructibility of vortex rings and taking the "ether" as the appropriate ideal fluid, Kelvin made vortex rings the basis of a primitive atomic theory. Kelvin conjectured the existence of nonaxisymmetric vortex rings whose "core" could be associated with thickened knot configurations in $\mathbb{R}^3$. (The differing topological structures of knots in $\mathbb{R}^3$ were supposed to classify the various atomic structures in Nature.) Indeed, on this basis, Kelvin's collaborator, Tait, was thus led to pioneering work in the mathematical theory of knots. Kelvin's theory declined with the overthrow of the "ether" concept. Nonetheless, in modern theoretical low-temperature physics the importance of vortex rings has strikingly reappeared in the theory of superfluidity and superconductivity since there ideal fluids approximate reality.

6.5 Critical Point Theory of Marston Morse in Hilbert Space

In order to study all the critical points of a given functional $\mathcal{G}(u)$, defined on a Hilbert space H, topological considerations are necessary to supplement our earlier discussion. This point has been clarified in both finite- and infinite-dimensional contexts by Marston Morse beginning with studies dating from 1925. To illustrate this point, we note that for a linear compact self-adjoint operator $C \in B(H, H)$, *orthogonality notions* can be used to develop a complete spectral theory once the analogue of (1.3.40(i)) is established. Moreover the *principle of superposition* for linear operators implies that for a self-adjoint Fredholm operator L, the solutions of the operator equation $Lu = f$ fill up finite-dimensional linear subspaces of H. In a nonlinear context, totally new ideas must be used to find analogues of these results. In this section we shall describe Morse's approach to this problem in a simple Hilbert space context. In Section 6.6 we describe the related theory of Ljusternik and Schnirelmann, while in Section 6.7 we illustrate the use of these theories in various contexts.

6.5A A sharpening of the steepest descent method

For any general study of critical points, it is necessary to supplement the notions of weak lower semicontinuity and coerciveness by more refined analytic and topological considerations. To this end, we first reconsider the method of steepest descent, introduced in Section 3.2. Suppose $F(x)$ is a smooth C^2 functional defined on a real Hilbert space H, and bounded from below on H. Then in Section 3.2 it was shown that the solutions $x(t, x_0)$ of the initial value problem

(6.5.1) $$dx/dt = -F'(x), \qquad x(0, x_0) = x_0,$$

exist for all $t \geqslant 0$, with $\lim x(t, x_0)$ as $t \to \infty$ a critical point of $F(x)$ provided that the critical points of $F(x)$ are isolated and that $F(x)$ satisfies the following *compactness condition* (mentioned earlier in (6.1.1′)).

(6.5.2) **Condition (C)** Any sequence $\{x_n\}$ in H with $|F(x_n)|$ bounded and $\|F'(x_n)\| \to 0$ has a convergent subsequence.

Under hypothesis (6.2.2), the discussion of Section 3.2 shows that $F(x)$ clearly attains its infimum on H. The following result shows the utility of (6.5.2) for the study of other types of critical points.

(6.5.3) **Theorem** Suppose that a C^2 functional $F(x)$ defined on H is bounded from below and satisfies (6.5.2), and has only isolated critical points. If $F(x)$ possesses two isolated relative minima y_1, y_2, then the functional $F(x)$ must possess a third critical point y_3, distinct from y_1 and y_2, which is not an isolated relative minimum.

Proof: Suppose that $F(x)$ does not have a third critical point. Then we shall show that H can be represented as the union of two open, disjoint subsets U_1 and U_2; which obviously contradicts the connectedness of H. To construct the sets U_i, suppose that $x(t, x_0)$ is the solution of (6.5.1). By (6.5.2), $x(t, x_0)$ exists for all $t \geqslant 0$ and $\lim x(t, x_0)$ as $t \to \infty$ is y_i $(i = 1, 2)$. Let $U_i = \{x_0 \mid \lim x(t, x_0) = y_i \text{ as } t \to \infty\}$ $(i = 1, 2)$. Clearly $H = U_1 \cup U_2$, while U_1 and U_2 are disjoint. To show that the sets U_i are open in H, we first note that each y_i, being a strict relative minimum, has a neighborhood W_i such that any solution $x(t, x_0)$ which enters W_i remains in W_i and, in fact, converges to y_i as $t \to \infty$. Indeed, for x_0 sufficiently near y_i, since $F(x(t, x_0))$ is a decreasing function of t, $x(t, x_0) \to y_i$. Thus, by virtue of the continuity of $x(t, x_0)$ with respect to the initial condition x_0, if $z_0 \in U_i$, then for $\epsilon > 0$ sufficiently small with $\|z_0 - \tilde{z}_0\| < \epsilon$, there is a T such that both $x(T, z_0)$ and $x(T, \tilde{z}_0)$ lie in W_i. Consequently $\tilde{z}_0 \in U_i$ if z_0 is. Therefore each U_i is open, and we have obtained the desired contradiction.

It is also immediate that the third critical point y_3 shown to exist by the above argument cannot be another relative minimum, for if it were and if $F(x)$ had no other critical points, then the argument just given would again lead to a contradiction.

6.5B Degenerate and nondegenerate critical points

To proceed with a deeper study of the critical points of a C^2 functional $F(x)$ defined on H, the following definitions are convenient. A critical point x_0 is called *nondegenerate* if the self-adjoint operator $F''(x_0)$ is invertible. Otherwise x_0 is termed *degenerate*. The *index* of a critical point x_0 of $F(x)$ is the maximal dimension on which the form $(F''(x_0)x, x)$ is negative definite. $F(x)$ is a Fredholm functional if $F'(x)$ is Fredholm.

The nondegenerate critical points of $F(x)$ have several important properties. First, by the inverse function theorem (3.1.5), the nondegenerate critical points of $F(x)$ are isolated. Next, the set of C^2 functionals $F(x)$ defined on a *bounded* subset of H such that the critical points of $F(x)$ are all nondegenerate form a dense subset of all such C^2 Fredholm functionals. Indeed, if $G(x)$ is a C^2 Fredholm functional some of whose critical points may be degenerate, consider the functional $\tilde{G}_p(x) = G(x) - (x, p)$. Clearly, for $\|p\|$ sufficiently small, $\|\tilde{G}_p(x) - G(x)\|_{C^2}$ can be made as small as desired. On the other hand, by (3.1.5), all the critical points of $\tilde{G}_p(x)$ are nondegenerate provided p is not a singular value for $G'(x)$. Since $G'(x)$ is a C^1 Fredholm operator of index zero, by (3.1.45), this set is nowhere dense in H. Furthermore, if a C^2 functional $F(x)$ has all its critical points nondegenerate on $F^{-1}[a, b]$ and $F(x)$ satisfies Condition (C) of (6.5.2), then there can be at most a finite number of critical points on $F^{-1}[a, b]$ for any $-\infty < a, b < \infty$. Indeed, otherwise there would be a sequence $\{x_n\}$ with $a \leqslant F(x_n) \leqslant b$ and $\|F'(x_n)\| = 0$, so that $\{x_n\}$ would have a convergent subsequence with limit $\bar{x}$. This would be a contradiction, for clearly $\bar{x}$ would be both a nonisolated and nondegenerate critical point of $F(x)$ with $\bar{x} \in F^{-1}[a, b]$.

Another interesting result is the following extension of Morse's lemma (1.6.1) to a Hilbert space context.

(6.5.4) **Theorem** Suppose $F(x)$ is a C^3 functional defined in the neighborhood of a nondegenerate critical point $x = 0$. Then there is a diffeomorphism h mapping a neighborhood U of $x = 0$ onto itself such that for $x \in U$, $F(x) - F(0) = \frac{1}{2}(F''(0)h(x), h(x))$.

Proof: Since $x = 0$ is a critical point, we have

$$F(x) - F(0) = \int_0^1 (F'(sx), x)\, ds,$$

while

$$F'(x) = \int_0^1 (d/dt)F'(tx)\,dt = \int_0^1 F''(tx)x\,dt.$$

Thus we can represent $F(x)$ in terms of F'' by writing

$$F(x) - F(0) = \int_0^1 \int_0^1 s(F''(stx)x, x)\,dt\,ds = (k(x)x, x) \quad \text{(say)},$$

where we may suppose that $k(x)$ is the self-adjoint operator defined by setting

$$(k(x)y, z) = \int_0^1 \int_0^1 s(F''(stx)y, z)\,dt\,ds.$$

Clearly with this definition, $\frac{1}{2}(F''(0)x, x) = (k(0)x, x)$. Now suppose that $B(x) = [k(0)]^{-1}k(x)$ and $C(x) = \sqrt{B(x)}$ which exists for $\|x\|$ sufficiently small, since $B(0) = I$ (the identity). Also note that, if $C^{\mathrm{T}}(x)$ denotes the adjoint to $C(x)$, then $C^{\mathrm{T}}(x)k(0) = k(0)C(x)$ since $B^{\mathrm{T}}(x)k(0) = k(0)B(x)$, and

$$[k^{-1}(0)C^{\mathrm{T}}(x)k(0)]^2 = k(0)^{-1}B^{\mathrm{T}}k(0) = B(x) = C^2(x).$$

Thus the result is obtained by the following simple computation:

$$\begin{aligned} F(x) - F(0) &= (k(x)x, x) = (k(0)C^2(x)x, x) \\ &= (C^{\mathrm{T}}k(0)C(x)x, x) \\ &= (k(0)h(x), h(x)) = \tfrac{1}{2}(F''(0)h(x), h(x)), \end{aligned}$$

where $h(x) = C(x)x$ is invertible for $\|x\|$ sufficiently small.

This result has the immediate consequence that after a local differentiable change of coordinates $y = Y(x)$ near the nondegenerate critical point $x = 0$ the functional $F(x)$ can be written in the form

$$F(y) = \|(I - P)y\|^2 - \|Py\|^2,$$

where P is the projection of H onto the linear subspace of H on which $F''(0)$ is negative definite.

We next observe that exactly as in the finite-dimensional case our discussion applies when the Hilbert space H is replaced by an infinite-dimensional manifold $\mathfrak{M}$ that locally approximates H. This fact is very useful for many differential geometric problems (see the notes at the end of the chapter). More explicitly,

Definition A manifold $\mathfrak{M}$ of class C^r modeled on a Hilbert space X (i.e., Hilbert manifold) is a collection of open sets $\{U_\alpha\}$ and mappings θ_α: $U_\alpha \to X$ such that:

(i) $\theta_\alpha\colon U_\alpha \to \theta_\alpha(U_\alpha)$ is a homeomorphism,
(ii) $\theta_\alpha \theta_\beta^{-1}\colon \theta_\beta(U_\alpha \cap U_\beta) \to \theta_\alpha(U_\alpha \cap U_\beta)$ is a smooth map of class C^r.

Definition *An atlas of class C^r for a set $\mathfrak{M}$ is a collection of open subsets $\{U_\alpha\}$ of $\mathfrak{M}$ and transition maps such that:*

(i) ζ_α is a homeomorphism of U_α onto $\zeta_\alpha(U_\alpha)$, a subset of a Hilbert space X;

(ii) the mappings are compatible in case $U_\alpha \cap U_\beta \neq \varnothing$, i.e., $\zeta_\alpha \zeta_\beta^{-1}\colon \zeta_\alpha(U_\alpha \cap U_\beta) \to \zeta_\beta(U_\alpha \cap U_\beta)$ is a homeomorphism of class C^r for each α, β;

(iii) the collection (U_α, ζ_α) is maximal with respect to properties (i) and (ii).

A *Banach manifold of class C^r* modeled on X is an atlas of class r defined on the set $\mathfrak{M}$ relative to the Banach space X. The members of the collection $\{(U_\alpha, \zeta_\alpha)\}$ are referred to as *charts*.

Most properties of a mapping f between spaces, defined by means of the derivative of f, can be extended to mappings between Banach manifolds via this definition. For example,

Definition Let $f\colon \mathfrak{M} \to \mathfrak{N}$ be a mapping defined on the manifolds $\mathfrak{M}$ and $\mathfrak{N}$. Then f is of class C^p, provided it is continuous and relative to charts at each $x \in \mathfrak{M}$ and $f(x) \in \mathfrak{N}$, f is C^p; i.e., $\zeta_\alpha f \theta_\beta^{-1}$ is a smooth map of class C^p between the Banach spaces $X_{\mathfrak{M}}$ and $X_{\mathfrak{N}}$.

These definitions enable us to carry over much of the local analysis of Chapter 3 to the study of mappings between Hilbert manifolds. In particular a tangent space, $T\mathfrak{M}_x$ to a Hilbert manifold $\mathfrak{M}$ at $x \in \mathfrak{M}$ is the set of all tangent vectors $\{p'(0)\}$ to $\mathfrak{M}$ relative to C^1 curves $p(t)$ passing through x at $t = 0$. Moreover, the differential of a mapping $f\colon \mathfrak{M} \to \mathfrak{N}$ between two Hilbert spaces is the mapping $df(x)\colon T\mathfrak{M}_x \to T\mathfrak{N}_{f(x)}$ defined by setting

$$df(x)(p'(0)) = \frac{d}{dt} f(p(t)) \,|_{t=0}$$

for every curve $p(t)$ and so is linear in the argument $p'(0)$. For a smooth functional $\mathcal{G}(x)$ defined on a Hilbert manifold $\mathfrak{M}$, the differential $d\mathcal{G}(x, p'(0))$ is linear in the argument $p'(0)$ and thus we write

$$d\mathcal{G}(x, y) = (\operatorname{grad} \mathcal{G}(x), y)$$

by virtue of the Riesz representation theorem for linear functionals. Furthermore, in terms of charts about x and $f(x)$, the differential $df(p'(0))$ can be computed as a derivative of a mapping. The critical points of a functional $\mathcal{G}(u)$ defined on a Hilbert manifold $\mathfrak{M}$ thus coincide with the points $x \in \mathfrak{M}$ such that $\operatorname{grad} \mathcal{G}(x) = 0$. Clearly notions such as critical point, degenerate and nondegenerate critical point, and Morse index, being invariant under coordinate changes, have a well-defined meaning on $\mathfrak{M}$. Many of the results we obtain in the remainder of this section can be adapted to apply to Hilbert manifolds.

In order to investigate the relations between the critical points of $F(x)$, we use singular homology. The basic properties of singular homology groups with coefficient group $\mathcal{G}$ can be summarized as follows:

(i) If $f: (X, A) \to (Y, B)$ is a continuous mapping, then there is a group homomorphism for each integer q, $f_*: H_q(X, A; \mathcal{G}) \to H_q(Y, B; \mathcal{G})$ with the following properties.

(a) If $f = i$, the identity i_* is the identity automorphism.
(b) If $g: (Y, B) \to (Z, C)$, then $(gf)_* = g_* f_*$.
(c) $df_* = f_* d$.
(d) (Homotopy property) If $f, g: (X, A) \to (Y, B)$ are homotopic, then $f_* = g_*$.

(ii) (Excision property) If U is an open subset of X with $\bar{U} \subset \operatorname{int} A$, the inclusion e: $(X - U, A - U) \to (X, A)$ induces the isomorphism

$$e_*: H_q(X - U, A - U; \mathcal{G}) \to H_q(X, A; \mathcal{G}') \qquad \text{(for each } q\text{)}.$$

(iii) (Exactness property) If $i: A \to X$ and $j: X \to (X, A)$ denote inclusion maps, then the following infinite sequence is exact, i.e., the image of each homomorphism equals the kernel of the next homomorphism

$$\cdots \to H_q(A; \mathcal{G}) \xrightarrow{i_*} H_q(X; \mathcal{G}) \xrightarrow{j_*} H_q(X, A; \mathcal{G}) \xrightarrow{d} H_{q-1}(A; \mathcal{G}) \to \cdots .$$

(iv) (Dimension property) If X is a space consisting of one point, then

$$H_p(X; \mathcal{G}) = \begin{cases} 0 & \text{for} \quad p \neq 0, \\ \mathcal{G} & \text{for} \quad p = 0. \end{cases}$$

6.5C Morse type numbers

In order to investigate the structure of all the critical points of a given functional, it is useful to establish a classification of critical points that is invariant under local differentiable changes of coordinates. The following classification (due to M. Morse) will be used in this section. First, consider the nondegenerate critical points of a smooth functional $F(x)$. We classify these by their index. The result (6.5.4) ensures that such a classification is invariant under local diffeomorphism. Assuming $F(x)$ possesses an isolated degenerate critical point x_0, we associate with x_0 a sequence of positive integers $(M_0(x_0), M_1(x_0), M_2(x_0), \ldots)$ called the type numbers of x_0. The integers $M_i(x_0)$, $i = 0, 1, \ldots,$ are a measure of the number of nondegenerate critical points of index i equivalent to x_0. These type numbers $M_i(x_0)$ are defined as the Betti numbers of the relative singular homology groups $H_i(F^{c+\epsilon} \cap U, F^{c-\epsilon} \cap U)$ with Z_2 coefficients, where $F(x_0) = c$, $F^d = \{x \mid F(x) \geqslant d\}$, and U is a small neighborhood of x_0 ($\epsilon > 0$ is sufficiently small). In order to justify these definitions, we prove

(6.5.5) **Theorem** Suppose $F(x)$ is a C^2 real-valued functional defined on a Hilbert space, which satisfies condition (6.5.2), and suppose $b > a$.

(i) If $F(x)$ has no critical points on the interval $[a, b]$, then the sets $F^b = \{x \mid F(x) \leqslant b\}$ and $F^a = \{x \mid F(x) \leqslant a\}$ are isotopic. Furthermore, the isotopy may be so chosen that the points of F^a are fixed, so that F^a is a deformation retract of F^b.

(ii) If $F(x) = c$ is an isolated critical value of $F(x)$ on which F has only a finite number of critical points $\sigma(c) = \{x_i\}$, then for any $\epsilon > 0$

$$H_q(F^{c+\epsilon}, F^{c-\epsilon}) \approx H_q(\mathring{F}^c \cap U \cup \sigma(c), \mathring{F}^c \cap U),$$

where $\mathring{F}^c = \{x \mid F(x) < c\}$ and U is any sufficiently small neighborhood of all the critical points $\sigma(c)$.

(iii) If $F(x)$ has a single nondegenerate critical point of index i on $[a, b]$, $H_q(F^b, F^a) = 0$, $q \neq i$, while $H_i(F^b, F^a) = G$ (the coefficient group of the homology theory).

Proof: (i): First we observe that since $F(x)$ satisfies Condition (C) for some $\epsilon_0 > 0$ sufficiently small, $F(x)$ has no critical points on $F^{-1}[a - \epsilon_0, b + \epsilon_0]$, and in fact there is a positive constant $d > 0$ such that $\inf_{F^{-1}[a-\epsilon_0, b+\epsilon_0]} \|F'(x)\| \geqslant d$. Otherwise, there would be a sequence $\{x_n\} \in H$ with $x_n \in F^{-1}[a - (1/n), b + (1/n)]$ and $F'(x_n) \to 0$ so that after passing to a subsequence $x_n \to \bar{x} \in F^{-1}[a, b]$, $F'(\bar{x}) = 0$.

Now to define the isotopy $\zeta_t(x)$ of F^b onto F^a, we use the method of steepest descent discussed in Section 3.2, and consider the truncated analogue of (6.5.1),

$$\frac{dx}{dt} = -\alpha(|F(x)|) \frac{F'(x)}{\|F'(x)\|^2}, \quad x(0) = x_0 \in F^{-1}[a, b], \tag{6.5.6}$$

where $\alpha(z)$ is a real-valued C^∞ nonnegative function such that $\alpha(z) = b - a$ for $a \leqslant z \leqslant b$ and $\alpha(z) = 0$ for $z \leqslant a - \epsilon_0$ or $z \geqslant b + \epsilon_0$. Since $\inf_{[a-\epsilon_0, b]} \|F'(x)\| \geqslant d > 0$, the right-hand side of the differential equation (6.5.6) is locally Lipschitz continuous and uniformly bounded. Furthermore, a simple argument using (3.1.27) shows that $x(t, x_0)$ exists for all $t \in (-\infty, \infty)$. Now

$$F(x(t, x_0)) - F(x_0) = \int_0^t \frac{d}{ds} F(x(s, x_0)) = -\alpha(F(t, x_0)).$$

Consequently, $x(t, x_0)$ leaves $F^{a-\epsilon_0}$ fixed (pointwise) and deforms F^b onto F^a. Thus setting $\zeta_t(x_0) = x(t, x_0)$ for $x_0 \in F^{-1}[a, b]$, we see that ζ_t is a deformation of F^b onto F^a. In fact, ζ_t is an isotopy of H onto itself since for any $x_0 \in H$ with $\zeta_{-t}(x_0) = x(-t, x_0)$,

$$\zeta_{-t}\zeta_t(x_0) = \zeta_{-t}(x(t, x_0)) = x(-t, x(t, x_0)) = x_0.$$

Actually, we can prove (by a slight modification of the above argument) that F^b is a deformation retract of F^a. Indeed, denote by $x(t, x_0)$, the solution of

$$\frac{dx}{dt} = -(F(x_0) - a) \frac{F'(x)}{\|F'(x)\|^2}, \quad x(0) = x_0 \in F^{-1}[a, b]. \tag{6.5.7}$$

Then the mapping $\zeta_t(x_0) = x(t, x_0)$ for $x_0 \in F^{-1}[a, b]$ and $\zeta_t(x_0) = x_0$ for $x_0 \in F(x) \leqslant a$. As before for $x_0 \in F^{-1}[a, b]$, $F(\zeta_t(x_0)) = F(x_0) - t\{F(x_0) - a\}$. Thus $F(\zeta_1(x_0)) = a$ and F^b is a deformation retract of F^a.

(ii): Now suppose that $F(x) = c$ is an isolated critical level of F on which F has a finite number of critical points $\{z_i\}$ $(i = 1, \ldots, N)$. We will show that F^c is a deformation retract of $F^{c+\epsilon}$ for some small $\epsilon > 0$, by using the deformation $\zeta_t(x_0)$ defined by (6.5.7). In particular, we show that $\lim_{t\uparrow 1} x(t, x_0)$ exists for $x_0 \in F^{-1}(c, c + \epsilon]$. First, suppose (for $i = 1, \ldots, N$), $\inf_{t\in[0,1)} \|x(t, x_0) - \{z_i\}\| > 0$, so that by Condition (C), $\|F'(x(t, x_0))\|$ is uniformly bounded above zero. Thus for any two values $0 < t_1, t_2 < 1$,

$$\|x(t_2, x_0) - x(t_1, x_0)\| \leqslant \int_{t_1}^{t_2} \left\| \frac{dx}{ds} \right\| ds$$
$$\leqslant K|F(x_0) - c - \epsilon|\,|t_2 - t_1|,$$

where K is a constant independent of t. For any sequence $\bar{t}_n \uparrow 1$, $\{x(\bar{t}_n, x_0)\}$ is a Cauchy sequence and $\lim_{t\uparrow 1} x(t, x_0)$ exists. Next, we suppose $\inf_{t\in[0,1)} \|x(t, x_0) - \{z_i\}\| = 0$, so that for some sequence $t_n \uparrow 1$ and some integer i, $\|x(t_n, x_0) - z_i\| \to 0$. Actually we prove $\lim_{t\uparrow 1} x(t, x_0) = z_i$, by supposing the contrary, and obtaining a contradiction. Indeed, if this limit does not exist, there are two spherical neighborhoods $S_1 \subset S_2$ of z_i such that for an infinite sequence of disjoint intervals $[t_j, t_{j+1}] \subset [0, 1)$ with $x(t, x_0) \subset S_2 - S_1$ for $t \in [t_j, t_{j+1}]$, there are absolute positive numbers $c, d > 0$ such that

$$\|x(t_{j+1}) - x(t_j)\| \geqslant c \quad \text{and} \quad \inf_{[t_j, t_{j+1}]} \|F'(x(t))\| \geqslant d. \tag{6.5.8}$$

Then (6.5.8) implies

$$c \leqslant \|x(t_{j+1}) - x(t_j)\| \leqslant \int_{t_j}^{t_{j+1}} \left\| \frac{dx}{ds} \right\| ds \leqslant |F(x_0) - c| \frac{|t_{j+1} - t_j|}{d}.$$

This is the desired contradiction since as $j \to \infty$, $|t_{j+1} - t_j| \to 0$. Finally, by excision

$$H_q(F^{c+\epsilon}, F^{c-\epsilon}) \approx H_q(\mathring{F}^c \cup \sigma \cap W, \mathring{F}^c \cap W),$$

where W is any neighborhood of σ that contains no other critical points of $F(x)$.

(iii): First, suppose $F(x)$ is a quadratic functional and $x = 0$ is the critical point of $F(x)$ of index i with $F(0) = 0$. Then there is a self-adjoint, invertible operator L of H into itself such that $F(x) = \frac{1}{2}(Lx, x)$. Clearly Condition (C) implies that L is compact, so that L has separable range. Thus, we may suppose without loss of generality that H is separable. Since $F(x)$ is negative definite on a closed subspace H_i of dimension i of H,

$H = H_i \oplus H_i^{\perp}$, and L is invariant on H_i and $H_i^{\perp}$. Therefore, if $\mathring{F}^0 = \{x \mid F(x) < 0\}$,

(6.5.9) $$H_q(H^{\epsilon}, H^{-\epsilon}) \approx H_q(F^0 \cup \{0\}, F^0).$$

Let Σ be the open unit ball in H and set $\Sigma_i = \Sigma \cap H_i$. We shall show that $(\Sigma_i \cup \{0\}, \Sigma_i)$ is a deformation retract of $(\mathring{F}^0 \cup \{0\}, \mathring{F}^0)$. Once this is established, (6.5.5(ii)) implies that

$$H_q(F^{\epsilon}, F^{-\epsilon}) \approx H_q(B_i \cup \{0\}, B_i) \approx H_q(B_i, B_i - \{0\}) \approx H_q(B_i, \partial B_i).$$

To show that $(\mathring{F}^0 \cup \{0\}, \mathring{F}^0)$ can be deformed into $(\Sigma_i \cup \{0\}, \Sigma_i)$, suppose that $F(x) < 0$ and $x = x_i + y$, where $x_i \in H_i$ and $y \in H_i^{\perp}$, and set $x(t) = x_i + (1-t)y$. Then

$$F(x(t)) = \tfrac{1}{2}(Lx(t), x(t)) = \tfrac{1}{2}(Lx_i, x_i) + \tfrac{1}{2}(1-t^2)(Ly, y) \leqslant \tfrac{1}{2}(Lx, x) < 0.$$

Consequently, $(\mathring{F}^0 \cup \{0\}, \mathring{F}^0)$ can be deformed into $(\Sigma_i \cup \{0\}, \Sigma_i)$, with $(\Sigma_i \cup \{0\}, \Sigma_i)$ left fixed.

Now the general case follows from (6.5.4) and (6.5.9). For, suppose x_0 is the only critical point of $F(x)$ on $F(x) = c$ and x_0 is nondegenerate. Then replacing $F(x)$ by $\tilde{F}(x) = F(x + x_0) - c$, $\tilde{F}(0) = 0$ and $x = 0$ is a nondegenerate critical point of index i of $\tilde{F}(x)$. Thus, since homology is unchanged under the homeomorphism h of (6.5.4),

$$\begin{aligned} H_q(\tilde{F}^{\epsilon}, \tilde{F}^{-\epsilon}) &= H_q(\tilde{F}^0 \cup 0 \cap W, \tilde{F}^0 \cap W) \\ &\approx H_q(\mathring{Q}^0 \cup 0, \mathring{Q}^0), \quad \text{where } Q \text{ is the quadratic part of } F. \\ &\approx H_q(B^i, \partial B^i) \\ &\approx G. \end{aligned}$$

6.5D Morse inequalities

(6.5.10) **Corollary** Suppose $F(x)$ is a C^2 real-valued functional defined on a Hilbert space H such that (i) $F(x)$ is bounded below, (ii) $F(x)$ satisfies the compactness condition (C) (6.5.2) on H, (iii) all the critical points of $F(x)$ are nondegenerate, (iv) $F(x)$ possesses only a finite number of critical points of fixed Morse index, then the following relations hold

$$\begin{aligned} M_0 &\geqslant 1, \\ M_1 - M_0 &\geqslant -1, \\ M_2 - M_1 + M_0 &\geqslant 1, \\ &\vdots \\ \sum_{i=0}^{\infty} (-1)^i M_i &= 1, \end{aligned}$$

where M_i denotes the number of critical points of Morse index i of $F(x)$.

Proof: This proof makes crucial use of the result (6.5.5) and the basic properties of singular homology theory. We first observe that the nondegeneracy hypothesis (iii) of $F(x)$ and the compactness condition (6.5.2) imply that for any real number b, the number of critical points of $F(x)$ on F^b is finite. This fact follows from the isolated character of nondegenerate critical points since an infinite number of nondegenerate critical points of F^b would be inconsistent whenever condition (6.5.2) is assumed. Then by our discussion of singular homology given above, as in Milnor (1963), for any subadditive integer invariant $S_\lambda(X, Y)$

$$S_\lambda(F^{a_n}, F^{a_0}) \leqslant \sum_{i=1}^{n} S_\lambda(F^{a_i}, F^{a_{i-1}})$$

with equality holding if this invariant is additive. Thus if we let $S_\lambda(X, Y)$ $(\lambda = 0, 1, 2, \ldots)$ denote the alternating sum of the Betti numbers of (X, Y), $\sum_{k=0}^{\lambda}(-1)^k R_{\lambda-k}(X, Y)$ (up to λ) (a subadditive invariant) and $S(X, Y)$ the full sum $\sum_{k=0}(-1)^k R_k(X, Y)$ (an additive invariant) we find, in each case by (6.5.5),

$$\sum_{k=0}^{k} (-1)^k R_{\lambda-k}(F^{a_n}, F^{a_0}) \leqslant \sum_{k=0}^{k} (-1)^k M_{\lambda-k},$$

$$\sum_{k=0}^{\lambda} (-1)^\lambda R_\lambda(F^{a_n}, F^{a_0}) \leqslant \sum_{i=0}^{\lambda} (-1)^\lambda M_\lambda.$$

Next we observe that for $b > \inf_H F(x)$ $(b \neq c_i)$, the set F^b can be decomposed as follows $b = a_n$,

$$a_0 < \inf_H F(x), \; F^{a_0} \subset F^{a_1} \subset F^{a_2} \cdots F^{a_n}$$

in such a way that each $\{F^{a_i} - F^{a_{i-1}}\}$ contains exactly one critical value. Next we observe that $F^{a_0} = \varnothing$, and by hypothesis (iv) we may assume, by taking a_n sufficiently large, that the set $H - F^{a_n}$ contains no critical points of index $\leqslant \lambda$, so that by choosing $b = a_n$ sufficiently large we may let $F^{a_n} = H$, thus the inequalities (6.5.10) follow since $R_\lambda(H) = 0$, $\lambda > 0$, whereas $R_0(H) = 1$.

The result (6.5.5) has a useful extension if $F(x)$ is defined only on a bounded domain of H, say $\Sigma_R = \{x \mid \|x\| < R\}$. In fact, we prove

(6.5.11) **Corollary** Suppose hypotheses (i)–(iii) of (6.5.5) are valid for the set $\Sigma_R = \{x \mid \|x\| < R\}$. Then if $(F'(x), x) > 0$ for each x on $\|x\| = R$, and M_i denotes the number of critical points of $F(x)$ of index i in Σ_R, the inequalities (6.5.10) hold.

Proof: We note that the condition $(F'(x), x) > 0$ implies that the solutions $x(t, x_0)$ of the equations of steepest descent used in the proof of

(6.5.10) remain in the set σ_R for all t. Since the homology groups of Σ_R coincide with those of H, the proof given in (6.5.10) carries over to this case. In this case hypothesis (iv) of (6.5.10) can be removed since by hypothesis there are no critical points of $F(x)$ on $\|x\|^2 = R$, and in Σ_R the number of critical points of $F(x)$ is finite.

Finally, we consider an extension of the inequalities (6.5.10) for functionals $F(x)$ that may possess degenerate critical points. To this end it is convenient to consider functionals defined on Hilbert manifolds $\mathfrak{M}$ since such sets often possess nontrivial topological properties. In such cases any smooth functional defined on $\mathfrak{M}$ that satisfies (6.5.2) will be shown to have an interesting critical point theory.

Before stating a result in this direction, we define an integer-valued measure of the critical points on an isolated critical value $F(x) = c$, by setting $M_i(c)$ equal to the Betti number of the relative homology groups with Z_2 coefficients, $H_i\,(F^{c+\epsilon}, F^{c-\epsilon})$, for $\epsilon > 0$ sufficiently small, where as usual, $F^d = \{x \mid x \in \mathfrak{M},\ F(x) \leqslant d\}$. Clearly as in (6.5.10), $F_i(c)$ is independent of ϵ if $F(x)$ satisfies condition (6.5.2) on the set $\{x \mid c - \epsilon \leqslant F(x) \leqslant c + \epsilon\}$.

We now state the following result that can be proved on the basis of (6.5.5(ii)). Note that it does not require the nondegeneracy of the critical points of $F(x)$.

(6.5.12) Suppose $F(x)$ is a C^2 real-valued functional on a complete, smooth, Hilbert manifold $\mathfrak{M}$ such that $F(x)$ is bounded from below, satisfies condition (6.5.2), and possesses isolated critical values $c_1, c_2, \ldots$. Then the following inequalities hold

$$(6.5.13) \qquad \sum_{\{c_j\}} M_i(c_j) \geqslant R_i(\mathfrak{M}),$$

where R_i is the ith Betti number of $\mathfrak{M}$.

This result can be proved by a simple modification of our previous results. See Rothe (1973).

6.5E Illustrations

To illustrate the results just given, consider the simple semilinear Dirichlet problem (6.1.30) discussed earlier, defined over a bounded domain $\Omega \subset \mathbb{R}^N$,

$$(6.5.14) \qquad \epsilon^2\,\Delta u + u - g(x)u^3 = 0, \qquad u|_{\partial\Omega} = 0.$$

Clearly the trivial solution $u_0(x) = 0$ exists for all ϵ, and its index is easily computed to be the sum of the multiplicities of all eigenvalues of the Laplacian Δ relative to Ω (subject to the null boundary condition) less than $1/\epsilon^2$. Indeed, the solutions of (6.5.14) are precisely the critical points of the functional $\mathcal{G}_\epsilon(u) = \int_\Omega \{\epsilon^2|\nabla u|^2 - u^2 + \frac{1}{2} g(x)u^4\}\, dx$ in the class $\mathring{W}_{1,2}(\Omega)$, and $(\mathcal{G}_\epsilon''(0)v, v) = \int_\Omega \{\epsilon^2|\nabla v|^2 - v^2\}\, dx$. We also note that $\mathcal{G}_\epsilon(u)$ is bounded from below on $\mathring{W}_{1,2}(\Omega)$ and (since $\mathcal{G}_\epsilon'(x)$ is a proper mapping for fixed ϵ) satisfies (6.5.2). Let the eigenvalues of (Δ, Ω) be denoted $0 < \lambda_1 \leqslant \lambda_2 \leqslant \cdots \leqslant \lambda_N \leqslant \cdots$, so that $u_0(x) \equiv 0$ is a nondegenerate critical point of $\mathcal{G}_\epsilon(u)$ if and only if $\epsilon^{-2} \neq \lambda_i$ $(i = 1, \ldots)$. Then, as mentioned in (6.1.31) for $1/\epsilon^2 \leqslant \lambda_1$, $u_0(x) \equiv 0$ is an absolute minimum of $\mathcal{G}_\epsilon(u)$ and the only solution of (6.5.14). However, for $1/\epsilon^2 > \lambda_1$, u_0 is no longer a relative minimum. As proven in (6.1.31) for $\lambda_1 < 1/\epsilon^2$, the infimum of $\mathcal{G}_\epsilon$ is attained at the unique positive function $u_1(x, \epsilon)$. By (6.1.31), for ϵ sufficiently small, say $\epsilon < \epsilon_0$, $\pm u_1(x, \epsilon)$ are nondegenerate minima for (6.5.14). *Consequently, for* $\epsilon < \min(\epsilon_0, 1/\lambda_2^2)$, *the inequalities* (6.5.10) *imply that the boundary value problem* (6.5.14) *has another pair of critical points* $\pm u^*(x, \epsilon)$ *which must necessarily change sign in* Ω. To prove this last statement, assume (6.5.14) has only the solutions $\pm u_1(x, \epsilon)$ and $u_0(x)$, for c sufficiently small. By virtue of (4.2.7) and (6.1.1) we may suppose $\pm u_1(x, \epsilon)$ are nondegenerate minima of the associated functional $\mathcal{G}_\epsilon(u)$, and moreover for $\epsilon^2 \neq \lambda_1^{-1}$ and less than λ_2^{-1} we may assume that the (Morse) index of $u_0(x)$ is at least 2. Consequently, by the inequalities (6.5.10) $M_0 \geqslant 2$, and thus $M_1 \geqslant M_0 - 1 \geqslant 1$. Consequently, either (6.5.14) possesses a nondegenerate critical point $u^*(x, \epsilon)$ of index 1 and thus distinct from $u_0(x)$ and $\pm u_1(x, \epsilon)$ or (6.5.14) has a degenerate critical point that must also be distinct from $u_0(x)$ or $\pm u_1(x, \epsilon)$. The oddness of u^3 then implies that $\pm u^*(x, \epsilon)$ satisfies (6.5.14). Actually in Section 6.7 we shall greatly improve this result by showing for $\lambda_i < \epsilon^{-2} \leqslant \lambda_{i+1}$, (6.5.14) has i *pairs* of distinct nonzero solutions $\pm u_1(x, \epsilon), \pm u_2(x, \epsilon), \ldots, \pm u_i(x, \epsilon)$. Of course to establish such a result it is important to find a *general critical point theory* that does not distinguish between degenerate and nondegenerate critical points.

As another useful application of the relations of Morse (6.5.10) we apply them to sharpen our results (5.4.29) and (6.3.25) by giving estimates for the number of solutions of operator equations of the form

$$\text{(6.5.15)} \qquad Lu - Nu = f, \qquad f \in H = \text{a Hilbert space},$$

where L is a self-adjoint Fredholm operator and $Nu \in C^1(H, H)$ is a uniformly bounded completely continuous gradient mapping with strictly convex antiderivative $\mathfrak{N}(u)$. Moreover, in accord with (5.4.29) we suppose $(N(ra + x_1), a) < (\phi(a), a)$, where

$$\phi(a) = \lim_{r\to\infty} PN(ra), \qquad a \in \{\operatorname{Ker} L \cap \|x\| = 1\}, \qquad \text{and} \qquad x_1 \perp \operatorname{Ker} L.$$

Then we prove

(6.5.16) **Theorem** Under the above assumptions, if $f \in \text{Range}(L - N)$, the functional $\mathcal{G}_f(u) = \frac{1}{2}(Lu, u) - \mathfrak{N}(u) - (f, u)$ satisfies the compactness condition (C) and apart from a possible exceptional set of first Baire category the number of solutions of (6.5.15) is finite. Moreover, in this case, if the dimension of the linear subspace T on which the quadratic form $(Lu, u) \leqslant 0$ is $j < \infty$, then the following Morse inequalities hold:

$$M_j \geqslant 1, \quad M_{j+1} - M_j \geqslant -1, \quad M_{j+2} - M_{j+1} + M_j \geqslant 1, \ldots$$

$$\sum_{i=0}^{\dim T} (-1)^i M_{j+i} = \pm 1.$$

Proof: To prove $\mathcal{G}_f(u)$ satisfies Condition (C) we shall prove (*) if $f \in \text{Range}\ (L + N)$ and $\|Lu_n + Nu_n - f\| \to 0$, then $\|u_n\|$ is uniformly bounded. Once this fact is established the complete continuity of N implies that, after possibly passing to a subsequence, Lu_n converges strongly and u_n converges weakly in $\bar{u}$, so that $u_n \to \bar{u}$ strongly, and Condition (C) follows. Now we prove (*); we decompose $H = \text{Ker}\ L \oplus H_1$ so that $u_n = z_n + y_n$, $z_n \in \text{Ker}\ L$, $y_n \in H_1$. Then $\|Ly_n\|$ is uniformly bounded since

$$\|Ly_n\| \leqslant \|Lu_n + Nu_n - f\| + \|Nu_n - f\| \leqslant \text{const.}$$

by the uniform boundedness of N. Since L is a Fredholm operator, $\|y_n\|$ is uniformly bounded. Next we show that $\|z_n\|$ (and consequently $\|u_n\|$) is uniformly bounded. To this end we use (5.4.29) to prove that since $f \in \text{Range}\ (L + N)$, $(f, a) < (\phi(a), a)$. Thus if we assume $\|z_n\| = \|r_n a_n\| \to \infty$, while $\|y_n\|$ remains bounded, $(N(z_n + y_n), a) \to (\phi(a), a) > (f, a)$. On the other hand, by assumption $\|Lu_n + Nu_n - f\| \to 0$, which implies $(N(z_n + y_n), a) \to (f, a)$, so $\|z_n\|$ stays uniformly bounded as required.

Next, we again apply (6.3.25) and Section 3.1D to assert that Range $(L + N)$ is open in H and $L + N$ is a C^1 nonlinear Fredholm operator of index 0, so apart from a possible exceptional set f of first Baire category the critical points of $L + N - f$ are nondegenerate and isolated. The complete continuity of N, the Fredholm property of L, and (*) then yield the finiteness of these solutions.

Finally, in order to verify the Morse inequalities (6.5.10) we show that $(\mathcal{G}_f''(u)x, x)$ is positive definite off the set T, where $(Lx, x) \leqslant \beta\|x\|^2$. Indeed for $x \in [T]^\perp$,

$$(\mathcal{G}_f''(u)x, x) = (Lx, x) - (N'(u)x, x)$$
$$\geqslant \beta\|x\|^2 - (\beta - \epsilon)\|x\|^2 = \epsilon\|x\|^2.$$

Consequently the Morse indices of a critical point $\bar{u}$ of $\mathcal{G}_f(u)$ must be less than dim T. On the other hand, $\mathcal{G}_f(\bar{u} + u) \to -\infty$ as $\|u\| \to \infty$ on the linear subspace $S = \{u \mid (Lu, u) \leqslant 0\}$. This is plain for $(Lu, u) < 0$ by the Fredholm property of L. Finally for $u = Ra \in \operatorname{Ker} L$ with $\|a\| = 1$, (6.4.29) implies for $f \in \operatorname{Range}(L - N)$ that as $R \to \infty$, $(f, a) < (N(\bar{u} + Ra), a)$. Consequently

$$\lim_{R \to \infty} \mathcal{G}_f(\bar{u} + Ra) = \lim_{R \to \infty} R\{(f, a) - (N(\bar{u} + Ra), a)\} = -\infty.$$

Moreover, on the set T, the strict convexity of $N(u)$ implies that $(\mathcal{G}_f''(\bar{u})w, w) < 0$. Consequently, if we restrict $\mathcal{G}_f(u)$ to $[T]^{\perp}$, we find $\mathcal{G}_f(u) \to \infty$ as $\|u\| \to \infty$, so that the result (6.5.10) implies that the Morse inequalities are valid. From (5.1.9) (the reduction lemma), we conclude that the only contribution due to studying the critical points on T is to increase the Morse indices of a critical point by the integer j, and consequently the Morse inequalities as described in the theorem hold.

6.6 The Critical Point Theory of Ljusternik and Schnirelmann

6.6A Heuristics

It is often important to study the critical points of a given smooth functional *independent of nondegeneracy considerations*. Such a theory of critical points was formulated in the years 1925–1947 by the Russian mathematicians L. Ljusternik and L. Schnirelmann. This theory is based on determining a topological analogue for the minimax principles which characterize the eigenvalues of a self-adjoint compact operator L. Indeed, as mentioned in (1.3.42), if the positive eigenvalues of L are denoted $\lambda_1^+, \lambda_2^+, \ldots,$ arranged in decreasing order, and counted according to multiplicity, then

(6.6.1) $$\lambda_n^+ = \sup_{[S_{n-1}]} \min_{x \in S_{n-1}} (Lx, x),$$

where S_{n-1} denotes the unit sphere in an arbitrary n-dimensional linear subspace Σ of H, and $[S_{n-1}]$ denotes the class of such spheres as Σ varies in H. Since the eigenvalues of L are precisely the critical values of the functional (Lx, x) on the unit sphere $\partial\Sigma_1 = \{x \mid \|x\| = 1\}$ of H, it is natural to extend (6.6.1) to general smooth functionals $F(x)$ by finding "topological" analogues for the sets S_{n-1} and $[S_{n-1}]$.

A fundamental result extending (6.6.1) to nonquadratic functionals can be found by supposing that $\mathfrak{M}$ is a Hilbert manifold and $n(A)$ is an integer-valued function defined on a class of closed subsets of $\mathfrak{M}$ with the properties:

(i) $n(A) = 1$ if A is a point of $\mathfrak{M}$; $n(\varnothing) = 0$;

(ii) $n(A) \geqslant n(B)$ if $A \supseteq B$;
(iii) $n(A \cup B) \leqslant n(A) + n(B)$;
(iv) $n(A) = n(A_t)$, where A_t is an isotopy of A;
(v) there is a neighborhood U of A such that $n(U) = n(A)$.

These properties are consistent since the trivial function $n(A) = 1$ if $A \neq \varnothing$ and $n(A) = 0$ if $A = \varnothing$ satisfies (i) – (v). Now we prove the following result (assuming, for the moment, the existence of an integer-valued function $n(A)$), which we shall use as a substitute for (6.6.1).

6.6B The minimax principle

In order to utilize the properties of $n(A)$ to study the critical points of a functional $F(x)$ defined on $\mathfrak{M}$ and set $\mathfrak{M}^\alpha = \{x \mid F(x) \geqslant \alpha, x \in \mathfrak{M}\}$ for real α, we prove

(6.6.2) **Theorem** Suppose that c is an isolated critical value for the C^1 real-valued functional $F(x)$ defined on a Hilbert manifold $\mathfrak{M}$ whose Fréchet derivative is Lipschitz continuous, and that $F(x)$ satisfies the compactness condition (6.5.2). Then for $\epsilon > 0$ sufficiently small, there is a neighborhood U of the set $K_c = \{x \mid x \in \mathfrak{M}, F(x) = c, \nabla F(x) = 0\}$ and a deformation $\{\zeta_t\}$ of $\mathfrak{M}^{c+\epsilon} - U_\epsilon$ such that $\zeta_1(\mathfrak{M}^{c+\epsilon} - U_\epsilon) \subseteq \mathfrak{M}^{c-\epsilon}$.

Proof: To define the deformation ζ_t, we consider the solution $x(t, x_0)$ of the system

$$(6.6.3) \qquad \frac{dx}{dt} = -\alpha(\|\nabla F(x)\|)\nabla F(x), \qquad x(0) = x_0,$$

where $\alpha(z)$ is any C^∞ function with $\alpha(z) = 1$ for $0 \leqslant z \leqslant 1$, $\alpha(z) = 2/z^2$ for $z \geqslant 2$, and such that $z^2\alpha(z)$ is monotone increasing for all $z \geqslant 0$. By (3.1.27), the solution of (6.6.3) exists for all t since $\alpha(\|\nabla F(x)\|)\|\nabla F(x)\|$ is uniformly bounded for $x \in \mathfrak{M}$.

Next, let U be a small neighborhood of the set K_c. Clearly K_c is compact by (6.5.2). We show that for some $\delta > 0$, U contains the set

$$V_\delta = \left\{x_0 \mid x_0 \in \mathfrak{M}, \quad |F(x_o) - c| < \delta, \quad \inf_{t \in [0,1]} \|\nabla F(t, x_0)\| < \delta\right\}.$$

Otherwise, there exists a sequence of points $y_n \in \mathfrak{M}$, $y_n \not\subseteq U$, and numbers $t_n \in [0, 1]$ such that $F(y_n) \to c$, $\nabla.F(x(t_n, y_n)) \to 0$ as $t_n \to t_*$. Now by Condition (C), since $|F(x(t_n, y_n))|$ is uniformly bounded, $\{x(t_n, y_n)\}$ has a convergent subsequence with limit $\bar{y}$. (Clearly $\bar{y}$ is a critical point of $F(x)$ on $\mathfrak{M}$.) After relabeling indices, we note that

$$y_n = x(-t_n, x(t_n, y_n)) \to x(-t^*, \bar{y}) = \bar{y},$$

i.e., y_n converges to $\bar{y}$, which contradicts the fact that $y_n \not\subseteq U$.

Finally, suppose that for δ sufficiently small, $F(x_0) \leqslant c + \frac{1}{2}\delta^2$ and $x_0 \not\in N_\delta$ a small spherical neighborhood of radius δ about K_c. We show that $F(x(1, x_0)) \leqslant c - \frac{1}{2}\delta^2$. Once this fact is established, setting $\epsilon = \delta^2/2$ and $U_\epsilon = N_\delta \cap \mathfrak{M}^{c+\epsilon}$, we find the desired deformation $\zeta_1(\mathfrak{M}^{c+\epsilon} - U_\epsilon) \subseteq \mathfrak{M}^{c-\epsilon}$ by setting $\zeta_t(x_0) = x(t, x_0)$ for $x_0 \in \mathfrak{M}$. Now for $x_0 \in F^{-1}[c - \delta^2/2, c + \delta^2/2]$ and $x_0 \not\in N_\delta$, we see that since $\alpha(z)z^2$ is monotone decreasing and $\|\nabla f(x_0)\| \geqslant \epsilon$,

$$F(x(1, x_0)) - F(x_0) = -\int_0^1 \alpha(\|\nabla F(x(t, x_0))\|)\|\nabla F(x_0, t)\|^2\, dt \leqslant -\epsilon^2.$$

Thus $F(\zeta_1(x_0)) \leqslant c + \epsilon^2/2 - \epsilon^2 = c - \epsilon^2/2$.

(6.6.4) **Minimax Theorem** Suppose $F(x)$ is a C^2 functional defined on a smooth Hilbert manifold $\mathfrak{M}$. If $F(x)$ satisfies condition (6.5.2) on $\mathfrak{M}$ and $[A]_i = \{A \mid A \subset \mathfrak{M}, n(A) \geqslant i\}$ is nonvacuous, where $n(A)$ satisfies conditions (i)–(v) above, then assuming the critical values of $F(x)$ are isolated:

(i) If finite

$$c_i = \inf_{[A]i} \sup_{x \in A} F(x) \tag{6.6.5}$$

is achieved and is a critical value of $F(x)$ relative to $\mathfrak{M}$.

(ii) If $c_i = c_{i+1} = \cdots = c_{i+j} = c$ is finite, then $n(K_c) \geqslant j + 1$, where $K_c = \{x \mid x \in H, F(x) = c$, and x is a critical point of $F(x)\}$.

(iii) If some $c_i = \infty$, then $\sup_K F(x) = \infty$, where K is the set of critical points of $F(x)$ on $\mathfrak{M}$.

(iv) (Ljusternik–Schnirelmann multiplicity theorem) Similarly, if finite, $c_i = \sup_{[A]_i} \inf_A F(x)$ is attained and is also a critical value of $F(x)$ relative to $\mathfrak{M}$. If $\tilde{c}_i = \cdots = \tilde{c}_{i+j} = \tilde{c}$ is finite, $n(K_{\tilde{c}}) \geqslant j + 1$. Furthermore, if some $\tilde{c}_i = -\infty$, then $\inf_K F(x) = -\infty$.

Proof: (i): If c_i is finite and not a critical value, then condition (6.5.2) implies that for some $\epsilon > 0$, $F^{-1}[c + \epsilon, c - \epsilon]$ contains no critical points. Consequently by (6.6.2), we may deform $\mathfrak{M}^{c+\epsilon} = \{x \mid F(x) \leqslant c + \epsilon\}$ into $\mathfrak{M}^{c-\epsilon} = \{x \mid F(x) \leqslant c - \epsilon\}$. Hence every $A \subset [A]_i$ with $A \subseteq \mathfrak{M}^{c+\epsilon}$ is deformed into a new set $A' \subseteq \mathfrak{M}^{c-\epsilon}$ and by Section 6.6A(iv), $n(A') \geqslant i$ so that $A' \in [A]_i$. Thus $\max_{A'} F(x) \leqslant c_i - \epsilon$, while $c_i = \inf_{[A]_i} \max_A F(x) \leqslant \max_{A'} F(x) = c_i - \epsilon$. This contradiction implies that c_i is a critical value of $F(x)$.

(ii): Next we suppose that $c_i = c_{i+1} = \cdots = c_{i+j} = c$, and prove

$$n(K_c) = n(\mathfrak{M}^{c+\epsilon}) - n(\mathfrak{M}^{c-\epsilon}) \qquad \text{for some } \epsilon > 0. \tag{6.6.6}$$

First by (v), there is a neighborhood $U \supset K_c$ with $n(K_c) = n(U)$. Secondly by (6.6.2), there is a deformation ζ_t of $\mathfrak{M}$ with $\zeta_t(\mathfrak{M}^{c+\epsilon} - U) \subseteq \mathfrak{M}^{c-\epsilon}$. Therefore by (6.6.2) and the properties of $n(A)$, $n(\mathfrak{M}^{c-\epsilon}) \geqslant n(\mathfrak{M}^{c+\epsilon} - U)$. Now by (6.6.2)

$$n(\mathfrak{M}^{c+\epsilon}) = n(\mathfrak{M}^{c+\epsilon} \cup U) \leqslant n(\mathfrak{M}^{c+\epsilon} - U) + n(U) \leqslant n(\mathfrak{M}^{c-\epsilon}) + n(K_c).$$

This establishes (6.6.6). Finally to prove (6.6.5), we observe that $n(\mathfrak{M}^{c+\epsilon}) \geqslant i + j$ since $\mathfrak{M}^{c+\epsilon}$ contains subsets A with $n(A) \geqslant i + j$. On the other hand, $c_i = c$ implies that $n(\mathfrak{M}^{c-\epsilon}) \leqslant i - 1$. Consequently (6.6.6) implies that $n(K_c) \geqslant j + 1$.

(iii): Suppose that $\sup_K F(x) = \alpha < \infty$, where K is the set of critical points of $F(x)$ on $\mathfrak{M}$. Then for some $\epsilon \geqslant 0$, since $\mathfrak{M}^{\alpha+\epsilon}$ is isotopic to $\mathfrak{M}$, $n(\mathfrak{M}^{\alpha+\epsilon}) = n(\mathfrak{M})$. Thus if $i \leqslant n(\mathfrak{M})$, $c_i = \inf_{[A]_i} \sup_{x \in A_i} F(x) \leqslant \alpha + \epsilon < \infty$; which contradicts the fact that $c_i = \infty$.

(iv): Repeat the proofs of (i)–(iii) but deform $\mathfrak{M}^{\bar{c}-\epsilon}$ into $\mathfrak{M}^{\bar{c}+\epsilon}$ for some $\epsilon > 0$.

As a simple application of formula (6.6.5) we prove:

(6.6.7) **Theorem** Suppose $\mathfrak{M}$ is a C^2 complete Hilbert manifold without boundary, $F(x)$ is a C^2 real-valued functional defined on $\mathfrak{M}$ which is bounded from below and satisfies condition (6.5.2). Then $F(x)$ possesses at least $n(\mathfrak{M})$ critical points, where $n(\mathfrak{M})$ is any integer-valued function defined on $\mathfrak{M}$ and satisfying properties (i)–(v).

Proof: Without loss of generality, we may suppose that the number of critical points of $F(x)$ on $\mathfrak{M}$ is finite. Then $F(x)$ has a finite number of critical values $c_0 \leqslant c_1 \leqslant \cdots \leqslant c_N$ on $\mathfrak{M}$ with $c_0 = \min_{\mathfrak{M}} F(x)$, so that we may write $\mathfrak{M} \supseteq \mathfrak{M}^{c_N} \supset \mathfrak{M}^{c_{N-1}} \supset \cdots \supset \mathfrak{M}^{c_0} \supset \varnothing$, where $\mathfrak{M}^{c_i} = \{x \mid x \in \mathfrak{M}, F(x) \leqslant c_i\}$. For some $\epsilon > 0$ independent of $i = 1, \ldots, N$, by the basic properties of $n(A)$, $n(K_{c_i}) \geqslant n(\mathfrak{M}^{c_i+\epsilon}) - n(\mathfrak{M}^{c_i-\epsilon})$. Summing this last inequality over i, and using the fact that, by (6.5.5(i)), the sets $\mathfrak{M}^{c_i-\epsilon}$ and $\mathfrak{M}^{c_i+\epsilon}$ are isotopic for each $i = 1, \ldots, N$ so that $n(\mathfrak{M}^{c_i-\epsilon}) = n(\mathfrak{M}^{c_{i-1}+\epsilon})$, we find (using the facts $\mathfrak{M}^{c_0-\epsilon} = \varnothing$ and $\mathfrak{M}^{c_N+\epsilon} = \mathfrak{M}$)

$$\sum_{i=0}^{N} n(K_{c_i}) = \sum_{i=0}^{N} n(\mathfrak{M}^{c_i+\epsilon}) - n(\mathfrak{M}^{c_i-\epsilon})$$

$$\geqslant n(\mathfrak{M}^{c_N+\epsilon}) - n(\mathfrak{M}^{c_0-\epsilon}) + \sum_{i=1}^{N} \{n(\mathfrak{M}^{c_i-\epsilon}) - n(\mathfrak{M}^{c_{i-1}+\epsilon})\}$$

$$\sum_{i=0}^{N} n(K_{c_i}) \geqslant n(\mathfrak{M}) - n(\varnothing) = n(\mathfrak{M}).$$

By our assumptions, K_{c_i} consists of a finite number of points $(x_1^i, \ldots, x_n^i)$

for each $i = 0, \ldots, N$, and by the above inequality, the total number of such points must be at least $n(\mathfrak{M})$. Indeed, otherwise, by property (iii) of $n(A)$,

$$\sum_{i=1}^{N} n(K_{c_i}) \leqslant \sum_{i=1}^{N} \sum_{j=1}^{M} n(x_j') < n(\mathfrak{M}),$$

which contradicts (6.6.8). Thus the theorem is proven.

6.6C Ljusternik–Schnirelmann category

We turn now to the problem of determining interesting and computable integer-valued functions $n(A)$ satisfying properties (i)–(v). It is thus interesting that a *maximal* (cf. (6.6.8) below) function $n(A)$ exists. This maximal function, called the category of a closed subset $A \subset \mathfrak{M}$ and denoted $\operatorname{cat}_{\mathfrak{M}}(A)$, is defined as follows:

Definition A closed subset A of a topological space X has category 1 (relative to X) if A is contractible (over X) to a point $\operatorname{cat}_X A = N$ if the least number of contractible closed subsets of X necessary to cover X is N. If no finite number of such sets suffice, we set $\operatorname{cat}_X A = \infty$.

(6.6.8) **Theorem** If $\mathfrak{M}$ is a complete Hilbert manifold, $\operatorname{cat}_{\mathfrak{M}}(A)$ defined above satisfies properties (i)–(v) and so is an admissible function $n(A)$.Furthermore, $\operatorname{cat}_{\mathfrak{M}}(A)$ is *maximal* in the sense that if $n(A)$ is any function defined on the closed subsets of $\mathfrak{M}$ possessing properties (i)–(iv), then $\operatorname{cat}_{\mathfrak{M}}(A) \geqslant n(A)$.

Proof: Properties (i)–(iii) are immediate. To prove (iv), suppose that $\zeta_t(A)$ is a deformation of A, and $\zeta_1(A) \subseteq \cup_{i=1}^{N} B_i$, where B_i is closed and contractible on $\mathfrak{M}$. If $A_i = \zeta_1^{-1}(B_i)$, then A_i is closed in A and hence in $\mathfrak{M}$. In addition, A is covered by $\cup_{i=1}^{N} A_i$, while A_i is contractible in X $(i = 1, \ldots, N)$ since $\zeta_t \mid_{A_i}$ is a deformation of A_i into B_i and B_i is contractible. To prove (v), it suffices to prove that any closed set A contractible in X has a neighborhood U with $\operatorname{cat}_X U = 1$. Let ζ_t be a deformation of A to a point p, and let V be a neighborhood of p with $\bar{V}$ contractible in $\mathfrak{M}$. By the homotopy extension theorem, ζ_t can be extended to a deformation $\tilde{\zeta}_t$ of X. Then $A = \zeta_1^{-1}(p) = \zeta_1^{-1}(V) = \tilde{\zeta}^{-1}(V)$. Let U be a neighborhood of A with $\bar{U} \subseteq \tilde{\zeta}^{-1}(V)$. Then $\operatorname{cat}_X(\bar{U}) \leqslant \operatorname{cat}(\tilde{\zeta}_1^{-1}(\bar{V})) \leqslant \operatorname{cat}_X \bar{V} = 1$ since $\bar{V}$ is contractible and closed in $\mathfrak{M}$.

Finally, we prove the maximality of $\operatorname{cat}_{\mathfrak{M}} A$. If $\operatorname{cat}_{\mathfrak{M}} A = 1$, A is deformable to a point p in $\mathfrak{M}$. Consequently, for any function $n(A)$ satisfying properties (i)–(v), $n(A) \leqslant n(p) = 1 = \operatorname{cat}_{\mathfrak{M}} A$. If $\operatorname{cat}_{\mathfrak{M}} A = N$

$< \infty$, $A = \bigcup_{i=1}^{N} A_i$, where each A_i is deformable to a point $p_i \in \mathfrak{M}$. Thus $n(A_i) = n(p_i) = 1$, so that

$$n(A) = n\left(\bigcup_{i=1}^{N} A_i\right) \leqslant \sum_{i=1}^{N} n(A_i) \leqslant N = \operatorname{cat}_{\mathfrak{M}} A.$$

Hence in all cases, $\operatorname{cat}_{\mathfrak{M}} A \geqslant n(A)$.

In order to utilize[1] the results of (6.6.4), it is clearly important to compute both the category of a given manifold $\mathfrak{M}$ as well as the number of distinct classes. $[A]_i = \{A \mid A \subset \mathfrak{M},\ \operatorname{cat}_{\mathfrak{M}} A \geqslant i\}$. Thus the following estimates, connecting the category of $\mathfrak{M}$ to other properties of $\mathfrak{M}$, are particularly important.

(6.6.9) Suppose A is a closed subset of a Hilbert manifold $\mathfrak{M}$. Then:

(a) $\operatorname{cat}_{\mathfrak{M}} A \leqslant \dim A + 1$ (where $\dim A$ denotes the dimension of the set A);

(b) $\operatorname{cat} \mathfrak{M} \geqslant$ cup length $\mathfrak{M}$ (see Appendix A for the definition of cup length);

(c) if $\dim \mathfrak{M} =$ cup length $\mathfrak{M}$, $\operatorname{cat} \mathfrak{M} = \dim \mathfrak{M} + 1$;

(d) let P^k denote the k-dimensional real projective space, and $P^\infty(X)$ denote the infinite-dimensional projective space obtained by identifying the antipodal points of the unit sphere $\{x \mid \|x\| = 1\}$ of a uniformly convex Banach space X, then

$$\operatorname{cat}_{P^n} P^k = k + 1 \qquad \text{for} \quad n \geqslant k,$$

and

$$\operatorname{cat}_{P^\infty(X)} P^k(X) = k + 1 \qquad \text{where} \quad P^k(X) \subset P^\infty(X).$$

Since these results are topological in nature, we omit their proof, and refer the reader to Schwartz (1969).

6.6D Application to nonlinear eigenvalue problems

The Ljusternik–Schnirelmann theory of critical points can be carried over with considerable success to manifolds $\mathfrak{M}$ modeled on reflexive Banach spaces. A simple, yet nonetheless important, instance of such an extension is the study of nonlinear eigenvalue problems mentioned in Section (6.3A) in which case $\mathfrak{M}$ is a hypersurface. Let $\mathcal{A}(x)$ and $\mathcal{B}(x)$ be real-valued C^2 functionals defined on the reflexive Banach space X, which we shall assume is uniformly convex. We wish to consider the nontrivial

[1]Another invariant $n(A)$ satisfying (i)–(v) of Section 6.6A is the "genus" function introduced by Krasnoselski (1964). However due to the maximality property (6.6.8) we use cat A here.

normalized solutions (x, λ) of the equation

$$(6.6.10) \qquad \mathcal{A}'(x) = \lambda \mathcal{B}'(x), \qquad \mathcal{A}(x) = \text{const.}$$

Clearly the solutions of (6.6.10) are contained in the set of critical points of the functional $\mathcal{B}(x)$ on the level set $\mathcal{A}_c = \{x \mid x \in X, \mathcal{A}(x) = c$ $(c = \text{const.})\}$. If we suppose that $\mathcal{A}(x)$ is a Fredholm functional, then, by virtue of (3.1.47), apart from a set of real numbers of measure zero, $\mathcal{A}_c$ is a Banach manifold modeled on the space X. We shall prove the following analogue of the spectral theorem (1.3.40) for linear self-adjoint compact operators defined on an infinite-dimensional real Hilbert space H.

(6.6.11) **Theorem** The equation (6.6.10) has a countably infinite number of normalized solutions (u_n, λ_n) with $u_n \in \mathcal{A}_c$, $|\lambda_n| \to \infty$, and $u_n \to 0$ weakly as $n \to \infty$ for every real number $c > 0$, provided the operators $\mathcal{A}'(x)$ and $\mathcal{B}'(x)$ satisfy the following conditions:

(i) $\mathcal{A}'(x)$ is a C^1 odd gradient operator with $\mathcal{A}'(0) = 0$, and for any $x \neq 0$, $(\mathcal{A}'(sx), x)$ is a strictly increasing function of the positive real variable s;

(ii) the functional $\mathcal{A}(x)$ is coercive, i.e., $\mathcal{A}(x) \to \infty$ as $\|x\| \to \infty$;

(iii) whenever $x_n \to x$ weakly in X and $\{\mathcal{A}x_n\}$ converges strongly in X^*, then $\mathcal{A}_{x_n} \to \mathcal{A}x$ and $x_n \to x$ strongly in X; and

(iv) $\mathcal{B}'(x)$ is a completely continuous gradient operator with $\mathcal{B}'(x) = 0$ if and only if $x = 0$.

Proof: Consider the sets $\mathcal{A}_c/\mathbb{Z}_2$ obtained by identifying antipodal points of the level sets $\mathcal{A}_c = \{x \mid x \in X, \mathcal{A}(x) = c\}$. Hypotheses (i)–(iii) show $\mathcal{A}_c$ is nonvacuous; indeed since $\mathcal{A}(x) = \int_0^1 (\mathcal{A}'(sx), x)\, ds$ so that $\mathcal{A}(kx)$ is a continuous function of k with range $[0, \infty)$. The coerciveness of $\mathcal{A}(x)$ implies that $\mathcal{A}_c$ is bounded in X, and consequently, since for $x \in \mathcal{A}_c$,

$$(\mathcal{A}'(x), x) \geqslant \int_0^1 (\mathcal{A}'(sx), x)\, ds = c, \quad \|\mathcal{A}'(x)\| \geqslant \frac{c}{\sup_{x \in \mathcal{A}_c} \|x\|} = \beta \geqslant 0.$$

Hypothesis (i) implies that each set $\mathcal{A}_c$ is starlike with respect to $x = 0$ since every ray through the origin, $\{tx \mid t \in \mathbb{R}^1, \|x\| = 1\}$, intersects $\mathcal{A}_c$ at precisely two points, $\pm t(x)x$. Thus there is a one-to-one mapping of $P^\infty(X) = \partial\Sigma_1/\mathbb{Z}_2$ (the unit sphere $\partial\Sigma_1$ of X with antipodal points identified) onto $\mathcal{A}_c/\mathbb{Z}_2$ defined by setting $f(x) = t(x)x$. To prove that $f(x)$ is continuous (and in fact differentiable), we note that for $x \in \mathcal{A}_c$, $c \neq 0$,

$$\frac{d}{ds}\mathcal{A}(sx) = (\mathcal{A}'(sx), x) > 0 \qquad \text{for} \quad s > 0.$$

Thus the implicit function theorem implies that $t(x)$ and, consequently,

$f(x)$ are continuous. Therefore, $\mathcal{C}_c/\mathbb{Z}_2$ is homeomorphic to $P^\infty(X)$ by means of the mapping f.

We now define a deformation of subsets of $\mathcal{C}_c/\mathbb{Z}_2$ which has some of the properties of the deformation along gradient lines described in (6.6.4). For this purpose, note first that since X is uniformly convex, the duality[1] map $J\colon X^* \to X$ is locally Lipschitz continuous. Then we consider the solution $x(t, x_0)$ of the initial value problem

$$\frac{dx}{dt} = v + a(x, v) J\,\mathcal{C}x, \qquad x(0) = x_0 \in \mathcal{C}_c,$$

where v is an element of X and $a(x, v) \in \mathbb{R}^1$ are to be chosen so that (i) $x(t, x_0) \in \mathcal{C}_c$ and (ii) $\mathcal{B}(x(t, x_0))$ is a decreasing function of t. We determine $a(x, v)$ by requiring that whenever $x(t, x_0)$ satisfies the initial value problem then $\mathcal{C}(x(t, x_0)) = c$. Thus $(\mathcal{C}'(x), v + a(x, v)J\,\mathcal{C}x) = 0$ and $a(x, v) = (\mathcal{C}'(x), v)/\|\mathcal{C}'(x)\|^2$. We determine v by supposing that x_0 is not a critical point of $\mathcal{B}(x)$ restricted to $\mathcal{C}_c$. Then

$$\begin{aligned}\mathcal{B}(x(t, x_0)) - \mathcal{B}(x_0) &= \int_0^t (\mathcal{B}'(x(t)), v + a(x, v)J\,\mathcal{C}'x)\, dt \\ &= \int_0^t \left(\mathcal{B}'(x(t)) - \frac{(\mathcal{B}'(x(t)), J\,\mathcal{C}'x)}{\|\mathcal{C}'(x)\|^2}\, \mathcal{C}'(x), v \right) dt.\end{aligned}$$

Setting

$$\nabla \mathcal{B}(x(t)) = \mathcal{B}'(x(t)) - \frac{(\mathcal{B}'(x(t)), J\,\mathcal{C}'x)}{\|\mathcal{C}'(x)\|^2} \qquad \mathcal{C}'(x) = \int_0^t (\nabla \mathcal{B}(x(t)), v)\, dt,$$

where $\nabla \mathcal{B}(x)$ denotes the gradient of $\mathcal{B}(x)$ restricted to $\mathcal{C}_c$. Thus choosing $v = -\nabla \mathcal{B}(x(t))$, we find that

$$\mathcal{B}(x(t, x_0)) - \mathcal{B}(x_0) = -\int_0^t \|\nabla \mathcal{B}(x(t, x_0))\|^2\, dt.$$

We now proceed to prove that the numbers

$$c_i^+ = \inf_{A \subset [A]_i} \sup_{x \in A} \mathcal{B}(x), \qquad c_i^- = \sup_{[A]_i} \inf_A \mathcal{B}(x)$$

are critical values of $\mathcal{B}(x)/\mathcal{C}_c$, where $[A]_i = \{A \mid A \in \mathcal{C}_c/\mathbb{Z}_2,\ \mathrm{cat}_{\mathcal{G}_c}(A) \geqslant i\}$. Clearly by (6.6.9), since $\mathcal{C}_c/\mathbb{Z}_2 \approx P^\infty(X)$, the classes $[A]_i$ are nonvacuous and form a strictly decreasing sequence $[A]_1 \supset [A]_2 \supset [A]_3$

[1] The duality mapping J is defined to be the Fréchet derivative of the functional $I(u) = \frac{1}{2}\|u\|^2$. For a uniformly convex Banach space, $I(u)$ is differentiable on the complement of the origin, and J satisfies the properties

$$(Ju, u) = \|Ju\|\,\|u\| \qquad \text{and} \qquad \|J(u)\| = \|u\|.$$

$\supset \cdots$. However, in order to repeat the argument in (6.6.4), we must verify the following analogue of Condition (C):

(*) If $c \neq 0$, $\epsilon > 0$ is sufficiently small, $u_n \in \mathcal{B}^{-1}(c+\epsilon, c-\epsilon)$, $u_n \in \mathcal{Q}_c$, and $\nabla \mathcal{B} x_n \to 0$, then $\{x_n\}$ has a convergent subsequence.

To verify (*), we may suppose (after possibly passing to a subsequence) that (a) $x_n \to \bar{x}$ weakly in X, (b) $\mathcal{Q}'(x_n)$ is weakly convergent, and (c) $\|\mathcal{Q}'(x_n)\|$ is uniformly bounded above zero and convergent. Then suppose

$$\text{(6.6.12)} \qquad \nabla \mathcal{B}(x_n) = \mathcal{B}'(x_n) - \frac{(\mathcal{B}'(x_n), J\,\mathcal{Q}'(x_n))}{\|\mathcal{Q}'(x_n)\|^2}\,\mathcal{Q}'(x_n) \to 0$$

Since $\mathcal{B}'$ is completely continuous, (c) implies that $(\mathcal{B}'(x_n), J\,\mathcal{Q}'(x_n)) \times \mathcal{Q}'(x_n)$ is strongly convergent. Consequently, we may suppose that $(\mathcal{B}'(x_n), J\,\mathcal{Q}'(x_n))$ is convergent to a real number β (say). The number $\beta \neq 0$ since otherwise (6.6.12) would imply that $\mathcal{B}'(x_n) \to 0$ and by hypothesis (iv) that $\bar{x} = 0$, which is impossible since by virtue of (iv), $\mathcal{B}^{-1}[c+\epsilon, c-\epsilon]$ is weakly closed and does not contain zero for $\epsilon > 0$ sufficiently small. Thus $\{\mathcal{Q}'(x_n)\}$ is strongly convergent, and by hypothesis (*), $x_n \to x$ strongly in X. Consequently the desired result is established.

Now we prove that each $c_i^+ \neq 0$ is a critical value of $\mathcal{B}(x)$ restricted to $\mathcal{Q}_c$. Indeed, otherwise by (*), for some $\epsilon > 0$, $\mathcal{B}^{-1}[c_i^+ + \epsilon, c_i^+ - \epsilon]$ contains no such critical points. Consequently, as in the proof of (6.6.2), we can find a deformation ζ_t and a set $\tilde{A} \subset [A]_i$ such that $\sup_A \mathcal{B}(x) = c_i^+ + \epsilon$, so that $\sup_{\zeta_1(A)} \mathcal{B}(x) \leqslant c_i - \epsilon$. This is the desired contradiction, since $\operatorname{cat}(\zeta_1(\tilde{A})) \geqslant i$, implies that $\zeta_1(\tilde{A}) \in [A]_i$ and

$$c_i^+ = \inf_{[A]_i} \sup_A \mathcal{B}(x) \leqslant \sup_{\zeta_1(\tilde{A})} \mathcal{B}(x) \leqslant c_i - \epsilon.$$

By reversing the procedure just given and deforming the sets $[A]_i$ so that $\mathcal{B}(x(t, x_0))$ is increasing along $x(t, x_0)$, we may show that the numbers

$$c_i^- = \sup_{[A]_i} \inf_{A \in [A]_i} \mathcal{B}(x),$$

if nonzero, are critical values of $\mathcal{B}(x)$ restricted to $\mathcal{Q}_c$.

We now show that under the given hypotheses, the sequence of critical points $\{x_n^\pm\}$ associated with critical values $c_n^\pm$ (a) satisfy the equation $\mathcal{Q}'(x_n^\pm) = \lambda_n^\pm \mathcal{B}'(x_n^\pm)$, where (b) $|\lambda_n^\pm| \to \infty$ and $x_n^\pm \to 0$ weakly as $n \to \infty$. The fact (a) follows immediately from the proof of (*) since each x_n satisfies the equation

$$\mathcal{B}'(x) = \left[\frac{(\mathcal{B}'(x), J\,\mathcal{Q}'(x))}{\|\mathcal{Q}'(x)\|^2}\right]\mathcal{Q}'(x) \qquad \text{and} \qquad \frac{(\mathcal{B}'(x), J\,\mathcal{Q}'(x))}{\|\mathcal{Q}'(x)\|^2} \neq 0$$

is finite.

6.7 Applications of the General Critical Point Theories

We now consider some applications of the critical point theories of Morse and Ljusternik and Schnirelmann described in the preceding two sections. In the first two subsections we prove some general results on nonlinear operator equations, and in the next two subsections we apply these results to some concrete problems of mathematical physics. Subsection E is devoted to a brief consideration of the differential geometric problem of studying geodesics on compact manifolds.

6.7A Application to bifurcation theory for gradient mappings

The bifurcation theory described in Chapter 4 can be supplemented by more global arguments. Consider, for example, the nonlinear eigenvalue problem

(6.7.1) $$u = \lambda\{Lu + Nu\}, \qquad \lambda \in \mathbb{R}^1,$$

defined on the real Hilbert space H. Here (a) L is a compact self-adjoint mapping of H into itself, and (b) $Nu = \mathcal{N}'(u)$ is an odd, higher order, completely continuous gradient mapping of H into itself with $N(0) = 0$ and

(6.7.2) $$\|Nu - Nv\| = O(\|u\| + \|v\|)\|u - v\| \quad \text{as} \quad \|u\|, \|v\| \to 0.$$

We shall be concerned with investigating the nontrivial solutions of (6.7.1) near $u = 0$, by means of the Ljusternik–Schnirelmann theory of critical points. Accordingly, we recall the approach to bifurcation theory discussed in Section 4.2 in which one selected an invariant I_f such that I_f (i) measures the solutions of (6.7.1), (ii) is invariant under small, suitably restricted perturbations, and (iii) can be approximately calculated by linearization. We shall show that the following *critical values* $c_n(R)$ *calculated by minimax principles*, are suitable invariants for each n and R sufficiently small:

(6.7.3) $$c_n(R) = \sup_{[V]_n} \inf_{V} \{ \tfrac{1}{2}(Lu, u) + \mathcal{N}(u)\}.$$

Here V is a symmetric subset on the sphere $\partial\Sigma_R = \{x \mid \|x\|^2 = R\}$ such that $\text{cat}(V, \partial\Sigma_R/\mathbb{Z}_2) \geq n$ and $[V]_n$ is the class of all such symmetric subsets of $\partial\Sigma_R$.

To proceed further, we observe that the numbers $c_n(R)$ satisfy properties (i)–(iii) for an invariant I_f. To begin, by (6.6.4), the numbers $c_n(R)$ are critical values of the function $\mathcal{F}(u) = \{\tfrac{1}{2}(Lu, u) + \mathcal{N}(u)\}$ on $\partial\Sigma_R$, so that for some number $\lambda_n(R)$ there is a solution $(u_n(R), \lambda_n(R))$ of (6.7.1) with

$\frac{1}{2}\|u_n(R)\|^2 = R$. Secondly, if $\tilde{c}_n(R) = \sup_{[V]_n} \inf_V \frac{1}{2}(Lu, u)$, we shall prove that

(6.7.4) $$|\tilde{c}_n(R) - c_n(R)| = o(R) \quad \text{as} \quad R \to 0.$$

Finally, we shall show that for $R = 1$, the critical points $\tilde{c}_n(R)$ of the quadratic isoperimetric problem defined above coincide with $\{\lambda_n^{-1}\}$, the eigenvalues of L ordered by decreasing magnitude and counted with multiplicities. As $R \to 0$, we shall also prove, for each n, that

(6.7.5) $$|\lambda_n(R) - \lambda_n| \to 0.$$

Assuming the truth of the above results, as $R \to 0$, for each n, the one-parameter family $(u_n(R), \lambda_n(R)) \to (0, \lambda_n)$; and thus represents a family of nontrivial solutions of (6.7.1) bifurcating from $(0, \lambda_n)$. Clearly this fact not only gives an alternative proof of (4.2.15) for equations of the type (6.7.1), but also yields interesting results for bifurcation near an eigenvalue of higher multiplicity λ_n for the linearized problem $u = \lambda Lu$. In fact, we shall prove the so-called "multiplicity preservation theorem"

(6.7.6) **Theorem** Suppose the hypotheses (a), (b) for the operators in equation (6.7.1) hold, and let λ_n be an eigenvalue of multiplicity N for the linear equation $u = \lambda Lu$. Then equation (6.7.1) has at least N distinct one-parameter families of nontrivial solutions $(u_{n+k}(R), \lambda_{n+k}(R)) \to (0, \lambda_n)$ for $k = 0, \ldots, (N-1)$ as $R \to 0$.

Proof: Assuming for the moment that the results (6.7.3)–(6.7.5) are known, the argument just presented above shows that the N distinct families of nontrivial solutions $(u_{n+k}(R), \lambda_{n+k}(R))$ exist for $k = 0, \ldots, N-1$; and since λ_n has multiplicity N, each family $\to (0, \lambda_n)$ as $R \to 0$. Thus it remains only to prove that these families are distinct. But this fact is an immediate consequence of the Ljusternik–Schnirelmann multiplicity theorem (6.6.4(iv)).

We now establish the equations (6.7.3)–(6.7.5), to complete the proof of the theorem. Clearly, this fact is accomplished by the following

Lemma A (The generalized minimax principle for quadratic functionals).

$$R\lambda_n^{-1} = \sup_{V_{n,R}} \min_{V} \tfrac{1}{2}(Lu, u)$$

where λ_n is the nth eigenvalue of $u = \lambda Lu$ (ordered by magnitude and counted according to multiplicity) and $\partial A_R \equiv \partial S_R$.

Lemma B $R\lambda_n^{-1} - c_n(R) = o(R)$.

Lemma C $|\lambda_n^{-1} - \lambda_n^{-1}(R)| \to 0$ as $R \to 0$.

Proof of Lemma A: Let S denote an n-dimensional subspace of H and $T_R = \{u \mid u \in S, \frac{1}{2}\|u\|^2 = R\}$, then we recall the following two facts:

(a) Let $P_R(n-1)$ be the set of elements obtained by identifying antipodal points of T_R and regarded as a subspace of $P_R(H)$. Then $\text{cat}(P_R(n-1), P_R^\infty(H)) = n$.

(b) The Courant–Fischer minimax principle can be rewritten [cf. (1.3.41)]

$$R\lambda_n^{-1} = \sup_{[T]n,\,R} \min_{T_R} \tfrac{1}{2}(Lu, u),$$

where T_R is defined as above and $[T]_{n,R}$ is the class of all such sets for n fixed. Now we consider the numbers

$$\tilde{c}_n(R) = \sup_{[V]n,\,R} \inf_{V} \tfrac{1}{2}(Lu, u).$$

By (a), $[T]_{n,R} \subset [A]_{n,R}$; so that $\tilde{c}_n(R) \geq R\lambda_n^{-1}$ for each n. Furthermore the numbers $\tilde{c}_n(R)$ are critical values of the function $\frac{1}{2}(Lu, u)$ on $P_R^\infty(H)$ and consequently on ∂S_R, so that $\tilde{c}_n(R) = R\lambda_{k(n)}^{-1}$ for some integer $k(n)$. To show that $\tilde{c}_n(R) = R\lambda_n^{-1}$ for each n, we proceed by induction. If $n = 1$, $\tilde{c}_n(R) = R\lambda_1^{-1}$ by definition. Suppose now that λ_1 is an eigenvalue of multiplicity exactly p, then $\lambda_1 \geq \lambda_{k(n)}$ for $n = 1, \ldots, p$. Hence $\lambda_1 = \lambda_{k(n)}$ $n = 1, \ldots, p$. Now we show $\tilde{c}_1(R) = \tilde{c}_2(R) = \cdots = \tilde{c}_p(R) \neq \tilde{c}_{p+1}(R)$. Indeed if $\tilde{c}_p(R) = \tilde{c}_{p+1}(R)$, then the critical set associated with the critical value would have dimension p on ∂S_R (6.6.4), which contradicts the fact that λ_1 is an eigenvalue of multiplicity p. Hence as an induction hypothesis we assume that the distinct eigenvalues $\lambda_{(1)}, \lambda_{(2)}, \ldots, \lambda_{(n-1)}$ are consistent with the distinct numbers $\tilde{c}_{(1)}(R), \tilde{c}_{(2)}(R), \ldots, \tilde{c}_{(n-1)}(R)$, with multiplicities included, by means of the relation $\tilde{c}_{(p)}(R) = R\lambda_{(p)}^{-1}$, $p = 1, \ldots, n-1$. Now suppose $\lambda_{(n)}$ is an eigenvalue of multiplicity exactly t, then we show $\tilde{c}_{(n)}(R) = \tilde{c}_{(n)+1}(R) = \cdots = \tilde{c}_{(n)+t}(R) = \lambda_{(n)}^{-1}R$. By our induction hypothesis clearly $\lambda_{n-1} < \lambda_{k(n+i)} \leq \lambda_n$ for $i = 1, 2, \ldots, t$. Thus $\lambda_{k(n+1)} = \lambda_n$, $i = 1, 2, \ldots, t$. Now suppose $\tilde{c}_{n+t+1}(R) = \lambda_n^{-1}R$, then the dimension of the critical set associated with the critical value of $\lambda_n^{-1}R$ exceeds $t-1$ on S_R by the above mentioned (6.6.4) again contradicting the fact that λ_n has multiplicity exactly t. Hence $\tilde{c}_{n+t+1}(R) \neq \lambda_n^{-1}R$ and hence the multiplicities of $\tilde{c}_{(n)}(R)$ and $\lambda_{(n)}$ agree. So the lemma is proven.

Proof of Lemma B: First we note that as $\mathfrak{N}(u) = \int_0^1(u, N(su))\,ds$, for small R and $u \in \partial S_R$,

$$|\mathfrak{N}(u)| \leq K(\|u\|)\|u\|^2 \quad \text{where} \quad K(\|u\|) \to 0 \quad \text{as} \quad \|u\| \to 0.$$

Hence $K_R = \sup_{S_R} |\mathfrak{N}(u)| = o(R)$. Now

$$c_n(R) = \sup_{[A]n,\,R} \inf_{A} \{\tfrac{1}{2}(Lu, u) + \mathfrak{N}(u)\},$$

and by Lemma A $R_{\lambda_n}^{-1} = \sup_{[A]_{n,R}} \inf_A \frac{1}{2}(Lu, u)$, so

$$|c_n(R) - R\lambda_n^{-1}| \leq |\sup_{[A]n,\,R} \inf_{A} \{\tfrac{1}{2}(Lu, u) + K_R\} - \sup_{[A]n,\,R} \inf_{A} \tfrac{1}{2}(Lu, u) \leq K_R = o(R).$$

Proof of Lemma C: Taking the inner product of (6.7.1) with $u_n(R)$ we obtain

$$R\lambda_n^{-1}(R) = \tfrac{1}{2}(Lu_n(R), u_n(R)) + \tfrac{1}{2}(Nu_n(R), u_n(R))$$

$$= c_n(R) + \{\tfrac{1}{2}(Nu_n(R), u_n(R)) - \mathfrak{N}(u_n(R))\}$$

$$= c_n(R) + o(R), \qquad \text{for small } R.$$

Hence by Lemma B,

$$R\lambda_n^{-1}(R) - R\lambda_n^{-1} = c_n(R) - \lambda_n^{-1}R + o(R)$$

So $|\lambda_n^{-1}(R) - \lambda_n^{-1}| = o(R)/R = o(1)$. Hence as $R \to 0$, $\lambda_n^{-1}(R) \to \lambda_n^{-1}$.

Conclusion of proofs: By the above results the set $(u_n(R), \lambda_n(R))$ for fixed n defines a one-parameter family of solutions, bifurcating from $(0, \lambda_M)$.

The result just obtained can itself be extended to a global context as follows:

(6.7.7) **Corollary** The families of solutions $(u_j(R), \lambda_j(R))$, discussed in (6.7.6), can be extended as solutions of (6.7.1) for all $R > 0$.

Proof: This result is an immediate consequence of (6.6.4), since for each j, the vectors $u_j(R)$ are critical points associated with the critical values

$$c_n(R) = \sup_{[V]_n} \inf_V \{ \tfrac{1}{2}(Lu, u) + \mathfrak{N}(u)\}.$$

Remarks: In order to apply Corollary (6.7.7) to continuation problems for bifurcation theory, it is necessary to investigate the continuity properties of $(u_j(R), \lambda_j(R))$ as a function of R. Definitive results in this direction have yet to be achieved and the problem is made more difficult by known examples where discontinuities arise. On the other hand our result (6.1.31) points to further affirmative results. (See Section 6.7C and Note F, p. 390.)

6.7B Multiple solutions of operator equations involving gradient mappings

Here we consider the problem of finding a lower bound for the number of solutions of the operator equation

$$x - Lx + N(x) = 0, \tag{6.7.8}$$

where L is a compact, self-adjoint mapping of a real Hilbert space H into itself, and $N(x) = \mathfrak{N}'(x)$ is a completely continuous gradient mapping of H into itself with $N(x)$ of higher order. We prove the following

(6.7.9) **Theorem** Suppose the above hypotheses are satisfied and at $x = 0$ the quadratic form $Q(x) = (x, x) - (Lx, x)$ has Morse index $q > 0$. Then the equation (6.7.8) will have at least q distinct solution pairs $\pm x_n$, $n = 1, 2, \ldots, q$, provided the following two conditions are satisfied:

(a) $F(x) = \frac{1}{2} Q(x) + \mathfrak{N}(x)$ is bounded below, and
(b) $F(x) \geqslant 0$ for $\|x\|$ sufficiently large.

Proof: We consider the manifold $\mathfrak{M}$ obtained by deleting the origin from the Hilbert space H and identifying the antipodal points of $H - \{0\}$. In addition, $\mathfrak{M}$ is a smooth manifold for each x bounded away from the origin. Clearly, $\mathfrak{M}$ contains sets of Ljusternik–Schnirelmann category $n = 1, 2, 3, \ldots$ since the real n-dimensional projective space $\mathfrak{P}_n(H) \subset \mathfrak{M}$ for each n. Now $F(x) = \frac{1}{2} Q(x) + \mathfrak{N}(x)$ is even in x and can thus be considered as a C^2 differentiable functional on $\mathfrak{M}$.

Clearly, the functional $F(x)$ satisfies Condition (C) on the set $\mathfrak{M}_{-\epsilon} = \{x \mid F(x) < -\epsilon\}$ for any $\epsilon > 0$ since if $x \in \mathfrak{M}_{-\epsilon}$ by hypothesis (b), $\|x\|$ is uniformly bounded so that if $F'(x_n) \to 0$ for $x_n \in \mathfrak{M}_{-\epsilon}$, then the sequence has a weakly convergent subsequence $\{x_{n_j}\}$ such that $x_{n_j} - Lx_{n_j} - Nx_{n_j} \to 0$. Now the complete continuity of L and N imply that $\{x_{n_j}\}$ is strongly convergent.

Now we consider the numbers defined by $c_n(\mathfrak{M}) = \inf_{[V]_n} \sup_V F(x)$, where V is a subset of $\mathfrak{M}$ such that $\text{cat}(V, \mathfrak{M}) \geq n$, and $[V]_n$ is the class of all such subsets. We show that for $\epsilon > 0$ and sufficiently small, all the numbers $c_o(\mathfrak{M}), c_1(\mathfrak{M}), \ldots, c_{q-1}(\mathfrak{M})$ are less than $-\epsilon$; so that any critical points of $F(x)$ lying on any of these levels is contained in $\mathfrak{M}_{-\epsilon}$.

To obtain the desired bound, we observe that since the index of the quadratic form $Q(x) = (x, x) - (Lx, x)$ is q, there is a q-dimensional subspace of H, H_q, and an absolute constant $c < 0$ such that for each $x \in H_q$, $Q(x) \leq c\|x\|^2$.

Consequently, identifying the antipodal points of the sphere of radius R in H_q, we obtain a set $\mathcal{P}_R$ that can be identified with $(q-1)$-dimensional real projective space. Thus $\text{cat}_{\mathfrak{M}_{-\epsilon}} \mathcal{P}_R \geq q$ for each $R > 0$, so that $\mathcal{P}_R \in [V]_n$ for each $n = 1, 2, \ldots, q$. On the other hand, for small R and any $x \in \mathcal{P}_R$,

$$F(x) = Q(x) + \mathfrak{N}(x) \leq c\|x\|^2 + o(\|x\|^2) \leq \tfrac{1}{2} cR^2 < 0. \tag{6.7.10}$$

Combining these two facts, we find that for R sufficiently small,

$$\inf_{[V]_n} \sup_{V} F(x) \leq \sup_{\mathcal{P}_R} F(x) \leq \tfrac{1}{2} cR^2 < 0.$$

Thus by the minimax theorem (6.6.4) the functional $F(x)$ has q pairs of critical points $(\pm x_n)$ $n = 1, \ldots, q$. such that $F(x_n) = c_n(\mathfrak{M})$ and these critical points satisfy the equation (6.7.8), as desired. Thus the theorem is established.

The result just obtained yields conditions that ensure the existence of finitely many distinct solutions of (6.7.8). We now mention an extension of this result, which provides criteria for a nonlinear operator equation (involving gradient mappings) to possess a countably infinite number of distinct solutions. We consider the operator equation

$$x = \mathfrak{N}'(x), \qquad \mathfrak{N}'(0) = 0, \tag{6.7.11}$$

where $\mathfrak{N}'(x)$ is an odd C^1 completely continuous gradient mapping of an infinite-dimensional Hilbert space H into itself, with the properties:

(6.7.12) $(\mathfrak{N}'(tx), x)$ for $x \neq 0$ is a strictly convex function of $t \geq 0$ such that $\mathfrak{N}(x) \leq \beta(\mathfrak{N}'(x), x)$, where β is some absolute constant less than $\frac{1}{2}$;

(6.7.13) $\lim_{t \to 0}(\mathfrak{N}'(tx), x)/t = 0$ uniformly over bounded subsets of $x \in H$;

(6.7.14) $\lim_{t\to\infty}(\mathfrak{N}'(tx), x)/t = \infty$ uniformly over subsets of H with $(\mathfrak{N}'(x), x)$ bounded above zero.

Then we note

Theorem Suppose the operator $\mathfrak{N}'(x)$ satisfies the above hypotheses, then the equation (6.7.11) possesses a countably infinite number of distinct solutions.

For a proof of this result see Ambrosetti (1973).

6.7C Global equilibrium states of a flexible elastic plate

We have already mentioned some connections between the critical point theories developed in this chapter and the problems of nonlinear elasticity. Indeed these problems provide the simplest nontrivial examples on this theory that can be checked by observation. Here we turn attention to the buckling problem for thin flexible elastic plates discussed earlier in Section 4.3B. As mentioned there, the equilibrium states of a clamped thin elastic plate B subject to compressible forces acting on the boundary of B are given as the solutions of the nonlinear operator equation

(6.7.15) $\quad u + Cu - \lambda Lu = 0, \qquad \lambda \in \mathbb{R}^1.$

Here, as earlier, λ is a measure of the force acting on $\partial\Omega$ and the operators L and C are bounded mappings of $\mathring{W}_{2,2}(\Omega)$ into itself, with L a linear compact self-adjoint mapping, and C a completely continuous gradient mapping, homogeneous of degree 3 and such that $(Cu, u) > 0$. Here we derive some information on the number of solutions of (6.7.15) as λ varies over the interval $(0, \infty)$. Throughout this section we use the definitions and notations of Section 4.3B. We prove (see Fig. 6.3)

(6.7.16) **Theorem** Suppose $(Lu, u) > 0$ for $u \neq 0$, and the eigenvalues of $u = \lambda Lu$ are ordered so that $0 < \lambda_1 \leqslant \lambda_2 \leqslant \lambda_3 \leqslant \cdots$ (with multiplicities included). Then:

(i) for fixed $\lambda \in (\lambda_{n-1}, \lambda_n]$, the equation (6.7.15) has at least $n-1$ distinct pairs of solutions $(\pm u_j)$, $j = 1, 2, \ldots, n-1$.

(ii) Let $R > 0$ be a fixed positive number, and $\mathfrak{M}_R = \{u \mid u \in \mathring{W}_{2,2}(\Omega),\ \frac{1}{2}\|u\|^2 + \frac{1}{4}(Cu, u) = R\}$. Then the equation (6.7.15) has a countably infinite number of distinct solutions $(u_n(R), \lambda_n(R))$ such that: (a) $u_n(R) \in \mathfrak{M}_R$ and $\lambda_n(R) > \lambda_n$; (b) $\lambda_n(R) \to \infty$ as $n \to \infty$; and (c) as $R \to 0$, $(u_n(R), \lambda_n(R)) \to (0, \lambda_n)$.

(iii) Suppose the multiplicity of λ_n is k and $\lambda_{n-1} < \lambda_n$, then near $(0, \lambda_n)$ there are at least k distinct one-parameter families of solutions $(u_{n+i}(\epsilon), \lambda_{n+i}(\epsilon))$, $i = 0, \ldots, k-1$, such that $(u_{n+i}(\epsilon), \lambda_{n+i}(\epsilon)) \to (0, \lambda_n)$ as $\epsilon \to 0$, where ϵ is proportional to $\|u_n(\epsilon)\|^2$.

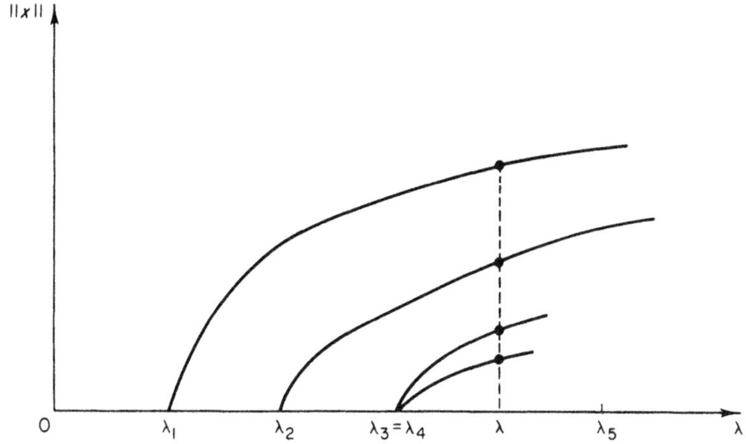

FIG. 6.3 Hypothesized configuration for the origin of four solution pairs for λ between λ_4 and λ_5. (The "branches" emanating from $\lambda = \lambda_i$ and $\|x\| = 0$ denote nontrivial solution pairs for the equation (6.7.15).)

Proof: The results (i)–(iii) are almost immediate consequences of the abstract theorems established earlier in this chapter.

Proof of (i): The desired conclusion is established by applying Theorem (6.7.9) to the equation (6.7.15). In the present case, we replace the operator L in (6.7.9) by λL for fixed $\lambda \in (\lambda_{n-1}, \lambda_n]$, and N in (6.7.9) by C. Thus $F(u) = \frac{1}{2}\|u\|^2 - \lambda(Lu, u)) + \frac{1}{4}(Cu, u)$. Clearly, the quadratic form referred to in (6.7.16) becomes $Q(u) = \|u\|^2 - \lambda(Lu, u)$; and since $\lambda \in (\lambda_{n-1}, \lambda_n]$, the index of $Q(u)$ is $n-1$. Thus (i) will be established once we show that $F(x) \to \infty$ as $\|x\| \to \infty$ since then the provisions of Theorem (6.7.9) will be satisfied. This coerciveness property has already been proven in Section 6.2B. A simple proof for the present case follows from the estimate

$$F(u) = \tfrac{1}{2}\|u\|^2 - \tfrac{1}{2}\lambda(\mathfrak{B}(u, f), u) + \tfrac{1}{4}\|\mathfrak{B}(u, u)\|^2$$
$$= \tfrac{1}{2}\|u\|^2 - \tfrac{1}{2}\lambda(\mathfrak{B}(u, u), f) + \tfrac{1}{4}\|\mathfrak{B}(u, u)\|^2.$$

Since $\frac{1}{2}\lambda(\mathfrak{B}(u, u), f) \leqslant \epsilon^2 \frac{1}{2}\lambda\|\mathfrak{B}(u, u)\|^2 + (\lambda/2\epsilon^2)\|f\|^2$ for any $\epsilon > 0$, we find that for $\lambda\epsilon^2 = \frac{1}{2}$, $F(u) \geqslant \frac{1}{2}\|u\|^2 - \lambda^2\|f\|^2$. Consequently, $F(u) \to \infty$ as $\|u\| \to \infty$ since λ is fixed.

Proof of (ii): The first part of the conclusion follows from Theorem (6.6.11). Indeed, setting $\mathcal{A}u = u + Cu$ and $\mathfrak{B}u = Lu$, we note that the hypotheses of Theorem (6.6.11) are easily verified. For example, to show that $f(s) = (\mathcal{A}'(su), u)$ is a strictly increasing function of $s \in [0, \infty)$ (for fixed $u \neq 0$), we observe that by virtue of the homogeneity of Cu,

$$f(s) = s\|u\|^2 + s^3(Cu, u),$$

so that

$$f'(s) = \|u\|^2 + 3s^2(Cu, u) > 0 \qquad \text{for} \quad u \neq 0.$$

On the other hand, to verify hypothesis (iii) of (6.6.11), we see that if $u_n \to u$ weakly and the sequence $\{\mathcal{Q}u_n\}$ converges strongly, then the complete continuity of Cu implies that $u_n \to u$ strongly.

The second part of the conclusion is obtained by using the fact that the solutions $\{u_n(R)\}$ are characterized as critical points associated with the critical values

$$c_n(R) = \sup_{[V]_n} \inf_{V} \tfrac{1}{2}(Lu, u),$$

where V is a symmetric subset of $\mathfrak{M}_R$ with $\text{cat}(V, \mathfrak{M}_R/\mathbb{Z}_2) \geqslant n$ and $[V]_n$ is the class of all subsets of $\mathfrak{M}_R$. Thus since $\frac{1}{2}\|u\|^2 \leqslant R$ as $R \to 0$, $\|u\| \to 0$; and consequently a slight variation of Lemmas (A)–(C) of (6.7.6) shows that as $R \to 0$, $(u_n(R), \lambda_n(R) \to (0, \lambda_n)$.

Proof of (iii): The result can be proved on the basis of Theorem (6.7.6), by converting the equation (6.7.15) into an equation of the form (6.7.1), by scaling. Indeed, setting $u = \sigma v$ in (6.7.15), where $\sigma \neq 0$ is a real number to be determined, we find that by virtue of the homogeneity of C, v satisfies the equation $v + \sigma^2 Cv = \lambda Lv$. Setting $\sigma^2 = \lambda$, we obtain

(6.7.17) $$v = \lambda(Lv - Cv) \qquad \text{for} \quad \lambda > 0.$$

Hence applying Theorem (6.7.6) to (6.7.17), we obtain the desired families of solutions bifurcating from $(0, \lambda_n)$.

Combined buckling–bending problem As mentioned earlier in Section 6.2B, the equilibrium states in this case are identical with the solutions of the inhomogeneous nonlinear operator equation (for fixed λ):

(6.7.18) $$u + Cu - \lambda Lu = f \qquad (f \neq 0).$$

The considerations of (i) do not apply to this equation since the associated functional

(6.7.19) $$\mathcal{I}_\lambda(u) = \tfrac{1}{2}\|u\|^2 + \tfrac{1}{4}(Cu, u) - \tfrac{1}{2}\lambda(Lu, u) - (f, u)$$

is no longer symmetric with respect to the antipodal map, because $f \neq 0$. However, we now show that the Morse critical point theory of (6.5.10) is applicable to the functional $\mathcal{I}_\lambda(u)$. First, we recapitulate some facts that will be needed in the ensuing argument.

(6.7.20) **Lemma** For fixed λ, the operator $A_\lambda(u) = u + Cu - \lambda Lu$ is a C^∞, proper nonlinear Fredholm mapping of index zero of $\mathring{W}_{2,2}(\Omega)$ into itself. Moreover, the singular values Φ_λ of $A_\lambda(u)$ form a closed subset that is nowhere dense in H.

Proof: The properness of $A_\lambda(u)$ was established in (2.7.18), whereas the C^∞ smoothness property follows immediately from the fact that the operator Cu is homogeneous of degree 3. In addition, that Φ_λ is nowhere dense in $\mathring{W}_{2,2}(\Omega)$ follows from Smale's extension (3.1.45) of Sard's theorem, once we show that $A_\lambda(u)$ is a nonlinear Fredholm operator of index zero. This fact follows immediately from (2.6.3) since $A_\lambda(u)$ can be represented as a smooth compact perturbation of the identity.

We now state an application of our discussion in (6.5D) to the solutions of (6.7.18).

(6.7.21) **Theorem** For each fixed λ and almost all $f \in \mathring{W}_{2,2}(\Omega)$ (i.e., $f \in \mathring{W}_{2,2}(\Omega) - \Phi_\lambda$), the solutions of (6.7.18) will be a finite number of nondegenerate critical points of $\mathcal{I}_\lambda(u)$. These critical points will satisfy the Morse inequalities (6.5.10) by virtue of the properness of $\mathcal{I}_\lambda'(u)$. More generally, for any fixed f, (6.7.18) is always solvable and the totality of solutions $\{w\}$ satisfies the a priori bound

$$(*) \qquad \|w\|_{2,2} \leqslant 2(+\lambda^2\|F_0\|^2 + \tfrac{1}{2}\|f\|^2)^{1/2}.$$

Furthermore, if $\mathcal{I}_\lambda(u)$ has at least two isolated relative minima, then $\mathcal{I}_\lambda(u)$ will possess a third critical point.

Proof: By virtue of Lemma (6.7.20), the points $f \in (\mathring{W}_{2,2}(\Omega) - \Phi_\lambda)$ are everywhere dense in $\mathring{W}_{2,2}(\Omega)$, and $A_\lambda^{-1}(f)$ must consist of nondegenerate critical points of $\mathcal{I}_\lambda(u)$. Thus the finiteness of this set $A_\lambda^{-1}(f)$ will follow from the properness of A_λ and the implicit function theorem. Indeed, if there were an infinite number of points in $A_\lambda^{-1}(f)$, these points would lie in a compact set and so have a convergent subsequence, contradicting the fact that nondegenerate critical points are isolated. That these critical points of $\mathcal{I}_\lambda(u)$ satisfy the Morse inequalities (6.5.10) when regarded as defined on $H = \mathring{W}_{2,2}(\Omega)$ then follows from the fact that the antiderivative of any proper gradient mapping defined on H automatically satisfies Condition (C) on H. The a priori bound $(*)$ also follows from simple considerations. Indeed, if w satisfies $\mathcal{I}_\lambda(u) = 0$, we find by virtue of (6.7.18), that

$$\|w\|^2 + (C(w), w) - \lambda(C(F_0, w), w) = (f, w).$$

Thus for any $\epsilon > 0$, the Cauchy–Schwarz inequality implies

$$\|w\|^2 + \|C(w, w)\|^2 - |\lambda|\epsilon\|C(w, w)\|^2 - \frac{|\lambda|}{\epsilon}\|F_0\|^2 \leqslant \|f\|\,\|w\|.$$

Setting $\epsilon = 1/|\lambda|$, we find

$$\|w\|^2 \leqslant \|f\|\,\|w\| + |\lambda|^2\|F_0\|^2,$$

from which $(*)$ follows.

The last statement in (6.7.20) follows from (6.5.3) and the fact that $\mathcal{I}_\lambda(u) \to \infty$ as $\|u\| \to \infty$.

6.7D Stationary states for some nonlinear wave equations

We seek some time-periodic, complex-valued solutions of the special form $u(x, t) = e^{i\lambda t}v(x)$ for nonlinear wave equations of the form

$$(6.7.22) \qquad u_{tt} = \Delta u - m^2 u + f(x, |u|^2)u$$

or

$$(6.7.23) \qquad -iu_t = \Delta u + f(x, |u|^2)u$$

defined on $\mathbb{R}^1 \times \mathbb{R}^N$. Here λ is a real number, $v(x)$ ($\not\equiv 0$) is a real-valued, smooth function vanishing exponentially at ∞, and $f(x, |u|^2)$ is a C^1 positive function of x that is odd in $|u|^2$. Such solutions are exact nonlinear analogues of stationary states for linear Schrödinger equations, and we therefore refer to these solutions as stationary states.

Substituting $u(x, t) = e^{i\lambda t}v(x)$ in either of the above wave equations, we find that $v(x)$ satisfies the semilinear elliptic equations

(6.7.24) $$(\Delta - \beta)v + f(x, v^2)v = 0, \qquad x \in \mathbb{R}^N;$$

where $\beta = \lambda^2 - m^2$ in the case of (6.7.22), and $\beta = \lambda$ in the case of (6.7.23).

Thus to investigate the stationary states of either (6.7.22) or (6.7.23), we shall restrict the function $f(x, y)$ and:

(a) determine those values of β for which (6.7.24) has nontrivial solutions in $L_2(\mathbb{R}^N)$,

(b) for fixed $\beta > 0$ prove that (6.7.24) has a countably infinite number of distinct solutions in $L_2(\mathbb{R}^N)$.

In answering both (a) and (b) we shall find special "nonlinear" phenomena.

We begin by considering the *question (a)* in case either:

(I) for all u,

$$0 < f(x, |u|^2) = g(x)\{|u|^\sigma\} \qquad \text{with} \quad 0 < \sigma < 4/(N-2),$$

where $g(x) \to 0$ as $|x| \to \infty$, or

(II) $f(x, |u|^2)$ is independent of x and $f(|u|^2) = g|u|^\sigma$, where g is a positive constant.

(6.7.25) **Theorem** Suppose $f(x, |u|^2)$ satisfies one of the conditions (I) or (II), then (6.7.24) has a solution $v(x) \in L_2(\mathbb{R}^N)$ with the provisos

(i) only if $\beta > 0$ in case (I); and
(ii) if and only if $\beta > 0$ and $\sigma < 4/(N-2)$, in case (II).

Proof: (i): This result is an immediate consequence of the results of Section 6.3.A and of the fact that the gradient mapping defined by

$$(\mathcal{N}'(v), \Phi) = \int_{\mathbb{R}^N} f(x, v^2)v\Phi, \qquad \Phi \in W_{1,2}(\mathbb{R}^N)$$

is a completely continuous operator mapping $W_{1,2}(\mathbb{R}^N)$ into itself. See p. 61.

(ii): We first establish the sufficiency of the restriction $\sigma < 4/(N-2)$ in case $f(x, v^2) = cv^\sigma$. To this end, observe that the results of this chapter cannot be applied directly since the operator corresponding to the term $c|u|^\sigma u$ is *not completely continuous* in $W_{1,2}(\mathbb{R}^N)$, although it is bounded and continuous. To circumvent this difficulty, we proceed by seeking only radially symmetric solutions of (6.7.24), i.e., solutions depending on $|x| = r$, and by making the change of variables $r^{(N-1)/2}v(r) = w(r)$. Then $w(r)$ vanishes at $r = 0$ and satisfies the equation

(†) $$\frac{d^2w}{dr^2} - \beta w + gr^{-\sigma}|w|^\sigma w = 0$$

for $\beta > 0$ and $0 < \sigma < 4/(N-2)$. This equation has a nontrivial solution w. Indeed, if we rewrite (†) as an operator equation in $H = \mathring{W}_{1,2}(0, \infty)$ with norm $\|w\|_H^2 = \int_0^\infty (w_r^2 + \beta^2 w^2)\, dr$, the operator defined by the duality method as

$$(\mathcal{N}'(w), \Phi) = \int_0^\infty g r^{-\sigma} |w|^\sigma w \Phi \qquad \text{with} \quad 0 < \sigma < 4/(N-2)$$

is a completely continuous mapping of H into itself (see p. 61). Consequently, by an application of (6.3.2) and the homogeneity of the term $g r^{-\sigma}|w|^\sigma w$, the Lagrange multiplier, referred to in (6.3.2), can be chosen to be unity. The fact that $w(r)$ and hence $v(r)$ decay exponentially at ∞ is a consequence of the principle of *amplification of decay* mentioned in Section 1.2.

To prove the necessity of the restrictions $0 < \gamma < 4/(N-2)$ and $\beta > 0$, we first recall that the necessity of $\gamma < 4/(N-2)$ for $\beta > 0$ was obtained by elementary means in Section 1.2. If $\beta < 0$, the impossibility of a suitable $v(x)$ follows from the results mentioned in Section 1.2(iv). Finally, if $\beta = 0$ and $\gamma = 4/(N-2)$, the nonexistence of a suitable $v(x)$ is obtained using the deeper properties of linear elliptic partial differential equations. Indeed, suppose $w(x)$ ($\not\equiv 0$) is a function decaying exponentially at ∞ and satisfying $\Delta w + |w|^{4/(N-2)} w = 0$ on $\mathbb{R}^3$. Then regarding w as a solution of the linear equation $\Delta w + g(x)w = 0$ with $g(x) = |w|^{4/(N-2)}$, the result follows by noting that after performing the Kelvin transformation of this equation, w has a pole of infinite order at the origin and contradicts the unique continuation property for the transformed equation $\Delta w + g(x)w = 0$.

Now we consider the problem (b), and suppose that the function

(6.7.26) $$f(x, v^2) = \sum_{i=1}^{p} g_i(x)|v|^{\sigma_i},$$

where $0 < \sigma_i < 4/(N-2)$, $g_i(x) \geqslant 0$ (and not all $g_i(x)$ are identically zero), where either:

(1) $g_i(x) \to 0$ as $|x| \to \infty$, or
(2) each $g_i(x)$ is a constant.

(6.7.27) **Theorem** Suppose the function $f(x, v^2)$ satisfies either of the hypotheses (1) or (2) above. Then the equation (6.7.24) possesses a countably infinite number of distinct solutions if $i = 1$. For $i > 1$, the solutions satisfy (6.7.24) apart from an eigenvalue factor.

Proof: The proof is based on an application of Theorem (6.6.11). We begin by supposing that $f(x, v^2)$ satisfies hypothesis (1). Then the operator defined implicitly by

$$(\mathcal{N}'(u), v) = \sum_{i=1}^{p} \int g_i(x)|u|^{\sigma_i} uv$$

is a completely continuous gradient mapping of $W_{1,2}(\mathbb{R}^N)$ into itself with

$$\mathfrak{N}(u) = \sum_{i=1}^{p} (1/(\sigma_i + 2)) \int_{\mathbb{R}^N} g_i(x)|u|^{\sigma_i+2}.$$

Since $\sigma_i > 0$, there is a constant $\alpha < \frac{1}{2}$ with $\mathfrak{N}(u) \leqslant \alpha(\mathfrak{N}'(u), u)$ for all $u \in W_{1,2}(\mathbb{R}^N)$. The reader will easily verify the remainder of the hypotheses of (6.6.11) for the operator $\mathfrak{N}'$.

If $f(x, v^2)$ satisfies hypothesis (2), we proceed as in the proof of Theorem (6.7.25) and consider the equation

$$\text{(6.7.28)} \qquad \frac{d^2w}{dr^2} - \beta w + \sum_{i=1}^{p} g_i r^{-\sigma_i}|w|^{\sigma_i} w = 0,$$

where $w(r) = r^{(N-1)/2}v(r)$ and $w(0) = 0$. As before, the operator $\mathfrak{N}'(w)$ defined by

$$(\mathfrak{N}'(w), \Phi) = \sum_{i=1}^{p} \int_0^\infty g_i r^{-\sigma_i}|w|^{\sigma_i} w\Phi$$

is a completely continuous gradient mapping of $\mathring{W}_{1,2}(0, \infty)$ into itself. Moreover, as in the above paragraph, the hypotheses necessary to apply Theorem (6.6.11) are then easily verified. Thus the theorem is established after rescaling the associated eigenvalue as in (ii) above.

6.7E Geodesics between two points of a compact Riemannian manifold

Let $(\mathfrak{M}^N, g)$ be a compact, smooth, Riemannian manifold of dimension N with metric tensor g. Then, in terms of local coordinates, the geodesics between two points a and b of $\mathfrak{M}^N$ are the solutions (passing through a, b) of the following second-order system of ordinary differential equations

$$\text{(6.7.29)} \qquad \frac{d^2x^k}{dt^2} + \sum_{i,j=1}^{N} \Gamma_{ij}^k(x) \frac{dx^i}{dt} \frac{dx^j}{dt} = 0, \qquad (k = 1, \ldots, N).$$

Alternately, these geodesics are the critical points of the arc length functional, $\int_a^b ds$, relative to all the smooth curves of $\mathfrak{M}^N$ passing through a, b. This latter characterization of geodesics and its variants are useful in studying the structure of the geodesics joining a and b. The simplest basic result is due to Hilbert.

(6.7.30) There is a geodesic of minimal length joining the two points a and b, on a smooth, compact, (connected) Riemannian manifold.

Proof: We consider the class of rectifiable curves $K_{a,b}$ on $(\mathfrak{M}^N, g)$ joining a and b. Since $\mathfrak{M}^N$ is connected, this class is nonvacuous. We parametrize the curves $\{c\}$ of $K_{a,b}$ by the parameter $\tau = s/L$, where L is the length of a curve in $K_{a,b}$ and s is the arc length measured from the initial point a. With this parametrization, the desired geodesics are critical points of the functional $\mathcal{G}(c) = \int_0^1 ds^2$. Clearly, by the properties of the Lebesgue integral, $\mathcal{G}(c)$ is lower semicontinuous with respect to uniform convergence in $K_{a,b}$. Let $\mathcal{G}_0 = \inf \mathcal{G}(c)$ over $K_{a,b}$. We will show that $\mathcal{G}_0$ is attained by an element c_0 of $K_{a,b}$. To this end, it suffices to show that the set of curves in $K_{a,b}$ with bounded length, is compact. Thus let $\{c_n(\tau)\}$ be a sequence of curves in $K_{a,b}$ with length $L(c_n) \leqslant M$ (say). Then for any $\tau_1, \tau_2 \in [0, 1]$ and fixed n,

$$d(c_n(\tau_1), c_n(\tau_2)) = L(c_n)|\tau_1 - \tau_2| \leqslant M|\tau_1 - \tau_2|, \tag{6.7.31}$$

where $d(x, y)$ denotes the Riemannian distance between x and y. Thus $\{c_n\}$, regarded as continuous mappings from $[0, 1] \to (\mathfrak{M}^N, g)$, are uniformly bounded and equicontinuous. By the Arzela–Ascoli Theorem, $\{c_n\}$ has a uniformly convergent subsequence whose limit c_0 will again be a rectifiable curve, since, by (6.7.31), $d(c_0(\tau_1), c_0(\tau_2)) \leqslant M|\tau_1 - \tau_2|$.

A deeper study of the geodesics joining a and b can be carried out by utilizing the results of 6.5 and 6.6. Indeed, suppose $(\mathfrak{M}^N, g)$ is isometrically embedded as a closed submanifold into a Euclidean space $\mathbb{R}^{k(N)}$ of sufficiently high dimension. Let $W_{1,2}([0, 1], (\mathfrak{M}^N, g))$ be the closed subset of the Hilbert space $H = W_{1,2}([0, 1], \mathbb{R}^{k(N)})$ consisting of those elements $c(\tau) \in H$ for which the image $c[0, 1] \subset \mathfrak{M}^N$. $W_{1,2}([0, 1], (\mathfrak{M}^N, g))$ is a Hilbert manifold, as is its closed subspace $\Omega(\mathfrak{M}^N; a, b)$ consisting of those elements $c(\tau)$ of $W_{1,2}([0, 1], (\mathfrak{M}^N, g))$ satisfying $c(0) = a$ and $c(1) = b$, with a, b fixed points of $\mathfrak{M}^N$. Now let $\langle c(\tau), c(\tau)\rangle$ denote the length of the vector $c(\tau)$. Then the critical points of the functional $\mathcal{G}(c) = \int_0^1 \langle \dot{c}(\tau), \dot{c}(\tau)\rangle \, d\tau = \int_c ds^2$ in the space $\Omega(\mathfrak{M}^N; a, b)$ coincide with the geodesics on $(\mathfrak{M}^N, g)$ joining a and b. If $\mathfrak{M}^N$ is homeomorphic to the sphere S^N, the result (6.5.12) can be used to show that *there are an infinite number of distinct geodesics joining a and b on $(\mathfrak{M}^N, g)$*. Indeed one can show that the functional $\mathcal{G}(c)$ satisfies the condition C, (6.5.2), on $\Omega(\mathfrak{M}^N; a, b)$. Then the result follows provided it can be shown that the Betti numbers of $\Omega(\mathfrak{M}^N; a, b)$ are different from zero for an infinite number of distinct integers. Indeed, it is known that the Betti numbers R_i of $\Omega(\mathfrak{M}^N; a, b)$ form a periodic sequence of length $(N-1)$ consisting of the numbers 1 when $i = 0 \pmod{(N-1)}$ and 0 otherwise. Note B at the end of this chapter mentions other interesting information on this topic. Since an adequate treatment of this topic is contained in other monographs, we refer the reader to the books of Morse (1934) and Schwartz (1969) and the article of Palais (1963) for further information.

NOTES

A The Dirichlet problem for parametric surfaces of constant mean curvature

Let Ω be a simply connected domain in the u-v plane with boundary $\partial\Omega$. We wish to determine a parametric surface $S = \{X(u, v) = x_1(u, v), x_2(u, v), x_3(u, v)\}$ of constant mean curvature M defined over Ω, which assumes prescribed smooth boundary values f on $\partial\Omega$. The system of partial differential equations defining S are

$$(*) \qquad \Delta X = 2M \frac{\partial X}{\partial u} \wedge \frac{\partial X}{\partial v} \quad \text{in } \Omega$$

$$(**) \qquad X \mid_{\partial\Omega} = f.$$

Here $\partial X/\partial u \wedge \partial X/\partial v$ denotes the vector product of $(\partial x_1/\partial u, \partial x_2/\partial u, \partial x_3/\partial u)$ and $(\partial x_1/\partial v, \partial x_2/\partial v, \partial x_3/\partial v)$. The problem can be simplified by setting $X(u, v) = F(u, v) + Y(u, v)$, where $F(u, v)$ is a harmonic 3-vector in Ω satisfying the boundary condition $(*)$. Then we seek a C^2 3-vector $Y(u, v)$ defined over Ω that satisfies $(*)$ and the homogeneous boundary condition $Y \mid_{\partial\Omega} = 0$. To determine the function Y, we suppose that $H < \frac{3}{2}R$ and minimize the functional

$$\mathcal{G}(Y) = \int\int_{\Omega} \left\{ |\nabla Y|^2 + \tfrac{4}{3} MY \cdot \left(\frac{\partial Y}{\partial u} \wedge \frac{\partial Y}{\partial v} \right) \right\} du\, dv$$

over the set $W_R = \mathring{W}_{1,2}(\Omega) \cap \Sigma_R$, where Σ_R denotes the continuous vector functions Y on $\bar{\Omega}$ with ess $\sup_{\bar{\Omega}} |Y(u, v)| \leq R$. Clearly, if $\tilde{Y} \in \mathring{W}_{1,2}(\Omega) \cap \Sigma_R$ and $|\tilde{Y}|_{L_\infty} < R$ attains the infimum of $\mathcal{G}(Y)$ over W_R, then for all test functions $\zeta \in W_{1,2}(\Omega) \cap L_\infty(\Omega)$,

$$\int\int_{\Omega} \left[\nabla Y \cdot \nabla \zeta + 2M \left(\frac{\partial Y}{\partial u} \wedge \frac{\partial Y}{\partial v} \right) \cdot \zeta \right] du\, dv = 0.$$

Consequently, by the regularity results mentioned in Section 1.5 Y will satisfy $(*)$ as well as the homogeneous boundary condition $Y|_{\partial\Omega} = 0$, provided we show that the desired infimum is attained at $\tilde{Y}$ with $|\tilde{Y}| < R$. Actually we prove

Theorem Let $|M| < 1$ and suppose that $\sup_{\partial\Omega} |f| \leq 1$. Then $(*)$ has a solution satisfying $(**)$ as well as the bound $\sup_{\Omega} |X| \leq 1$.

The result is obtained in two steps by first showing that the minimization problem mentioned above has a solution $\tilde{Y}(x)$, and secondly by proving a bound for $\tilde{Y}(x)$ for $|M| < 1$. For the proof we refer the reader to the paper of Hildebrandt and Widman (1971).

B Marston Morse's results on geodesics between two points P, Q on a compact Riemannian manifold (M, g)

The geodesics on S^N with its metric of constant curvature are easily found for two points (P, Q) of S^N that are not antipodal; these geodesics are in one-to-one correspondence with the integers, and when ordered by length and denoted $g_0, g_1, \ldots, g_n, \ldots$, the Morse index of g_n is easily found to be $(N-1)n$; here we can associate the integer n with the number of times g_n contains the antipode of P in its interior. This result has the following generalization for the geodesics between the two fixed points P, Q of $(\mathfrak{M}, g)$ with dim $\mathfrak{M} = N$:

(i) The arc length functional J has critical points if and only if the critical point corresponds to a geodesic of $(\mathfrak{M}, g)$ parametrized by arc length; and the geodesics between P

and Q are all nondegenerate if and only if Q is not a conjugate point of P, and moreover form a set of measure zero of points of $(\mathfrak{M}, g)$.

(ii) Provided the points P and Q are not conjugate relative to $(\mathfrak{M}, g)$, the critical points of J all have finite Morse index, and this index is exactly the number of conjugate points (counted with multiplicity) of one endpoint in the interior of the associated geodesic.

C Use of the calculus of variations in the large for calculations of homotopy groups

The results of Section 6.5 on infinite-dimensional Morse theory were used in Section 6.7 to obtain results about the existence of critical points from topological information. Actually the converse procedure of utilizing knowledge of the critical points of a standard functional (say the arc length functional) defined on a compact manifold $\mathfrak{M}$ to determine facts about the topology of $\mathfrak{M}$ has proved remarkably successful. Many interesting results concerning homotopy groups can be achieved in this way. Thus by utilizing the results of Note B the first part of the Freudenthal suspension theorem (1.6.8) concerning the homotopy groups of spheres can be proven. Much information on the homotopy theory of the classical Lie groups was obtained in this way by Bott and Samelson. For the details of these results we refer the interested reader to the monograph of Milnor (1963).

A typical result in this connection is the following periodicity theorem for the homology of the loop space $\Omega(S^N)$:

$$(*) \qquad H_q(\Omega(S^N)) \approx \begin{cases} \mathbb{Z} & q \equiv 0 \pmod{N-1}, \\ 0 & \text{otherwise}. \end{cases}$$

This periodicity phenomenon is also true for the homotopy groups of many Lie groups as was discovered by Bott. Moreover, from $(*)$ one can compute the cohomology ring $H_*(\Omega(S^N))$ mentioned in Appendix A and conclude from the general critical point theories of Sections 6.5 and 6.6 that for every Riemannian metric on S^N and two distinct points $P, Q \in S^N$, there are an infinite number of geodesics joining P and Q.

D Applications of Ljusternik–Schnirelmann type invariants to equivariant mappings

In Section 6.6 topological invariants (such as the category of a set) were used to investigate the existence of critical points of an even functional. Such topological invariants can also be used to study mapping properties of other equivariant operators. An example is the following generalization of the Borsuk–Ulam theorem [Holm and Spanier, 1971].

Theorem Let C be any compact mapping defined on the unit sphere $\partial\Sigma = \{x \mid x \in X, \|x\| = 1\}$ of a Banach space X, such that for $f = I + Cf(\partial\Sigma)$ lies in a subspace of codimension k of X, then

$$\dim\{x \mid f(x) = f(-x), x \in \partial\Sigma\} \geqslant k-1.$$

E Bibliographic notes

Section 6.1: The results contained here are generalizations of the so-called direct method of the calculus of variations. These methods can be traced back to Lebesgue (1907) and Hilbert (1900). The notion of lower semicontinuity in numerous forms was investigated in the works of Tonelli (1921, 1923). The compactness Condition (C) stated in the text was first stated in Palais and Smale (1964). The illustration of the minimization methods contained in Theorem 6.1.8 can be found in Berger and Schecter (1977), while (6.1.20) can be found in the book Ladyhenskaya and Uralsteva (1968). The results in equation (6.1.30) are contained in the paper Berger and Fraenkel (1970).

Section 6.2: The result (6.2.5) is adapted from Berger (1975) where a more complete discussion of its relation to algebraic manifolds is given. Our discussion of flexible plates and shells can be found in Berger (1967, 1971). Plateau's problem is well discussed in Nitsche (1974). Our proof is a modification of Garabedian (1964). The result (6.2.28) is contained in an unpublished paper of Berger and Wightman.

Section 6.3: The results of this section can be found in Berger (1973) and Berger and Schechter (1977).

Section 6.4: Our discussion of large amplitude periodic solutions of Hamiltonian systems is based on Berger (1971b), and its application to the perturbed Kepler problem is based on unpublished work of Berger and Arensdorf. Our discussion of metrics of prescribed curvature is based on Berger (1975) and Yamabe (1960). These results have been extended by Kazdan and Warner (1975) and by Moser (1973). Our discussion of steady vortex rings can be found in Fraenkel and Berger (1974).

Sections 6.5 and 6.6: Our discussion of Morse theory in Hilbert space is adapted from Rothe (1973) and Smale (1964). A more detailed reference is the paper Palais (1963). The illustration (6.5.16) can be found in Berger and Podolak (1977). Our discussion of the critical point theory of Ljusternik and Schnirelmann can be found in papers by Schwartz (1964) and Palais (1966), and in the book of Ljusternik (1966). A reference for the application of this result to nonlinear eigenvalue problems is Browder (1965). Proof of (6.6.11) is due to Amann (1972).

Section 6.7: The application of the Ljusternik–Schnirelmann theory to bifurcation theory is due to Berger (1970). Later papers on the subject are due to Bohme (1973) and Riddell (1975). The idea contained in (6.7.9) is due to Clark (1973). For the result (6.7.11) see Ambrosetti (1973). The application of general critical point theories to nonlinear elasticity can be found in the papers of Berger (1974). Our discussion of nonlinear stationary states is based on Berger (1972).

F On the multiplicity preservation theorem (6.7.6)

This theorem reiterates the importance of utilizing global topological methods in studying degenerate critical points as contrasted with the algebraic methods of Thom's catastrophe theory. The result has particular importance for the study of "nonlinear" normal modes of a Hamiltonian system near a singular point as described in Chapter 4. In that case, the violation of the Liapunov irrationality conditions associated with jth normal mode (say) of the linearized system implies the eigenvalue λ_j associated with this mode is not simple. The number of such violations can be interpreted as the multiplicity of λ_j, and the preservation theorem implies that the jth normal mode is not destroyed by the nonlinear Hamiltonian perturbation but merely distorted with the preservation of periodicity. This approach has been described in the papers Berger (1969, 1970a). Alternate finite-dimensional approaches to this problem have been carried out by Weinstein and Moser. As mentioned in the text, the virtue of the multiplicity preservation theorem is that it provides a method of "continuing" these nonlinear normal modes to large amplitude. Further results await research on the continuity of the eigenvalue "branches." It can be shown, for example, that the period of the solutions described in Theorem (6.4.2) tends with decreasing amplitude to the smallest nonzero period of a linearized system (if nontrivial).

APPENDIX A

ON DIFFERENTIABLE MANIFOLDS

A set $\mathfrak{M}$ is a *manifold* of dimension N if $\mathfrak{M}$ is a Hausdorff topological space and each point $x \in \mathfrak{M}$ has a neighborhood W_x that is homeomorphic to an open subset of $\mathbb{R}^N$. The set $\mathfrak{M}$ is called a *differentiable manifold* (of class C^k) if $\mathfrak{M}$ can be covered by a family of open sets O_α (called coordinate patches) each homeomorphic (by a C^k map h_α) to an open set in $\mathbb{R}^N$ and such that on the intersection of any two such coordinate patches the change of coordinate mappings $h_\alpha h_\beta^{-1}$: $h_\beta(O_\alpha \cap O_\beta) \to \mathbb{R}^N$ is a smooth map of class C^k. A differentiable manifold $\mathfrak{M}^N$ is called orientable if it can be covered by coordinate patches O_α as above such that the change of coordinate mappings $h_\alpha h_\beta^{-1}$ have positive Jacobian. Connected one-dimensional differentiable manifolds can be easily described.

Any connected differentiable manifold of dimension 1 is diffeomorphic to a circle or some open interval of real numbers.

A subset V of a differentiable manifold $\mathfrak{M}^N$ is an r-dimensional submanifold of $\mathfrak{M}^N$ if there is a family of coordinate patches $\{O_\alpha\}$ covering $\mathfrak{M}^N$ such that $(O_\alpha \cap V)$ is a system of coordinate patches covering V and if $x = (x_1, x_2, \ldots, x_N)$ are local coordinates in O_α, then

$$O_\alpha \cap V = \{x \mid x_{r+1} = x_{r+2} = \cdots = x_N = 0\}.$$

For the solution of problems in differential geometry, it is essential to study calculus on manifolds. For some considerations, this can generally be achieved by introducing *local coordinates* $(x_1, \ldots, x_N)$ (regarded as a point of $\mathbb{R}^N$) in a coordinate patch O_α of M^N. Indeed, for $x \in O_\alpha$ each map h_α: $O_\alpha \to \mathbb{R}^N$ can be written $h_\alpha(x) = (\zeta_1(x), \zeta_2(x), \ldots, \zeta_n(x)) = (x_1, x_2, \ldots, x_N)$.

Moreover, $x = h_\alpha^{-1}(x_1, \ldots, x_N)$. It is also useful to introduce the notion of a *partition of unity* on M^N (i.e., a locally finite covering $\mathcal{V}$ of M and a collection of real-valued nonnegative smooth functions f_V, $V \in \mathcal{V}$ on M such that the support of f_V is contained in V and $\sum_{V \in \mathcal{V}} f_V(x) = 1$). Thus to define the integral of a real-valued function g on M^N with respect to a volume element dV, let (V_j, f_j) be any partition of unity on M^N, and set

$$\int_M g\, dV = \sum_j \int_{V_j} f_j g\, dV.$$

Now each term on the right can be evaluated by using local coordinates since the V_j can be chosen subordinate to the coordinate patches O_α. Moreover, this definition is independent of the partition of unity used since if (W_k, ϕ_k) is any other partition of unity (subordinate to O_α),

$$\sum_j \int_{V_j} f_j g\, dV = \sum_{j,k} \int_{V_j \cap W_k} f_j \phi_k g\, dV = \sum_k \int_{W_k} \phi_k g\, dV.$$

The use of differential forms is essential for the solution of many problems in differential geometry (cf. Section 1.2A). Local coordinates are useful in defining differential forms on a manifold M^N. Indeed, one can begin by defining differential forms on an open set $\Omega \subset \mathbb{R}^N$ and then extending the definition to M^N.

We define differential p-forms of class C^k on an open set Ω of $\mathbb{R}^N$ (denoted $\bigwedge_p^{(k)}(\Omega)$) to be the formal expressions

$$\omega = \sum_{i_1 < i_2 \cdots i_p} c_{i_1 i_2 \cdots i_p}(x)\, dx_{i_1} \wedge dx_{i_2} \wedge \cdots \wedge dx_{i_N}$$

where the functions $c_{i_1 \cdots i_p}(x) \in C^k(\Omega)$ and the integers $i_1, \ldots, i_p$ lie between 1 and N. Two such differential forms may be added componentwise. One also defines the exterior product of two forms $\omega \in \bigwedge_p^{(k)}(\Omega)$ and $\omega' \in \bigwedge_q^{(k)}(\Omega)$ to be $(\omega \wedge \omega') \in \bigwedge_{p+q}^{(k)}(\Omega)$ in the following two steps:

(i) For any permutation σ of the indices $i_1, \ldots, i_k$,

$$dx_{\sigma(i_1)} dx_{\sigma(i_2)} \cdots dx_{\sigma(i_k)} = \operatorname{sgn}(\sigma)\, dx_{i_1} \wedge dx_{i_2} \wedge \cdots \wedge dx_{i_k},$$

where $\operatorname{sgn}(\sigma)$ is the sign of the permutation σ.

(ii) If ω is defined as above and

$$\omega' = \sum_{j_1 < j_2 < \cdots < j_q} b_{j_1 j_2 \cdots j_q} dx_{j_1} \wedge dx_{j_2} \wedge \cdots \wedge dx_{j_q},$$

then

$$\omega \wedge \omega' = \sum c_{i_1 \cdots i_p}(x) b_{j_1 \cdots j_q}(x) (dx_{i_1} \wedge \cdots \wedge dx_{i_p} \wedge dx_{j_1} \wedge \cdots \wedge dx_{j_q}),$$

where the summation is taken over all indices $i_1, \ldots, i_p, j_1, \ldots, j_q$. If $\bigwedge^k(\Omega) = \sum_p \bigwedge_p^{(k)}(\Omega)$, then the $\bigwedge^k(\Omega)$ form a ring under addition and exterior multiplication.

The exterior derivative of a form $\omega \in \bigwedge_p^{(k)}(\Omega)$ defined by (1.6.6) is a form

$d\omega \in \bigwedge_{p+1}^{(K-1)}(\Omega)$ defined by setting $df = \sum_i (\partial f/\partial x_i)\, dx_i$ for any function f and

$$d\omega = \sum_{i_1 < i_2 < \cdots < i_p} d(c_{i_1 \cdots i_p})\, dx_{i_1} \wedge dx_{i_2} \wedge \cdots \wedge dx_{i_p}.$$

In fact, the operator d has the following additional properties:

$$\text{(1)} \qquad \begin{aligned} &\text{(i)} \quad d(\omega + \omega_1) = d\omega + d\omega_1, \\ &\text{(ii)} \quad d(\omega \wedge \omega') = d\omega \wedge \omega' + (-1)^p \omega \wedge d\omega, \\ &\text{(iii)} \quad d^2\omega = d(d\omega) = 0. \end{aligned}$$

Differential forms transform very elegantly under changes of variables. Indeed, if $f = (f_1, \ldots, f_N)$ is a smooth mapping of U onto V, we can define a mapping $f_*\colon \bigwedge_p^{(k)}(\Omega) \to \bigwedge_p^{(k)}(V)$. Indeed, let ω be defined by (1.6.6), then we set

$$f_*(\omega) = \sum_{i_1 < i_2 < \cdots < i_p} c_{i_1 \cdots i_p}(f(x))\, df_{i_1} \wedge \cdots \wedge df_{i_p}.$$

Now the mapping f_* has the crucial properties

$$\text{(2)} \qquad \begin{aligned} &\text{(i)} \quad f_*(\omega + \omega_1) = f_*(\omega) + f_*(\omega_1) \\ &\text{(ii)} \quad f_*(\omega \wedge \omega') = (f_*\omega) \wedge (f_*\omega') \\ &\text{(iii)} \quad d(f_*\omega) = f_* d\omega \\ &\text{(iv)} \quad \text{if } f\colon U \to V \text{ and } g\colon V \to W, \text{ then } (g \circ f)_* = f_* \circ g_*. \end{aligned}$$

Thus, for example, the exterior derivative of a differential form is independent of the coordinate system in which it is computed.

We can define p-differential forms on a manifold $\mathfrak{M}$ of class C^k, denoted $\bigwedge_p^1(\mathfrak{M})$, in a natural way by using the previous discussion and the properties of the coordinate mappings h_α. Then ω is a p-form of class of C^s on $\mathfrak{M}$ if on U, $\omega(x)$ can be written in terms of local coordinates as

$$\omega(x) = \sum_{i_1 < \cdots < i_p} \zeta_{i_1 \cdots i_p}(x)\, dh_{\alpha_{i_1}} \wedge dh_{\alpha_{i_2}} \wedge \cdots \wedge dh_{\alpha_{i_p}}.$$

This definition is independent of the local coordinates used. Furthermore, as above, exterior multiplication and differentiation can be defined on $\bigwedge_p^{(k)}(\mathfrak{M})$ in such a way that they possess properties (1) and (2).

De Rham cohomology groups

Let $\mathfrak{M}$ be a smooth N-dimensional manifold. Suppose $\bigwedge_p(\mathfrak{M}, \mathfrak{N})$ denotes the vector space of smooth differential forms of degree p defined on $\mathfrak{M}$ that vanish in a neighborhood $\mathfrak{N} \subset \mathfrak{M}$. Then the exterior differentiation operator

$$d: \wedge_p(\mathfrak{M}, \mathfrak{N}) \to \wedge_{p+1}(\mathfrak{M}, \mathfrak{N}).$$

With $Z^p(\mathfrak{M}, \mathfrak{N}) = \text{Ker } d$ and $B^p(\mathfrak{M}, \mathfrak{N}) = \text{Range } d$, we define the pth de Rham cohomology group of $(\mathfrak{M}, \mathfrak{N})$ as the quotient

$$H^p(\mathfrak{M}, \mathfrak{N}) = Z^p(\mathfrak{M}, \mathfrak{N})/B^p(\mathfrak{M}, \mathfrak{N}).$$

The following theorem holds.

Theorem The De Rham cohomology groups are topological invariants of the pair $(\mathfrak{M}, \mathfrak{N})$.

To investigate the notion of cup length mentioned in (6.6.9) we use the De Rham cohomology. Actually a ring structure can be defined on $\sum_{p=0}^{\dim} H^p(\mathfrak{M}, \mathfrak{N})$ by virtue of the existence of the exterior product of two differential forms defined on $\mathfrak{M}$. In fact, if $\omega_1 \in \wedge_p(\mathfrak{M}, \mathfrak{N})$ and $\omega_2 \in \wedge_q(\mathfrak{M}, \mathfrak{N})$ are such that $d\omega_1 = d\omega_2 = 0$, then

$$d(\omega_1 \wedge \omega_2) = d\omega_1 \wedge \omega_2 + (-1)^p \omega_1 \wedge d\omega_2 = 0;$$

while if we suppose that $d\omega_1 = 0$ and $\omega_2 = d\omega$, then

$$d(\omega_1 \wedge \omega_2) = \pm \omega_1 \wedge \omega_2.$$

Thus $\omega_1 \wedge \omega_2 \in \wedge_{p+q}(\mathfrak{M}, \mathfrak{N})$, so that $\sum_{p=0}^{N} \wedge_p(\mathfrak{M}, \mathfrak{N})$ is an associative algebra under exterior multiplication with $Z(\mathfrak{M}, \mathfrak{N}) = \sum_{p=0}^{N} Z^p(\mathfrak{M}, \mathfrak{N})$ a subalgebra, and $B(\mathfrak{M}, \mathfrak{N}) = \sum_{p=0}^{N} B^p(\mathfrak{M}, \mathfrak{N})$ an ideal in $Z(\mathfrak{M}, \mathfrak{N})$. Consequently,

$$\sum_p H^p(\mathfrak{M}, \mathfrak{N}) = \sum_p Z^p(\mathfrak{M}, \mathfrak{N})/B^p(\mathfrak{M}, \mathfrak{N})$$

$$\approx Z(\mathfrak{M}, \mathfrak{N})/B(\mathfrak{M}, \mathfrak{N})$$

is also an associative algebra. Hence there is a product ($\cup$, the cup product) between cohomology groups with

$$H^p(\mathfrak{M}, \mathfrak{N}) \cup H^q(\mathfrak{M}, \mathfrak{N}) \to H^{p+q}(\mathfrak{M}, \mathfrak{N}).$$

Then the integer cup length is the maximal number of nonzero elements of $H^*(\mathfrak{M})$ whose cup product does not vanish.

Complex manifolds

We now turn to a consideration of complex structures on differentiable manifolds. A manifold $\mathfrak{M}$ of even dimension $2N$ is called a *complex manifold* if it is possible to cover $\mathfrak{M}$ by coordinate patches in each of which the local $2N$ real coordinates may be expressed as distinguished complex coordinates $z_1, \ldots, z_N$ in such a way that coordinate mappings between local coordinates of overlapping patches can be given by complex analytic functions in $z_1, \ldots z_N$.

Differential forms defining such complex manifolds have special features. Thus (allowing complex coordinates) $\bigwedge_1^{(k)}(\mathfrak{M})$ can be decomposed into two spaces of forms: one spanned by $dz_1, \ldots, dz_N$ called type (1, 0), and the other spanned by the complex conjugates $d\bar{z}_1, d\bar{z}_2, \ldots, d\bar{z}_N$ of type (0, 1) (i.e., $\bigwedge_1^{(k)}(\mathfrak{M}) = \bigwedge_{1,0}^{(k)}(\mathfrak{M}) + \bigwedge_{0,1}^{(k)}(\mathfrak{M})$). Note also that this splitting is invariant under analytic changes of coordinates.

The exterior differentiation operator d then decomposes in a natural way into the sum $d = \partial + \bar{\partial}$, where:

(i) for a function $f = f(z, \bar{z})$, with the summation convention understood,

$$\partial f = (\partial f/\partial z_k)dz_k \qquad \text{and} \qquad \bar{\partial} f = (\partial f/\partial \bar{z}_k)d\bar{z}_k;$$

(ii) $$\partial\left(c\, dz_{i_1} \wedge \cdots \wedge dz_{i_p} \wedge d\bar{z}_{j_1} \wedge \cdots \wedge d\bar{z}_{j_p}\right) = (\partial c)dz_{i_1} \wedge \cdots dz_{i_p} \wedge d\bar{z}_{j_1} \wedge \cdots \wedge d\bar{z}_{j_p}$$

and a similar expression for $\bar{\partial}$.

Thus ∂ maps $\bigwedge_{p,q}(\mathfrak{M}) \rightarrow \bigwedge_{p+1,q}(\mathfrak{M})$ and $\bar{\partial}$ maps $\bigwedge_{p,q}(\mathfrak{M}) \rightarrow \bigwedge_{p,q+1}(\mathfrak{M})$, with $\partial^2 = \bar{\partial}^2 = \partial\bar{\partial} + \bar{\partial}\partial = 0$.

APPENDIX B

ON THE HODGE–KODAIRA DECOMPOSITION FOR DIFFERENTIAL FORMS

On any smooth compact manifold $\mathfrak{M}$ we may introduce a Riemannian metric

$$ds^2 = \sum_{i,j=1}^{N} g_{ij}(x)\ dx_i\ \ dx_j.$$

Relative to this metric, we may define an L_2 inner product $\langle\ ,\ \rangle$ on differential p-forms defined on $\mathfrak{M}$, $\bigwedge_p(\mathfrak{M})$, and a formal adjoint d^{T} of the exterior differential operator d by setting

$$\langle d^{\mathrm{T}}\omega, \omega_1\rangle = \langle \omega, d\omega_1\rangle.$$

Then we define the Laplace–Beltrami operator

$$\Delta = dd^{\mathrm{T}} + d^{\mathrm{T}}d \qquad \text{on} \quad \bigwedge_p(\mathfrak{M}).$$

Clearly Δ maps $\bigwedge_p(\mathfrak{M})$ into itself, is self-adjoint, commutes with d and d^{T}, and $\Delta\omega = 0$ if and only if $d\omega = d^{\mathrm{T}}\omega = 0$. A form ω with $\Delta\omega = 0$ is called harmonic.

We now mention the analogue for compact manifolds of the decomposition of a smooth vector field of compact support defined on $\mathbb{R}^N$.

Theorem If $\omega \in \bigwedge_p^1(\mathfrak{M})$, there is a $(p-1)$-form ω_1, a $(p+1)$-form ω_2, and an harmonic p-form H such that

$$\omega = d\omega_1 + d^{\mathrm{T}}\omega_2 + H.$$

The forms $d\omega_1$, $d^{\mathrm{T}}\omega_2$, and H are unique in the sense that if $\omega = d\bar{\omega}_1 + d^{\mathrm{T}}\bar{\omega}_2 + \bar{H}$, then $d\bar{\omega}_1 = d\omega_1$, $d^{\mathrm{T}}\omega_2 = d^{\mathrm{T}}\bar{\omega}_2$, and $H = \bar{H}$. Moreover the dimension of the vector space of harmonic p forms is finite dimensional and equal to the pth Betti number R_p of $\mathfrak{M}$.

In this connection, the following inequality is of great importance.

Suppose $\omega \in \bigwedge_p^\infty(\mathfrak{M})$ is such that the harmonic part of ω, $H(\omega) = 0$, then there is an absolute constant $c > 0$ such that

$$c\langle \omega, \omega \rangle \leqslant \langle d\omega, d\omega \rangle + \langle d^{\mathrm{T}}\omega, d^{\mathrm{T}}\omega \rangle.$$

We now define Green's operator $G(\omega)$ for any $\omega \in \bigwedge_p^\infty(\mathfrak{M})$ as the solution u of

$$\Delta u = \omega - H(\omega).$$

The solution of this equation is unique, and in the Hilbert space $\mathcal{H}^p(\mathfrak{M})$ obtained by completing $\bigwedge_p(\mathfrak{M})$ relative to the norm

$$\|\omega\|^2 = \langle d\omega, d\omega \rangle + \langle d^{\mathrm{T}}\omega, d^{\mathrm{T}}\omega \rangle,$$

the solution u exists (in the generalized sense), and by the ellipticity of Δ, in the classical sense. Thus we find for $\omega \in \bigwedge_p(\mathfrak{M})$, the Hodge–Kodaira decomposition

$$\omega = dd^{\mathrm{T}}G(\omega) + d^{\mathrm{T}}dG(\omega) + H(\omega).$$

On a complex manifold $\mathfrak{M}$ one may introduce a real analytic Riemannian metric g that in terms of the distinguished complex analytic coordinates can be written as

$$ds^2 = 2\sum_{j,k} g_{jk}\, dz_j\, d\bar{z}_k.$$

Such a metric is called Hermitian. A Hermitian metric may be introduced on any compact complex manifold.

A Hermitian metric on $\mathfrak{M}$

$$g = 2\sum_{j,k} g_{jk}\, dz_j \wedge d\bar{z}_k$$

is called Kähler, if $dg = 0$. In that case, the complex Laplacean $\square$ is written $\square f = \sum g^{i\bar{j}}\, \partial^2 f/\partial z_i\, \partial\bar{z}_j$, and the following result holds

If $(\mathfrak{M}, g)$ is a Kähler manifold, $\square$ defined as above satisfies $\square = \frac{1}{2}\Delta$, where Δ is the Laplace–Beltrami operator defined on the real $2N$-dimensional Riemannian manifold $(\mathfrak{M}, g)$.

In the general Hermitian case, we may again define the Hodge–Kodaira decomposition by following the argument used in the Riemannian case, with $\square$ replacing Δ. Indeed, if the $(0, q)$-forms are denoted $\bigwedge_{0,q}(\mathfrak{M})$, and the harmonic (scalar or vector valued) $(0, q)$-forms are those $(0, q)$-forms ω for which $\square\omega = 0$, then we have a Green's function G such that

$$\omega = \bar{\partial}\bar{\partial}^{\mathrm{T}}G(\omega) + \bar{\partial}^{\mathrm{T}}\bar{\partial}G(\omega) + H(\omega),$$

where $H(\omega)$ is the harmonic part of ω and $G(\omega)$ is the unique solution u of $\square u = \omega - H(\omega)$.

We note an L_2 scalar product $\langle\ ,\ \rangle$ for forms of the same type and correspondingly, a formal adjoint $\bar{\partial}^{\mathrm{T}}$ of $\bar{\partial}$ defined by

$$\langle \bar{\partial}^{\mathrm{T}}\omega, \omega_1\rangle = \langle \omega, \bar{\partial}\omega_1\rangle.$$

$\bar{\partial}^{\mathrm{T}}$ maps $\wedge_{p,q}(\mathfrak{M}) \to \wedge_{p,q-1}(\mathfrak{M})$ and $(\bar{\partial}^{\mathrm{T}})^2 = 0$. In addition, the complex Laplacean

$$\square = \bar{\partial}\bar{\partial}^{\mathrm{T}} + \bar{\partial}^{\mathrm{T}}\bar{\partial}$$

preserves $\wedge_{p,q}(\mathfrak{M})$, commutes with $\bar{\partial}$ and $\bar{\partial}^{\mathrm{T}}$, and is a strongly elliptic partial differential operator on forms of any fixed type.

For a discussion and proof of these results we refer the reader to the book by Kodaira and Morrow (1971).

REFERENCES

Agmon, S. (1965). "Lectures on Elliptic Boundary Value Problems." Van Nostrand-Reinhold, Princeton, New Jersey.

Agmon, S., Douglis, A., and Nirenberg, L. (1959). Estimates near the boundary for solutions of elliptic partial differential equations satisfying general boundary conditions, I, *Comm. Pure Appl. Math.* **12**, 623–727.

Agmon, S., Douglis, A., and Nirenberg, L. (1964). Estimates near the boundary for solutions of elliptic partial differential equations satisfying general boundary conditions, II, *Comm. Pure Appl. Math.* **17**, 35–92.

Alber, S. (1970). The topology of functional manifolds and the calculus of variations in the large, *Russian Math Surveys* **25**, 51–117.

Alexiewicz, A., and Orlicz, W. (1954). Analytic operations in real Banach spaces, *Studia Math.* **14**, 57–78.

Ambrosetti, A. (1973). Esistenza di infinite solutioni per problemi non lineari, *Atti Accad. Naz. Lincei Mem. Cl. Sci. Fis. Mat. Natur. Sez. I* **52**, 660–667.

Ambrosetti, A., and Prodi, G. (1972). On the inversion of some differentiable mappings with singularities between Banach spaces, *Annali di Math.* **93**, 231–246.

Amann, H. (1972). Ljusternik–Schnirelmann theory and nonlinear eigenvalue problems, *Math. Ann.* **199**, 55–72.

Amann, H. (1976). Fixed point theorems and nonlinear eigenvalue problems in ordered Banach spaces, *SIAM Rev.* **18**, 620–709.

Amann, H. (1976). "Nonlinear Operators in Ordered Banach Spaces and Some Applications to Nonlinear Boundary Value Problems" (Lect. Notes in Math.). Springer-Verlag, New York.

Appel, P. (1921). "Mecanique Rationnelle," Vol. IV. Gauthier-Villars, Paris.

Banach, S. (1920). Thesis, published in *Fund. Math.* **3**, 133–181.

Banach, S., and Mazur, S. (1934). Uber mehrdeutige stetige abbildungen, *Studia Math.* **5**, 174–178.

Bartle, R. (1953). Singular points in functional equations, *Trans. Amer. Math. Soc.* **75**, 366–384.

Batchelor, G. (1967). "Introduction to Fluid Dynamics." Cambridge Univ. Press, London and New York.

Benjamin, T. B. (1971). A unified theory of conjugate flows, *Philos. Trans. Roy. Soc.* **269**, 587–647.

Berger, M. S. (1965). An eigenvalue problem for non-linear elliptic partial differential equations. *Trans. Amer. Math. Soc.* **120**, 145–184.

Berger, M. S. (1967). On Von Kármán's equations and the buckling of a thin elastic plate, I, *Comm. Pure Appl. Math.* **20**, 687–719.

Berger, M. S. (1969). A bifurcation theory for real solutions of non-linear elliptic partial differential equations. *In* "Bifurcation Theory and Nonlinear Eigenvalues" (J. Keller and S. Antman, eds.), pp. 113–216. Benjamin, Reading, Massachusetts.

Berger, M. S. (1970a) On multiple solutions of non-linear operator equations arising from the calculus of variations, *Amer. Math. Soc. Proc. Symp.* **17**, 10–27.

Berger, M. S. (1970b). On stationary states for a nonlinear wave equation, *J. Math. Phys.* **11**, 2906–2912.

Berger, M. S. (1971a). On Riemannian structures of prescribed Gaussian curvature for compact 2-manifolds, *J. Differential Geometry* **5**, 325–332.

Berger, M. S. (1971b). On a family of periodic solutions of Hamiltonian systems, *J. Differential Equations* **10**, 17–26.

Berger, M. S. (1971c). Periodic solutions of second order dynamical systems and isoperimetric varational problems, *Amer. J. Math.* **93**, 1–10.

Berger, M. S. (1972). On the existence and structure of stationary states for a nonlinear Klein-Gordon equation, *J. Functional Analysis* **9**, 249–261.

Berger, M. S. (1973). Applications of global analysis to specific non-linear eigenvalue problems, *Rocky Mountain J. Math.* **3**, 319–354.

Berger, M. S. (1974). New applications of the calculus of variations in the large to non-linear elasticity, *Comm. Math. Phys.* **35**, 141–150.

Berger, M. S. (1975). Constant scalar curvature metrics for complex manifolds, *Proc. Amer. Math. Soc. Inst. Differential Geometry* **27**, 153–170.

Berger, M. S., and Berger, M. S. (1968). "Perspectives in Nonlinearity." Benjamin, New York.

Berger, M. S., and Fife, P. (1968). On Von Kármán's equations and the buckling of a thin elastic plate, II, Plate with general boundary conditions, *Comm. Pure Appl. Math.* **12**, 227–247.

Berger, M. S., and Fraenkel, L. E. (1970). On the asymptotic integration of a nonlinear Dirichlet problem, *J. Math. Mech.* **19**, 553–585.

Berger, M. S., and Fraenkel, L. E. (1971). Singular perturbations of non-linear operator equations, *J. Math. Mech.* **20**, 623–631.

Berger, M. S., and Fraenkel, L. E. (1976). Applications of the calculus of variations in the large to free boundary problems of continuum mechanics, *Proc. Conf. Appl. Functional Anal. to Continuum Mech.* (Lect. Notes in Math. **503**), pp. 186–193. Springer-Verlag, New York.

Berger, M. S., and Meyers, N. G. (1971). Generalized differentiation and utility functions. *In* "Preference Utility and Demand." Harcourt, New York.

Berger, M. S., and Plastock, R. (1977). On proper non-linear Fredholm operators (to appear).

Berger, M. S., and Podolak, E. (1974). On nonlinear Fredholm operator equations, *Bull. Amer. Math. Soc.* **80**, 861–864.

Berger, M. S., and Podolak, E. (1975). On the solutions of a non-linear Dirichlet problem, *Indiana J. Math.* **24**, 837–846.

Berger, M. S., and Podolak, E. (1977). On the homotopy groups of spheres and nonlinear Fredhold operator equations (to appear).

Berger, M. S., and Schechter, M. (1972). Embedding theorems and quasilinear elliptic boundary value problems for unbounded domains, *Trans. Amer. Math. Soc.* **172**, 261–278.

Berger, M. S., and Schechter, M. (1977). On the solvability of semilinear gradient operator equations, *Advances in Math.* (to appear).

Berger, M. S., and Westreich, D. (1973). A convergent iteration scheme for bifurcation theory on Banach space, *J. Math. Anal. Appl.* **43**, 136–144.

Bers, L. (1957). Topology (Lect. Notes). Courant Inst., New York.

Bers, L., John, F., and Schechter, M. (1964). "Partial Differential Equations." Wiley, New York.

Beurling, A., and Livingston, A. E. (1962). A theorem on duality mappings in Banach spaces, *Ark. Mat.* **4**, 405–411.

Birkhoff, G. D. (1927). "Dynamical Systems" (Colloq. Publ. **9**), Amer. Math. Soc., Providence, Rhode Island.

Birkhoff, G., and Zarantonello, E. (1957). "Jet, Wakes and Cavities." Academic Press, New York.

Böhme, R. (1971). Nichtlineare störung der isolierten eignewerte selbstadjungierter operatoren, *Math. Z.* **123**, 61–92.

Böhme, R. (1972). Die lösung der verzweigungsgleichungen für nichtlineare eigenwertprobleme, *Math. Z.* **127**, 105–126.

Bourbaki, N. (1949). "Elements de Mathematique." Hermann, Paris.

Brézis, H. (1968). Equations et inequations non lineaires dans les espaces vectoriels en dualité, *Ann. Inst. Fourier (Grenoble)* **18**, 115–175.

Brézis, H. (1973). "Operateurs Maximaux Monotones et Semi-Groupes de Contractions Dans les Espaces de Hilbert" (Notas de Mat. **50**). American Elsevier, New York.

Brout, R. (1965). "Phase Transitions." Benjamin, Reading, Massachusetts.

Browder, F. E. (1954). Covering spaces, fibre spaces and local homeomorphisms, *Duke Math. J.* **21**, 329–336.

Browder, F. E. (1965). Infinite dimensional manifolds and non-linear elliptic eigenvalue problems, *Ann. of Math.* **82**, 459–477.

Browder, F. E. (1968). Nonlinear eigenvalue problems and Galerkin approximations, *Bull. Amer. Math. Soc.* **74**, 651–656.

Browder, F. E. (1976). Nonlinear operators in Banach spaces, *Proc. Symp. P. M.* **18**, pt. 2. Amer. Math. Soc., Providence, Rhode Island.

Cacciopoli, R. (1931). Problemi non lineari in analisis funzionale, *Rend. Sem. Nat. Roma* **1**, 13–22.

Cartan, H. (1940). Sur les matrices holomorphes de variables complexes, *J. Math. Pures Appl.* **19**, 1–26.

Cartan, H. (1966). Some applications of the new theory of Banach analytic spaces, *J. London Math. Soc.* **41**, 70–78.

Cartan, H. (1970). "Differential Forms." Hermann, Paris.

Cartan, H. (1971). "Differential Calculus." Hermann, Paris.

Cauchy, A. (1847). Méthode genéral pour la resolution des systemes d'equations simultanées, *C. R. Acad. Sci. Paris* **25**.

Cesari, L., and Kannan, R., (eds.) (1976). "Nonlinear Functional Analysis and Its Applications." Dekker, New York.

Clark, D. (1973). A variant of the Ljusternik–Schnirelmann theory, *Indiana J. Math.* **22**, 65–74.

Courant, R. (1950). "Dirichlet's Principle." Wiley (Interscience), New York.

Crandall, M. G., and Rabinowitz, P. H. (1971). Bifurcation from simple eigenvalues, *J. Functional Analysis* **8**, 321–340.

Cronin, J. (1953). Analytic functional mappings, *Ann. of Math.* **58**, 175–181.

Cronin, J. (1964). "Fixed Points and Topological Degree in Nonlinear Analysis" (Math. Surveys **11**). Amer. Math. Soc., Providence, Rhode Island.

Cronin, J. (1972). Eigenvalues of some nonlinear operators, *J. Math. Anal.* **38**, 659–667.

Cronin, J. (1973). Equations with bounded nonlinearities, *J. Differential Equations* **14**, 581–596.

Dancer, E. N. (1971). Bifurcation theory in real Banach spaces, *J. London Math. Soc.* **23** (3), 699–734.

De Villiers, J. M. (1973). A uniform asymptotic expansion of the positive solution of a non-linear Dirichlet problem, *Proc. London Math. Soc.* **27**, 701–722.

Dieudonné, J. (1960). "Foundations of Modern Analysis." Academic Press, New York. Enlarged and Corrected edition, 1969.
Deprit, A., and Henrard, J. (1968). A manifold of periodic orbits, *Advances in Astron. Astrophys.* **6**, 2–124.
Douady, A. (1965). Le Probleme des Modules. Sem. Coll. du France.
Dunford, N., and Schwartz, J. (1958). "Linear Operators," Part I; (1963), Part II. Wiley, New York.
Eells, J. (1966). A setting for global analysis, *Bull. Amer. Math. Soc.* **72**, 751–807.
Eells, J. (1970). Fredholm structures, *Symp. Nonlinear Functional Anal.* **18**, 62–85. Amer. Math. Soc., Providence, Rhode Island.
Ekeland, I., and Temam, R. (1976). "Convex Analysis and Variational Problems." North-Holland Publ., Amsterdam.
Elworthy, K., and Tromba, A. (1970). Differential structures and Fredholm maps, *Proc. Symp. Global Anal.* **15**, 45–94.
Einstein, A. (1955). "The Meaning of Relativity." Princeton Univ. Press, Princeton, New Jersey.
Faddeev, L., and Zakharov, V. (1971). Korteweg–de Vries equation as a completely integrable Hamiltonian system, *J. Math. Phys.* **12**, 1548–1551.
Federer, H. (1969). "Geometric Measure Theory." Springer-Verlag, New York.
Fife, P. (1973). Semilinear elliptic boundary value problems with small parameters, *Arch. Rational Mech. Anal.* **52**, 205–232.
Förster, O. (1975). Power series methods in deformation theorems (Lect. notes, AMS Summer Inst. Several Complex Variables).
Fraenkel, L. E. (1962). Laminar flow in symmetrical channels with slightly curved walls, II, *Proc. Roy. Soc.* **A272**, 406–428.
Fraenkel, L. E. (1973). On a theory of laminar flow in channels of a certain class, *Proc. Cambridge Philos. Soc.* **73**, 361–390.
Fraenkel, L. E., and Berger, M. S. (1974). On the global theory of vortex rings in an ideal fluid, *Acta Math.* **32**, 13–51.
Fréchet, M. (1906). Sur quelque points du calcul fonctionnel, *Rend. Circ. Mat. Palermo* **22**, 1–74.
Friedman, A. (1969). "Partial Differential Equations." Holt, New York.
Friedrichs, K. O. (1955). Asymptotic phenomena in mathematical physics, *Bull. Amer. Math. Soc.* **59**, 485–504.
Fucik, S., Necas, J., and Soucek, V. (1973). "Spectral Analysis of Nonlinear Operators" (Lect. Notes in Math. **343**). Springer-Verlag, New York.
Fujita, H. (1961). On the existence and regularity of the steady solutions of the Navier–Stokes equation, *J. Fac. Sci. Univ. Toyko* **9**, 59–102.
Garabedian, P. (1964). "Partial Differential Equations." Wiley, New York.
Gateaux, R. (1922). Sur les fonctionnelles continues et les fonctionnelles analytique, *Bull. Soc. Math. Fr.* **50**, 1–21.
Geba, K. (1964). Algebraic topology methods in the theory of compact fields, *Fund. Math.* **54**, 177–209.
Geba, K., and Granas, A. (1973). Infinite dimensional cohomology theories, *J. Math. Pure Appl.* **52**, 145–270.
Goldring, T. (1977). Thesis, Yeshiva Univ., New York.
Gordon, W. B. (1971). *J. Differential Equations* **10**, 324–335.
Gordon, W. B. (1972). *Amer. Math. Monthly* **79**, 755–759.
Gortler, H., Kirchgassner, K., and Sorger, P. (1963). Branching solutions of the Benard problem, "Problems of Continuum Mechanics." SIAM, Providence, Rhode Island.

Granas, A. (1961). Introduction to Topology of Functional Spaces (Univ. of Chicago lecture notes).

Granas, A. (1969). Topics in Infinite Dimensional Topology. Sem. Coll. de France.

Graves, L. (1950). Some mapping theorems, *Duke Math. J.* **17**, 111–114.

Gross, L. (1964). Classical analysis on Hilbert space. *In* "Analysis on Function Space" (T. Martin and I. Segal, eds.), pp. 51–68. MIT Press, Cambridge Massachusetts.

Hadamard, J. (1904). Sur les equations fonctiouelles, C. R. *Acad. Sci. Paris Sér. A–B* **136**, 351.

Hale, J. K. (1969). "Ordinary Differential Equations." Wiley (Interscience), New York.

Hartman, P. (1967). On homotopic harmonic maps, *Canad. J. Math.* **19**, 673–687.

Heissenberg, W. (1967). Nonlinear problems in physics, *Phys. Today* (May), pp. 27–33.

Hilbert, D. (1900). Address, Internat. Congr. Math., Paris.

Hildebrandt, S., and Widman, K. O. (1971). On the Dirichlet problem for surfaces of constant mean curvature, *Math. Ann.*

Hille, E. (1948). "Functional Analysis and Semigroups" (Colloq. Publ. **31**). Amer. Math. Soc., Providence, Rhode Island.

Hille, E., and Phillips, R. (1957). "Functional Analysis and Semigroups" (Colloq. Publ. **31**). Amer. Math. Soc., Providence, Rhode Island.

Hilton, P. (1953). "Introduction to Homotopy Theory." Cambridge Univ. Press, London and New York.

Holm, P., and Spanier, E. H. (1971). Involutions and Fredholm maps, *Topology* **10**, 203–218.

Hopf, E. (1951). Über die Anfangswertaufgabe für die hydro gleichungen, *Math. Nachr.* **4**, 213–231.

Hormander, L. (1966). "An Introduction to Several Complex Variables." Van Nostrand-Reinhold, Princeton, New Jersey.

Hu, S. (1959). "Homotopy Theory." Academic Press, New York.

Ize, G. (1975). Bifurcation Theory for Fredholm Operators. Thesis, New York Univ.

John, F. (1968). On quasi-isometric mappings, I, *Comm. Pure Appl. Math.* **21**, 77–110.

Joseph, D. (1976). "Theory of Stability of Viscuous Fluids." Springer-Verlag, New York.

Judovitch, V. I. (1967). Free convection and bifurcation, *Appl. Math. Mech., Trans. PMM* **31**, 101–111.

Judovitch, V. I. (1966). Secondary flows and fluid instability between rotating cylinders, *Appl. Math. Mech., Trans. PMM* **30**, 688–698.

Karlin, S. (1968). "Total Positivity." Stanford Univ. Press, Palo Alto, California.

Kazdan, J., and Warner, F. (1974). Curvature functions for compact 2-manifolds, *Ann. of Math.* **99**, 14–47.

Kazdan, J., and Warner, F. (1975). Scalar curvature and conformed deformation, *J. Differential Geometry* **10**, 113–134.

Keller, J. B., and Antman, S., (eds.) (1969). "Bifurcation Theory and Nonlinear Eigenvalue Problems." Benjamin, Reading, Massachusetts.

Kelvin, and Tait, P. (1879). "Treatise on Natual Philosophy." Cambridge Univ. Press, London and New York.

Kirchgässner, K., and Sorger, P. (1969). Branching analysis for the Taylor problem, *Quart. J. Mech. Appl. Math.* **22**, 183–210.

Kodaira, K., and Morrow, J. (1971). "Complex Manifolds." Holt, New York.

Kolomogrov, A. (1954). Theorie generale des systemes dynamique et mechanique classique, *Proc. Internat. Congr. Math., Amsterdam.*

Krasnoselski, M. A. (1965). "Topological Methods in the Theory of Nonlinear Integral Equations." Pergamon, Oxford.

Krasnoselski, M. A., *et al.* (1972). "Approximate Solution of Operator Equations." Noordhoff, Groningen.

Krasovskii, J. P. (1961). On the theory of steady-state waves of finite amplitude, *Z. Vycisl. Mat. Mat. Fiz.* **1**, 836–855 (in Russian).
Kupka, I. (1965). Counterexample to the Morse–Sard theorem in the case of infinite-dimensional manifolds, *Proc. Amer. Math. Soc.* **16**, 954–957.
Kuranishi, M. (1965). New proof for the existence of locally complete families of complex structures, *Proc. Cont. Compl. Analy.* Springer-Verlag, Berlin and New York.
Ladyhenskaya, O. (1969). "Mathematical Theory of Viscous and Compressible Flow" (2nd ed.). Gordon & Breach, New York.
Ladyhenskaya, O., and Uralsteva, N. (1968). "Linear and Quasilinear, Elliptic Partial Differential Equations." Academic Press, New York.
Landau, L. (1937). On the theory of phase transitions, *Phys. Z. Sov. Univ.* **11**, 26–39.
Landau, L. (1944). On the problem of turbulence, *Dokl. Akad. Nauk USSR* **44**, 339–342.
Lang, S. (1972). "Differentiable Manifolds." Addison-Wesley, Reading, Massachusetts.
Landesman, E., and Lazar, A. (1970). Nonlinear perturbations of linear eigenvalue problems at resonance, *J. Math. Mech.* **19**, 609–623.
Lax, P. (1968). Integrals of nonlinear equations of evolution and solitary waves, *Comm. Pure Appl. Math.* **21**, 467–490.
Lebesgue, H. (1907). Sur le probleme de Dirichlet, *Rend. Circ. Math. Palermo* **24**, 371–402.
Leray, J. (1933). Etude de diverses equations integrales non lineaires, *J. Math.* **12**, 1–82.
Leray, J. (1952). "La Theorie des Points Fixes et ses Applications en Analyse." pp. 202–208. Amer. Math. Soc., Providence, Rhode Island.
Leray, J., and Schauder, J. (1934). Topologie et equations fonctionnelles, *Ann. Sci. École. Norm. Sup.* **51**, 45–78.
Levi-Civita, T. (1925). Determination rigoureuse des ondes permanentes d'ampleur finie, *Math. Ann.* **93**, 264–314.
Liapunov, A. (1906). Sur les figures d'equilibrium, *Acad. Nauk St. Petersberg* pp. 1–225.
Liapunov, A. (1907). Problem general de la stabilite du mouvement, *Ann. Fac. Sci. Univ. Toulouse* **17**, 203–474.
Lichtenstein, L. (1931). Vorlesungen über einige Klassen nichtlinearen Intergralgleichungen und Intergrodifferentialgleichungen nebst Anwendungen, Berlin.
Lichtenstein, L. (1933). "Gleichgewichtsfiguren Rotierender Flussigkeiten." Springer-Verlag, Berlin.
Lions, J. L. (1969). "Quelques Method de Resolution des Problemes aux Limites Non-lineaires." Dunod, Paris.
Littman, W. (1967). A connection between α-capacity and m–p polarity, *Bull. Amer. Math. Soc.* **73**, 862–866.
Ljusternik, L. (1966). "Topology of the Calculus of Variations in the Large" (Amer. Math. Soc. Transl. **16**), Amer. Math. Soc., Providence, Rhode Island.
Ljusternik, L., and Schnirelmann, L. (1930). "Method Topologique Dans Les Problémes Variationelles" (Actualités Sci. Indust. **188**). Hermann, Paris.
Ljusternik, L., and Sobolev, V. (1961). "Elements of Functional Analysis." Ungar, New York.
Loginov, B. V., and Trenogin, V. A. (1972). The use of group properties to determine multi-parameter families of solutions of nonlinear equations, *Math. Sb.* **14**, 438–452.
Malgrange, B. (1969). Sur l'integrabilité des structures presque complexes, *Symp. Math.* **2**. Academic Press, New York.
Michal, A. (1958). "Le calcul Differential dans les Espaces de Banach." Gauthier-Villars, Paris.
Milnor, J. (1963). "Morse Theory." Princeton Univ. Press, Princeton, New Jersey.
Milnor, J. (1965). "Topology from the Differentiable Viewpoint." Univ. of Virginia Press, Charlottesville, Virginia.

Minty, G. J. (1962). Monotone (nonlinear) operators in a Hilbert space, *Duke Math. J.* **29**, 341–346.

Morrey, C. Jr. (1966). "Multiple Integrals in the Calculus of Variations." Springer-Verlag, New York

Morse, M. (1934). "The Calculus of Variations in the Large" (Colloq. Publ. **18**), Amer. Math. Soc., Providence, Rhode Island.

Morse, M., and Cairns, S. (1969). "Critical Point Theory in Global Analysis." Academic Press, New York.

Moser, J. (1966). A rapidly convergent iteration method and nonlinear partial differential equations, I, *Ann. Scuola Norm. Sup. Pisa* **20**, 226–315; II, 449–535.

Moser, J. (1971). A sharp form of an inequality of N. Trudinger, *Indiana Univ. Math. J.* **20**, 1077–1092.

Moser, J. (1973a). On a nonlinear problem in differential geometry. *In* "Dynamical Systems" (M. Peixoto, ed.). Academic Press, New York.

Moser, J. (1973b). "Stable and Random Motions in Dynamical Systems." Princeton Univ. Press, Princeton, New Jersey.

Moser, J. (1976). Periodic orbits near an equilibrium and a theorem by Alan Weinstein. *Comm. Pure Appl. Math.* **29**, 727–747.

Nash, J. (1956). The imbedding problem for Riemannian manifolds, *Ann. of Math.* **63**, 20–63.

Nevanlinna, F., and Nevanlinna, R. (1957). "Absolute Analysis." Springer-Verlag, Berlin.

Newlander, A., and Nirenberg, L. (1957). Complex coordinates in almost complex manifolds, *Ann. of Math.* **65**, 391–404.

Nirenberg, L. (1959). On elliptic partial differential equations, *Ann. Scuola Norm. Sup. Pisa* **13**, 115–162.

Nirenberg, L. (1964). Partial differential equations with application in geometry. *In* "Lectures in Modern Math" (T. Saaty, ed.), Vol. 2. Wiley, New York.

Nirenberg, L. (1971). An application of generalized degree to a class of nonlinear problems, *3rd Colloq. Anal. Fonct., Liege Centre Belge de Recherches Math.* pp. 57–73.

Nirenberg, L. (1972). An abstract form of the Cauchy–Kowalewski theorem, *J. Differential Geometry* **6**, 561–576.

Nirenberg, L. (1973). Lectures on linear partial differential equations, Reg. Conf. 17. Amer. Math. Soc., Providence, Rhode Island.

Nirenberg, L. (1974). Topics in Nonlinear Functional Analysis (Lecture Notes). New York. Univ.

Nitsche, J. C. C. (1974). "Minimal Surfaces." Springer-Verlag, New York.

Nussbaum, R. (1972). Some asymptotic fixed point theorems, *Trans. Amer. Math. Soc.* **171**, 349–375.

Palais, R. (1963). Morse theory on Hilbert manifolds, *Topology* **2**, 299–340.

Palais, R. (1966). Ljusternik–Schnirelmann theory on Banach manifolds, *Topology* **5**, 115–132.

Palais, R. (1967). "Foundations of Global Nonlinear Analysis." Benjamin, Reading, Massachusetts.

Palais, R., and Smale, S. (1964). A generalized Morse theory, *Bull. Amer. Math. Soc.* **70**, 165–171.

Petryshyn, W. V. (1970). Nonlinear equations involving non-compact operators, *Proc. Symp. Pure Math.* **18**, pt. 1, 206–233. Amer. Math. Soc., Providence, Rhode Island.

Pimbley, G. H. (1969). "Eigenfunction Branches of Nonlinear Operators and Their Bifurcations." Springer-Verlag, New York.

Pitcher, E. (1958). Inequalities of critical point theory, *Bull. Amer. Math. Soc.* **64**, 1–30.

Plastock, R. (1972). Thesis, Yeshiva Univ., New York.

Plastock, R. (1974). Homeomorphisms between Banach spaces, *Trans. Amer. Math. Soc.* **200**, 169–183.

Podolak, E. (1974). Thesis, Yeshiva Univ., New York.

Podolak, E. (1976). On asymptotic nonlinearities, *Indiana J. Math.*.

Podolak, E. (1977). On the range of operator equations with an asymptotically linear term, *Indiana J. Math.* (to appear).

Pohozaev, S. I. (1967). The solvability of nonlinear equations with odd operators, *Funkcional. Anal. Prilozen.* **1**, 66–73.

Pohozaev, S. I. (1968). The set of critical values of a functional, *Math. USSR* **4**, 93–98.

Poincaré, H. (1885). Les figures equilibrium, *Acta Math.* **7**, 259–302.

Poincaré, H. (1890). Les fonctions fuchsiennes et l'equation, *Oeurves* **1**, 512–591.

Poincaré, H. (1892). "Les Methodes Nouvelles de la Mechanique Celeste," Vols. 1–3. Gauthier-Villiars, Paris.

Poincaré, H. (1905). Sur les lignes geodesiques, *Trans. Amer. Math. Soc.* **6**, 237–274.

Prodi, G. (1967). Problemi di diramazione per equazioni funzionali, *Boll. Un. Mat. Ital.* **22**, 413–433.

Protter, M., and Weinberger, H. (1967). "Maximum Principles in Partial Differential Equations." Prentice-Hall, Englewood Cliffs, New Jersey.

Rabinowitz, P. H. (1971). Some global results for nonlinear eigenvalue problems, *J. Functional Analysis* **7**, 487–513.

Rabinowitz, P. H. (1973). Some aspects of nonlinear eigenvalue problems, *Rocky Mountain J. Math.* **3**, 161–202.

Rabinowitz, P. H. (1974). Pairs of positive solutions for nonlinear elliptic partial differential equations, *Indiana Univ. Math. J.* **23**, 173–186.

Rabinowitz, P. H. (1975). Theorie du Degree Topologique et Applications (Lecture Notes).

Riddell, R. C. (1975). Nonlinear eigenvalue problems and spherical fibrations of Banach spaces, *J. Functional Analysis* **18**, 213–270.

Riesz, R., and Nagy, B. (1952). Lecons d'analyse fonctionelle, *Akad. Kiado Budapest.*

Rosen G. (1969). "Formulations of Classical and Quantum Dynamical Theory," Academic Press, New York.

Rosenbloom, P. C. (1956). The method of steepest descent, *Symp. Appl. Math., Providence, Rhode Island* **6**, 127–176.

Rosenbloom, P. C. (1961). The majorant method, *Proc. Symp. Pure Math.* **4**, 51–72.

Rothe, E. (1951). A relation between the type numbers and the index, *Math. Nachr.* **4**, 12–27.

Rothe, E. (1953). Gradient mappings, *Bull. Amer. Math. Soc.* **59**, 5–19.

Rothe, E. (1973). Morse theory in Hilbert space, *Rocky Mountain J. Math.* **3**, No. 2, 251–274.

Ruelle, D., and Takens, F. (1971). On the nature of turbulence, *Comm. Math. Phys.* **20**, 167–192.

Sather, D. (1973). Branching of solutions of nonlinear equations in Hilbert space, *Rocky Mountain Math. J.* **3**, 203–250.

Sattinger, D. H. (1971). Stability of bifurcating solutions by Leray–Schauder degree, *Arch. Rational Mech. Anal.* **43**, 154–166.

Sattinger, D. H. (1973). "Topics in Stability and Bifurcation Theory." (Lect. Notes in Math. **309**). Springer-Verlag, New York.

Schauder, J. (1927). Zur theorie stetiger abbildungen in funktionalraumen, *Math. Z.* **26**, 417–431.

Schauder, J. (1929). Invarianz des gebietes in funktionalraumen, *Studia Math.* **1**, 123–139.

Schauder, J. (1930). Der fixpunktsatz in funktionalraumen, *Studia Math.* **2**, 171–180.

Schechter, M. (1971). "Principles of Functional Analysis." Academic Press, New York.

Schmidt, E. (1908). Zur theorie der linearen und nichtlinearen integralgleichungen, III, *Math. Ann.* **65**, 370–399.

Schwartz, J. (1963). Compact analytic mapping of *B*-spaces, *Comm. Pure Appl. Math.* **16**, 253–260.

Schwartz, J. (1964). Generalizing the Ljusternik–Schnirelmann theory of critical points, *Comm. Pure Appl. Math.* **17**, 807–815.

Schwartz, J. (1969). "Nonlinear Functional Analysis." Gordon and Breach, New York.

Seifert, H., and Threfall, W. (1938). "Variations in Grossen" (reprint). Chelsea, New York.

Sergeraert, F. (1972). Un theoreme de fonctions implicites sur certains espaces de frechet et quelques applications, *Ann. École Norm. Sup.* **5**, 599–660.

Serrin, J. (1969). The problem of Dirichlet for quasilinear elliptic differential equations with many independent variables, *Philos. Trans. Roy. Soc. London* **264**, 413–496.

Segal, I. E. (1963). The global Cauchy problem for a relativistic scalar field with power interaction, *Bull. Soc. Math. France* **91**, 129–135.

Segal, I. E. (1966). Nonlinear relativistic partial differential equations, *Proc. Internat. Congr. Math., Moscow* pp. 681–690.

Siegel, C. L., and Moser, J. (1971). "Lectures on Celestial Mechanics." Springer-Verlag, New York.

Smale, S. (1964). Morse theory and a nonlinear generalization of the Dirichlet problem, *Ann. of Math.* **17**, 307–315.

Smale, S. (1965). An infinite dimensional version of Sard's theorem, *Amer. J. Math.* **87**, 861–867.

Smirnov, V. (1964). "Course in Higher Math." Addison–Wesley, Reading, Massachusetts.

Sobolev, S. L. (1938). Sur un theorems d'analyse fonctionnelle, *Math. Shor.* **45**, 471–496.

Sobolev, S. L. (1950). "Applications of Functional Analysis in Mathematical Physics" (Amer. Math. Soc. Transl.). Providence, Rhode Island.

Stromgren, (1932). *Bull. Astron.* **9**, 87–130.

Spanier, E. (1966). "Algebraic Topology." McGraw-Hill, New York.

Srubshchik, L. S. (1964). On the asymptotic integration of a system of nonlinear equations of plate theory, *Appl. Math. Mech., Trans. PMM* **27**, 335–349.

Stakgold, I. (1971). Branching solutions of nonlinear equations, *SIAM Rev.* **13**, 289–332.

Sternberg, S. (1969). "Celestial Mechanics," Part II. Benjamin, Reading, Massachusetts.

Svarc, A. S. (1964). *Dokl. Akad. Nauk USSR* **154**, 61–63.

Szebehely, V. (1967). "Theory of Orbits." Academic Press, New York.

Takens, F. (1972). Some remarks on the Böhme–Berger bifurcation theorem, *Math. Z.* **129**, 359–364.

Taylor, G. I. (1923). *Proc. Roy. Soc.* **A104** (Sci. Papers **4**), 112–147.

Temam, R. (1975). On the Euler equations of incompressible perfect fluids, *J. Functional Analysis* **20**, 32–43.

Ter-Krikorov, A. M. (1969). On the asymptotic character of the motion of a conservative system acted on by an aperiodic perturbing force, *Appl. Math. Mech., Trans. PMM* **33**, 730–736.

Thom, R. (1975). "Structural Stability and Morphogenesis." Benjamin, Reading, Massachusetts.

Titza, R. (1951). On the general theory of phase transitions. *In* "Phase Transitions in Solids," Chapter 1. Wiley, New York.

Toda, H. (1961). "Composition Methods in Homotopy Groups of Spheres" (Ann. Math Studies **49**). Princeton Univ. Press, Princeton, New Jersey.

Tonelli, L. (1921–1923). "Fondamenti di Calculo Delle Variazioni," Vols. I and II. Zanichelli, Bologna, Italy.

Treves, F. (1970). An abstract nonlinear Cauchy–Kowalewski theorem, *Trans. Amer. Math. Soc.* **150**, 77–92.

Trudinger, N. (1967). On imbedding into Orlicz spaces and some applications, *J. Math. Phys.* **17**, 473–484.

Vainberg, M. (1964). "Variational Methods for the Study of Nonlinear Operators." Holden-Day, San Francisco, California.

Vainberg, M., and Tregogin, V. (1974). "Theory of Branching of Solutions of Nonlinear Equations." Noordhoff, Leyden, The Netherlands.

Velte, W. (1964). Stabilitatsverhalten und verzweigung stationares logsungen der Navier–Stokes-schen gleichungen, *Arch. Rational Mech. Anal.* **16**, 97–125.

Visik, M. I. (1963). Quasilinear strongly elliptic systems of differential equations in divergence form, *Trans. Moscow Math. Soc.* **12**, 140–208.

Volmir, A. S. (1967). "Flexible Plates and Shells" (transl. from Russian by Air Force Systems Command, Wright Patterson A.F.B., Ohio).

Volterra, V. (1930). "Theory of Functionals." Blackie, London.

Von Kármán, T. (1910). "Festigkeitsprobleme Ency. der Math. Wiss," Vol. 4. Teubner, Leipzig.

Von Kármán, T. (1940). The engineer grapples with nonlinear problems, *Bull. Amer. Math. Soc.* **46**, 615–683.

von Neumann, J. (1949). "Collected Works," Vol. 6. "Recent Theories of Turbulence," pp. 437–472. Macmillan, New York.

Wallace, A. (1970). "Algebraic Topology." Benjamin, Reading, Massachusetts.

Weinstein, A. (1973). Lagrangian submanifolds and Hamiltonian systems, *Ann. of Math.* **98**, 377–410.

Westreich, D. (1972). Banach space bifurcation theory, *Trans. Amer. Math. Soc.* **171**, 135–156.

Westreich, D. (1973). Bifurcation at eigenvalues of odd multiplicity, *Proc. Amer. Math. Soc.* **41**, 609–614.

Wehausen, J. (1969). Free surface flows. *In* "Research Frontiers in Fluid Mechanics," Chapter 18. Wiley, New York.

Wightman, A. (1974). Constructive field theory, Introduction to the problems. *In* "Fundamental Problems" (B. Kursonoglu, ed.). Gordon and Breach, New York.

Williams, S. A. (1972). A sharp sufficient condition for solution of a nonlinear problem, *J. Differential Equations* **16**, 580–586.

Wintner, A. (1947). "Analytical Founcations of Celestial Mechanics." Princeton Univ. Press, Princeton, New Jersey.

Yamabe, H. (1960). On the deformation of Riemanniam structures on compact manifolds, *Osaka Math. J.* **12**, 21–37.

Yoshida, K. (1965). "Functional Analysis." Springer-Verlag, Berlin.

Zehnder, E. J. (1974). A remark about Newton's method, *Comm. Pure Appl. Math.* **27**, 361–366.

Zygmund, A. (1934). "Trigonometric Series" (reprint). Dover, New York.

INDEX

Pure and Applied Mathematics

A Series of Monographs and Textbooks

Editors **Samuel Eilenberg and Hyman Bass**

Columbia University, New York

RECENT TITLES

ROBERT A. ADAMS. Sobolev Spaces

JOHN J. BENEDETTO. Spectral Synthesis

D. V. WIDDER. The Heat Equation

IRVING EZRA SEGAL. Mathematical Cosmology and Extragalactic Astronomy

J. DIEUDONNÉ. Treatise on Analysis: Volume II, enlarged and corrected printing; Volume IV; Volume V; Volume VI

WERNER GREUB, STEPHEN HALPERIN, AND RAY VANSTONE. Connections, Curvature, and Cohomology: Volume III, Cohomology of Principal Bundles and Homogeneous Spaces

I. MARTIN ISAACS. Character Theory of Finite Groups

JAMES R. BROWN. Ergodic Theory and Topological Dynamics

C. TRUESDELL. A First Course in Rational Continuum Mechanics: Volume 1, General Concepts

GEORGE GRATZER. General Lattice Theory

K. D. STROYAN AND W. A. J. LUXEMBURG. Introduction to the Theory of Infinitesimals

B. M. PUTTASWAMAIAH AND JOHN D. DIXON. Modular Representations of Finite Groups

MELVYN BERGER. Nonlinearity and Functional Analysis: Lectures on Nonlinear Problems in Mathematical Analysis

CHARALAMBOS D. ALIPRANTIS AND OWEN BURKINSHAW. Locally Solid Riesz Spaces

JAN MIKUSINSKI. The Bochner Integral

THOMAS JECH. Set Theory

CARL L. DEVITO. Functional Analysis

MICHIEL HAZEWINKEL. Formal Groups and Applications

SIGURDUR HELGASON. Differential Geometry, Lie Groups, and Symmetric Spaces

ROBERT B. BURCKEL. An Introduction To Classical Complex Analysis: Volume 1

C. TRUESDELL AND R. G. MUNCASTER. Fundamentals of Maxwell's Kinetic Theory of a Simple Monatomic Gas: Treated as a Branch of Rational Mechanics

JOSEPH J. ROTMAN. An Introduction to Homological Algebra

LOUIS HALLE ROWEN. Polynominal Identities in Ring Theory

ROBERT B. BURCKEL. An Introduction To Classical Complex Analysis: Volume 2

BARRY SIMON. Functional Integration and Quantum Physics

DRAGOS M. CVETLOVIC, MICHAEL DOOB, AND HORST SACHS. Spectra of Graphs

Printed and bound by CPI Group (UK) Ltd, Croydon, CR0 4YY
03/10/2024
01040418-0009